T0260800

The Evolution of Biological Information

The Evolution of Biological Information

HOW EVOLUTION CREATES COMPLEXITY, FROM VIRUSES TO BRAINS

CHRISTOPH ADAMI

PRINCETON UNIVERSITY PRESS
PRINCETON & OXFORD

Published by Princeton University Press
41 William Street, Princeton, New Jersey 08540
99 Banbury Road, Oxford OX2 6JX

press.princeton.edu

All Rights Reserved

Library of Congress Cataloging-in-Publication Data
Names: Adami, Christoph, author.
Title: The evolution of biological information : how evolution creates
    complexity, from viruses to brains / Christoph Adami.
Description: Princeton : Princeton University Press, [2024] | Includes
    bibliographical references and index.
Identifiers: LCCN 2022060838 (print) | LCCN 2022060839 (ebook) | ISBN
    9780691241142 (hardback) | ISBN 9780691241159 (ebook)
Subjects: LCSH: Information theory in biology. | Evolution (Biology) |
    BISAC: SCIENCE / Life Sciences / Evolution | COMPUTERS / Information
    Theory
Classification: LCC QH507 .A336 2024 (print) | LCC QH507 (ebook) | DDC
    570—dc23/eng/20230505
LC record available at https://lccn.loc.gov/2022060838
LC ebook record available at https://lccn.loc.gov/2022060839

British Library Cataloging-in-Publication Data is available

Editorial: Alison Kalett and Hallie Schaeffer
Cover Design: Heather Hansen
Production: Danielle Amatucci
Publicity: William Pagdatoon

This book has been composed in Arno Pro

10 9 8 7 6 5 4 3 2 1

CONTENTS

PREFACE

Darwin's theory of evolution occupies a unique place in the history of science. Among the great discoveries of the last five hundred years set in motion by Copernicus and Galilei, Darwin's theory about the common origin of all forms of life and their evolution represents a major leap forward in our understanding of the world. Alongside those discoveries in astronomy and physics that concern the inanimate universe, Darwin's theory stands out because it concerns the origins of the investigator *himself*. Discoveries about the place of humankind in time and space have always necessitated the removal of a mental and emotional barrier, both in the originator of the idea and in the public that receives them. Darwin's proposition is unique in history because it obliterates the last vestiges of human hubris and declares us kin with bacterial slime and leaves of grass. Loren Eiseley (1958) poignantly remarks:

> It is my genuine belief that no greater act of the human intellect, no greater gesture of humility on the part of man has been or will be made in the long history of science. The marvel lies not in the fact that bones from the caves and river gravels were recognized with trepidation and doubt as beings from the half-world of the past; the miracle, considering the nature of the human ego, occurs in the circumstance that we were able to recognize them at all, or to see in these remote half-fearsome creatures our long-forgotten fathers who had cherished our seed through the ages of ice and loneliness before a single lighted city flickered out of the darkness of the planet's nighttime face.

While the theory of evolution by no means materialized out of the blue (as I shall discuss) and while an assortment of naturalists and scientists formulated ideas similar to the central tenets of Darwinism, only Darwin himself wrote about them fully conscious of their world-shaking implications.[1]

---

1. In a January 11, 1844, letter to England's most eminent botanist, J. D. Hooker, Darwin reveals his apprehensions: "At last gleams of light have come, and I am almost convinced (quite contrary to the opinion I started with) that species are not (it is like confessing a murder) immutable." (F. Darwin, 1887, p. 384)

Among the celebrated theories that constitute the framework of our knowledge about the world, Darwinism is unique in another way. At the time of its inception, the theory of evolution was based solely on observation and logical deduction, not on empirical facts obtained by experimentation. It is perhaps not unreasonable to see Darwin as akin to a sleuthing detective, weighing the evidence available to him while formulating and rejecting hypothesis after hypothesis until hitting on the one explanation that is consistent with all available facts. While evolutionary theory can be couched in abstract terms just as any theory in the physical sciences, its predictions could (at the time) not be submitted to experimental tests, not even after the molecular revolution that brought with it the discovery of the genetic code and the molecular mechanisms giving rise to variation. This is because *macroevolution* (that is, speciation, adaptation, and innovation) would have taken too long for the meticulous experimental approach that characterizes progress in the physical sciences. Macroevolutionary timescales are usually measured in the millions of years, or in the hundreds at the very best.[2]

This weakness of evolution as a scientific theory must be recognized as one of the two major elements that have prevented Darwinism from being fully accepted by both scientists (most of whom accepted it almost instantly) and the public alike. The other element that constantly gnaws at the foundations of evolutionary theory is the controversy about the explanation (or nonexplanation) of life's complexity. The controversies are varied, battles are fought both within the ranks of scientists that would not doubt for a second the validity of the central tenets of Darwinism, and without, by an incredulous public and by entrenched creationists.

Explaining how Darwinian evolution can account for the complexity of life (biocomplexity) thus emerges as one of the last remaining major problems in evolutionary biology.[3] Whether or not complexity increases in evolution, and if it did, what the mechanisms are that fuel this growth, is a thorny question because complexity itself is historically a vague concept. If different scientists understand complexity in different ways, it is no wonder that an agreement over this issue has not been reached.

These two factors, which impede a full acceptance of evolutionary theory as sufficient to explain all forms of life on Earth, are moreover related: If experiments could be conducted in which complexity visibly evolves from

---

2. But occasionally, macroevolutionary changes can be observed to occur in the span of decades; see, for example, chapter 4.

3. The other one, namely the emergence and maintenance of the sexual mode of reproduction, has stirred less interest in the general public, while it still is the source of intense research among evolutionary biologists.

simplicity, the controversy would surely shift from "Does It?" to "How Does It?" In this book I shall pull together two strands of research that meet head-on the perceived vulnerabilities of evolution as a complete and satisfying theory of organic origin, diversity, and complexity. The first strand is the field of experimental evolution, a discipline that few could have imagined in Darwin's days, but that today has matured into a quantitative science with the power of *falsification* (the hallmark of scientific theories) in the last twenty years. The other is the theoretical development of a concept of complexity, rooted in mathematics and the theory of information, but germane to biology. No mathematical concept of complexity has hitherto satisfied both mathematicians and biologists alike. Both scientists and nonscientists have an intuitive notion of what constitutes complexity; we "know it when we see it." The theoretical concept that I introduce in this book seems to satisfy our intuition every time it is subjected to the test, which bodes well for its acceptance as a measure for biocomplexity. Moreover, it proves to be both practical and universal. Practical, because it implies a recipe to attach a *number* to the complexity of any class of organisms (allowing in principle a comparison between species), and universal because it does not refer at all to nucleic acids or proteins, or any other particular feature of this world and the forms of life that populate it. Rather, it is based on the universal concepts of automata and information theory, which are abstract.

A mathematical description of the mechanisms that are responsible for the evolution and growth of complexity, and experimental evidence buttressing such a description, should go a long way to eliminate those doubts that are anchored around the startling complexity of life and the seeming inability of scientific theory to account for it. I will try in this book to convince the reader that it can be accounted for, both in abstract terms and mathematical formulae, and that the mechanisms responsible for the growth of complexity can be investigated experimentally and be tested and retested.

But information theory can do more for biology than just provide for a measure of complexity. In hindsight, *everything* in biology uses information in one form or another, be it for communication (between cells, or organisms) or for prediction (via molecular as well as neural circuits). As a consequence, we must think of information theory as the unifying framework that allows us to understand complex adaptive systems far from equilibrium, with biological life being the prime example.

From the preceding comments, it should be clear that this is not a conventional book about evolution. It is not a "whodunnit" in which the complicated relationship between adapted forms is revealed through elaborate genetic or behavioral experiments and observation. Rather, it treats evolutionary theory as an empirical science, in which abstract concepts, mathematical models, and dedicated experiments play the role they have been playing in the physical sciences in the last few hundred years. While I try to keep the mathematical sophistication to a minimum, a reader who wants to make the most of this book should be prepared to follow basic algebra. Indeed, the concept of genomic complexity—its acquisition and evolution—is so firmly rooted in the theory of information that it would be impossible to bypass an exposition of the framework due to Shannon (1948). Furthermore, molecular evolution theory (due to Eigen 1971, Eigen and Schuster 1979), the theory of self-replicating macromolecules under mutation and selection, is a kinetic theory in which the time-dependence of concentrations of molecules are key. Thus, basic notions from calculus will be required to follow those sections. Yet, I have strived to explain the concepts that are introduced mathematically also in intuitive language, so that the dynamics of evolution, and the circumstances surrounding the evolution of complexity, should appear more clearly to every serious reader interested in biocomplexity.

Another somewhat less conventional feature of this book is its extensive use of the methods of computation. Virtually all the physical sciences have branches nowadays that are almost entirely computational in nature: the power of modern computers to take initial data and, armed with a set of equations that model the system under investigation, grind through to the consequences has revolutionized every facet of modern science. Even within biology, the computer has taken major inroads, in particular in the analysis of bioinformatics data, and the modeling of cellular processes and development. *Genetic algorithms*, a method to search for rare bit patterns that encode solutions to complex problems (usually in engineering) are inspired by the Darwinian idea of inheritance with variation coupled with selection of the fittest. But this is not, by far, the limit of how computers can aid in the study of the evolutionary process. Computational evolutionary biology involves building models of worlds in which the Darwinian principles are explored. A particular branch of computational evolutionary biology involves not only creating such an artificial world, but implanting in it not a simulation of evolution, but the *actual process itself*. Because it is possible to create a form of life that can inhabit and thrive in an artificial world (Ray 1992; Adami 1998), it has become possible to conduct dedicated experiments that can explore fundamental aspects of evolution as they affect an alien form of life (sometimes called "digital life") (Harvey 1999; Zimmer 2001; Lenski 2001). Because such experiments can only study those aspects of the evolutionary process that are *independent* of the form of life it affects, we cannot, of course, hope to gain insight from these experiments about phenomena that are intimately tied to the type of chemistry used by the organism. But the beauty of digital life experiments lies in their ability to make predictions about evolutionary mechanisms and phenomena affecting *all forms of life anywhere in the universe*. When designing experiments to test evolutionary theory, then, we must judiciously choose the experimental organism to use, weighing its advantages and idiosyncrasies (there is, after all, no "universal" organism anywhere on this world and likely others), from the gallery available to us: viruses, bacteria, yeast, or fruit flies (to name but a few), or indeed digital ones.

The foundation laid by Darwin, and the edifice of modern evolutionary theory constructed in the twentieth century, is not shaken by the experimental, computational, and information-theoretic approach outlined here. I imagine it strengthened in those quarters where the structure was perceived to be weak or vulnerable, and its expanse increased, to an ever more dazzling, towering achievement within humanity's endeavor to understand the world around us, and ourselves.

ACKNOWLEDGEMENTS

This book has had a long gestation period. I started it around 2002, thinking it would be the follow-up to *Introduction to Artificial Life*, my 1998 book that I was still using as the basis of the course I taught at the time at the California Institute of Technology. Originally, I thought the book would focus more on evolutionary biology seen through the lens of digital evolution, since so many new results had appeared since 1998. But gradually, the book evolved. The focus shifted to understanding the emergence of complexity using tools from information theory, a theory that gradually grew to become the backbone of the book. After I moved to the Keck Graduate Institute in Claremont (California), the book languished until I took a sabbatical at Michigan State University (MSU) in 2010. There, reunited with the Caltech team of Titus Brown and Charles Ofria, who wrote the first (and last) versions of Avida (the digital life platform), as well as with Rich Lenski (who had collaborated on many of the big digital life experiments), work intensified again—only to succumb once more to the more pressing needs of grant proposals and manuscripts. But my move to Michigan State University after the sabbatical influenced the book in another way: Lenski's long-term evolution experiment would be afforded a much more central place in the book, since I had learned so much from his experimental approach to evolutionary questions.

My plans to finish the book during the 2010 sabbatical thwarted, I pinned my hope on the next sabbatical, which I took at Arizona State University in 2018. I am very grateful to Paul Davies, Ann Barker, and George Poste for giving me the opportunity to spend that year under the Arizona sun, and the book grew tremendously during that year. I also owe thanks to Olaya Rendueles and Olivier Gascuel at the Institut Pasteur in Paris, where I spent a mini-sabbatical during that year.

After returning to MSU, the book kept growing, despite my efforts to contain the scope. In the end, it was the global pandemic that began in 2020 that made it possible for me to finally finish the book, 18 years after I wrote the first 150 pages. There are, of course, a large number of people to whom I am indebted, and without whose support, collaboration, and insight this book would never exist. At Caltech, Steve Koonin welcomed me into his group and turned my career from nuclear physics to complex systems and digital life. Francis Arnold (who sat in on many of the classes when I taught "Introduction to Artificial Life") has been a steadfast supporter and collaborator, cosupervising two star Ph.D. students. I also received tremendous support at Caltech from Christof Koch and David Baltimore. After moving to the Keck Graduate Institute, I gained a new set of colleagues and full support and the friendship of Shelly Schuster, David Galas, and Greg Dewey, along with fantastic colleagues and collaborators Animesh Ray, Alpan Raval, Angelika Niemz, and Herbert Sauro.

My move to MSU would not have been possible without the combined efforts of Rich Lenski, Titus Brown, Charles Ofria, and Walt Esselman, who as department chair at the time proved that his word was as good as he said it was. In the end, this book carries the mark of MSU the most: it was there where I turned my focus increasingly to the evolution of intelligence and cognition. And without a doubt I owe a debt of gratitude to my current Chair Vic DiRita, who never questioned me on how research on the evolution of intelligence (or on the quantum physics of black holes, for that matter) was an appropriate use of my time in the Department of Microbiology and Molecular Genetics.

With respect to my entering the fray of artificial intelligence research, one person cannot go unmentioned: it was Arend Hintze who started this work with me when he joined me as a postdoc at the Keck Graduate Institute in 2006, moved to MSU with me during the sabbatical and stayed on, ultimately to lead his own group there. During those years, our families became close friends. Furthermore, I am indebted to Jeff Hawkins, whose book *On Intelligence* changed how I thought about the brain, and who graciously spent a weekend talking to me about his theory during SciFoo in 2007.

In these last twenty years I've had the privilege to interact with a large number of students, postdocs, collaborators, and friends and colleagues. All of those have touched this book in one way or another. I wish to thank Larissa Albantakis, David Arnosti, Mark Bedau, Steve Benner, Anton Bernatskiy, Jesse Bloom, Cliff Bohm, Josh Bongard, Sebastian Bonhöffer, Paulo Campos, Sam Chapman, Nicolas Chaumont, Stephanie Chow, Nitash C G, Jacob Clifford, Jeff Clune, Lee Cronin, Evan Dorn, Vic DiRita, D. Allan Drummond, Fred Dyer, Jeffrey Edlund, Santi Elena, Josh Franklin, Murray Gell-Mann, James Glazier, Nigel Goldenfeld, Virgil Griffith, Aditi Gupta, Dimitris Illiopoulos, Betül Kaçar, Stuart Kauffman, Laura Kirby, Doug Kirkpatrick, Dave Knoester, Eugene Koonin, Donna Koslowsky, Thomas LaBar, Dante Lauretta, Joel Lehmann, Lars Marstaller, Chris Marx, Joanna Masel, Devin McAuley, Masoud Mirmomeni, Dule Misevic, Randy Olson, Bjørn Østman, Anurag Pakanati, Anthony Pargellis, Ted Pavlic, Josh Plotkin, Daniel

Polani, Anselmo Pontes, Bill Punch, Jifeng Qian, Vinny Ragusa, Steen Rasmussen, Tom Ray, Matt Rupp, Tom Schneider, Jory Schossau, Adrian Serohijos, Eugene Shakhnovich, Eric Smith, Paul Sternberg, Darya Sydykova, Jack Szostak, Tracy Teal, Ali Tehrani, Giulio Tononi, Greg VerSteeg, Daniel Wagenaar, Sara Walker, Jialan Wang, Claus Wilke, and Mike Wiser.

Finally, I would like to acknowledge the influence that MSU's BEACON Center for the Study of Evolution in Action had on this book, and on my career in general. The NSF-funded Science and Technology Center opened its doors in 2010, welcoming me for the inaugural BEACON-funded sabbatical. At BEACON, I found a group of researchers who were passionate about evolutionary biology, and unafraid to move into unfamiliar directions, and collaborate with a theorist like me. I am grateful to the leadership of BEACON, from the Director Erik Goodman to Rich Lenski, Charles Ofria, Kay Holekamp, and Rob Pennock on the Executive Committee, along with the Managing Director Danielle Whittaker, as well as Connie James and Judy Brown-Clarke. In a real way, this book carries BEACON's essence in its DNA.

Last (but not least) I would like to thank my family: my partner in life, Taylor, and our daughter, Julia, who has heard about "the book" during all of her life; the book that took time away that rightfully belonged to her.

I dedicate this book to two men who have shaped my thinking and offered their unwavering support to me since my time as a graduate student, until the last of their days. I still think about them—and what they have taught me—nearly every day.

Hans A. Bethe (1906–2005)
Gerald E. (Gerry) Brown (1926–2013)
Okemos, May 2022

The Evolution of Biological Information

1

Principles and Origins
of Darwinism

There is grandeur in this view of life, with its several powers, having been originally breathed into a few forms or into one; and that, whilst this planet has gone
cycling on according to the fixed law of gravity, from so simple a beginning endless forms most beautiful and most wonderful have been, and are being, evolved.

—CHARLES DARWIN, *ON THE ORIGIN OF SPECIES* (1859)

Nothing in biology makes sense except in the light of evolution.

—TH. DOBZHANSKY (1973)

The basic principles of Darwinian theory as outlined by Darwin in his abstract
of a book that became the book of reference, the *Origin of Species* (Darwin
1859), are deceptively simple. Indeed, those principles can (in their most
elementary form) be summarized in a single sentence: *"Inheritance with modification, coupled with natural selection, leads to the evolution of species."* We should
not, however, be deceived by the simplicity of evolution's basic mechanism.
After all, we take it for granted that scientific theories that can be summarized
by a single formula can give rise to centuries of research, to work out its consequences in real (rather than idealized) settings. A framework of ideas such
as Darwinism can never constitute the endpoint of inquiry into the origin and
complexity of organic forms, but is rather the very point of departure. Within
an extraordinarily complicated environment (made so complicated in part
because of the organic forms in it), the Darwinian mechanism leads to such
a vast diversity of seemingly unrelated consequences that a single scientist can
spend their entire scientific career studying the mechanism's ramifications for
a *single* species out of many millions.

Darwinism, as implied in the header quote, is what explains biology. Its
claim is that it not only explains the complexity and variation in all the existing

1

forms of life, but that the same mechanism allows an extrapolation backward in time to our, and all other terrestrial forms of life's, beginning. This is a magnificent and confident claim, and such a theory must therefore expect to be challenged strongly and repeatedly (as it has been and continues to be). This is the natural state of affairs for all scientific theories and so it is with evolution, except that challenges to established theories (for example, testing their applicability in extreme circumstances) usually does not imply a challenge to the very foundations and structure of the theory itself. In other words, theories that have withstood many decades of attempts at falsification are unlikely to be ultimately shown wrong *in their entirety*, but only in details. Thus, anti-Darwinian enthusiasts should keep in mind that they are as likely to disprove the Darwinian principles as Newton and Einstein will be shown to have been *completely wrong* about gravity.

While today's reader is sure to be already acquainted with the main principles of Darwinism, it is important to start by spelling them out as succinctly and clearly as possible. Each element will be treated in much more detail throughout the book. We shall be guided by the single italicized sentence at the beginning of this chapter and begin by fleshing out the terms that appear in it. After this exposition, we will explore the impact of each of the elements of the triad in a simple simulation of evolution, to show that each must be present for the process to work.

1.1   Principles of Darwinian Theory

1.1.1   *Inheritance*

That certain traits are inherited from parent to offspring is obvious to anyone who observes plants and animals, but this observation alone (like most of the components of Darwinism on their own) is unable to shed light on the origin of species and the evolution of complexity. A trait is an "observable feature" of an organism and does not necessarily have to be inherited (it can also be acquired as a response of the organism to the environment). Furthermore, a number of traits can be due to a single gene, or several genes can affect the character of a single trait. This explains (together with the complications engendered by sexual reproduction) why understanding the inheritance of traits has not led immediately to the discovery of the first central element of Darwinism: the *reproduction* of the organism, and the concomitant *replication* of information, in the form of the organism's genetic material. Indeed, inheritance is a *consequence* of reproduction, while the replication of genes is both a consequence and a necessity for reproduction. This (backward) inference from inheritance to reproduction to replication appears trivial from the vantage point gained by the discovery of the genetic code, but is far from obvious prior to that discovery.

From a purely mechanistic point of view, we can thus distill inheritance to the *replication of genes*, or, even more abstractly, to the *copying of information*. As we shall discuss at length in later chapters, the replication of genes, which encode the necessary information to grow the organism and increase the chances for its survival in the environment in which it lives, is the ordering force that preserves the continuity of lineages. We should keep in mind that only the faithful replication of an organism's genes is required for Darwinian evolution, *not* the faithful reproduction of the organism itself. As we shall see later in this chapter, however, a close correlation of the organism's phenotype (the sum of traits and characters) with its genotype (the sum of its genetic information) is required for selection to work properly.

1.1.2 Variation

If replication was perfect, all offspring would be identical to their parents, and therefore all members of such a population would be indistinguishable. Because selection (discussed below) implies a concept of ranking, selection would be impossible in the absence of variation. This variation, however, must occur at a genetic (that is, inheritable) level, because while selection can act on acquired characters, such selection does not give rise to evolution. Thus, variation must occur on the genotypic level: on the information stored in an organism's genome.

Perfect (error-free) replication of information also has another drawback. While it is ideal for protecting the information coded in the genes, it is counterproductive if new information needs to be discovered and incorporated into the genes. The importance of genetic variations is best understood by again taking a purely mechanistic, information-based view of evolution. If the genome alone contains the information about how to make an organism that best survives in the given environment, how does this information *get there*? Since acquired characteristics—changes to an organism's phenotype due to interactions with its environment, such as damage, injury, or wear and tear—do not change the genes, they cannot be inherited. For information to enter the genome, changes must occur in the genomic sequence itself. We thus need a force that works in the opposite direction to the replication process that keeps genes intact: this is the process of *mutation*. A mutation is an alteration of the genetic material (the genetic sequence) that is potentially transmitted to the next generation. In a sense, mutations are the natural consequence of a physical world: they reflect the difficulty of keeping an ordered state (the sequence) intact while it is being manipulated, and exposed to numerous potentially corrupting agents. For example, mutations are a natural by-product of replication simply because it is impossible to perform perfect replication using imperfect machinery. The replication of information (the replication of DNA for organisms based on terrestrial biochemistry) is a physical process that involves the duplication of the carriers of information in sequence. Because this process takes place in a physical environment, there will always be errors associated with this process (the process is "noisy"), and these errors give rise to an alteration of the original sequence: a mutation.

Even though point mutations (that is, replacements of one letter in the sequence by another) are the simplest way to account for genetic variation, they are by no means the only ones that occur. In retrospect, nature has taken advantage of essentially all possible ways in which information can be changed, including deletion and insertion of a letter, deletions and insertions of whole sequences of code, inversions, shuffling, and so on. One of the most well-known sources of variation in evolution is the genetic recombination of code during sexual reproduction. No matter the origin of the mutation, however, because the code defines the organism, variations in the genotype can give rise to variations in the phenotype. It is this variation that the next element of Darwinism acts upon: selection.

1.1.3   Selection and adaptation

Among the primary concepts of Darwinian evolution, selection and adaptation are perhaps those most often misunderstood. Natural selection as a mechanism is now part of our vernacular and occupies, for good reason, a central place in Darwinian theory. Natural selection is what happens if some organisms are better at surviving and/or reproducing than others. From what we saw earlier, this clearly implies that there must be some agent of variation, as otherwise all organisms would be the same, and some could not be better than others. If one type of organism is better at survival/reproduction than another, then the relative numbers of these two types must necessarily change. If, at the same time, the total number of organisms in this competition is fixed (either due to a finite amount of resources in the niche, or due to finite space), then it is clear that a constantly changing ratio of numbers between two competing species will result in the inferior species being driven into extinction. This is, in a nutshell, the mechanism of natural selection, but its consequences, as well as its subtle variations, are far from trivial.

   To begin with, the previous sentence implies that natural selection acts on organisms that are "better at survival and/or reproduction." What exactly does that mean? This question addresses the concept of "fitness" within evolutionary biology, which is a central concept and deserves a brief discussion here (and a more detailed one later on). Much has been made from the apparent tautology that declares those organisms as the fittest who end up surviving. More precisely, this type of criticism has been leveled at the statement that "survival of the fittest" is a tautology if the fittest is defined as that which survives. There is no tautology, of course, because we do not, in biology, define "fit" as "one who survives." Fitness, instead, is a concept meant to characterize a *lineage*, not a single organism (even though the word is often used to describe individuals of that lineage). A lineage is a set of organisms that are tied together by their genes, that is, they all share the same genetic characteristics because of shared heritage. Any particular organism representing a lineage may be subject to random occurrences that may cause it to lose out in a competition with a representative of a less fit lineage purely through chance. This does not persuade us to change our fitness assessment of this lineage. Instead, on average, the representatives of the lineage that is fitter will outcompete the representatives of the less fit lineage, but any single competition may go either way.

This being said, the fitness of an organism is not always easy to estimate. Technically, the word "fitness" implies "adaptation," namely a lineage that "fits" its environment well. In evolutionary biology, fitness is defined as "expected reproductive success," where "reproductive success" implies success both in reproduction and survival. The reason we must emphasize our expectation is that, as we already saw, in a natural world expected success does not always equate with realized success. The phrase "survival of the fittest" is, therefore, really a poor rendition of the natural selection concept within Darwinian evolution. Selection is simply a mechanism by which the frequency of particular types of organisms are changed *depending on what genes they have,* and the mechanism is such that those genes that *increase* the carrier's relative numbers are precisely those that will carry the day. The logic of selection is so unassailable that it sometimes seems like an utter triviality. That this is not so is exemplified by the stupendous variety of mechanisms and technical complications that accompany natural selection, from sexual selection to a dependence on mutation rates, to neutral evolution and the selection for robustness.

Adaptation is perhaps even more misunderstood than selection. There is no doubt that adaptation is perhaps the most stunning result of Darwinian evolution, and can be observed in minute details of function in every organism inhabiting Earth. In the following chapters, we will largely do away with the concept of adaptation because it is too vague for a quantitative analysis. Some features of living organisms are easily identified as adaptations, namely traits that clearly further the reproductive and survival chances of a species. Other traits are not so easily interpreted, and the fitness value of any particular gene or trait cannot be linked directly to its adaptedness. Still, there is no doubt that adaptation occurs, when through natural selection those organisms are favored whose particular (well-adapted) trait allows them to exploit their environment in a more efficient manner. Adaptation, thus, is "adaptation to one's environment." Because the ability of an organism to exploit and thrive in its environment is directly related to the genes that code for such prowess, we can say that a well-adapted organism is in possession of a good amount of *information* about its environment, stored in its genes. Because the concept of information is a quantitative one, we shall use it in place of adaptation in most of what follows. We should keep in mind, however, that it is not guaranteed a priori that information is a good proxy for "adaptedness" or function, or even fitness for that matter. We shall have to examine this assumption in detail.

1.1.4 Putting it together

To get a better picture of how these three elements work together to generate evolution, it is instructive to put them all together in a computer simulation. We will keep this simulation as simple as possible so as to involve only the processes discussed above in their purest form, while making sure that we can turn off any of the elements independently to observe the dynamics that ensue. The simplicity of the simulation of course implies that it is not intended to simulate any actual evolving organism. Rather, its purpose is to illuminate the *interaction* between the elements, and to test their respective necessity.

The goal of our little simulation is to optimize the fitness of a population of alphabetic strings. The alphabet could be anything as long as it is finite. It could be binary (bits), quaternary (like DNA and RNA), base 20 (such as with amino acids), or base 26 (English lowercase alphabet). Here, we arbitrarily choose an alphabet consisting of the first twenty letters, (from a to t). Also, we will fix the length of any sequence to one hundred letters. To enact selection, we can construct a simple fitness landscape by arbitrarily declaring one particular (randomly chosen) sequence of letters the most fit, and stipulating that you lose fitness the more mutations away you are from that sequence. The number of point mutations it takes to get from one sequence to another is called *Hamming distance* in the mathematical literature, so in this case the fitness is based on a sequence's Hamming distance to the optimum.

Clearly, this fitness landscape does not resemble anything like what we would encounter in natural systems. In particular, no natural fitness peak is this cleanly designed from the outside, and even more importantly, the Hamming distance fitness implies that each site in the string of length 100 contributes *independently* to the fitness of the string. As a consequence, there are only one hundred different fitness values in this landscape, and the order in which the beneficial mutations are acquired is inessential. As we will see as the chapters unfold, this is so far removed from realistic fitness landscapes for nucleotides or proteins or even genes that such a simulation is little more than a caricature of the evolutionary process. Indeed, if all mutations were to depend on each other instead, that is, if the fitness effect of one mutation at one site depends on the state of all other sites, then a string of length 100 can encode up to D^{100} different fitness values, where D is the size of the alphabet. For proteins ($D = 20$), the difference in the "richness" of the fitness landscape amounts to about a factor 10^{128}! Mutations that depend on each other are called *epistatic*, and we will see that the interaction between mutations is the single most important factor in the emergence of complexity via Darwinian evolution. To some extent, the simulation we study below can be viewed as representing evolution with all its interesting bits (namely epistasis) stripped off. Its only purpose is to illustrate the combined effects of replication (inheritance), mutation (variation), and selection. Any fitness landscape suffices for this purpose, as long as it is not completely flat, that is, if there are any fitness differences at all.

Evolution occurs on sequences within a population, so in this simulation we shall observe the fate of a population of fixed size (here, 200), in competition with each other. Later, we will relax even this condition, to see what happens to evolution in the absence of competition (by allowing the population to grow indefinitely). Mutations are implemented so that each generation, an arbitrary string will suffer on average one mutation per replication cycle. This means that oftentimes they will suffer no mutations, more likely only one, and in rarer cases two or more mutations. This mechanism can be applied even if sequences do not replicate. The replication of these sequences is implemented in a probabilistic manner, so that those sequences that are ranked the highest according to the fitness criterion discussed above are accorded multiple offspring, while the sequence with the smallest score is assured not to leave any descendants.

If all this is put together, the algorithm effectively implements a parallel search (parallel because the search occurs in a population) for the optimum sequence. Algorithms just like that are indeed often used in engineering and other applications, and are termed *Genetic Algorithms* (see, e.g., Mitchell 1996 for an introduction or Michalewicz 1999 for a more advanced exposition).

Figure 1.1 shows a typical result of such a simulation when all elements of Darwin's triad are present. The mean fitness (solid line) of a population of 200 random sequences is steadily increasing, and the optimum fitness is found after 71 generations (the dashed line is the fitness of the best-in-population). Also note the population diversity (dotted line), which here is the logarithm of the number of different types of sequences n_s in the population (where we use

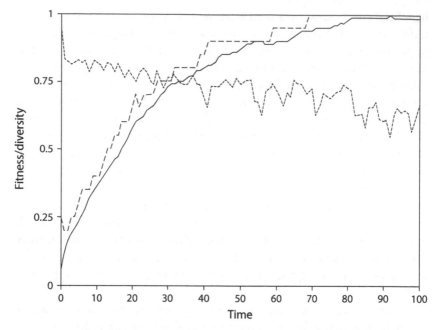

FIGURE 1.1. Mean fitness (solid line), fitness-of-best (dashed line), and population diversity (dotted line) in a simulation of evolution with mutation, reproduction, and selection. Fitness is measured in arbitrary units between one (optimum) and zero (worst), while time is measured in generations. Diversity is measured as the logarithm of the number of different sequences n_s, to the base of the population size $\log_{200}(n_s)$, which also lies between zero (no diversity) and 1 (all sequences different).

the population size as the base). It starts at approximately its maximal value 1 and declines to a steady state that remains below the maximum.

Let's first consider the same exact process, but in the absence of mutations. We start with a random population, so there is plenty of variation to begin with, but none is added as time goes on. Because the population size is so much smaller than the possible number of sequences, the chance that the fitness peak is accidentally already in the population is astronomically small. The best-of-population fitness is constant throughout since it is given by the highest fitness individual present (by chance) at the beginning, while the mean fitness of the population quickly increases (see Fig. 1.2[a]) because selection is working. The best sequence in the population quickly gains in numbers at the detriment of the less fit ones. At the same time, you can see the population diversity plummet drastically, because less fit variants are replaced by copies of the fitter variant, all of them identical but far from the maximum

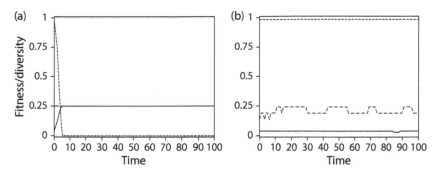

FIGURE 1.2. (a) Simulation of evolution with reproduction and selection, but without mutation. (b) Simulation of evolution with selection and mutation, but without reproduction. Legend as in Figure 1.1.

fitness possible). After nine generations, all two hundred individuals in the population are identical, and nothing else will ever happen here.

Next, we study the importance of reproduction. We can perform the same simulation, including a ranking of organisms according to their Hamming distance to the optimum, but now this ranking does not affect a sequence's reproduction rate (they do not reproduce at all). Mutations continue to occur, so in principle the optimum sequence could still be found because the sequences are immortal in this setting; however, the probability of this happening here is exponentially small. In Figure 1.2(b), we can see that the fitness of the best organism in the population is fluctuating (the fitness is taking what is known as a *random walk*), and the mean fitness mirrors that. Population diversity is maximal and unchanged from the initial diversity, since replication is the only process that can appreciably reduce the diversity. It is possible, of course, that random mutations create several copies of the same exact sequence by accident, thus lowering the population diversity. However, the probability of this occurring is again exponentially small, and such a state would be replaced by a more probable one in the next instant.

Finally, we consider the case where we have both mutation and reproduction, but no selection. To turn selection off, we can simply rank all sequences equally, independently of their Hamming distance to the optimal sequence. As a consequence, each individual is guaranteed exactly one offspring, regardless of the sequence of instructions. This case is interesting because even though there is no selection, random fluctuations can give rise to differences in reproductive ability, and sometimes certain mutations can become quite common in the population even though they have the same fitness as all others. (This case is known as "neutral evolution," and will be treated in detail in chapter 6.)

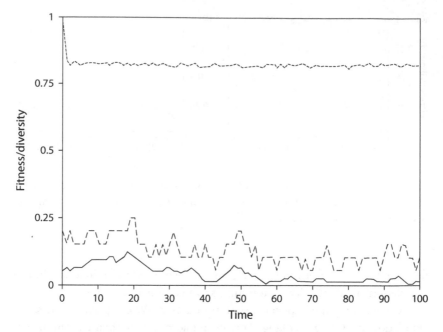

FIGURE 1.3. Simulation of evolution with reproduction and mutation, but without selection (neutral evolution). Legend as in Figure 1.1.

As a consequence, the dynamics are quite a bit different from the case we treated just before (no replication). The population diversity is not maximal, and the average and best fitness fluctuate more strongly (see Fig. 1.3). In each of the three cases where one of the necessary elements is absent, it is patently obvious that evolution does not occur even though two of the required three elements are present. Such is the interaction of the three elements of the Darwinian triad: all for one, and each for all!

1.1.5 Speciation

The species and its origin, while clearly a central concept in Darwinian theory, is not actually a central element of the Darwinian mechanism (the first three in this chapter are all that is needed), but rather one of its consequences. Still, it deserves to be treated in this quick tour of the principles because of its pivotal role in evolutionary biology.

Species are all around us, and are (usually) easily identified by eye as those members of a population that share certain phenotypic (meaning here, man-ifested) properties of an organism. That there is a "species problem" (this is what Darwin told his friends and colleagues he was working on before the publication of his *Origins*) takes a little thinking, because at first glance one might think that it is only too obvious that "populations of closely related organism that are mostly similar," namely species, will form as a consequence of the mechanisms described above. After all, a mutated organism is necessarily directly related to its progenitor if the mutation happened during the reproduction process. Furthermore, the probability that a mutation creates a dramatically different organism (one that would be classified as a different species) is expected to be exceedingly low. However, these obvious observations are precisely those that lead to the species question. If organisms naturally form populations of closely related specimens, why do new species arise at all? And how can it be that the process of species formation has led to types so dramatically different that it is well nigh inconceivable that they were once siblings, in particular while relatives of the original stock still exist today largely unchanged? What, then, drives the changes that species undergo, this fragmentation of populations into distinct groups, and why do they not coalesce into a muddled amalgam of types that blend one into another, with intermediate organisms everywhere between bacteria and the giraffe?

For sexual organisms, the standard species definition is that all organisms that can produce fertile offspring are considered as belonging to the same species and are different species if they cannot. This idea is called the "biological species concept" (Coyne and Orr 2004). It is a very sensible way of defining species because groups that cannot produce offspring with each other are effectively genetically isolated, as no gene mixing can occur between them. As a consequence, two groups that are isolated in this manner will evolve independently and become more and more different. On the other hand, it is not a perfect criterion because examples exist of distinct species that can produce fertile hybrids. In any case, defining species in this manner does not solve the species problem, as we now have to understand how it can happen that one species breaks into two or more "proto-species," who then gradually lose the ability to interbreed.

There are two main ways in which we can imagine that this breakup happens. First, it is possible that a species is accidentally separated into two groups due to a geographic partition, say, one group crosses a river while another does not. If subsequently the river grows so large that it renders any other crossing impossible, the groups are *reproductively isolated* and can evolve independently without mixing of genes *as if* they were different species. After some time, the different evolutionary paths taken by the respective groups is likely to have resulted in changes that make interbreeding biologically impossible (not only practically) so that the species will remain separate even if the river dries up and the groups are reunited. This process is called *allopatric speciation* in the literature ("allopatry" translates literally to "having different fatherlands").

Another (much more controversial) possibility is that groups can drift apart *without* geographic separation, a process termed *sympatric speciation*. As the name suggests, this is a mode of speciation where the future species occupy the *same* homeland. The difficulty with this mode is that it is hard to understand how the small genetic changes that occur within a population can give rise to separate species if the organisms are able to interbreed. The genetic mixing implied by interbreeding should wash out any genetic differences that may have arisen. Thus, interbreeding is a force that opposes speciation. To understand speciation in sympatry, we would have to assume that small genetic changes can cause some reproductive isolation that ultimately stops the gene flow between the incipient species (there is actually direct evidence for such an effect, see for example Uy et al. 2009). Most theories of sympatric speciation invoke local adaptation to different resources (so-called microhabitats). But because adaptation to different local resources (for example, changing your diet via a genetic mutation) does not prevent such differently adapted groups to interbreed, we usually have to assume that the change in diet must be directly associated with mate choice behavior also. In other words, the change in diet has to turn the prospective partners off sufficiently so that interbreeding is prevented.

It should be clear that one of the difficulties in testing theories of speciation is that it is rare that the process can be observed in real time. One of the most laudable exceptions is perhaps the decades-long work of the Grants (Grant and Grant 1989; Grant and Grant 2008), but other examples exist such as speciation in flowering plants (Soltis and Soltis 1989), sticklebacks (Colosimo et al. 2005), and cichlid fish (Schliewen et al. 1994; Seehausen et al. 2008).

The species concept can also be applied to asexual organisms, albeit in a different form, naturally. Bacteria and viruses, for example, do occur in distinct groups rather than in genetically fluid amalgams even though they do not reproduce sexually. For asexual organisms, allopatric and sympatric modes of speciation can occur, but the difference is not so profound because asexual species do not mix genetic material to begin with (I am ignoring lateral gene transfer here for the purpose of simplicity). Thus, within asexual organisms, a new species can be born simply by one (or several) propitious mutations. The difficulty for the bacterial species concept that arises in this case is somewhat different. If new species can arise within asexual organisms with every beneficial mutation, why do we not see have an almost infinite number of them, one corresponding to each such mutation? The answer to this question naturally lies in selection: when such a beneficial mutation sweeps a population, the inferior kind is driven to extinction. This seems to imply that evolution in asexual populations simply proceeds by one species supplanting another. Where, then, do all the different bacterial and viral species come from? This

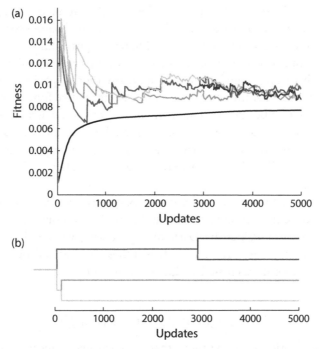

FIGURE 1.4. Evolution of an ecosystem of different species that specialize for survival on different limited resources. Each species discovers and then specializes on a different resource. (a) Lines in different shades of gray show relative fitness of the different species, splitting off ancestral species. Overall fitness of ecosystem in black at the bottom of the figure. (b) Ancestral reconstruction (phylogeny) of the simulated species (shades of gray the same as those in panel (a)). Adapted from Østman et al. (2014).

question seems to find its explanation in a resource-based sympatric process, where different species can coexist because they all "make their living" in a different manner, which means that they do not directly compete against each other anymore. A typical example that shows the emergence of new species (in a computational simulation, see Østman et al. 2014 for more details) in a resource-limited environment is shown in Figure 1.4. As new species emerge, the relative fitness of each depends on the frequency of that species in the population, as well as the frequency of others in some cases. When rare (that is, when it just emerges), a new species has a competitive advantage because the resource it relies upon is very abundant—nobody else relies on it yet. As a consequence, the emerging species has no trouble invading the existing type (see, for example, the new species in black that splits off the dark-gray species just before generation 3,000 in Fig. 1.4).

Ultimately, the relative fitness (here the overall reproductive rate) of each species must match those of any other species in the same ecosystem, as otherwise the equilibrium among the different types would be disrupted. In the case of speciation via adaptive radiation in a resource-limited environment, this equilibrium is stable because any time a species increases in number beyond what the ecosystem can carry, its relative fitness must drop, adjusting the number down. The mathematics of species equilibrium in a global ecosystem should give us pause, of course, because once the connection between resource limits and species fitness is broken, unimpeded replication can catastrophically exhaust a resource, leading ultimately to the demise of that species.

It is clear then that a theory of biocomplexity must also address the species problem. Not only is it important to understand how a diversity of species is created and maintained by Darwinian processes, we should also strive to understand how the *interaction* between these species and the ecological networks they form are created, maintained, and nurtured.

1.2   Origin of Darwinian Thought

Successful theories, meaning those that are particularly good at explaining observations, often seem so obvious that it is impossible to imagine a time when the world was looked at without this piece of knowledge. The idea that the Earth is round and not flat is one good example, perhaps the heliocentric worldview somewhat less so (but only because it is somewhat less obvious). With respect to evolution, its general principles have so permeated our everyday thinking that to delve into pre-Darwinian thought processes might seem like an exercise without merit. Yet, if you think about the simplicity of the main elements of Darwinism, it could appear like a preposterous accident that they have been discovered comparatively late.

To fully appreciate Darwin's insight, and perhaps to get a better gut feeling for these seemingly innocuous "three principles," we are going to take a little detour back in time to search out the roots of Darwinist ideas, to get a feel for the mindset of the era into which they were flung, and to follow the fits and starts of other scientists, who got little pieces of the story, but did not solve the puzzle.

1.2.1   Eighteenth century

The origins of Darwinian thought can be found scattered among the writing of naturalists, theologians, and geologists of the eighteenth century. To do justice to the sometimes timid attempts at making sense of the natural world during this time, it is important to understand that the century was steeped in religious dogma, which did not allow any opinion to be held that contradicted the Bible's creation story or its affirmation of the fixity of the Earth and the animal life that inhabits it.

In those times, you would be hard-pressed, therefore, to find someone to openly consider the question of "the origin of species," because this question was considered solved by a singular act of creation. Within this era, however, a few people were willing to ask questions which, perhaps, could be construed as heretic, but which nevertheless were openly dedicated to the "Glory of God." Because at the time it was modern to consider Nature as "the other great book" through which one could discover God,[1] some of the foremost naturalists of the time were in fact reverends and priests.

Two church doctrines governed all discussions about biological diversity in the eighteenth century. The first is the idea that all species were created at once, independently from each other, and arranged into the famous "great chain of being" with God, angels, and then humans at its top (see Fig. 1.5). This concept is actually an old philosophical one dating back to Plato and Aristotle, and has inspired thinking about the world order up until the nineteenth century.[2]

The second doctrine prescribed the age of the Earth, namely about 6,000 years. Both doctrines essentially prevented any thinking about *time*, thereby locking the universe, the Earth, and its inhabitants into a static stranglehold from which only the adventurous thinker could free himself.

The questions that at that time were encouraged were mostly concerned with classification. The first name to mention here is of course that of Carl Linnaeus (1707–1778), the Swedish botanist (and son of a pastor) who instead of following in his father's footsteps became obsessed with collecting and studying plants. He became known as the first *taxonomist* (meaning one who studies the general principles of classification of biological organisms) largely through his main work, the *Systema Naturae* (first published in 1735) which went through many editions (Linnaeus 1766).

---

1. Thomas Browne (1643) famously wrote: "Thus there are two Books from whence I collect my Divinity, besides that written by God, another of His servant Nature, that universal and publick Manuscript that lies expans'd unto the Eyes of all: those that never saw Him in the one, have discovered Him in the other."

2. Alexander Pope, in his *Essay on Man* (1733) pronounces:

"Vast chain of Being! which from God Began,
Natures aethereal, human, angel, man,
Beast, bird, fish, insect, what no eye can see,
No glass can reach; from Infinite to thee(. . .)"

FIGURE 1.5. Depiction of the great chain of being, from *Rhetorica Christiana* (1579) by Diego de Valadés.

Linnaeus's work was important not just because of its attempt to put order into the seemingly unbounded variety of plants and animals, but also because it started a fashion trend of sorts: to find hitherto undiscovered species and have them named after the discoverer. It was only through this combined effort of classification and discovery that the people of the eighteenth century would slowly acquire a grasp of what kind of life was out there sharing the planet with them. This was, after all, a time when tales of monsters with

FIGURE 1.6. Portrait of Carl Linnaeus by Alexander Roslin (Nationalmuseum Stockholm).

several heads, giants, mermaids, men with tails, etc., were widely believed by the literate public.

Because this effort of classification was ostensibly one of cataloguing God's creation, the idea of extinct species, or even *novel* ones, was still not one anybody would openly entertain. But, because it is clear that without a serious classification effort it would hardly have been possible even to make a claim about extinct or recent species, we can see that Linnaeus, while steeped in his time, was preparing the world for far greater discoveries.

Among those who conformed to the general belief system walked a few who dared to heretically question some of those most deeply held beliefs. Around the time of publication of Linnaeus's book, another famous tome was being read and discussed, this one by Benoit de Maillet (1656–1738), a French nobleman, later appointed Consul General of King Louis XIVth in Cairo.

De Maillet was an amateur as far as geology and natural history was concerned, but he showed a shrewd sense of discovery and deduction. Armed

FIGURE 1.7. A portrait of Benoit de Maillet (from de Maillet and Mascrier 1735).

with those, he attempted to make sense of the evidence available to him, link it, and construct a view of the universe. What is remarkable is that his theory of the universe does not involve a God, and envisions an Earth that is subject to external natural forces that change it, as opposed to catastrophes willed by a creator. He thus implicitly questioned one of the most important of church doctrines, namely that of the fixity of the Earth.

In his book *Telliamed* (de Maillet 1750), first published anonymously and widely read only after his death, de Maillet staunchly opposes the biblical deluge myth and suggests a much older age of the Earth. He deduces both opinions from observations, which he insists should hold preponderance over beliefs handed down from generations. In this sense, de Maillet was a radical revolutionary. His attack on religious dogma is particularly vitriolic in the following passage, where he speaks about the type of people who might reject his theory without giving due consideration to his facts:

The Case is not the same with another Class of Persons, to whom this Idea of Novelty and Singularity will perhaps appear a just Reason for condemning the Work; I mean those persons remarkable for their excessive Scruples and Delicacies in point of Religion. I grant indeed, we cannot too much respect this Delicacy, when it is enlightened and guided by reason; but it is equally certain, that this excessive Zeal sometimes only proceeds from Ignorance and Meanness of Spirit, since it often degenerates into false Prejudices, and a barbarous and ridiculous Blindness; that without giving a Shock to Religion, we may boldly attack ill-grounded Scruples, which are only the Effects of an inexcusable Superstition; and that if we were obliged to support the pure and salutary Ideas of the former, we are equally bound to oppose the Propagation of the stupid opinions set on Foot by the latter (. . .)

Even though the theory of the universe in *Telliamed* is being advocated by an Eastern philosopher (by the name of Telliamed) who is being interviewed by the God-fearing author, this ruse is only thinly disguised given that *Telliamed* is the author's last name spelled backward!

This is not to say that de Maillet anticipated Darwinism in any real sense. While he advocated the possibility that species could transform, and some species would evolve from sea dwellers to land animals, he also believed the stories of giants, and mermaids and mermen. So while he mixed legends and observations in support of his theories of the universe, he was unequivocal about the interpretation of fossil shells and animals discovered in strata far higher than the current sea level: species can go extinct, new ones can emerge, and the Earth is subject to constant forces of erosion that shape its appearance. Because de Maillet was born in 1656, we should really see in him a precocious precursor of the enlightenment that was to follow in the eighteenth and nineteenth centuries.

Perhaps the most important figure in natural history in the middle of the eighteenth century was the Comte de Buffon (1707–1788), more precisely George-Louis Leclerc, Comte de Buffon. Buffon's influence on the history of science would likely have been far greater had he not become obsessed with the idea of cataloging all existing knowledge in the fields of natural history, geology, and anthropology, which he did in his forty-four-volume *Histoire Naturelle* (Buffon 1749–1804).

Within the volumes depicting animal life, however, can be heard the voice of the scientist who attempts to make sense of all this variety. And, scattered among the many volumes that he wrote can be found essentially all the elements that are necessary to put together the theory of evolution. In particular, he advocated a theory of "degeneration" (by which he essentially meant change and progress) to link one species to another. He also clearly saw the evidence of fossils and therefore maintained that not only are there extinct species, but that new types emerge continuously. Finally, using

FIGURE 1.8. George-Louis Leclerc, Comte de Buffon, detail of a painting by Francois-Hubert Drouais at Musée Buffon.

geological arguments he refuted the church's claim of a 6,000-year-old Earth and proposed a much longer one (while still far off from what we know today). According to the sign of the times, though, he made these claims not as forcefully as Darwin would later dare, but somewhat timidly, followed immediately by affirmations of the church's general doctrines of thinking, often thereby contradicting himself in subsequent sentences.

While Buffon's *Histoire Naturelle* was widely read and admired in his time, he was less well known as an original thinker in natural history, simply because he never portrayed his thoughts as theory. It is entirely likely that he thought it best to throw in such ruminations to break the monotony of the animal descriptions that make up the bulk of his work. Still, in retrospect we can see that every one of the elements that Darwin used to synthesize his theory were already available at the middle of the eighteenth century, for somebody who had the perspicacity to appreciate their importance and the courage to put them together.

Inspired by the ideas of Buffon, Erasmus Darwin (1731–1802) forms a link to the nineteenth century. A respected physician and poet, Erasmus—who is the grandfather of Charles Darwin—had a serious interest in natural history and strongly entertained the idea that *species can change without limits*. In particular, he was convinced that the similarities between the different forms of life were due to ancestral relationships, and were changing and adapting through time. Still, these ruminations of the grandfather of the famous grandson (Erasmus Darwin died seven years before Charles was born) were not formulated as a coherent theory, but rather were observations about nature contained in his books *The Botanic Garden* (E. Darwin 1791) and *The Temple of Nature* (E. Darwin 1802), both in verse. His vision of such a theory, however, cannot be denied, as we can read in his *Botanic Garden*:

> As all the families both of plants and animals appear in a state of perpetual improvement or degeneracy, it becomes a subject of importance to detect the causes of these mutations.

Before venturing into the nineteenth century, two more influences on Darwin should briefly be mentioned. Thomas Malthus's (1766–1834) *Essay on the Principle of Population* (Malthus 1798) is often cited as having given Darwin the inspirational spark for his theory by emphasizing the "struggle of existence" going on everywhere, among animals and plants as well as humans. The main point that Malthus tried to make in his essay concerned the danger of overpopulation in the face of limited resources, in particular among the poor. Indeed, Darwin himself remarks in his autobiography edited by his son Francis (F. Darwin 1887, p. 68):

> In October 1838, that is, fifteen months after I had begun my systematic inquiry, I happened to read for amusement Malthus on *Population*, and being well prepared to appreciate the struggle for existence which everywhere goes on from long-continued observation of the habits of animals and plants, it at once struck me that under these circumstances favourable variations would tend to be preserved, and unfavourable ones to be destroyed. The results of this would be the formation of a new species. Here, then I had at last got a theory by which to work.

It seems somewhat odd, though, that it would be in Malthus's essay that Darwin first heard about the idea of a struggle of existence (see Eiseley 1958, 180 for a discussion of this point), as several authors had already extensively discussed it by the time Darwin read Malthus, in particular his good friend Lyell (whom we meet later), and the Reverend William Paley (1743–1805), whose writings concerning the "Evidence for Christianity" were required reading at

FIGURE 1.9. Erasmus Darwin by Joseph Wright of Derby (1792), at the Derby Museum and National Gallery.

Cambridge University, where Darwin took his B.A. In fact, Darwin professed to be fascinated by the logical deductive approach taken by Paley, almost as if theology was a subbranch of mathematics. In his book *Natural Theology* (Paley 1802) (famous for its comparison of complex life to a finely tuned watch, and the argument that just as the watch needs a watchmaker, life would need a creator) the Anglican priest Paley attempted to prove the existence of God by observing and showcasing astonishing details of adaptation. While we now know Paley's solution to the puzzling complexity of organic life to be wrong, he was certainly right to be puzzled, and his zeal to document the extent of the intricate complications and adaptations of life make him a worthy naturalist on the cusp of the nineteenth century.

1.2.2 Nineteenth century

The eighteenth century was marked more by the recognition that natural laws govern the movements of planets, and a growing awareness of the universe around us through astronomical discoveries, than a leap in our understanding

FIGURE 1.10. Thomas Robert Malthus, Mezzotint by John Linnell. From Wellcome Collection (CC BY 4.0).

of the natural world. But the groundwork had been laid, and in short succession several important new elements and ideas emerged.

In the year Erasmus Darwin passed away, the German physician (later professor of medicine and mathematics at the University of Bremen) Gottfried Reinhold Treviranus published his ruminations about the origin of species in a book entitled *Biologie, oder Philosophie der lebenden Natur*[3] (Treviranus 1802). In this book he set forth a theory of the "transmutation of species," arguing that each species had the potential to change in response to changes in the environment, and that it is this capacity that lies at the origin of the observed diversity of species. He writes:

> In every living being there exists a capacity of endless diversity of form; each possesses the power of adapting its organization to the variations of the external world, and it is this power, called into activity by cosmic changes, which has enabled the simple zoophytes in the primitive world to

3. Biology, or Philosophy of the Animate Nature

FIGURE 1.11. Reverend William Paley, painting by George Romney (National Portrait Gallery, London).

climb to higher and higher stages of organization, and has brought endless variety into nature.

While it is clear that Treviranus does not pretend to know the origin of the power to change that he sees inherent in every organism, he does glimpse lines of descent (in the form of "lines of transmutations") that span from the simplest forms all the way to us.

The idea of "adaptation" as the core of species diversity was (presumably independently) taken up by Jean-Baptiste Lamarck (1744–1829), who thought that he had (unlike Treviranus) hit upon the origin of the power to change: the use and disuse of organs in response to environmental changes. Lamarck echoed the poetic ruminations of Erasmus Darwin in a more scientific manner and, while influenced by Buffon just like Erasmus Darwin, was a man more on the cusp of the new century. After a career in the army, Lamarck took a post at the National Museum of Natural History in Paris, where he attempted to classify "insects and worms" (for which he coined the

FIGURE 1.12. Gottfried Reinhold Treviranus (1776–1837).

word "invertebrates"). For him, it was clear that species could change, but he also speculated that this change was not random, but instead occurred as a response to the environment. In particular, he figured that the influence of the environment was *direct*: if the leaves on an acacia tree are so high to be almost out of reach (to quote a famous example), then those giraffes that stretch their necks highest will survive preferentially. However, contrary to the genetic and heritable origins of an elongated neck that constitutes Darwinian evolution, Lamarck reckoned that a giraffe that had acquired a longer neck due to a lifetime of straining to reach the leaves would bear offspring with just as long a neck. In other words, his theory of evolution—and the system he outlined in *Philosophie Zoologique* (Lamarck 1809) certainly qualified as such—was based on the inheritance of such *acquired* characteristics.

Apart from invoking this particular mechanism to create diversity (after all, the laws of inheritance were only discovered quite a bit later by Mendel) Lamarck's theory reads curiously like that laid down later by Charles Darwin, but with *adaptation* at its core. Unlike Darwin, who professed that while there seems to be a general trend from the simple to the complex in evolution (a concept we will examine in detail in later chapters) and who could see

FIGURE 1.13. Jean-Baptiste de Monet, Chevalier de Lamarck. Portrait by Jules Pizzetta in *Galerie des naturistes*, Paris: Ed. Hennuyer (1893).

adaptation giving rise to more *or less* complex organisms, Lamarck believed fervently that evolution would produce *only* advancements, that nature was constantly improving organisms. Note, however, that because *all* animals, according to Lamarck's view, would "improve" their organs through use (and lose functions of others through disuse) natural selection through competition for limited resources is not an important element in this system. While being very much inspired by Lamarck's book, Darwin would later dismiss it as an "error" for its failure to recognize the importance of natural selection.

Even though Lamarck's *Philosophie Zoologique* should have shocked and amazed his contemporaries as the first system to explain the diversity and evolution of species, it went largely unnoticed. Lamarck's success was constantly being undermined by his much more successful colleague Cuvier (whom we meet momentarily), and he died blind, destitute, and forgotten.

Georges Cuvier (1769–1832) plays a starring role in any exploration of the origins of Darwinian thought because he was the first one to pay attention to the details of anatomy in a scientific manner. Cuvier worked and taught at the National Museum of Natural History alongside Lamarck and developed a system of comparative anatomy that would allow him to reconstruct the skeleton of entire organisms from the examination of a few or even a single bone. This system was based on correlations and similarities between bones that he had studied, and a belief that each organism's manner of living and precise function

FIGURE 1.14. George Cuvier (1831). Engraving by George T. Doo after a painting by W. H. Pickergill.

would affect each and every bone in its body; that the functioning of one organ would affect all others in such a manner that the entire *plan* of the organism was defined by them. While such a holistic view of organism structure appears somewhat mystic by today's standards, there is no denying that Cuvier took his skills in comparative anatomy to heights resembling magic.

For example, an ongoing excavation of the rock formation of the Paris Basin provided him with heaps of bones of large and extinct animals, which he proceeded to resurrect (so to speak), classify, and name. The importance of this skill for the development of a theory of evolution can be felt from retelling just two of his exploits. In another example, he examined a slab of granite containing a fossil that was described in 1726 and that was thought to contain the remains of the skeleton of a man who lived before the floods. This "relique" was soon put into its place when Cuvier demonstrated that the fossil was really that of a giant (now extinct) salamander. Clearly, this was *not* the eighteenth century anymore! In 1798, he published a study of elephant bones that proved that not only were the African and Indian elephants different species, but that they were also quite different from the fossil mammoths discovered in Europe and Siberia. Thus, he established evolution as a *fact*, and proved without the possibility of doubt that certain animal species were indeed extinct (as has been suggested by others before, of course).

More generally, Cuvier's precise analysis of anatomy led to the total destruction of the idea of a greater chain of being, that all organisms could be arranged in an unbroken chain reaching from the simplest all the way to God. Instead, he showed that a number of different body types had been evolving for ages completely in parallel, and that there was no conceivable way in which they could be put into a sequence. Thus, evolution did not proceed as the ladder or chain that was imagined during antiquity, but instead rose up more like a bush, with innumerable twigs, some prospering, some not. This type of insight, and the precision and diligence with which it was obtained, proved to be inspirational for Darwin, along with the momentous changes that occurred in the field of geology.

When discussing the thoughts that influenced Darwin and moved him on to the path to the *Origins*, most agree that the work of the geologist Charles Lyell (1797–1875) takes the crown. To gauge the importance of Lyell's work, we should remind ourselves again of the thinking with respect to geology that was current in this era. Up until the time that Lyell published his most important work, the three-volume *Principles of Geology* (Lyell 1830–1832), it was almost universally accepted that the present state of the Earth was a result of the biblical deluge, and that the forces that shape the Earth were due to divine interventions. This theory of *catastrophism* was popular despite the work of James Hutton (1726–1797), who in the eighteenth century had argued against catastrophism by invoking an eternal cycle of natural forces to which the Earth's features are exposed, building and reshaping surfaces, mountains, and seas. Hutton's *uniformitarianism* was the first blueprint of a physical theory of the Earth, in which processes of erosion and rebuilding are the main characters, and the surface of the Earth itself serves as the memory of bygone days. In particular, it was Hutton who boldly exclaimed that this process of erosion and regeneration showed that time was unlimited, "that we find no vestige of a beginning, no prospect of an end" (Hutton 1795). On the other hand, Hutton was a devout Christian who believed that the Earth was formed in this manner by God to allow human habitation, and the influence of his ideas gradually waned.

Enter Lyell, who after a brief career as a lawyer decided to devote himself full-time to his geological studies, while marshaling his talents of persuasion. The publication of the *Principles* irrevocably cemented the concept of unlimited time and the operation of natural forces into geology. Lyell's book was extraordinarily popular not only within scientific circles but also with

FIGURE 1.15. Charles Lyell, drawn by H. Maguire (National Library of Medicine).

the public at large, and contributed to a change in attitude among what would later constitute Darwin's audience. Darwin read the first edition of the *Principles* while on board the *Beagle* and was deeply impressed. Lyell's version of uniformitarianism, sans the mystic and obscure overtones of Hutton's theory, argued for a natural origin of the features of the Earth, while Darwin himself was searching for a natural origin of the species inhabiting it. The style of argumentation and the abundance of evidence presented to make his case deeply resonated within Darwin, and he appears to have copied some of Lyell's style in writing the *Origins*, which is dedicated to Lyell.

Armed with so much geological evidence and knowledge of the fossils, it may appear somewhat curious that Lyell did not discover evolution himself, or even strongly believe in its reality until very late in life. After all, Lyell wrote abundantly about the diversity of flora and fauna, as well as its distribution in relation to geology and geographic location; indeed, he was even the first to use the words "struggle for existence"! The reason for Lyell's failure to see the

FIGURE 1.16. James Hutton. Detail of a painting by Sir Henry Raeburn (1776) (Scottish National Portrait Gallery).

natural forces operating on the level of species can be found in his deep belief in the principles of uniformitarianism. This view argues for constant natural forces shaping and reshaping the Earth, but it does not call for *progress*. Indeed, right around the time that Darwin prepared the *Origin* for publication, Lyell was advocating a theory of *nonprogressionism* for the species problem, which mirrored his uniformitarianism. On the one hand, it is astounding how close Lyell came to anticipating Darwin when he was describing the mechanisms by which old species are being replaced by new ones in their struggle for existence in an environment with limited food supply. On the other hand, these thoughts are permeated with the concept of a cyclic nature of uniformitarian forces and do not anticipate the evolution of radically new forms of life and their common descent. Instead, the species that come and go and succeed each other in Lyell's view are all somewhat similar variations on a theme, coming and going eternally.

It is of course impossible to close a chapter on the evolution of Darwinian thought without mentioning the influence of Alfred Russel Wallace (1823–1913), a naturalist and explorer from Wales, whose forays into

FIGURE 1.17. Alfred Russel Wallace. London Stereoscopic & Photographic Company.

evolutionary theory famously prompted Darwin to rush the publication of his *Origins*. Wallace was interested in the causes of the different *distributions* of species around the globe, a pursuit now known as *biogeography*. Observing different species and the relationship between their abundance and fecundity, he observed that it was the availability of food, rather than any other factor, that predicted the abundance of any particular species. From this he deduced that *varieties* of species (members of a species that differ—heritably—in small characteristics) that are better at exploiting the available resources would increase in number and ultimately supplant the original species. Consequently, he argued, varieties could become *permanent* in this manner, in contradiction to the then-current belief that varieties were "unstable," and revert to the ancestral type unless carefully maintained by breeding (Wallace 1858).

FIGURE 1.18. Charles Darwin (ca. 1854). Photograph from the frontispiece of (Darwin 1887).

In a way, Wallace's observation short-circuited the long and arduous work of Darwin and may be seen as demonstrating that the time had arrived where the idea of evolution—including all of its consequences—was inevitable. Yet, without Darwin's championship of Wallace it is doubtful that Wallace's idea would have had a great impact, and neither was Wallace prepared to defend the theory as Darwin had been, due to the preceding twenty years of work. Consequently, the credit for the theory duly goes to Darwin, who worked tirelessly on his "dangerous idea" (Dennett 1995), but had to be coaxed into revealing it to the public out of fear of being scooped.

1.3 Summary

Darwinian evolution, from a process-oriented point of view, relies on three elements acting simultaneously on a population of sequences, composed of elements drawn from an alphabet. Sequences have to be replicated to guarantee *inheritance*, an agent of *modification* has to act in order to ensure variety among the sequences, and the environment should be *selective* with respect to the particular sequences, ranking them according to their expected fitness. This expected fitness is a number that is attached to each sequence that could in principle be measured, but it is not synonymous with survival. It is simply an attribute of the sequence that we expect to be a good *predictor* of survival, just as the level of mercury in a thermometer is used to predict the temperature (but we would not say that the mercury level is the *same* as the temperature, and that measuring temperature therefore is tautological). Besides selection, *competition* for limited resources is an important component of evolution. Because unlimited resources are only a theoretical possibility, competition is always present at least in the long term.

Darwinian evolution over time gives rise to closely related groups of organisms that we may call *species*. The process of speciation, over geological time scales, has given rise to all types of organisms living on Earth today, forming an unbroken link between today's biocomplexity and the first forms of life over 3.5 billion years ago. In this manner, Darwin's insight has explained the origin of species, even though the true origin of the mutations that produce the variance he observed (changes in an organism's germ cell DNA) would only be discovered much later.

While evolution today seems as simple an idea as putting two and two together, the very thought that species could originate, change, and go extinct was, for the longest time, heretical and therefore uttering it was bordering on suicide. In the Western hemisphere, Roman Catholic doctrine had created an intellectual prison from which it would prove fiendishly difficult to escape. From revolutionary ideas couched in mystic language to scientific observations reluctantly juxtaposed with church doctrine, from geological and stratigraphic discoveries over the meticulous comparison of fossil bones, the strands that would make Darwin's theory slowly emerged in the eighteenth and nineteenth century, ready for him to weave into a satisfying explanation of the origin and diversity of species.

It took the initiative of a few brave people who were not afraid to possibly die at the stake, so that the rest of the world could first learn that the Earth is not the center of the universe, nor does the Sun revolve around it. This humiliation of humanity by removing it from its exalted place in the universe was a necessary precursor to the demolition of Earth as a place fabricated in time and space by a deity, and this allowed the blossoming of scientific methods and made thoughts about the mutability of the species ("it is like confessing a murder!") finally possible.

2

Information Theory in Biology

It has not escaped our notice that the specific pairing we have postulated immediately suggests a possible copying mechanism for the genetic material.

—J. D. WATSON AND F. H. C. CRICK (1953b)

The quest for a reductionist view of life has yielded perhaps the most fundamental insight in all of biology, namely that the basis of life—certainly all life
on Earth, but most likely even beyond—is information. This insight, almost
commonplace today in light of the ever-accelerating rate at which the genomes
of whole organisms are sequenced and deposited in databanks, emerged only
slowly from meticulous work (see Box 2.1).

The crucial role of DNA as the carrier of inherited information was
cemented by the discovery of DNA's structure by Watson and Crick (1953b,
1953a) because the structure immediately suggested a mechanism of *duplication* of the information (see the quote heading this chapter) as well as its
modification by mutation, thus unifying the ideas of inheritance and evolution. Because information—in its precise mathematical form—plays a crucial
role in our understanding of complexity, its evolution and the concepts of
adaptation and *progress*, we will spend time in this chapter to introduce those
aspects of the *theory of information* that are relevant for evolutionary biology,
and proceed to apply them to molecular sequence data. While the theory of
information finds many other applications in biology, for example in neurobiology (Borst and Theunissen 1999; Bialek et al. 2001), we will not dwell on
those here (but will take them up more formally in chapter 9).

In the last quarter-century, we have gotten used to language describing
the cellular machinery as involved in the copying, processing, and transmission of information (Nurse 2008), and yet such descriptions are not usually
accompanied by a quantitative analysis. How much information is encoded in
one organism's genome? How does it compare to the amount of information
carried by another, and how much is shared between them? How much

Box 2.1. The Emergence of the Concept of Genetic Information

That cells contain self-replicating "elements" that are responsible for distinct inheritable characters was, of course, first noted by Mendel (1865). The path toward recognizing these elements as carriers of information was circuitous, however. Bateson and Saunders (1902) found that these discrete units could take on two or more states (the "alleles"), just like the random variables of probability and statistics introduced below. It was also necessary to realize that the *carriers* of information were logically and physically separate from their *expression* (that is, from the characters). This insight belonged to the Danish botanist Wilhelm Johannsen (1909), who coined the word "gene." But perhaps the most lasting impression (besides, of course, the work of Watson and Crick) was made by the geneticist Thomas Hunt Morgan along with his students Alfred Sturtevant and Calvin Bridges at the California Institute of Technology, who were able not only to draw "chromosome maps" of a fruit fly's genes, but also to show that their linear order on the map corresponds to their location on the chromosome (Morgan et al. 1925) (see Fig. 2.1).

These maps implied that the genetic information for particular characters had specific locations within a chromosome, and that modification (mutations) of these locations could destroy the function associated with the information. One of the last pieces of the information puzzle was the recognition that DNA controls the synthesis of proteins (Beadle 1946), and that it represents messages that encode protein structure via a simple translation code (Nirenberg et al. 1965). Finally, Maynard Smith (1970) observed that proteins—and by analogy also the DNA messages that code for them—could be viewed as elements of a coding space of random variables. By the time Jacques Monod argued that the function of biological molecules is *symbolic* (Monod 1971), the first comprehensive book applying some concepts from Claude Shannon's theory of information directly to biomolecules was about to appear (Gatlin 1972).

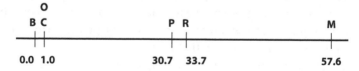

FIGURE 2.1. A figure showing the arrangement of genes on a chromosome and their relation to a map of genes in the fruit fly *Drosophila*. Adapted from Sturtevant (1913).

information is transmitted from generation to generation, gained in an evolutionary adaptation, or lost due to extinction? It is possible, in principle, to answer all of these questions if the information stored in genomes—preferably in each gene—can be measured. As we will see later in this chapter, this is indeed possible but can turn out to be exceedingly difficult in practice. Indeed, the quantitative framework to attempt such a monumental task has been around since 1948, when Claude Shannon, an engineer at Bell Laboratories in New Jersey, published a pair of papers that started one of the most successful frameworks in all of science: the theory of information (Shannon 1948).

Shannon's theory has had a tremendous influence in engineering disciplines because it not only quantifies the concepts of entropy and information, but also characterizes information transmission channels and provides bounds on how well information can be protected from noise. The concepts of error correction and channel capacity introduced by Shannon are put to practical use every day in any digital or analogue communication, and virtually every digital music player. In this chapter, we take a look at the basic constructions of information theory (entropy and information) and apply them to molecular sequences. While the exposition is mathematical, the concepts themselves are actually very intuitive. After all, Shannon's information concept corresponds precisely to what we mean when we use the word "information" in natural language. For this reason, all mathematical expositions are followed by explicit examples wherever possible.

One of the first things to keep in mind is that the concept of information is inherently *probabilistic*. Most—though not all—statements and laws of information theory apply to *averages* but cannot predict individual events. This is of course true for most statements in the theory of *probability and statistics*, and indeed information theory is based on both. For example, statistics tells us that a true coin will land on heads or tails half the time each. While this prediction is as firm as any we can make in statistics, it does not make *any* prediction about the outcome of the next throw. Coins and dice are standard examples that stand in for the mathematical concept of a *random variable*, and we will encounter them again. But we will also use *monomers* as random variables: nucleotide "variables" that can take on four states, or amino acid variables that can take on twenty states, and treat them essentially as we would a coin or a die.

2.1 Random Variables and Probabilities

A (discrete) random variable X is a mathematical object defined to take on a finite number s of specified *states* $x_1, \ldots, x_s$ with specified probabilities $p_1, \ldots, p_s$. While technically the states x_i must carry numerical values (so that we can speak about the average and variance of a random variable), we will often refer to the states themselves, such as heads or tails. The space of the possible states of X is often called an *alphabet*, and the set of p_i is called the *probability distribution* $p(x)$.

In general, there are many different realizations of any particular random variable, and the set of these variables is called an *ensemble*. Therefore, when we ask questions about X, we are really asking questions about the ensemble from which particular realizations of X (taking on the states $x_1, \ldots, x_s$) are drawn from.

We should always bear in mind that the mathematical random variable is a description—sometimes accurate, sometimes not—of a physical object. In other words, the random variable X with two states $x_1 = $ heads and $x_2 = $ tails and $p_1 = p_2 = 0.5$ is not the same thing as a coin. A coin is a much more complicated object: it might deviate from being true, and it might have more than two states—perhaps it can land on its edge once in a while. When we describe the statistical properties of a physical object using a mathematical random variable, we should be mindful that we are making a number of important assumptions, the most important of which is a tacit agreement to distinguish only a given number of possible states. Because this assumption is tacit, it is often overlooked that this is a *crucial* restriction. For a coin, for example, we could have counted many more different states than, say, $x_3 = $ edge. What about the markings on either side of the coin? Could we not measure the angle the head or tail markings make with, say, geographic north for each throw, and use it to label more states? If we distinguish n angles for each side of the coin, we would have a total of $2n$ states (plus the edge states). Clearly, this reasoning can be continued to absurdity ("let us count the number of excited states of the nuclei that make up the atoms that make up the metal of which the coins are made. . ."), and we see why it is necessary to first agree (between sender and receiver), which states to consider (or in other words, what measurement device to use) when discussing the possible states of a physical object (Adami 2011; Adami 2016). This agreement is tantamount to assigning a mathematical random variable to a physical object.

Just as the number of states are never given a priori, for a physical object the probabilities $p_1, \ldots, p_s$ are usually also not known with arbitrary precision. They are commonly obtained by experiment, for example by repeating coin flip experiments, and approximating the probabilities with the frequencies $M(x_i)$ of outcomes given a finite number M of trials. So for a coin, we might find

$$p_1 \approx \frac{M(x_1)}{M}, \quad p_2 \approx \frac{M(x_2)}{M}, \quad M(x_1) + M(x_2) = M. \qquad (2.1)$$

Clearly, only in the limit $M \to \infty$ will our estimate of the probabilities reflect the dynamics of the physical object precisely.

Let us apply these concepts directly to DNA. Suppose we are given a sequence

AGAGCGCCTGCTTTGCACGCAGGAGGTCTGCGGTTCGATCCCGCATAGCTCCACCA

$$(2.2)$$

The letters A, C, G, and T stand for the four nucleotides that are the "building blocks" of DNA. A stands for Adenine, C for Cystein, G is Guanine, and T stands for Thymine. There are several ways to interpret such a sequence of letters probabilistically. First, we can say that this is but one of 4^{56} possible DNA sequences of length $L = 56$ nucleotides, that is, we could view this sequence as one possible state of a random variable that can take on 4^{56} possible states. Of course we do not know how likely this sequence is, but if we assume that all states of this random variable are taken on with equal probability (which we would have to, in the absence of any evidence to the contrary), then this would imply that each nucleotide position takes on any of its four possible states with equal probability $p_i = 1/4$. However, the sequence itself provides us with information to narrow down these probabilities.

Instead of treating the entire sequence of length 56 as a random variable, we could also view this sequence as the record of fifty-six independent experiments checking the state of one random variable X that can take on the four states $x = $ A, T, G, C only. Viewed this way, the record above would allow us to infer that for this sequence

$$p_A \approx \frac{10}{56} \approx 0.18, \; p_T \approx \frac{11}{56} \approx 0.20,$$

$$p_G \approx \frac{16}{56} \approx 0.29, \; p_C \approx \frac{19}{56} \approx 0.34. \qquad (2.3)$$

Note that these estimates are quite uncertain (by about 19 percent for p_C, higher for the others) and only give us limited knowledge about the nature of the gene, because the deviation of the probabilities above from the random unbiased assumption $p_i = 1/4$ for all nucleotides is not very significant. However, for longer sequences, such an analysis can reveal a real bias in nucleotide usage. For example, while random sequences would show an equal usage of GC pairs and AT pairs, the genomes of many organisms show a distinct bias.[1] For example, the soil bacterium *Streptomyces coelicolor* has a

---

1. Because DNA occurs as a double-stranded molecule in organisms, a direct analysis of the frequencies of G and C in a cellular extract always reveals the same number—an observation called *Chargaff's rule* (Chargaff 1951)—so that usually only the G+C content is reported.

GC bias of 71.17%, that is $p_G + p_C = 0.7117$ (Bentley et al. 2002), while the common baker's yeast *Saccharomyces cerevisiae* is AT-rich in comparison: $p_G + p_C = 0.387$ (Dujon 1996). Such deviations from the unbiased estimate $p_G + p_C = 0.5$ may be adaptive in nature via the bias's biochemical consequences. For example, because G is paired with C using three hydrogen bonds rather than the two in AT pairs, GC base pairs can withstand higher temperatures before dissolving. However, nonadaptive processes can also change the GC content, as we will discuss in more detail in section 5.3.

Probabilistic analyses *along* a sequence or gene are an important tool for classifying whole genomes, single genes, or even regions within a gene. A large body of work has accumulated that uses stochastic methods to look for patterns and long-range correlations within sequences that can shed light on, for example, whether or not a sequence is likely to code for a gene or not (Fickett 1982) (including information-theoretic ones; Grosse et al. 2000). But there is another way of analyzing nucleotide frequencies that can reveal far more information, by looking at the probability to find a particular nucleotide at a particular *site* in the sequence. Thus, rather than considering the state of a single random variable X and using the values along the sequence at sites 1 to L as independent outcomes, we could define L random variables X_i, one for each site, so that X can be written as the joint variable

$$X = X_1 X_2 \cdots X_L. \tag{2.4}$$

To gain statistics to narrow down the probabilities p_i for each site (each random variable X_i), we now need a set of sequences (the aforementioned *ensemble*) so that we can count the values X_i takes on *across* sequences. Suppose, for example, that we have access to the sequences of the same gene of evolutionarily related organisms (this set constitutes our ensemble). Because of the balance of mutation and selection going on at each site, we expect those sites that are important for the survival of the organism to be much more conserved (meaning unvarying) than those that are irrelevant. To wit, because DNA codes for amino acids in triplets of nucleotides (called "codons," see section 2.3.1) the third position of many codons is unimportant in the determination of the corresponding amino acid, so that a mutation at this position would be *selectively neutral*. Consequently, we expect such nucleotides not to be informative.

Let us take a look at thirty-two more sequences of the type displayed above, and align them (see Table 2.1). Note the extra symbols '-' and '.' in this alignment. The symbol '-' denotes a "missing" symbol, that is, a deletion in that sequence with respect to the other sequences. On the other hand, the symbol '.' signals that the identity of the nucleotide at that position is not known.

Table 2.1. Alignment of the last fifty-six nucleotides of DNA sequences coding for tRNA genes of *E. coli* (with different anticodons)

```
AGAGCGCCTGCTTTGCACGCAGGAGGTCTGCGGTTCGATCCCGCATAGCTCCACCA
AGAGCGCTTGCATGGCATGCAAGAGGTCAGCGGTTCGATCCCGCTTAGCTCCACCA
TATGTAGCGGATTGCAAATCCGTCTA-GTCCGGTTCGACTCCGGAACGCGCCTCCA
AGAATACCTGCCTGTCACGCAGGGGGTCGCGGGTTCGAGTCCCGTCCGTTCCGCCA
AGGACACCGCCCTTTCACGGCGGTAA-CAGGGGTTCGAATCCCCTAGGGGACGCCA
AGAGCAGGGGATTGAAAATCCCCGTGTCCTTGGTTCGATTCCGAGTCCGGGCACCA
ATTACCTCAGCCTTCCAAGCTGATGA-TGCGGGTTCGATTCCCGCTGCCCGCTCCA
AGAGCACGACCTTGCCAAGGTCGGGGTCGCGAGTTCGAGTCTCGTTTCCCGCTCCA
AGAACGAGAGCTTCCCAAGCTCTATA-CGAGGGTTCGATTCCCTTCGCCCGCTCCA
AGAGCCCTGGATTGTGATTCCAGTTGTCGTGGGTTCGAATCCCATTAGCCACCCCA
AGAGCGCACCCCTGATAAGGGTGAGGTCGGTGGTTCAAGTCCACTCAGGCCTACCA
AGAGCAGGCGACTCATAATCGCTTGGTCGCTGGTTCAAGTCCAGCAGGGGCCACCA
AGAGCAGTTGACTTTTAATCAATTGGTCGCAGGTTCGAATCCTGCACGACCCACCA
AGAGCACATCACTCATAATGATGGGGTCACAGGTTCGAATCCCGTCGTAGCCACCA
AGAACGGCGGACTGTTAATCCGTATGTCACTGGTTCGAGTCCAGTCAGAGGAGCCA
AGCGCAACTGGTTTGGGACCAGTGGGTCGGAGGTTCGAATCCTCTCTCGCCGACCA
AGCGCACTTCGTTCGGGACGAAGGGGTCGGAGGTTCGAATCCTCTATCACCGACCA
AGCGCACCGTCATGGGGTGTCGGGGGTCGGAGGTTCAAATCCTCTCGTGCCGACCA
AAGGCACCGGTTTTTGATACCGGCATTCCCTGGTTCGAATCCAGGTACCCCAGCCA
AAGGCACCGGATTCTGATTCCGGCATTCCGAGGTTCGAATCCTCGTACCCCAGCCA
AGAGCGCTGCCCTCCGGAGGCAGAGGTCTCAGGTTCGAATCCTGTCGGGCGCGCCA
AGAGCAACGACCTTCTAAGTCGTGGGCCGCAGGTTCGAATCCTGCAGGGCGCGCCA
AGAGCAACGACCTTCTAAGTCGTGGGCCGCAGGTTCGAATCCTGCAGGGCGCGCCA
AGAGTACTCGGCTACGAACCGAGCGGTCGGAGGTTCGAATCCTCCCGGATGCACCA
ATAACGAGCCCCTCCTAAGGGCTAAT-TGCAGGTTCGATTCCTGCAGGGGACACCA
AGAGCGCACCCTTGGTAGGGGTGGGGTCCCCAGTTCGACTCTGGGTATCAGCACCA
AGAGCGCACCCTTGGTAAGGGTGAGGTCGGCAGTTCGAATCTGCCTATCAGCACCA
AGAGCAACTGACTTGTAATCAGTAGGTCACCAGTTCGATTCCGGTA.TCGGCACCA
AGAGCAGCGCATTCGTAATGCGAAGGTCGTAGGTTCGACTCCTATTATCGGCACCA
AGAGCGCACCCTTGGTAAGGGTGAGGTCCCCAGTTCGACTCTGGGTATCAGCACCA
AGAGCACCGGTCTCCAAAACCGGGTGTTGGGAGTTCGAGTCTCTCCGCCCCTGCCA
AGCTCGTCGGGCTCATAACCCGAAGATCGTCGGTTCAAATCCGGCCCCCGCAACCA
AGCTCGTCGGGCTCATAACCCGAAGGTCGTCGGTTCAAATCCGGCCCCCGCAACCA
```

Here, we will treat each of these symbols as if there was no data for the random variable at that position.[2] Inserts and deletes, as well as unresolved nucleotides or residues, are quite common in realistic sequence alignments.

2. Another possibility is to treat deletions as the fifth state (besides A, C, G, and T) of the random variable, because deletions are often crucial to the functioning of a sequence.

Let us for example examine position 8, that is, random variable X_8. Here we find the probabilities

$$p_8(A) = 5/33, \; p_8(C) = 17/33, \; p_8(G) = 5/33, \; p_8(T) = 6/33, \qquad (2.5)$$

which, even though using thirty-three sequences does not afford us good statistics, are significantly different from the unbiased assumption $p_i = 1/4$.

More drastic deviations are found for X_{33} to X_{36}, for example: only a single nucleotide can be found at these positions. Indeed, the set of sequences presented above represents the last 56 nucleotides (counting from 5' to 3') of the transfer RNA (tRNA) gene of the E. coli bacterium (see Sprinzl et al. 1998 for the sequences). Transfer RNA molecules serve as the intermediary between messenger RNA (ribonucleic acids) and amino acids in the synthesis of proteins from messenger RNA strands, and as a consequence their structure must be very precise. The positions that are absolutely conserved are therefore *essential*: mutating any of these positions to any other nucleotide results in a nonfunctioning tRNA molecule, which has severe fitness consequences for the organism. Hence, such mutations do not show up in the DNA of living bacteria. Note that because the sequences in Table 2.1 come from tRNAs with different anticodons (the anticodon specifies which amino acid the particular tRNA attaches to the forming sequence), that region (positions 14–16) does not appear conserved in this alignment.

Let us use these random variables to introduce more concepts in probability theory. As we already saw, it is possible to fashion new random variables through the union of others. For example, $Z = X_8 X_9$ is a well-defined random variable that can take on the sixteen different symbols AA, AC, AG, ..., TT with probabilities $p_{AA}, p_{AC}, p_{AG}, \cdots, p_{TT}$. Such probabilities are also called *joint probabilities*, and they can of course also be extracted from the above sequence alignment.

In general, joint random variables (like Z) can be constructed from random variables that can each take on a different number of states. Let X be random variable taking on s states x_i with probability $p_i = p(x_i) = P(X = x_i)$. Here we see three different types of notation for the probability that X takes on state x_i: $P(X = x_i)$ is often abbreviated as $p(x_i)$, which if the context allows for it, is often simply written as p_i. Y is a random variable taking on s' states y_j with probabilities $p(y_j)$. Then $Z = XY$ is a random variable that takes on the $s \times s'$ states $x_i y_j$ with probabilities

$$p(x_i, y_j) = P(X = x_i \text{ and } Y = y_j). \qquad (2.6)$$

It is not too difficult to work out that

$$\sum_j p(x_i, y_j) = p(x_i).\qquad(2.7)$$

This "summing over states of other variables" is sometimes called "marginalization" in statistics, and the probability (2.7) is then a "marginal probability."

Another important concept is the *conditional probability*, defined for random variables X, Y and $Z = XY$ as

$$p(x_i|y_j) = \frac{p(x_i, y_j)}{p(y_j)},\qquad(2.8)$$

and read as "p of x_i given y_j." This is the probability to find random variable X in state x_i if it is known that *at the same time* variable Y is in state y_j. Of course, we can also ask what is the probability $p(y_j|x_i)$ to find random variable Y in state y_j when we know that (at the same time) variable X is in state x_i. Obviously we can write this probability as

$$p(y_j|x_i) = \frac{p(x_i, y_j)}{p(x_i)},\qquad(2.9)$$

and as a consequence the two conditional probabilities can be related via

$$p(x_i|y_j) = \frac{p(y_j|x_i)p(x_i)}{p(y_j)}.\qquad(2.10)$$

The relation (2.10) is also known as "Bayes' theorem" after the reverend Thomas Bayes (1701–1767), one of the founders of the theory of probability (Bayes 1763).

The concept of conditional probabilities is perhaps easiest illustrated using binary random variables, but we will return to the nucleotide sequences shortly. Imagine two coins described by binary random variables A and B that can take on the states H and T. If they are true coins, then throwing them both together should represent the variable $C = AB$ that takes on the states HH, HT, TH, and TT. If we collect the statistics of coin A for fixed outcome of coin B (that is, given H or given T), then we should still see H and T appear with equal probability. This is because the variables A and B are independent: $p(a_i, b_j) = p(a_i)p(b_j)$.

Now imagine two "magic coins," where one of the coins always lands in the opposite state of the other (we can also see those two variables as the "facing up" and "facing down" sides of a single coin, which would make the variables less magical). For these variables, we should find $p(\text{HH}) = 0$, $p(\text{HT}) = 0.5, p(\text{TH}) = 0.5$, and $p(\text{TT}) = 0$. In this case, the conditional probabilities are (when written in terms of a matrix)

$$p(A|B) = \begin{pmatrix} p(\text{H}|\text{H}) & p(\text{H}|\text{T}) \\ p(\text{T}|\text{H}) & p(\text{T}|\text{T}) \end{pmatrix} = \begin{pmatrix} \frac{p(\text{HH})}{p(\text{H})} & \frac{p(\text{HT})}{p(\text{T})} \\ \frac{p(\text{TH})}{p(\text{H})} & \frac{p(\text{TT})}{p(\text{T})} \end{pmatrix} = \begin{pmatrix} 0 & 1 \\ 1 & 0 \end{pmatrix}.$$

$$(2.11)$$

The concept of a conditional probability is central—not just in the statistics of sequences but all of probability theory—because it characterizes the relative state of physical objects, that is, "if-then" statements (but it does not imply causation). From a fundamental point of view, all probability distributions that are not uniform (that is, probabilities that are not all equal for the different states the variable can take on) are tacitly conditional on some knowledge. Conditional probabilities such as $p(X = x_i | Y = y_j)$ simply make the condition explicit.

The above matrix of conditional probabilities states the fact that "if I find Y in state H then I *know for certain* that I will find X in state T, and vice versa." Thus, conditional probabilities are at the heart of our ability to make predictions, and therefore at the heart of the concept of information.

Let us examine conditional probabilities in the aligned tRNA sequence data displayed above. We can ask, for example, whether knowing the state of any particular random variable, say X_8, allows us to say anything interesting about the state of another variable, meaning, anything that we would not already know in the absence of knowledge of the state of X_8. Let us see whether the probabilities characterizing X_{21}, say, depend on knowing X_8. Collecting frequencies for X_{21} from the alignment in Table 2.1 gives approximately

$$p_{21}(\text{A}) = 0.24, \ p_{21}(\text{C}) = 0.46, \ p_{21}(\text{G}) = 0.21, \ p_{21}(\text{T}) = 0.09, \qquad (2.12)$$

while

$$p(X_{21}|X_8) = \begin{pmatrix} p(\text{A}|\text{A}) & p(\text{A}|\text{C}) & p(\text{A}|\text{G}) & p(\text{A}|\text{T}) \\ p(\text{C}|\text{A}) & p(\text{C}|\text{C}) & p(\text{C}|\text{G}) & p(\text{C}|\text{T}) \\ p(\text{G}|\text{A}) & p(\text{G}|\text{C}) & p(\text{G}|\text{G}) & p(\text{G}|\text{T}) \\ p(\text{T}|\text{A}) & p(\text{T}|\text{C}) & p(\text{T}|\text{G}) & p(\text{T}|\text{T}) \end{pmatrix} \qquad (2.13)$$

$$
= \begin{pmatrix} 0.2 & 0.235 & 0 & 0.5 \\ 0 & 0.706 & 0.2 & 0.333 \\ 0.8 & 0 & 0.4 & 0.167 \\ 0 & 0.059 & 0.4 & 0 \end{pmatrix}. \tag{2.14}
$$

We first notice that the probability to observe A, C, G, or T at position 21 depends very much on what is found at position 8. Indeed, it is clear that each row of matrix (2.14) represents a different probability distribution, and moreover these distributions are all quite distinct from the unconditional distribution (2.12).

We can now examine the probabilities in Equation (2.14) in more detail. Our ability to predict the state of X_{21} given the state of X_8 would be good when any of the numbers in any of the columns is significantly larger or smaller than 0.25, which is the probability to correctly guess the state of a nucleotide by chance. In fact, only a very few of the probabilities are even above 0.5, but there are a number of probabilities that vanish, which is also predictive. However, there appears to be no particular trend in how one of the sites predicts the other.

We will learn below how to quantify the amount of knowledge gained about one random variable by revealing the value of another, but in the meantime, let us look at another pair of sites, this time X_8 and X_{22}. We can obtain the *unconditional* probabilities for X_{22} from the alignment (as before, we are rounding the probabilities)

$$
p_{22}(A) = 0.18, \; p_{22}(C) = 0.15, \; p_{22}(G) = 0.52, \; p_{22}(T) = 0.15, \tag{2.15}
$$

but the conditional probability matrix is very different from (2.14):

$$
p(X_{22}|X_8) = \begin{matrix} & \begin{matrix} A & C & G & T \end{matrix} \\ \begin{matrix} A \\ C \\ G \\ T \end{matrix} & \begin{pmatrix} 0 & 0 & 0 & 1 \\ 0 & 0 & 1 & 0 \\ 0 & 1 & 0 & 0 \\ 1 & 0 & 0 & 0 \end{pmatrix} \end{matrix}. \tag{2.16}
$$

This matrix is interpreted as follows: If we find T at position 8, we can be certain ($p = 1$) to find A at position 22, while if we see a G at position 8, we are sure to encounter C at 22, and so forth. Very obviously, these are the associations implied by Watson-Crick pairing, so this conditional probability matrix suggests to us that positions 8 and 22 are in fact paired within the molecule. We will see later that this turns out to be true, and that this type of sequence analysis can be used to detect secondary structures in RNA molecules that are due to Watson-Crick pairings with high accuracy.

2.2 Entropy and Information

One of the cornerstones of Shannon's theory of information is the introduction of a quantitative measure of how well an observer, armed with a set of probabilities p_i describing a random variable X, can make predictions about the future state of X. He called this measure (of the uncertainty about X) the *entropy* of X: $H(X)$. The name entropy for the function $H(X)$ that will be introduced below is the same as that used for the thermodynamical concept "entropy," and indeed the two are intimately related (see, e.g., Adami 2006a; Adami 2011; Adami 2016). The function $H(X)$ is a simple function of the probability distribution. For a random variable X taking on s different states x_i with probabilities $p_i \equiv P(X = x_i)$, Shannon defined

$$H(X) = -\sum_{i=1}^{s} p_i \log p_i. \tag{2.17}$$

Before we discuss the interpretation of this quantity, let us study its normalization. $H(X)$ is not normalized unless we specify a basis for the logarithm. In computer science and engineering, the base "2" is most often chosen (the dual logarithm), which gives $H(X)$ the units "bits." In physics, the base "e" is the standard choice, implying $H(X)$ is counted in "nats." For us, it will be advantageous to keep the units open and adjust them depending on the random variable used. For nucleotides, for example, we might use base 4, while taking logarithms to base 20 for amino acids. In those cases, we will simply use the unit "mer" for the entropy, where a nucleotide mer is worth two bits, while an amino acid mer translates to $\log_2(20) \approx 4.32$ bits. Using those units, the entropy of a random n-mer is, conveniently, n mers.

If $H(X)$ measures my uncertainty about X, what is the largest uncertainty that I can have? As the earlier examples have taught us, this should occur if all possible outcomes x_i occur with equal probability. Then

$$H(X) = -\sum_{i=1}^{s} p_i \log p_i = -\sum_{i=1}^{s} \frac{1}{s} \log \frac{1}{s} = \log s. \tag{2.18}$$

Indeed, we can show that $p_i = 1/s$ maximizes $H(X)$ (this proof is left as Exercise **2.2**).

An example is sufficient to illustrate the idea of maximal uncertainty. Imagine the random variable to be a binary string of length 5. There are, then,

$2^5 = 32$ possible states. In the absence of any information about the probability distribution, all states must be equally likely, and the entropy is $H = \log 32$. If we take the base of the logarithm to be 2, then $H(X) = 5$ bits. This is sensible: the string consists of five variables, each of which is binary and could therefore store one bit. The string made of five binary digits each can thus maximally store five bits. We can say then that the uncertainty of a random variable is just the amount of information it *could* hold, that is, the uncertainty is *potential information*. We will return to this interpretation in more detail when we introduce the concept of information more formally. But let us also look at another possible way to normalize our uncertainty about X. We could use the possible number of states of the variable X as the base of the logarithm. In this case

$$H_{\max}(X) = \log_{32} 32 = 1. \tag{2.19}$$

This example illustrates that all uncertainties can be normalized in such a way that their maximal value is 1. In a sense, we did the same thing when we chose base 2 earlier: we just made sure then that the entropy of *each element* of the string is bounded by 1, and we did this also when we chose the base to be the number of nucleotides or amino acids.

What is the smallest value of $H(X)$? The intuition gained when studying probabilities earlier tells us that this should occur if we know that one particular state $x_\star$ is taken on with certainty $[p_\star \equiv P(X = x_\star) = 1]$ while all other states are ruled out (have vanishing probability). In that case

$$H(X) = -\sum_{i=1}^{s} p_i \log p_i = -p_\star \log p_\star - \sum_{i \neq \star}^{s} p_i \log p_i = -1 \log 1 - \sum_{i \neq \star}^{s} 0 \log 0.$$

The first term is clearly zero, but to evaluate the second term we have to rely on the (mathematically well-founded) convention that $0 \times \log 0 \equiv 0$, so that $H_{\min} = 0$. So, we can now state that for a random variable X taking on s states x_i with probabilities p_i, the uncertainty $H(X)$ is bounded as

$$0 \leq H(X) \leq \log s. \tag{2.20}$$

We can immediately apply these definitions to biological random variables. Let us go back to the sequence (2.2). If we think of the sequence as a record of fifty-six trials for a nucleotide random variable X, then we can use the probabilities (2.3) to estimate our uncertainty, that is, how much we do *not* know, about it. Let us first choose the dual logarithm:

$$H_2(X) = -0.18 \log_2 0.18 - 0.2 \log_2 0.2 - 0.29 \log_2 0.29$$
$$- 0.34 \log_2 0.34 = 1.96, \tag{2.21}$$

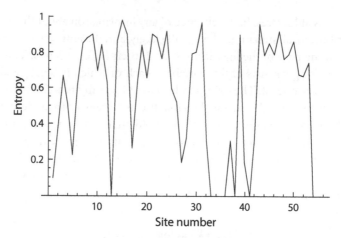

FIGURE 2.2. Entropy per site for the fifty-six nucleotides from the alignment in Table 2.1, with logarithms to base 4 (entropy measured in mers).

that is, the entropy of this variable is approximately 1.96 bits. The maximal entropy, of course, is 2 bits, which implies that—given the limited knowledge of fifty-six trials—the sequence appears nearly random. If we choose base 4 instead (the number of possible states), then

$$H_4(X) = H_2(X)/2 = 0.98 \text{ mers},\qquad(2.22)$$

close to its maximal value $H_{\max} = 1$ mer. We will use base 4 in the rest of this section, if not indicated otherwise. So while the sequence (2.2) appears nearly random when looking at the entropy of a single site (and assuming that the sequence is just reflecting different trials), we know from looking at Table 2.1 that it is very unlikely that the sequence is random.

Let us instead examine the entropies *per site*, that is, let us study the entropy for each random variable X_i separately. Above, we already obtained the probability distribution for sites $i = 8, 21$ and 22 from the alignment in Table 2.1. We can do this for all random variables X_i and plug the probabilities into formula (2.17). Using the base 4 logarithm, Figure 2.2 shows the entropy as a function of site number. This figure shows how different the uncertainties are for each site, ranging from zero (the sites 33–36 mentioned earlier, which are maximally informative), to nearly maximal (sites 15, 31, and 43 for example, that are nearly completely uninformative). Each of these sites has been undergoing mutation and selection for billions of years, so we can therefore safely assume that the values we see here are a direct consequence of the function of the molecule within the organism.

We will now study whether the correlations between sites that we detected by examining conditional probabilities are reflected in the entropies. Nothing appears to be particular about the entropies of sites 8 and 22 in the entropy plot of Figure 2.2, except perhaps that they are exactly the same. As emphasized already, to quantify the *relative* states of the sites, we have to look at conditional probabilities and their associated entropies. And to quantify how much information one site has about another, we finally introduce Shannon's information concept.

2.2.1 Information

Information, as discussed earlier, is that something that allows an observer (that is, one who is in possession of this information) to make predictions about the state of another system (which could be the same system only at a later time) that are better than chance. To quantify information, we then have to consider the relative states of two systems: the possible states of the observer X (or more precisely, the possible states of the observer's measurement device X), and the possible states of the observed, Y. Both will be treated as random variables, of course, and without loss of generality we assume that X can take on states $x_1, \ldots, x_s$ with probabilities $p(x_1), \ldots, p(x_s)$ whereas Y can take on states $y_1, \ldots, y_{s'}$ with probabilities $p(y_1), \ldots, p(y_{s'})$. We would like to characterize the amount of information that X has about Y. (We can just as well, and will, calculate the amount of information that Y has about X.) Note that it does not make sense to talk about the amount of information X or Y possess without specifying what this information is *about*, because information is used to predict the state of something other than one self (the state of "self" at a later time is considered a different system here). An observer X can possess information about Y while not knowing anything about Z. Information that is useless, that is, that cannot be used to make predictions because it is not known *what* this information is about, is really nothing but entropy, instead.

First, let us write down the entropies of X and Y:

$$H(X) = -\sum_{i=1}^{s} p(x_i) \log p(x_i), \quad H(Y) = -\sum_{i=1}^{s'} p(y_i) \log p(y_i). \quad (2.23)$$

The uncertainty $H(Y)$ represents *how much there is possibly to know about Y*. The entropy $H(X)$, on the other hand, represents how much X *can* possibly know about Y. It may look odd at first that the amount that X can know is limited by its entropy, but remember that the entropy is determined in large part by the number of possible states that X can take on. It is impossible for X

to make more distinct predictions than it has states. The reason why "what there is to know about Y" is limited is because *we ourselves* determine the amount of potential information by defining the states that we would like to resolve (the states y_i), and by agreeing to do so with a measurement device with states x_i (Adami 2016). We usually tacitly imply that the measurement device X is perfectly suited to resolving the states y_i, in other words, that there are probabilities $p(x_i|y_j)$ that are close to unity (while others are zero). While this may be true for measurement devices designed by engineers, we will see later that biology has plenty of use for measurement devices (sensors) that are significantly less reliable.

We can quantify the amount of information that X has about Y with the help of Shannon's *conditional entropy* concept. So, for example, let's say X knows (with certainty) that Y is in state y_j. The (remaining) entropy of X in this case is calculated using the conditional probabilities $p(x_i|y_j)$ as

$$H(X|Y=y_j) = -\sum_{i=1}^{s} p(x_i|y_j) \log p(x_i|y_j). \qquad (2.24)$$

If the probabilities to observe the states x_i are unchanged whether we know what state Y is in or not, then X is not a good measurement device for Y (and then $H(X|Y=y_j) = H(X)$).

From the above expression, we can calculate the *average conditional entropy*, that is, the remaining entropy of X when X knows what state Y is in, independent of the state. This is naturally just the average of Equation (2.24) over the states y_j

$$H(X|Y) = \sum_{j=1}^{s'} p(y_j)H(X|Y=y_j) = -\sum_{ij} p(x_i, y_j) \log p(x_i|y_j), \quad (2.25)$$

where the last equality holds because of Equation (2.8).

In order to calculate conditional entropies for our set of sequences from Table 2.1, all we need then are conditional and joint probability distributions. For example, to calculate $H(X_{21}|X_8)$, the entropy of site 21 given that we know what is at site number 8, we need $p(X_{21}|X_8)$, which we already obtained in (2.14), and $p(X_{21}, X_8)$, which we can obtain from (2.14) as long as we know $p(X_8)$ by using (2.8) again. This gives

$$H(X_{21}|X_8) = 0.58 \text{ mers}. \qquad (2.26)$$

As we noted earlier, knowing the state of site 8 does affect our ability to predict the state of site 21, and indeed the conditional uncertainty (2.26) is considerably smaller than $H(X_{21}) = 0.9$, that is, knowing X_8 does reduce our uncertainty about X_{21} somewhat. But let us now calculate the entropy that remains for site 22 if we know site 8. This calculation gives

$$H(X_{22}|X_8) = 0, \qquad (2.27)$$

that is, knowing site 8 tells us *all* we can possibly know about site 22: they are perfectly correlated and one will perfectly predict the state of the other.

From these considerations, the definition of information becomes obvious: it is just the reduction in uncertainty due to knowledge. So, if Y is again our system whose entropy we seek to decrease using measurement device X, then the information X has about Y is just[3]

$$I(X:Y) = H(Y) - H(Y|X). \qquad (2.28)$$

Note that this definition of information (also sometimes called "shared" or "mutual" entropy) is symmetric: $I(X:Y) = I(Y:X)$, and therefore besides expression (2.28), we can also write

$$I(X:Y) = H(X) - H(X|Y). \qquad (2.29)$$

Thus, "what X knows about Y, Y also knows about X." This is particularly clear for the example of information through Watson-Crick pairing that we looked at above. We can calculate the information site 8 has about site 22 as

$$I(X_{22}:X_8) = H(X_{22}) - H(X_{22}|X_8) = H(X_{22}) \qquad (2.30)$$

because $H(X_{22}|X_8) = 0$ as we found above in Equation (2.27). At the same time,

$$I(X_8:X_{22}) = H(X_8) - H(X_8|X_{22}) = H(X_8) \qquad (2.31)$$

because $H(X_8|X_{22})$ also vanishes. Then, $H(X_8)$ and $H(X_{22})$ have to be equal, a fact we noted earlier already. We can repeat this analysis for the other pair we looked at:

$$I(X_8:X_{21}) = H(X_{21}) - H(X_{21}|X_8) = 0.9 - 0.58$$

$$= 0.32 \text{ mers.} \qquad (2.32)$$

A simple diagram (Fig. 2.3) helps to see how entropies are distributed among shared and conditional entropies. As we discussed earlier, if two sites share most of their entropy, we can conclude that they bind to each other in a Watson-Crick pair. And because this binding is responsible for the structure of the molecule, information theory can help us to determine the molecule's

---

3. In this book, we use a colon between pairs of random variables in Shannon's information measure. Other books use a semi-colon.

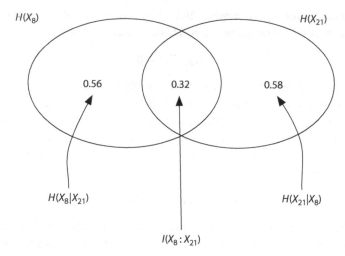

FIGURE 2.3. Venn diagram of entropies (in mers) between sites 8 and 21 of the sequence alignment (2.1).

secondary structure, that is, how the molecule is arranged as a chain in two dimensions. (This secondary structure folds into a three-dimensional structure, and information theory can also give hints on how this happens, as we will discuss later.)

To discover all Watson-Crick pairs in this molecule, we must calculate the mutual entropy between *all pairs*, $I(X_i : X_j)$. We do this here only for the fifty-six positions in Table 2.1, and show the result in a two-dimensional grid where darker entries represent higher information content, in Figure 2.4. Note the two diagonal strips of length 6 and 4 in the (symmetric) plot. They represent *stacks* of Watson-Crick pairs, a major component of RNA secondary structure. (Note also that because $I(i : i) = H(i)$, the diagonal contains the per-site entropy, which is, however, suppressed in this figure.) The stacks, betrayed by their mutual entropy, allow us to infer the secondary structure of the molecule, as done in Figure 2.5. In this sketch, nucleotides are drawn in three different shades of gray: white for high entropy, light gray for intermediate entropy, and dark gray for low or vanishing entropy (they can be read off the entropic profile in Fig. 2.2).

Treating the full-length (76-nucleotide) sequences in this manner reveals the secondary structure of the tRNA molecule, as depicted in Figure 2.6. Note that while the data suggests that six nucleotides pair in the anticodon stack (the rightmost stack in Fig. 2.5), a more thorough analysis reveals that the first pair (32–38), while showing some correlation, does not bind in such a manner. An analysis of 1415 sequences of the yeast tRNA for the amino

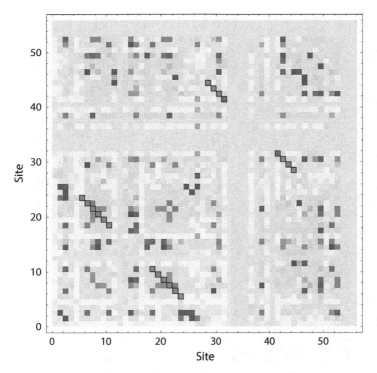

FIGURE 2.4. Mutual entropy (information) between sites of the sequence alignment 2.1. Darker squares indicate high information content. Two diagonal lines of high information stand out, and predict the secondary structure seen in Figure 2.5. (Because information is symmetric, the figure shows four diagonal lines.)

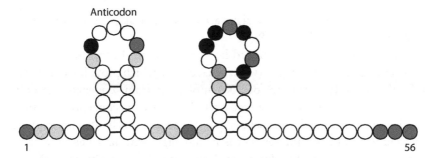

FIGURE 2.5. Sketch of secondary structure of the alignment of the fifty-six-base fragment of Table 2.1. The shades of the residues indicate the degree of conservation, where dark is strongly conserved and light is variable. The anticodon region is variable in this alignment because different tRNA species are mixed together.

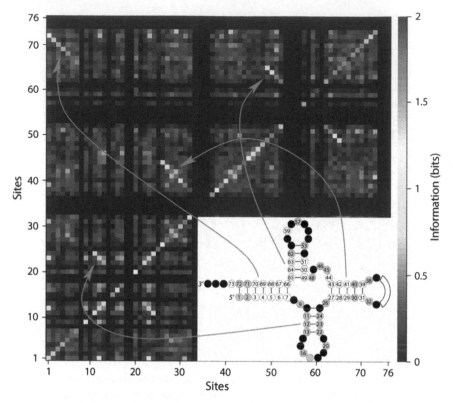

FIGURE 2.6. Mutual entropy (information) between sites for an alignment of thirty-three *E. coli* tRNA sequences (Adami 2004). The inset shows the secondary structure of the molecule, with arrows pointing toward the information corresponding to the stacks of Watson-Crick pairs.

acid phenylalanine can also reveal covariances between bases due to tertiary structure (Eddy and Durbin 1994).

The previous treatment showed how information theory can reveal the correlations due to binding between nucleotides, which provides a means to infer the secondary structure of RNA molecules from sequence data only. However, this application served mostly to introduce the concepts of entropy and information in molecular sequences; its usefulness is limited because the method works only if the sequences in the alignment have been undergoing mutation and selection for a very long time. If the sequences in the set used to build the alignment have a recent common ancestor, mutations have not yet had the chance to randomize those positions that are functionally unimportant, and most of those sites will appear to have very low or vanishing entropy even though they are *potentially* variable. Methods to obtain secondary structure information in such cases do, however, exist, as described by Eddy and Durbin (1994).

2.3 Information Content of Genes

In this section, we will use information theory in another manner. It is intuitively clear that living organisms *know* some very important facts about the world in which they thrive. We use the word "know" here strictly in the sense of "having information" (without assuming any mental faculties). Furthermore, while colloquially we say "information is stored in the sequence," strictly speaking, information requires an ensemble of sequences (see Box 2.2).

For example, the gene for the protein hemoglobin, which binds oxygen and transports it from the lungs to the body, clearly stores information about the presence of oxygen in the atmosphere. But of course, the protein knows a lot more: it knows the atomic configuration of oxygen, where the oxygen is supposed to go, and how to release it when it arrives at its target. Furthermore, ithe molecule "knows" how to interact with a number of other molecules, and maybe even more importantly, how *not* to interact with most others. This knowledge, as we will see in the next chapter, has been acquired in eons of evolution, and is manifested in the *functional* characteristics of the hemoglobin molecule. But how much information is this? How many bits?

We will see that this question can be answered in principle, for every molecule of every organism, even though this will turn out to be exceedingly difficult in practice (see also Box 10.1). Even worse, it is becoming more and more clear that a good fraction of what an organism knows about its environment is stored not in its genes, but in the regulatory apparatus that controls them. However, as I will show in detail in section 5.1.3, the information content is a good proxy for complexity, so trying to estimate such a number is going to be worth our time. We will return to a discussion of this topic soon enough; in the meantime we shall first study the information content of proteins, followed by an application to DNA and RNA sequences.

2.3.1 Information content of proteins

Proteins are sequences of amino acids that (mostly) fold into stable structures and make up the bulk of organic matter. The twenty amino acids that almost all forms of life use are encoded by triplets of nucleotides called *codons*. The genetic code that specifies how each particular triplet is translated into an amino acid is almost universal across all forms of life (there are a few exceptions, however). Because the sixty-four possible combinations of nucleotides

Box 2.2. Information in Single Sequences?

According to Shannon's definition Equations (2.28) or (2.29), information is a correlational property of two ensembles, which implies that it is not possible to determine how much information is stored in a single sequence, or a single piece of text. Yet, we seem to discuss here the information stored "in a sequence of DNA" or "in a protein" (and later on I will estimate how much information is stored in the human genome). Oftentimes, people talk about how much information is contained in a single message. How are these statements compatible? I will try to illustrate this by invoking an extremely simple world: the game of "Cluedo" (known as "Clue" in North America). In this game, originally developed in England in 1949, the player must deduce the identity of a murderer, along with the murder weapon and the room in which the murder was committed, via a series of inferences and eliminations. For our purposes, I will simplify the game so that only the murderer's identity must be guessed. In the classic game there are six possibilities: the characters Professor Plum, Colonel Mustard, Mrs. White, Miss Scarlet, Mrs. Peacock, and Mr. Green. In each game, any of these characters is equally likely to be the culprit, so the entropy of Y (the world whose state we are attempting to infer) is $H(Y) = \log(6)$. (If we take logarithms to the base 2, this is approximately 2.6 bits.) Suppose as a player, I am handed a strip of paper with the text Murderer is Male. Does this message not carry $\log_2 3$ bits of information, as it narrows the number of suspects from six to three? If I am handed the words Colonel Mustard, does this sequence not carry $\log_2 6$ bits of information? While revealing the clue does indeed produce *specific information* (see Exercise **2.7**), whether the sequence *is* information cannot be determined, because even if in that one instance my prediction (armed with the tip) turns out to be correct, we do not know whether the success occurred by chance or not. To assess this, we have to repeat the game many times, with the perpetrator randomly chosen among the set of six each time, so that the *average* specific information in Colonel Mustard or Murderer is Male can be determined. But you might argue that according to Equation (2.34), the information content of biomolecular sequences is determined not with respect to an ensemble, but a very specific realization of a world, namely $E = e$. Is that not the same as there being only one possible murderer (say, it is always Colonel Mustard) in which case a single sequence can carry information after all? Again the answer is no, but for a different reason. In molecular sequences, many *possible* messages are conceivably encoded at the same time, but we cannot from inspection of the sequence determine which of them is responsible for the fitness benefit, that is, which of them makes the valuable prediction. In our example, imagine that all six of the possible clues are encoded on a sequence. In a population that undergoes mutation and selection, only one of the six messages will remain unaltered, namely the one that contains the information that gives rise to the superior prediction. Only the ensemble created by the evolutionary process could reveal that information, even if there was only a single world.

Table 2.2. Assignment of nucleotides to amino acids in the genetic code

Ala (A)	GCT, GCC, GCA, GCG	Leu (L)	TTA, TTG, CTT, CTC, CTA, CTG
Arg (R)	CGT, CGC, CGA, CGG, AGA, AGG	Lys (K)	AAA, AAG
Asn (N)	AAT, AAC	Met (M)	ATG
Asp (D)	GAT, GAC	Phe (F)	TTT, TTC
Cys (C)	TGT, TGC	Pro (P)	CCT, CCC, CCA, CCG
Gln (Q)	CAA, CAG	Ser (S)	TCT, TCC, TCA, TCG, AGT, AGC
Glu (E)	GAA, GAG	Thr (T)	ACT, ACC, ACA, ACG
Gly (G)	GGT, GGC, GGA, GGG	Trp (W)	TGG
His (H)	CAT, CAC	Tyr (Y)	TAT, TAC
Ile (I)	ATT, ATC, ATA	Val (V)	GTT, GTC, GTA, GTG
START	ATG	STOP	TAA, TGA, TAG

code for only twenty amino acids,[4] the genetic code has significant redundancy. Thus, measuring information content on the level of nucleotides could potentially be very different from the information content measured on the level of amino acids. For example, because the third position in any particular codon is often neutral (often only the first two specify the amino acid; see Table 2.2), the information content of the third position in a codon is, on average, significantly smaller than that of the others. However, we will see that even for amino acids whose third position should be neutral, the third nucleotide *can* in fact store information: it is just not information about which amino acid to use.

A good starting point is to ask how much information a protein of L residues (L amino acids) could store in principle; that is, what is a protein's *maximal* information content? We already have the tools to answer this basic question; we just need to go back to the definition Equation (2.28) of the information one random variable X has about another, Y. To be specific, let X be our protein random variable (henceforth referred to as P), while Y represents the potential environments the protein could find itself in (we refer to those environments in the following as E). The information our protein has about its worlds E would then be

$$I(P : E) = H(P) - H(P|E), \tag{2.33}$$

4. More precisely, sixty-one codons code for twenty amino acids, as the stop codons are not translated.

that is, it is the entropy of P in the absence of a selective environment, minus the entropy of P *given* all potential environments. But this formula immediately reveals trouble: we usually can only know the entropy of a protein for one particular environment, say, $E = e$: that in which it has evolved (we follow the convention introduced earlier to represent random variables by capital letters, and particular values the variable can take on by lowercase letters). There is no conceivable way in which we could average over all *possible* environments, but neither are we interested in those: we care about the information our protein has about our (namely, its) world only. This information can be written as the difference between an entropy and a conditional (rather than *average* conditional) entropy:

$$I(P : E = e) = H(P) - H(P|E = e). \tag{2.34}$$

This difference is also sometimes called a *specific information*. While in principle specific information can be negative (see Exercise **2.7**) this will never be the case in our application to the information content of biomolecules because the first term in Equation (2.34) is always the maximal entropy.[5]

Let's see what we need to calculate the information content Equation (2.34). The first term is easy: as just discussed, in the absence of a selective environment, all protein sequences are equally likely, and therefore $H(P) = H_{max}$, which we calculate below. The second term refers to the remaining entropy given its selective environment [the conditional entropy (2.24)], and all the work in obtaining $I(P : E = e)$ is buried in that term.

But let us return to the first term $H(P)$. To calculate this, we first have to decide whether to use the nucleotide sequences that code for the protein as our random variables, or to use the amino acid sequences directly. For $H(P)$ as well as for $H(P|e)$, there is an important difference between these choices, due to constraints imposed by the genetic code. For ensembles that are subject to perfect mutation-selection balance, that is, where each codon appears precisely with the probability assigned to it by the genetic code, the difference $H(P) - H(P|e)$ should not depend on this choice. We will see below, however, that small deviations do exist, suggesting that information is encoded in genetic sequences that goes beyond the amino acid sequence (codon bias is just one such source of information).

Let us consider the amino acid sequence first. The uncertainty $H(P)$ represents how much we do not know about protein P in the absence of any

---

5. There is an alternative definition of specific information as a relative entropy between an unconditional and a conditional probability distribution. However, this definition is problematic as it lacks a fundamental additivity property; see DeWeese and Meister (1999).

environment constraining its sequence. Each residue ought to appear with equal probability in such a (maximally unadapted) protein, implying a maximal entropy $H = \log s$, where s is the total number of possible states of the protein, according to Equation (2.20). For a protein of L residues, each taking on one of twenty possible states, we have $s = L^{20}$, and therefore, if we take logarithms to base 20,

$$H(P) = \log_{20} 20^L = L \text{ mers.} \tag{2.35}$$

To calculate an L-mer's information content, we need the remaining uncertainty given the environment, $H(P|e)$, which we extract from an aligned ensemble of sequences that evolved in this environment e. Ideally, this is an ensemble (preferably infinite) of proteins that was subject to mutation and selection, without changing its structure or function, for preferably an infinite amount of time! In practice, of course, we have to make do with finite ensembles of molecules that are subject to mutation and selection (while functionally unchanged) for only a finite amount of time. This imposes severe limitations that we address in detail in section 3.2. We will largely ignore those here to get a feel for the method. As an example, let us try to measure the information content of the *homeobox* protein: a DNA binding protein that regulates development in animals and plants. Most of these proteins are about 57–62 amino acids long and consist of three helices that bind to DNA as depicted in Figure 2.7. These proteins are very similar across different animals, but there is also significant variation within each group. To calculate $H(P|e)$, we need to estimate the probability to find a polymer that folds into the homeodomain structure, among all possible polymers. Suppose for a moment that we could enumerate all protein structures and create a random variable P that takes on any particular structure i among all possible ones. In that case we need to estimate the probability $p(X = \text{homeodomain}|e)$ from available data. Because the number of possible protein structures is very large (on the order of thousands), this probability cannot be obtained accurately from any data set, because of the finite size corrections to entropy estimates described in Box 2.3.

Fortunately, there is an avenue to estimate this probability if the frequencies of amino acids at any particular position in the sequence are independent from those of others. For example, let us write the protein random variable in terms of the site random variables P_i as we did earlier for DNA sequences in Equation (2.4):

$$P = P_1 P_2 \ldots P_L \tag{2.36}$$

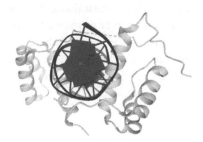

FIGURE 2.7. Two views of the three-dimensional structure of the homeobox protein (light gray), involved in the overall determination of an animal's body plan, bound to DNA (dark gray). Homeobox proteins are transcription factors that initiate development by binding to specific DNA sites.

for an L-mer. Previously, we used correlations between sites (in the form of terms $I(P_i : P_j)$) to deduce the secondary structure of ribozymes (ribozymes are enzymes made out of RNA). But because proteins do not form ladders, the correlations between amino acids are much weaker, and often it is assumed that they vanish altogether (Swofford et al. 1996). However, these correlations cannot be assumed to be negligible in general, and we will return to correct our results at the end of this section. In the following, we will ignore correlations as a first approximation and assume[6]

$$I(P_i{:}P_j) = 0 \quad i \neq j. \tag{2.37}$$

If all correlations between more than two residues also vanish, then the entropy of P is simply given by the sum of entropy of each of the sites:

$$H(P|e) = \sum_{i=1}^{L} H(P_i|e). \tag{2.38}$$

In that case, it is sufficient for the estimate of $H(P|e)$ to only know the per-site entropy, which we can obtain from an alignment of structurally similar homeodomain proteins. Such alignments can be found in sequence databases such as the *Pfam* ("Protein families") database (Finn et al. 2006). To be specific, let us align the sequences of homeodomain proteins of the rodent family (*Rodentia*). At the time of this writing, Pfam lists 810 such sequences with 57 residues

---

6. Here and in what follows, when we write $I(P_i{:}P_j)$ we always imply that the corresponding entropies $H(P_i)$ are conditional on the environment variable e, because were they not the information $I(P_i{:}P_j)$ would always vanish trivially.

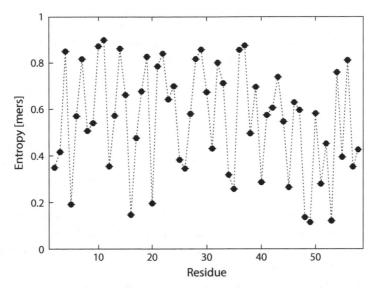

FIGURE 2.8. Entropic profile of the $L = 57$ amino acid rodent homeodomain, obtained from 810 sequences in Pfam (accessed February 3, 2011). Error of the mean is smaller than the data points shown. Residues are numbered 2–58 as is common for this domain (Billeter et al. 1993).

in its core, and they are similar enough to each other to have the same structure[7] (39 percent similarity), but different enough to allow us an estimate of the information content.

When estimating entropies from finite ensembles, care must be taken to correct the entropy estimates for a bias that arises when the ensemble is small. For example, it is clear that estimating the frequency of twenty symbols from a collection of only, say, twenty sequences, is not only highly inaccurate but also biased, because some symbols simply will not appear at all even though they would be present in a larger collection. Box 2.3 introduces the mathematics of correcting for the bias, and also for estimating the variance of the entropy estimate. In the following, we always apply the bias correction and show the standard deviation according to Equation (2.41).

Calculating the per-site entropies gives the entropy profile shown in Figure 2.8, which curiously reveals almost alternating positions of high and low entropy. We can calculate the information content by summing up the entropies in Figure 2.8, to find (remember that the unconstrained $H(P) = H_{max} = 57$):

7. Generally, two proteins that share more than 30 percent sequence similarity have a very high likelihood (better than 90 percent) to share the same structure (Brenner et al. 1998).

Box 2.3. Bias and Variance of Entropy Estimates

In this chapter and subsequent ones, we need to calculate the entropy of a random variable X that can take on s states with probabilities p_i $(i = 1, \cdots, s)$, based on the distribution of events x_i that occur with frequency n_i, from a total of N "trials." As the number of trials tends to infinity, we can rest assured that our estimates for the probabilities become more and more accurate: $n_i/N \to p_i$ as $N \to \infty$. But how does our estimate $H[n_1/N, \cdots, n_s/N] = - \sum_i \frac{n_i}{N} \log \frac{n_i}{N}$ of the entropy relate to the asymptotic value $H[p_i] = - \sum_i p_i \log p_i$ reached for $N \to \infty$? For many random variables, their expectation value does not depend on the size of the sample used to extract an estimate, only the variance does. Not so for entropy: it is a *biased* random variable (Miller and Madow 1954; see also Basharin 1959). This means that small sample sizes can seriously mislead our estimate, and we need to correct for this bias. Let $H[p_i]$ stand for the asymptotic value $H[p_i] = - \sum_i^s p_i \log_D p_i$ while $\langle H[n_i/N] \rangle$ stands for the biased expectation value obtained by estimating the p_i using the frequencies n_i/N. The distribution of each n_i is binomial: either an event is of type i (with probability p_i) or it is not (with probability $1 - p_i$). This means we can calculate both the expectation value and the variance of our estimator n_i/N: $\langle n_i/N \rangle = p_i$ as already advertised, and $\text{var}(n_i/N) = p_i(1 - p_i)/N$. These can now be used to calculate the expectation value and variance of the entropy, to first order in $1/N$ (meaning we neglect corrections proportional to $1/N^2$ below, because these terms will be much smaller than the leading terms for N sufficiently large, see Exercise **2.8**). If we take our logarithms to base D (units: mers) as usual, then

$$\langle H[n_i/N] \rangle = H[p_i] - \frac{s-1}{2N \ln(D)}, \tag{2.39}$$

$$\text{var}(H[n_i/N]) = \frac{1}{N} \left(\sum_{i=1}^{s} p_i \log_D^2 p_i - H^2[p_i] \right). \tag{2.40}$$

This means that we need to add a term $(s-1)/[2N \ln(D)]$ to our estimate to correct for the bias, so that ultimately

$$H[p_i] = \langle H[n_i/N] \rangle + \frac{s-1}{2N \ln(D)} \pm \sqrt{\frac{1}{N} \left(\sum_{i=1}^{s} p_i \log_D^2 p_i - H^2[p_i] \right)}. \tag{2.41}$$

Nowadays, there are several computational methods that can correct for bias in more sophisticated ways than the Miller-Madow correction (2.41), for example the "NSB Method" (Nemenman et al. 2002) and the "James-Stein estimator" (Hausser and Strimmer 2009), and any study that estimates entropy or information from small data sets must use them.

$$I(P_{\text{Rodentia}}) = H_{\max}(P) - \sum_{i=1}^{L} H(P_i|\text{rodentia})$$

$$= 25.29 \pm 0.09 \text{ mers.} \tag{2.42}$$

It is interesting to ask whether animal families that appear to be more complex have a higher information content in this gene. An alignment of 903 sequences of primate homeobox proteins (also from Pfam) results in an entropic profile remarkably similar to that of Figure 2.8. The information content turns out to be

$$I(P_{\text{Primates}}) = 25.43 \pm 0.08 \text{ mers,} \tag{2.43}$$

in other words, identical to that of the rodents within statistical error. Because we believe that the protein carries out essentially the same function in rodents as in primates (both mammals), perhaps we should not be too surprised about this result. But just because the total information content is the same does not imply that information is coded in a similar manner. For example, as each residue encodes a piece of the information, it is in principle possible that the information is distributed across the molecule in different ways between rodents and primates. We can study subtle differences in information encoding by subtracting one entropic profile from the other, looking at the difference between entropies per-site for aligned sites. This way we might notice if the entropy of some sites has increased while some others might have decreased, in the evolution from rodents to primates. We can see this difference plot in Figure 2.9, which suggests that some recoding did indeed take place. There are differences in particular at the interfaces between the helix motifs, but the DNA recognition motif (helix 3) seems to be the same within error. Still, it appears that information theory can track some overall effects in the adaptation of proteins.

Before we return to the question of whether it is better to use the DNA or amino acid sequences to study information content, we should check our assumption regarding the absence of interactions between sites—the assumption that $I(P_i : P_j) = 0$ $i \neq j$ for the homeobox proteins. We obtained these correlations earlier for RNA and can use the same method to reveal the matrix of informations, which we show in Figure 2.10.

The entropy of the protein $P = P_1 P_2 \ldots P_L$ can be written generally as

$$H(P) = \sum_{i=1}^{L} H(P_i) - \sum_{i<j} I(P_i : P_j) + H_{\text{corr}} \tag{2.44}$$

where H_{corr} represents higher-order (three or more) shared entropy, which we neglect. In principle, those higher-order contributions can be written down using Fano's entropy and information decomposition theorems (Adami and

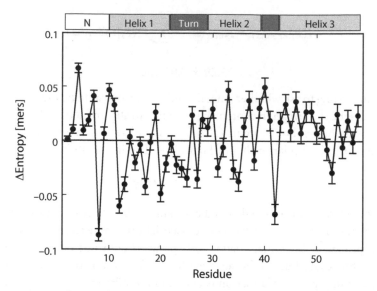

FIGURE 2.9. Difference between the entropic profile of the homeobox protein of rodents (shown in Fig. 2.8) and that of primates (the latter from 903 sequences in Pfam, accessed February 3, 2011). ΔEntropy = $H_{\text{rodents}} - H_{\text{primates}}$. Error bars are the error of the mean of the difference, using the average of the number of sequences. The shaded boxes indicate structural domains as determined for the fly version of this gene. (N refers to the protein's "N-terminus.")

C G 2022). The off-diagonal entropies $I(P_i : P_j)$ appearing in Equation (2.10) are each small (mostly between 0.01 and 0.05, with some as large as 0.2), but each one is also subject to a significant bias correction, and comes with larger error bars than the single entropy estimates. As a consequence, the first order correction due to mutual entropies in Equation (2.44) cannot reliably be estimated, and we must neglect them in what follows. However, we will take these up again in the next chapter, when we deal with much larger data sets.

We mentioned at the beginning of this section that choosing DNA or protein sequences in the calculation of the information content $I = H(P) - H(P|e)$ can make a difference. In fact, these differences can be quite subtle, so we will spend a little time discussing those, before we look specifically at the information content of nucleotide sequences.

2.3.2   Information content in amino acids vs. nucleotides

It is easy to see that it makes a difference whether we simply sum up the entropies *per nucleotide* site or *per codon*, that is, per triplet of nucleotides. Let us define a *codon random variable* as the product of three nucleotide random

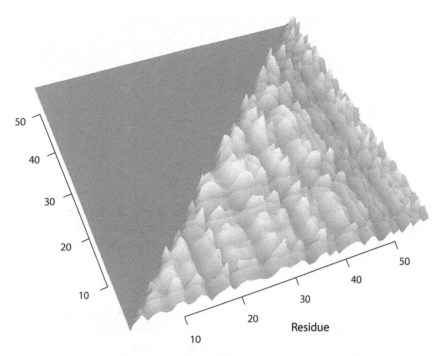

FIGURE 2.10. Mutual entropies $I(P_i : P_j)(i < j)$ among amino acids in the rodent homeodomain proteins from the Pfam data set. The diagonal representing the entropies in Fig. 2.8 has been removed for clarity.

variables $D_1D_2D_3$. Here, the order of the variables is important, as the genetic code implies a relationship between amino acids and nucleotides in which the three positions are treated unequally. As the translation Table 2.2 shows, several amino acids are encoded using the first two nucleotides only, while the third one is free to vary (Ala, Gly, Pro, Thr, Val). The mapping defined by Table 2.2 implies a relationship between the codon entropy $H(D_1D_2D_3)$ for a particular amino acid, and the simple sum $H(D_1) + H(D_2) + H(D_3)$. In fact, it is possible to calculate how much entropy is shared between each pair of codon positions (between D_1 and D_2, D_1 and D_3, and D_2 and D_3), as well as between all three. We write the latter quantity as $I(D_1:D_2:D_3)$ without defining it explicitly here (something that is done in all introductory textbooks for information theory, such as Cover and Thomas 1991 and Ash 1965) but it is defined implicitly in the Venn diagram shown in Figure 2.11, as well as in Exercise **2.6**.

It is not difficult but a little tedious to show that for three random variables

$$H(D_1D_2D_3) = \sum_{i=1}^{3} H(D_i) - \sum_{i<j}^{3} I(D_i : D_i) + I(D_1 : D_2 : D_3). \quad (2.45)$$

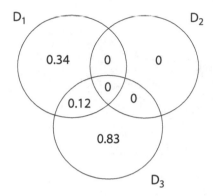

FIGURE 2.11. Entropy Venn diagram showing the correlations between nucleotides in a codon that codes for the amino acid Arginine, with logarithms taken to base 4 (nucleotide mers). As there is no entropy in the second codon position, it is not correlated to any of the other positions.

By comparing with Equation (2.44), it is clear that the single term $I(D_1{:}D_2{:}D_3)$ represents all the higher-order correlations that you can get for three variables. The information-theoretic relationship between any three variables can be visualized using a Venn diagram as in Figure 2.11. The entries in this diagram can be filled in by assuming that each codon that codes for an amino acid appears equally likely in an organism's gene (this is not true in practice; see the discussion of codon bias below). If we assume this for simplicity's sake, meaning that the three codons CGT, CGC, CGA, CGG, AGA, and AGG each encode Arginine equally often (see Table 2.2), we obtain the numbers shown in Figure 2.11.

We see immediately, for example, that there are no correlations between the first and second position and the second and third position $(I(D_1 : D_2) = I(D_2 : D_3) = 0)$, because the entropy of the second position vanishes (every Arginine codon has a G at the second position). Also, the entropy of the third position is almost maximal $H(D_3) = 0.95$, but some of that entropy is shared with the first position, implying that the sum of entropies $\sum_i H(D_i)$ actually *overestimates* the entropy by double-counting the significant correlation between nucleotide positions 1 and 3 due to the genetic code alone. As a consequence, when estimating the entropy of protein-coding sequences, we will measure entropies on the nucleotide level by counting the frequency of triplets rather than the frequency of single nucleotides.

If we measure the nucleotide entropies by measuring the frequency of codons, we should find that some of the nucleotide entropy is shared with the amino acid entropy (in an amount equal to the amino acid entropy, because specifying the codon unambiguously determines the residue). The shared amount, on the other hand, has a piece that is shared with the environment (true information), and a piece that is not (neutral positions). This situation is best illustrated with another Venn diagram, using the homeodomain example from above. Let us first calculate the information content of a DNA sequence by defining it just as for proteins: only the base of the logarithm differs. If $X = X_1 X_2 \ldots X_\ell$ (with $\ell = 3L$) is the sequence of DNA random variables X_i that, in triplets, codes for the protein sequence P, then the unconditional entropy $H(X)$ for nucleotides coding for the 57 residue homeodomain protein is $57 \times 3 \times 2$ bits $= 342$ bits, which we can translate to $57 \times \log_{20}(64)$, or approximately 79.1 (amino acid) mers (compared to 57 mers). This may appear curious at first: the maximal entropy for nucleotides is much larger than that for amino acids. However, we should have expected this: there is much redundancy in the genetic code. Note that because we are calculating an unconditional (marginal) entropy here, it does not matter whether we sum up single nucleotide or codon entropies: they are each maximal. In what follows, we will take the logarithm to base 20 even if we consider nucleotide codons (that can have 64 states) so as to be able to compare nucleotide-based with amino acid–based entropies. This implies that in principle the entropy at each codon can be larger than 1, even though it never will be in practice.

The information content for the homeodomain DNA sequence (as opposed to the amino acid sequence that we calculated earlier) is then

$$I(X) = H(X) - H(X|e) = 79.1 + \sum_{i=1}^{57} \sum_{j=1}^{64} p_j(i) \log_{20} p_j(i) \text{ mers}, \qquad (2.46)$$

where $p_j(i)$ is the probability to find codon j at position i in the sequence X, in the notation introduced in (2.5). The second term in (2.46) is just the conditional entropy of the DNA sequence given the environment $H(X|e)$ (with a minus sign). Using these definitions, we can analyze how information is shared between nucleotides and amino acids, and the environment. Consider the entropy Venn diagram in Figure 2.12. The solid circles represent the entropy of codons $H(X)$ on the left, and the entropy of amino acids $H(P)$ on the right. The unconditional protein entropy $H(P)$ is of course just 57 mers. As before, we would like to know how much of that is information; we would like to split

$$H(P) = H(P|e) + I. \qquad (2.47)$$

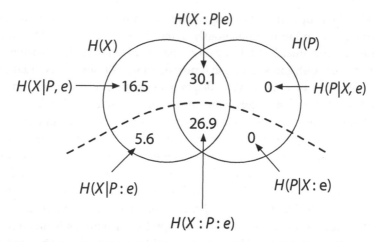

FIGURE 2.12. Entropy Venn diagram relating codon-based (X) and amino acid–based (P) entropies (mediated by the genetic code), divided into entropy given the particular environment, and information. Note that we draw the separation of entropies into a conditional piece and information with a dashed line (rather than representing the environment as a circle as for example in 2.11), because the conditional entropies are only conditional on a single state of the environment, e. To construct this diagram, I used 379 sequences of human homeodomain sequences (part of the primate set used earlier), for which I had both nucleotide and amino acid sequences.

Of course, a similar split also exists on the nucleotide level, as implied by (2.46). Gathering all the entropies, we can fill in the values in Figure 2.12. This figure allows us to draw a number of important conclusions about how information is coded in proteins. First, it is obvious that $H(P|X) = 0$, as knowing the codon trivially specifies the residue, and this is true whether or not the environment is specified ($H(P|Xe) = 0$ also). Second, the entire entropy of proteins is shared with the codons, but only a fraction of that is information. Most interestingly, there appears to be information stored in codons that is *not* shared with the amino acids; in other words, there is information coded in DNA that is not transmitted in translation.

Using the definitions in Figure 2.12, we see that the protein information

$$I(P) = H(P|X:e) + H(X:P:e) \qquad (2.48)$$

actually only consists of the piece that is shared with the codons, whereas the codon information

$$I(X) = H(X|P:e) + H(X:P:e) \qquad (2.49)$$

has the additional piece $H(X|P:e)$ that is nonvanishing. It is this piece that makes the difference in measuring information based on nucleotide codons and amino acids: it measures information stored in codons that is *not* translated into proteins. And while it is not especially large, we could ask what this information is about.

One possible origin of information in codons is *codon usage bias*. A codon bias can develop in an organism, for example, if some of the tRNA molecules necessary for translating codons have been lost during evolution. In that case, mistranslations can occur, which can be avoided by changing the codon to one for which the tRNA is available: the preferred codon. If the gene is highly expressed, then an evolutionary pressure for *translation accuracy* can lead to this conservation of untranslated information (Akashi 2003). Another possibility is that the set of sequences used to produce the entropy estimates is not fully equilibrated, that is, that an insufficient amount of time has elapsed for mutations to have tried all the possible combinations of codons. Often enough, this is a key limitation, but less so perhaps for this set of human homeodomain proteins, because the set comprises many *paralogs* (homologs that evolved via a gene duplication event, rather than a speciation event), which therefore would have had sufficient time to diverge.

Thus, we see that measuring information using protein or nucleic acid sequences poses different challenges and allows insights on different fronts. While proteins reveal only information that is translated, nucleotide sequences can reveal other informational biases, and could reveal splice sites or even DNA binding sites hidden in exons.

2.3.3 Information content of DNA and RNA aptamers

After this detour into protein information, let us return to DNA multiple sequence alignments such as the one shown in Table 2.1. Is such an alignment sufficient to estimate the information content of the gene that codes for the sequence? Just as before with proteins, let us describe the nucleotide random variable X of a sequence of L nucleotides as a joint random variable $X = X_1 X_2 \cdots X_L$ [recall Eq. (2.4)].

According to our construction, the information content of this sequence about a particular environment e is given by

$$I(X:e) = H_{\max}(X) - H(X|e) = L + \sum_j p_j \log p_j, \qquad (2.50)$$

where the sum runs over all the possible 4^L genotypes, and the p_j represents the probability to find each genotype j in an infinite population of sequences encoding the same information about e, at mutation-selection balance. Needless to say, the alignment in Table 2.1, or any other alignment for that matter, is insufficient to calculate this information: we have to resort to approximations to estimate at least the second term in (2.50). The most drastic assumption we can make is that the probabilities for each site $p(x_i)$ are independent of one another, which is the approximation we used earlier for proteins, in Equation (2.37), and later for nucleotides coding for these proteins, in Equation (2.46). Thus, suppose we approximate

$$H(X|e) \approx \sum_i H(X_i|e), \qquad (2.51)$$

where $H(X_i|e)$ is the conditional entropy of site i. These entropies *can* be obtained from an alignment as we saw earlier; indeed we plotted them against the site number in Figure 2.2 using the probabilities gleaned from Table 2.1. If we were to add these contributions as in Equation (2.51), we would estimate the information content of the 56 mer that is a part of the sequence that codes for an *E. coli* tRNA molecule as

$$I = 56 - 30.82 = 25.18 \text{ mers.} \qquad (2.52)$$

However, we already know that the sites X_i are not independent; this is precisely what allowed us to predict the secondary structure of the RNA molecule in the first place! A more accurate estimate of the information content takes all two-base correlations into account:

$$I \approx L - \sum_i H(X_i|e) + \sum_{i<j} H(X_i : X_j|e). \qquad (2.53)$$

We cannot estimate this "pair-correlation" correction [the second term in Eq. (2.53)] using the data from the alignment for the same reason that we could not calculate this correction for the protein example earlier: the sample size (here thirty-three sequences) is just too small to overcome the bias and error in the entropy estimate of the pair-correlation term, as we can see using the formulae of Box 2.3.

Each correction term in Equation (2.53) is of the form (dropping the notation that indicates these entropies are conditional on e, as we will assume this throughout):

$$H(X_i : X_j) = H(X_i) + H(X_j) - H(X_i, X_j). \qquad (2.54)$$

If the bias terms are independent, then the bias for the mutual entropy is

$$\Delta H(X_i : X_j) = 2\Delta H(X_i) - \Delta H(X_i, X_j). \qquad (2.55)$$

For the present alignment of thirty-three sequences, $\Delta H(X_i) = \frac{3}{2 \times 33 \ln 4}$, while $\Delta H(X_i, X_j) = \frac{15}{2 \times 33 \ln 4}$ because there are sixteen possible symbols for the joint entropy. The total bias is then

$$\Delta H(X_i : X_j) = \frac{6}{66 \ln 4} - \frac{15}{66 \ln 4} \approx -0.1. \qquad (2.56)$$

This does not appear large by itself, but as there are 1540 terms in the second sum in Equation (2.53), the bias correction will be of the same order of magnitude as the correlation correction, invalidating its use with this data set. However, this correction can in principle be carried out if the alignment has sufficient statistics, and we will perform this correction in the following chapter on data obtained for HIV sequence data, but with a more sophisticated method.

For sequences that code for RNA structures, we can use a simple trick to estimate the correction: we can assume that only one of the nucleotides that is involved in a stem should count toward the entropy, because its binding partner in the stem will be perfectly (or nearly perfectly) correlated to it. Indeed, this is what we found for X_{22} and X_8. There are twenty nucleotides involved in pairs (as we can see from Fig. 2.5), leading to an overcounting of the entropy by ten mers. Thus, a corrected information estimate of the structure is

$$I = 56 - (30.82 - 10) \text{ mers} = 35.18 \text{ mers.} \qquad (2.57)$$

Performing this correction for the entire tRNA molecule reveals an information content of about 44 mers for the 73-mer (this is the seventy-six-nucleotide sequence minus the anticodon, which specifies each particular amino acid and thus is not subject to mutational variation by definition; see Adami and Cerf 2000).

Is there a relation between information content and fitness, or even information content and structure? Intuitively, we might surmise that because information is that which allows *prediction*, sequences with higher information content should "know more" about the job they have to do, for example binding to a target molecule. This question has been studied by Carothers et al. (2004), who evolved RNA aptamer sequences (aptamers are polymers with only a few nucleotides) to bind to a specific target starting from a random library of RNA sequences using a molecular evolution technique called SELEX (Ellington and Szostak 1990; Tuerk and Gold 1990) (Systematic Evolution of Ligands by EXponential enrichment). These RNA sequences (between thirty and sixty-nine mers long) were evolved to bind to the nucleotide Guanosine triphosphate (GTP) in separate experiments, and the sequence's information content was measured using the method outlined

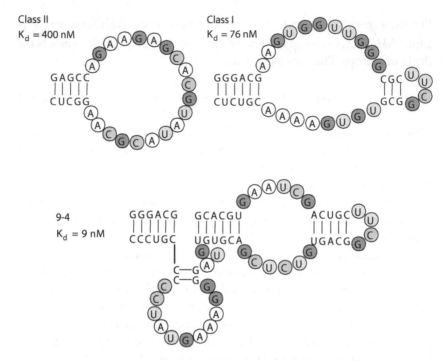

FIGURE 2.13. Secondary structure of RNA aptamers that evolved to bind GTP (adapted from Carothers et al. 2004). The three different classes of ribozymes have different binding affinities K_d (lower number indicates higher affinity). The structurally more complex ribozymes (like class 9-4) have higher information content and lower K_d.

above. Figure 2.13 shows the secondary structure of three different classes of evolved aptamers each with a different number of bulges or loops.

The affinity for evolved RNA aptamers to bind to their GTP substrate can be measured in a binding assay, where RNA sequences are incubated with GTP molecules and subsequently filtered through a membrane that has pores that are small enough so that only unbound molecules but not the bound complexes can pass. As a consequence, the ratio of the concentrations of substances on one side of the filter (the filtrate) to that on the other side (the retentate) is equal to the ratio of free ligand concentration [L] to concentration of free ligand and bound complex [A·L], where [A] represents the concentration of the RNA aptamer. As the concentration of RNA that is exposed to the GTP substrate is increased, the fraction that is bound to it also increases, as seen in Figure 2.14. The RNA concentration that results in half of the RNA to bind to the GTP target is called the *apparent dissociation constant* K_d. Thus, lower K_ds characterize molecules with higher affinity

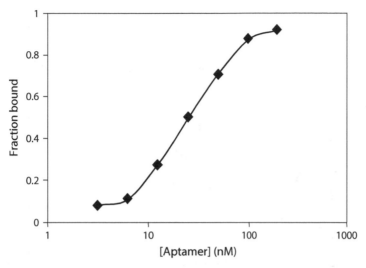

FIGURE 2.14. Fraction of bound RNA as a function of aptamer concentration (in nanomoles) in an ultrafiltration assay (adapted from Davis and Szostak 2002). This fraction reached 1/2 at an RNA concentration of 25 nM, defining a $K_d = 25$ nM for this particular aptamer.

to the substrate: they require a smaller concentration in order to bind GTP efficiently.

Figure 2.13 shows each RNA secondary structure of the evolved aptamer with its measured K_d. First, we notice that the more complicated the structure (here we can simply count the number of loops or bulges), the lower the K_d; that is, the higher the affinity of the molecule to its target. And interestingly, this correlates well with information content too: the simplest of the structures (called "Class II" in Fig. 2.13) has an information content of thirty-eight bits as measured by Carothers et al. (2004). The intermediate Class I structure comes in at forty-five bits, while the complex 9-4 structure, with the highest binding affinity of 6 nM, weighs in at a hefty sixty-five bits. Indeed, the authors found a roughly linear relationship between information content and binding affinity, as can be seen in Figure 2.15, which recapitulates the measured affinities and information content for eleven different structures. While there are some outliers that do not conform to the linear law, it seems as if you need about ten bits of information to improve your binding affinity ten-fold.

Let us return to the interpretation of information as that quantity that allows us to make predictions about random variables with an accuracy better than chance. If we are given the information content of a sequence, what can we say about the likelihood that we will encounter that sequence in a random

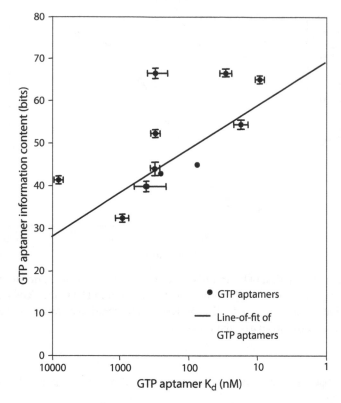

FIGURE 2.15. Aptamer information content plotted against binding affinity K_d for the eleven structures evolved by Carothers et al. (2004).

pool? In general, the probability to correctly predict the state of a random variable X with entropy $H(X)$ measured in bits is given by

$$P(X = x_i) = 2^{-H(X)}, \tag{2.58}$$

for a typical x_i. This is equivalent to noting that it takes on average $H(X)$ discerning yes/no questions to ascertain the identity of X. In a random pool of sequences of length L coded in an alphabet of size D, the probability to correctly predict the state of any random sequence is accordingly

$$P_{\text{random}} = D^{-H_{\text{max}}} = \frac{1}{D^L}, \tag{2.59}$$

because the entropy of an unspecified ensemble is maximal: $H_{\text{max}} = L$ (we are taking logarithms to the base D here). This implies that it would take an average of L "which of the D states?" questions (the D-dimensional analogue of the binary "yes/no" questions) to ascertain the molecule's identity. The pool of *functional* molecules, on the contrary, is much smaller. Among all those molecules that are functional (with information I), the probability to correctly specify any one of them is

$$P(X = x_i | x_i \text{ functional}) = D^{-H(X|e)}, \qquad (2.60)$$

that is, we need $H(X|e)$ questions on average to specify the identity of any one of the equivalent functional molecules. From these two probabilities, we can obtain the probability P_{func} that a molecule is functional with information I, since

$$P_{\text{random}} = P(X = x_i | x_i \text{ functional}) \times P_{\text{func}} \qquad (2.61)$$

that is, the probability to correctly specify any random molecule is equal to the probability to correctly specify any functional molecule, times the probability that the molecule is functional. Inserting Equations (2.59) and (2.60) into Equation (2.61) then yields

$$P_{\text{func}} = \frac{P_{\text{random}}}{P(X = x_i | x_i \text{ functional})} = \frac{D^{-L}}{D^{-H(X|e)}} = D^{-L+H(X|e)} = D^{-I}. \quad (2.62)$$

The interpretation of this formula is quite elegant (Carothers et al. 2004; Szostak 2003): it takes I mers to completely specify a sequence with information content I in a random pool. Or, in other words, armed with information I (the knowledge that the molecular sequence is functional with information I), it takes far fewer "which of the D states?" questions to specify the molecule: information allows you to make predictions with an accuracy that exceeds the accuracy of a chance prediction.

We can also turn the problem of predictability on its head. We asked earlier whether we could predict the identity of a molecule with information content I with accuracy better than chance. But can the molecules themselves predict *their* environment more accurately than chance? Can we say that the molecule has less uncertainty about its environment if it stores I mers of information? This question is at first sight ambiguous because we purposefully avoided talking about an *ensemble* of environments when we defined the information content about a particular environment. Fortunately, for the case of binding targets, such an analysis can be carried out by studying whether sequences that have information about GTP (that bind to GTP with high affinity) easily discriminate between different targets. Carothers et al., in a follow-up study (Carothers et al. 2006), have provided a partial answer.

We could couch the question in the following manner: Are molecules with information about GTP specific to GTP, or do they bind to other targets as well? Instead of testing the specificity of the eleven evolved aptamers to a group of targets that are very different from GTP, Carothers et al. instead chose to test the specificity of their molecules to chemical *analogues* of GTP. Information theory would predict that molecules with information about GTP would not bind to targets that are different from GTP, but should not be particularly choosy when it comes to targets similar to GTP. And indeed, while the evolved aptamers were most specific to GTP, they showed a sizable (if reduced) affinity to the chemical analogues also. These authors also asked whether more information about GTP would be correlated with higher specificity to GTP, that is, whether molecules that bind GTP more effectively (have more information about GTP) would also discriminate better between targets that are different from GTP. The surprising result is that there is no such correlation: molecules with different information content were neither more nor less specific to the analogues. Instead, the authors found that the molecules with different GTP specificity differed significantly in the free energy of formation, that is, they had different *thermostabilities*, which implies different stabilities of the tertiary structure of the molecule.

We can conclude from this observation that molecules store information not just about their target; there are other constraints that need to be satisfied. While all the aptamers that are specific to GTP store information about their target in terms of the binding contacts (complementarity of the RNA binding pocket to the ligand), they can differ in thermostability, which is a "piece of information" that is unrelated to the functional group complementarity.

2.4   Information Channels and Communication

Evolution, it bears repeating, is a process that increases the fit of a population to the environment within which it makes its living. The information that is accumulating in the genome of these organisms is information *about* that environment (see section 3.1 for an in-depth look at how this happens), and that information helps the organism predict what it will encounter in the future. But just fitting the environment as it is at one point in time is not enough: the environment is constantly changing, and this change occurs on many different time scales, and at microscopic to global scales. A population that is truly fit anticipates changes and reduces surprises. How exactly do you do that?

Generally speaking, there are two ways to thrive in a changing environment. One way is to make yourself immune to the changes, that is, you insulate yourself from the environment in such a way that no external changes will affect you. This strategy (generally called "robustness," and discussed in detail in chapter 6) is particularly effective if the changes in the environment are unpredictable, but it also comes with some drawbacks. Most obviously, insulating you from changes means that you forgo *taking advantage* of changes. After all, a change does not necessarily imply that one would be worse off in the changed world.

The other way to deal with changes is to anticipate them, that is, to have the right response for any possible change. This would allow you to mitigate (or eliminate) any costs to you when the environmental change would be harmful to you, but also take advantage of changes that are beneficial: the best of both worlds! To be able to do this, two conditions must be met. First: changes must be recognized, that is, the organism has to have sensors that can change state in response to a change in the environment. Second, the organism must be able to itself change in some way, so as to become maximally fit given the observed change. All these mechanisms have information at its heart, as we'll see. Mind you, if you do not have sensors, you can still adapt to a changing environment, but you have to do it via "bet hedging." This strategy, described in Box 2.4, relies on evolving a switching program that matches the rate at which the environment changes (this "matching" does not imply that the rates should be equal). But clearly, *sensing* a changed environment and reacting to it by switching to a persistent state in response would be a much more efficient strategy. Indeed, such strategies do exist (Harms et al. 2016) for some bet hedgers, so perhaps we should ask why a sensing strategy is not more common among bet hedgers.

We encounter perhaps the most obvious and common example of sensing in response to changes in the cellular metabolism, in particular bacterial metabolism. Bacteria have an unmatched ability to thrive in very different environments (this is one of the many differences to eukaryotic cells, which usually require fairly constant and defined environments instead). This bacterial versatility can be traced back in part to their ability to adapt to the environment by sensing what carbon source is prevalent, and turning on the right genes in response. The textbook example for such a sensing module is the so-called "lac operon." How this module works was first described by the French biologists François Jacob and Jacques Monod (Jacob and Monod 1961), work that ultimately garnered them the Nobel Prize in 1965.

The lac operon is a set of genes that allows *E. coli* (and other related enteric bacteria) to use the sugar lactose to grow. *E. coli* prefers glucose as a carbon source, which is understandable because glucose is a simple sugar (a monosaccharide, so-called because it consists of a single molecule of $C_6H_{12}O_6$). The world in which *E. coli* lives has plenty of other sugars one could live on, and one of them is lactose. To digest lactose (which is a di-saccharide, a not-so-simple sugar) requires an enzyme to cleave the double-sugar into two

Box 2.4. Bet Hedging: Adapting to Change without Sensing

Bet hedging in evolutionary biology is, as the name implies, a probabilistic response to a changing environment. In finance, a "hedging" strategy would be to buy different securities (for example) that would pay off in different future scenarios. In biology the idea is very similar: it allows an organism to "place bets" on what kind of an environment it is most likely to encounter. In this manner, the organism does indeed attempt to predict the future, but it does so without sensing the state of the environment. A typical case of bet hedging is the probabilistic germination of plants (Philippi and Seger 1989). If a plant produces a large number of seeds in a good year but only a small number of seeds in a bad year, then delaying germination is a good strategy for bad years because ungerminated seeds will survive to the next year and receive another chance, as it were. The optimal seed germination probability depends on how likely bad years are: it is better to germinate later the more likely the bad years. In this case, the information about the frequency of bad years enters the genome, via the process of evolution, encoded into the germination probability. Another interesting case of bet hedging to deal with changing environments is the evolution of "persisters" in bacteria. Exposure to antibiotics is an ongoing threat for bacteria. While we are used to antibiotics as medications we take to counter bacterial infections, most of the antibiotics in use today were initially "invented" by bacteria or fungi.

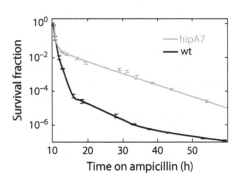

To protect itself against antibiotics, a bacterium can enter a "persistent" state in which it grows very slowly (if at all). Because most antibiotics target genes involved in growth, such a strategy is similar to the germination-delay strategy of plants, betting on surviving long enough to see another day. The probability to switch to the persistent state depends on the likelihood of antibiotic exposure. The bacterium *Escherichia coli* (our favorite commensal since it is also a darling of experimental evolution research and the star of section 4.2), for example, has the capacity to switch to persistence also, but there are also mutants

FIGURE 2.16. Fraction of surviving bacteria exposed to the antibiotic ampicillin. The light gray curve shows the mutant hipA7 that switches to the persister type at a much higher rate than the wild type (black curve). Adapted from Balaban et al. (2004).

that switch at a rate that is a thousand times higher (see Fig. 2.16 and Balaban et al. 2004; Kussell et al. 2005). Those types that switch at a higher rate are usually found in environments in which the likelihood to encounter antibiotics is much higher.

(a)

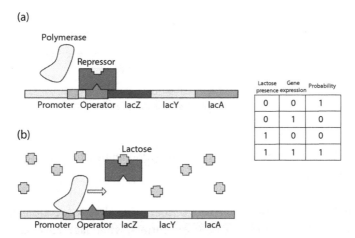

Lactose presence	Gene expression	Probability
0	0	1
0	1	0
1	0	0
1	1	1

FIGURE 2.17. Schematic view of the lac operon. (a) When lactose is absent, the lac repressor (dark gray) binds to the operator DNA sequence, thus blocking the RNA polymerase (light gray) to bind to the promoter sequence and then transcribe the operon. (b) When lactose is present, the molecule (gray crosses) binds to the lac repressor, which changes its shape in response. With a changed shape, the repressor cannot bind to the operator, allowing the polymerase to bind and then proceed to transcribe the three genes lacZ, lacY, and lacA. The computational logic is displayed in the upper right.

glucose molecules. Of course, *E. coli* has such an enzyme in its arsenal, called β-galactosidase (literally, sugar-cutter), but expressing it all the time just to make sure you are ready when there is lactose around is costly.[8] It would thus help the cell to only express β-galactosidase (encoded by the lacZ sequence in Fig. 2.17) when lactose is actually present. To do this, the cell needs a lactose sensor that is coupled to the gene expression machinery in such a way that the gene is only turned on if lactose is actually present. The lac operon is precisely that machine. The way it works is described schematically in Figure 2.17.

The three genes encoded in the lac operon are all transcribed whenever the RNA polymerase binds to its promoter sequence. It just so happens that when lactose is not present, a molecule called *lac repressor* binds to a sequence just downstream of the lac promoter, and prevents the polymerase to bind there and proceed with its job of transcription. If lactose is present, however, the lactose molecule binds to the repressor which, in response, changes shape. The morphed repressor cannot bind to the operator sequence anymore, and thus frees the polymerase to create the mRNA for the three proteins encoded by the

---

8. What is costly about expressing the lac operon unconditionally has been controversial for a long time (Stoebel et al. 2008), but it has now been established that it is in fact the expression of the permease lacY that creates a significant fitness deficit (Eames and Kortemme 2012).

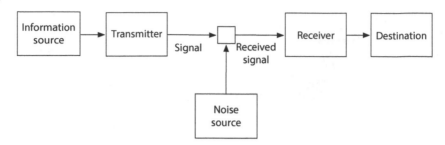

FIGURE 2.18. Schematic diagram of a general communication system (after Shannon 1948).

genes lacZ, lacY, and lacA. The lacZ sequence, as already mentioned, makes the enzyme β-galactosidase (the sugar-cutter), while lacY encodes a membrane protein that actually pumps the lactose molecule inside the cell. The role of lacA is much less clear. You can delete it and the *E. coli* metabolism works just fine, but it is unlikely that lacA is an evolutionary spandrel (Gould and Lewontin 1979; see also Box 10.1). Very likely we just have not encountered the situation where it is essential.

2.4.1 Noiseless channels

From an information-theoretic point of view, the lac operon creates an *information transmission channel* (Rhee et al. 2012) that allows the transmission of one bit per use of the channel. The presence or absence of lactose informs the cell whether or not the gene should be expressed. The information is formed by making sure that the middle two cases in the logic table on the top right in Figure 2.17 do not occur. Let's look at the mathematical definition of channel information and capacity, so that we can discuss information transmission more quantitatively.

The definition of a channel, along with the calculation of how much information can be sent through such a channel with arbitrary accuracy (the channel capacity) is no doubt the central achievement of Shannon's theory (Shannon 1948). Figure 2.18 is the standard depiction of a noisy channel, and it is essentially the figure that appeared already in the original publication (it is Shannon's Figure 1). In this view of the information transmission channel, information coming from some source is first encoded by the transmitter, and then sent over the channel. If the channel is noisy, then the code may be corrupted during transmission. The receiver will then attempt to correct for the errors, and then decode the information so that it is in its original form at the destination. Before Shannon's work, it was generally believed that it was not possible to protect messages from arbitrary levels of noise, but Shannon proved that this was not true. He showed that unless the noise is so extreme so as to completely randomize the message, it is possible to protect messages so that they can be deciphered with arbitrary accuracy, as long as one is willing to sacrifice the rate at which the information is transmitted. We will not delve into the theory of channels here at any depth, as there are many excellent textbooks that cover this topic. Instead, we will only discuss the main features of noisy discrete and continuous value *biological* communication channels.

Shannon defined the capacity of a channel to be the maximum rate at which information can be transmitted with arbitrary accuracy. Let's first discuss discrete noiseless channels, for which the lac operon as discussed above is a good example. For those, Shannon defined the capacity

$$C = \lim_{T \to \infty} \frac{\log_2 n(T)}{T}, \tag{2.63}$$

where $n(T)$ is the number of different signal states that can occur while operating the channel for a duration T. The way we looked at information transmission in the lac operon, each "message" was the presence or absence of lactose ($n = 2$), measured over an arbitrary length of time, say ΔT. If we looked at two events in a row, there would be $n = 2 \times 2$ possible events (low-low, low-high, high-low, high-high), but the duration is $2\Delta T$. In the limit $T = n\Delta T \to \infty$ then, the capacity is just

$$C = \lim_{n \to \infty} \frac{\log_2 2^{n\Delta T}}{n\Delta T} = 1, \tag{2.64}$$

that is, $C = 1$ bit, as expected.

Clearly this description of the channel is highly idealized. Are there really only two levels of expression, low or high? Of course not: information channels in biology are almost never isolated single-bit channels, but rather are part of a complex set of interlinked information channels. For starters, while having a binary (on/off) channel for controlling the lac gene expression clearly gives a fitness advantage, it is still rather coarse because the environment is more complex than that. For example, suppose lactose is present but there is also plenty of glucose. In that case, it would help *not* to turn on the lac gene because it would be wasteful. And indeed, bacteria have figured this out long ago.

Figure 2.19 shows that just "in front" (or "upstream" in the parlance of genetics in which the "stream" is the direction of transcription) of the promoter site is a binding site for the CAP protein. CAP stands for "catabolite activator protein," which is a pretty good name because the protein helps in activating the lactose gene (which, after all, catabolizes, i.e., breaks down, lactose). The idea is that if glucose is present, this helper protein will not be

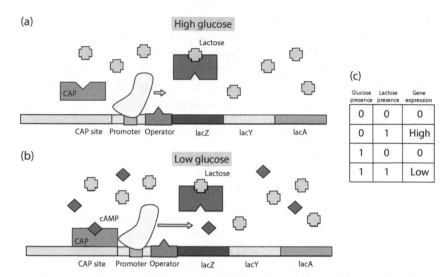

FIGURE 2.19. Regulation of lac expression as a function of glucose and lactose levels. (a) When glucose is present, the CAP protein does not bind to its site, unable to help the polymerase stick to its binding site, leading to low levels of lac expression. (b) When glucose levels are low, cAMP levels are high. A cAMP molecule (diamonds) binds to the CAP protein, changing its shape so that it can bind to the CAP site. There, it helps the RNA polymerase bind to its site, leading to strong expression of the lac operon. (c) With two binary signal states, it is possible to encode four different output states. In this case, the logic creates three different levels: 0, low, and high.

there, and lactose will only be activated at a low level (Fig. 2.19[a]; note the small arrow). But when glucose is absent, CAP goes to work and significantly amplifies lac expression (Fig. 2.19[b]; note the long arrow). As a consequence, lac expression is more nuanced, involving three levels rather than two: see Fig. 2.19(c). How does CAP do this? First, CAP can only bind when it is activated, just the opposite of the repressor protein that cannot bind when activated. But unlike the repressor protein, it needs to be activated in the *absence* of its sensor molecule. This means it cannot be activated by glucose itself.

Evolution, of course, figured out how to solve this conundrum. When glucose levels are low, the cell synthesizes a molecule called "cAMP," which stands for "cyclic AMP." We'll encounter AMP (adenosine monophosphate) again in section 7.1: it is one of the fundamental molecules in a cell's metabolism. Of the three energy carriers AMP, ADP, and ATP, AMP is the lowest energy state: it is transformed to ADP and then ATP by adding more and more energy (via the addition of phosphoryl groups to the molecule, a process called "phosphorylation"). However, cAMP is not produced directly from AMP; instead it is produced from ATP via an enzyme called a "cyclase." The reason for this circuitous signaling likely lies in detaching the role that AMP, ADP, and ATP play in metabolism from the signaling function. Indeed, cAMP is called a "second messenger," that is, a signaling molecule (much like allolactose is used in the lac operon). If there are plenty of cAMP molecules, this signals to the cell that it is "hungry," so to speak.

We can now see that in the absence of glucose there are plenty of cAMP molecules, and *they* play the role of activators for CAP (cAMP is shown as diamonds in Fig. 2.19[b]). When CAP is activated it can bind to DNA and help the polymerase bind to its promoter sequence, and as a consequence transcription levels are high. In retrospect we now realize that the picture shown in Figure 2.17 was incomplete: it showed the RNA polymerase transcribing the lac gene when lactose is present, but without the help of the CAP protein, the polymerase "falls off" its site often enough that transcription levels are low.

What is the capacity of the channel in this case? The answer to this question is easy: it is $\log_2 3 \approx 1.585$ bits per use of the channel, even though it seems that there are four possible messages (the four possible glucose/lactose configurations). However, the cell only responds with three different states, and according to the capacity formula (2.63), this means the capacity is $\log_2 3$. But you might object to this description in terms of the table in Figure 2.17 and remark that molecular binding is never so accurate that expression is all or nothing (so that the probabilities are not one and zero, but perhaps more like 0.9 and 0.1), so that the information that the cell receives about lactose must be less than one bit. And indeed this would be correct: the channel is not noiseless. But it turns out that this reduced capacity is not a liability to the cell; instead it turns out that the system *would not work* if the logic table would actually be like the one shown in Figure 2.17. Let's first find out why, and then recalculate the capacity of the channel.

It has not escaped the attention of workers in the field that if lactose molecules cannot enter the cell by being pumped through the membrane via the permease (encoded by lacY), then those molecules cannot activate the gene that codes for the permease to pump the molecules: it is a lactose catch-22. It turns out, however, that the lac operon is sometimes transcribed even in the absence of lactose (Choi et al. 2008). These rare stochastic events are sufficient to maintain a low concentration of permeases in the cell that can become membrane-bound sentinels for any lactose molecules the cell may encounter. The leakiness of the channel reduces the capacity for information transmission to below one bit per usage of the channel, but it is the leakage that makes the channel functional in the first place.[9]

9. There is even an additional wrinkle to this story. The molecule that binds to the repressor is actually not lactose, but allolactose, which is produced from lactose by β-galactosidase via transglycosylation. Thus, the cell uses both the lacY *and* the lacZ gene products stochastically to drive its own expression.

2.4.2   Binary noisy channels

Let's look at the simple "lactose-only" channel (Fig. 2.17) in the presence of noise first. If activation of the lac operon is stochastic, this means that the probability of gene expression in the absence of lactose is not 0, but something small, say ϵ. When lactose is present, on the other hand, let us imagine that the gene is not always turned on, but instead the probability that this happens is $1 - \epsilon$. Shannon taught us how to calculate the capacity of a noisy channel. If we define the sender and receiver random variables as S (for "sender") and R ("receiver") respectively, the capacity is

$$C = \max I(R : S), \qquad (2.65)$$

where $I(R : S) = H(R) - H(R|S)$ is the shared entropy between sender and receiver, defined abstractly earlier in Equation (2.28). The maximization is to be carried out over the probability distribution of the input to the channel, as we'll discuss in a moment.

It is immediately clear that the noisiness of the channel is reflected in the conditional entropy $H(R|S)$, which is the uncertainty we have about the message on the receiver's end. In a noiseless channel, this entropy is zero: lactose low means gene off, lactose high means gene on, without fail. Incidentally, because the entropy of the signal source $H(R)$ here is just the logarithm of the number of states, the capacity (2.65) in the noiseless case just turns into (2.63), as it should. But if $H(R|S)$ is non-zero, the capacity is reduced. Let's calculate it, for the simple model in which the gene is activated with a small probability ϵ even if lactose is not present.

According to Equation (2.25), the conditional entropy is

$$H(R|S) = - \sum_{s=\text{high, low}} \sum_{r=\text{on, off}} p(S = s, R = r) \log_2 p(R = r | S = s). \quad (2.66)$$

The conditional probability $p(R|S)$ is the *channel matrix*, and according to our simple model, it is

$$p(r|s) = \begin{array}{c} \\ \text{on} \\ \text{off} \end{array} \begin{array}{cc} \text{low} & \text{high} \\ \left(\begin{array}{cc} \epsilon & 1 - \epsilon \\ 1 - \epsilon & \epsilon \end{array} \right). \end{array} \qquad (2.67)$$

To calculate the conditional entropy (2.66), we still need the joint probability distribution $p(r, s)$. As long as we know the probability distribution $p(s)$ (that is, how likely we will be seeing low or high lactose levels) we can get the probability $p(r, s)$ using Bayes' theorem, namely $p(r, s) = p(r|s)p(s)$. Say $p(\text{low}) = q$ and $p(\text{high}) = 1 - q$. Then (see Exercise **2.9**) the entropy is

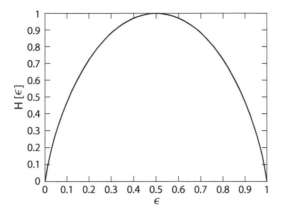

FIGURE 2.20. Binary entropy function $H[\epsilon]$ as a function of the error probability ϵ.

actually independent of q and reads

$$H(R|S) = -\epsilon \log_2 \epsilon - (1 - \epsilon) \log_2 (1 - \epsilon) \equiv H[\epsilon], \qquad (2.68)$$

where I defined the binary entropy function $H[\epsilon]$. This function is plotted in Figure 2.20, which shows that it is maximal at $\epsilon = 0.5$, while it vanishes at $\epsilon = 0$ and at $\epsilon = 1$. This is completely expected for a "one-bit" uncertainty function, because it simply reflects the uncertainty you have about a single bit if an error probability of ϵ was applied to that bit. The uncertainty is maximal if $\epsilon = 0.5$: half the time your bit is flipped, and half the time it is not. If $\epsilon = 0$ then of course you have no uncertainty (given the input), and you also can be sure about the input when $\epsilon = 1$, because if you are certain that the bit was flipped, then all you have to do in order to know what the message was is to flip your outcome. Using this result, we can now calculate the capacity as a function of ϵ using the formula (2.65). We know that $I(R : S) = H(R) - H[\epsilon]$, but to obtain the capacity we need to maximize over the input distribution $p(s)$, not the output distribution $p(r)$. As we found out earlier, the conditional entropy $H(R|S)$ is independent of q (see Exercise 2.8), which is due to the fact that the channel is symmetric (the same probability ϵ accounts for the error in how likely it is that the gene is turned on when lactose is absent, and how likely it is that the gene is off when lactose is present). In that case, it turns out that $H(R|S) = H(S|R)$, and we can write

$$I(R : S) = H(S) - H[\epsilon]. \qquad (2.69)$$

Note that in nonsymmetric channels, $H(R|S)$ could depend on the input probability distribution. With the information written in terms of the input

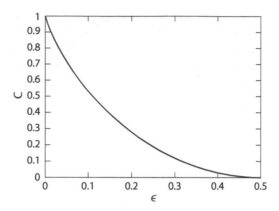

FIGURE 2.21. Capacity of the lac-operon symmetric channel as a function of the error probability ϵ.

entropy (which is maximal at $q = 0.5$ because $H(S) = H[q]$), we see that the capacity is

$$C = \max_q I(R:S) = 1 - H[\epsilon], \tag{2.70}$$

which is plotted in Figure 2.21. The capacity vanishes at $\epsilon = 0.5$, a probability that represents a level of noise that makes information transmission impossible. Indeed, at this noise level, rather than sending information, the process has become akin to a coin flip. No amount of error correction can extract information from a perfect coin flip.

But it is also clear that for small error probabilities, a sizable fraction of the information can be transmitted. For example, for a 5 percent error, the capacity only drops to about 0.71 bits per use of the channel. Now, it is clear that this level of noise in the channel will be detrimental to the cell. As a consequence, there is a selective pressure to correct for such errors. In engineering, a considerable amount of work is devoted to finding codes that allow for efficient error correction, and the same is true in biological channels. In fact, a lot of the complexity of the channels that we ignored here is geared toward making sure that the cell can obtain the full bit of information, even in the presence of noise. Indeed, that the capacity is 0.71 (say) does not imply that the cell only gets a fraction of the information. It means that *per use* of the channel, you get to transmit 0.71 bits. Error correction is achieved by embedding the message in a longer code (for example, using the channel n times for a code of length n). The rate drops when using such codes, but it still allows for sending the full information over the channel, only slower than what is possible in the absence of noise.

The central takeaway from the lac operon story is simple from an information-centric point of view. The lactose molecule is a signal to the cell about what kind of environment it is currently in. The signal is relayed to the central cellular machinery via a cascade: first it is transported into the cell via the permease, then the signal molecule binds to its receptor, the lac repressor molecule. It is at this point that the signal molecule becomes information, because it engenders a change in the repressor's 3D structure (an *allosteric* change), which in turn affects whether the repressor binds to DNA and obstructs the polymerase. Information after all is, as the English biologist William Bateson famously wrote, "differences that make a difference" (Bateson 1979). Thus, if the difference (whether or not lactose is present) did not make a difference (change the shape of the repressor and influence gene expression), then the lactose molecule would not represent information, it would simply represent entropy: a lot of differences without consequences.

2.4.3 Continuous value channels

Discrete binary channels are often used to demonstrate the information-theoretic principles behind communication because it is easy to calculate the capacity of such channels. In biology, information transmission is usually quite a bit messier than the lac operon story.

And while sensing of molecules is best described in terms of discrete variables because the molecules that *bind* the signal can usually only take on a finite number of discrete states (like the lac repressor in Fig. 2.17), the response to such signals can be quite complex. For example, a common channel in biology is represented by the activation of a gene via a transcription factor. Typically, the transcription factor has a particular concentration in the cell and will bind to a site on the DNA that it is complementary to. In response to this binding, a gene is being transcribed at a particular level. In the previous section we looked at binding and gene expression as a binary choice, but in reality gene expression is a stochastic process (Elowitz et al. 2002). The "noise" in gene expression comes from a variety of sources, both intrinsic to the process of gene expression (DNA binding), as well as extrinsic sources (such as fluctuations in the signal molecules).

Here we will focus on the consequences of having a continuous probability distribution of signal molecules $p(s)$ (extrinsic noise) and a model of gene expression that is probabilistic (intrinsic noise). The two sources of noise will give rise to a distribution of response molecules described by $p(r)$, where r is the concentration of the gene product (we will study the binding of transcription factors to their sites in terms of thermodynamics and information theory in much more detail in section 3.4). The capacity of the channel can then be written as[10]

$$C = \max_{p(s)} I(S:R), \qquad (2.71)$$

that is, it is given by the shared entropy between signal S and response R, maximized over the input probability distribution $p(s)$. This expression generalizes the capacity for the simple binary channel, Equation (2.70).

In the simplest case, the signal molecule's abundance (for example, a transcription factor) is described by a Gaussian variable S with a given mean and variance, while the response variable reflects that abundance, but with added Gaussian noise (see Box 2.5). This situation is shown in Figure 2.22, where the only effect of the channel is to add noise to the sensed variable.

Channels with Gaussian noise are called *Gaussian channels* and play a major role in engineering (where they are sometimes called "AWGN channels," where AWGN stands for "Additive White Gaussian Noise"), and the quantum version of that channel plays a role in quantum communication (Holevo and Werner 2001) and even black hole physics (Bradler and Adami 2015).

For Gaussian channels, the capacity is given by the simple formula (2.77), because we have assumed here that the input distribution is Gaussian along with the noise. When S is not normal, the transmitted information $I(S:R)$ is less than the capacity.

Let us now consider more general channels, where the input distribution is not necessarily Gaussian, and is transformed by a function $r(s)$, shown as the solid black line in Figure 2.22. Since the transmitted information is then less than the capacity, we can ask what evolution can do to increase the amount of information transmitted per use of the channel. Because the optimization of the mutual entropy $I(S:R)$ is obtained by changing the input distribution $p(s)$, we must therefore ask, "What input distribution optimizes (maximizes) the transmitted information?," given all other constraints of the channel, of course. This is a question that can be answered mathematically (Tkačik et al. 2008b; Tkačik et al. 2008a) (see in particular the review Tkačik and Walczak 2011), and the result is surprisingly simple.

The derivation of the result is somewhat tricky so we'll skip most of the details. First, we must generalize the channel matrix from the expression

$$p(r|s) = \frac{1}{\sqrt{2\pi\sigma^2}} e^{-\frac{(r-s)^2}{2\sigma^2}} \qquad (2.72)$$

10. Technically, the maximization is really a search for the supremum, because while the max function returns the highest number within a set, the supremum of the set bounds the set, and does not have to be a part of the set.

Box 2.5. Capacity of Gaussian Channels

The simplest example of a continuous variable channel is given by the Gaussian channel. Let us define a signal variable S and a response variable R, where $R = S + \mathcal{N}$ and $\mathcal{N}$ represents Gaussian noise with zero mean and variance σ^2. We will assume here that the signal variable *itself* is distributed in a Gaussian manner around its mean $\bar{s}$ with variance σ_s^2, that is

$$p(s) = \frac{1}{\sqrt{2\pi\sigma_s^2}} e^{-\frac{(s-\bar{s})^2}{2\sigma_s^2}}. \tag{2.73}$$

Let us calculate the information that S conveys about R, that is, $I(S:R)$, in this special case. We can write this information as

$$I(S:R) = H(R) - H(R|S). \tag{2.74}$$

Now, since $R = S + \mathcal{N}$, $H(R|S)$ is really the same as $H(S + \mathcal{N}|S)$, which in turn is just $H(\mathcal{N}|S)$ since $H(S|S) = 0$. So, $H(\mathcal{N}|S)$ is the entropy of the Gaussian noise given the signal. But since the noise is actually independent of the signal, it is clear that $H(R|S) = H(\mathcal{N})$. This is also intuitively clear: if the response is just the signal plus noise, then given the signal the only remaining entropy must be the entropy of the noise (this is not true anymore if the response is a more complicated function of the signal).

It is easy to calculate the entropy of a Gaussian variable with variance σ^2, it is simply (see Exercise **2.10a**)

$$H(\mathcal{N}) = \frac{1}{2}\log_2(2\pi e\sigma^2). \tag{2.75}$$

To calculate the information (2.74), we need to calculate $H(R)$. The variable R is a convolution of two Gaussian random variables, since $p(r) = \int p(r|s)p(s)\,ds$, and $p(s)$ is Gaussian by definition (2.73) and $p(r|s)$ is just the noise distribution of $\mathcal{N}$. We then find that (see Exercise **2.10b**)

$$H(R) = -\int p(r)\log p(r)\,dr = \frac{1}{2}\log_2\left(2\pi e(\sigma^2 + \sigma_s^2)\right), \tag{2.76}$$

that is, the entropy of a distribution with variance $\sigma^2 + \sigma_s^2$. As a consequence, the information shared between signal and response is simply

$$I(S:R) = \frac{1}{2}\log_2\left(1 + \frac{\sigma_s^2}{\sigma^2}\right). \tag{2.77}$$

This is the standard result for the capacity of a Gaussian channel (Cover and Thomas 1991), where $\frac{\sigma_s^2}{\sigma^2}$ is the signal-to-noise ratio.

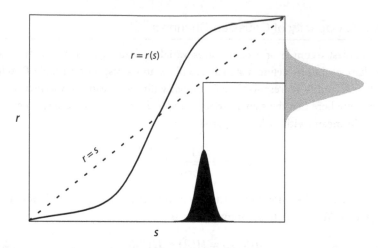

FIGURE 2.22. Information transmission channel where the Gaussian input distribution $p(s)$ (black) is transformed into the output response distribution $p(r)$ (gray). In the simplest case, the transfer function is the identity (dashed line, $r = s$) and the only effect of the channel is to increase the width of the distribution. In the more general case, the input is transformed into the output with a function $r(s)$ (solid line).

that holds for the case described in Box 2.5 to

$$p(r|s) = \frac{1}{\sqrt{2\pi\sigma_r^2}} e^{-\frac{(r-\bar{r}(s))^2}{2\sigma_r^2}}, \qquad (2.78)$$

where $\bar{r}(s)$ is the function that transforms the mean signal expression level s into the mean response level $\bar{r}(s)$. Here, the variance of the distribution is given by σ_r^2, which can differ from the noise variance σ^2 because in this case the noise can depend on the signal.

Next, we need to optimize expression (2.74). This is best done by a variational procedure, using Lagrange parameters to take care of the constraints on the channel. In the limit of small noise (an expansion in inverse powers of σ), the result is

$$p_{\text{opt}}(s) = \frac{d\bar{r}(s)}{ds}. \qquad (2.79)$$

Using this distribution to calculate the response distribution yields

$$p_{\text{opt}}(r) = \int p(r|s)p_{\text{opt}}(s)ds = \int p(r|s)\frac{d\bar{r}(s)}{ds}ds \propto \int p(r|s)dr, \qquad (2.80)$$

but the last integral is simply 1, as the distribution (2.78) is properly normalized. Thus, the capacity is optimized when the output distribution is uniform (a constant). A little bit of calculation shows that this constant distribution is given by (Tkačik et al. 2008b; Tkačik et al. 2008a)

$$p_{\text{opt}}(\bar{r}) = \frac{1}{Z}\frac{1}{\sigma_r}, \tag{2.81}$$

where (recall that σ_r in general depends on r)

$$Z = \int \frac{1}{\sigma_r}\,\mathrm{d}r \tag{2.82}$$

normalizes the probability distribution (2.81). In particular, we can calculate the optimal information as

$$I_{\text{opt}}(R:S) = \log_2\left(\frac{Z}{\sqrt{2\pi e}}\right). \tag{2.83}$$

This optimization of information transmission, which works by adapting the input distribution in such a manner that the output distribution has a near-uniform distribution is, as it turns out, a well-known trick that has been used in optimizing image processing for example, where it is called *histogram equalization*. In fact, this trick was discovered much earlier by Laughlin (1981), who showed that the response function in a fly's retina has almost perfectly adapted to the contrast variations in a fly's environment so as to produce a near-uniform output distribution. Even better, analyzing a particular gene expression channel, namely the expression of the "gap" gene hunchback (a gene that is crucial in fly development, that is, the formation of the fly body) as a response to the morphogen signal bicoid (Tkačik et al. 2008b) shows a close-to-optimal distribution $p_{\text{opt}}(r)$, with about 1.7 bits of information processed per use of the channel.

2.4.4 Information transmission between animals and plants

To close out this chapter, we take a look at how information theory can characterize communication between whole organisms, as opposed to cells. After all, communication takes place on all scales, from molecules all the way to the edge of the solar system.[11] An instructive example of communication in the service of fitness is the communication going on between a plant and an animal. While one may think that plants and animals have little to talk about, the

---

11. As far as we know today, the largest distance covered by two-way communication is between Earth and the Voyager 1 spacecraft, launched on September 5, 1977, which is currently about 150 AU (astronomical units) from Earth, traveling at approximately 3.6 AU/year. (One AU is the distance between Earth and the sun, about 150 million kilometers.)

language being used is chemical in nature, and all the "talking" is being done by the plant.

Plants are under attack daily by herbivores that munch on their leaves, and it seems there is little that they can do to protect themselves from that threat. However, if given a chance, evolution always finds a way, even if a plant is unlikely to evolve a swatter to remove the herbivores from their leaves. Perhaps they can ask someone else to do it for them? One ingenious way to lessen the impact of a herbivore has evolved in a number of species of flowering plants, such as cotton, tobacco, and maize plants. The herbivores that prefer the leaves of these plants themselves have predators of course, and these predators have particular tastes for some, but not all the herbivores. For example, the tobacco budworm (*Heliothis virescens*) is the larval stage of a moth, and feasts on tobacco leaves, cotton leaves, as well as many other plants. Enter the wasp *Cardiochiles nigriceps*. It is a parasitoid: it lays its eggs in the larvae of other arthropods, and the tobacco budworm is its favorite target. However, there are other herbivores on the typical tobacco or cotton plant, such as the corn earworm *Helicoverpa zea*. The two caterpillars *H. virescens* and *H. zea* are very difficult to distinguish by eye (and the wasp will readily attack both), but there is a problem: the wasp eggs only develop in *H. virescens* (Lewis and Brazzel 1966). There would thus be a significant fitness benefit for the wasp if it did not mix up the two constantly.

Enter the signaling system of the plant. It turns out that when an herbivore bites into the leaf of the plant, the oral secretions that go along with the feeding trigger—in the leaf of the plant—the synthesis of certain terpenoids. Terpenoids are a class of organic chemical that are derived from *terpenes*, and as the name suggests (think turpentine), they are highly aromatic. It so happens that when *H. virescens* or *H. zea* bites the plant, *different* terpenoids are synthesized, and as a consequence, the identity of the herbivore can be determined by smell alone. In an ingenious set of experiments, De Moraes et al. have shown that *C. nigriceps* has evolved to take advantage of this opportunity (De Moraes et al. 1998), while the plant has gained an effective herbivore remover, by hailing it with the tell-tale perfume.

We can quantify this signaling in terms of information theory. In general, there are many different herbivores that can infest the plant, but to quantify how much information is in the generated odor, for our purposes we will only distinguish the smell of *H. virescence*, the smell of *H. zea*, and the smell of nothing (the undamaged plant). To test whether *C. nigriceps* can really distinguish their host *H. virescence* from other species, the team exposed a plant bitten by *H. virescence*, one bitten by the similar *H. zea* (in which, however, *C. nigriceps* eggs cannot mature), or an undamaged plant, to *C. nigriceps* in the field, for an hour in each experiment. During those trials, the number of times a

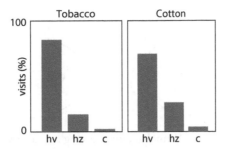

FIGURE 2.23. Percentage of times the wasp visited a tobacco or cotton plant bitten by *H. virescence* (hv), by *H. zea* (hz), or a an undamaged control plant (c). Data from De Moraes et al. (1998).

female *C. nigriceps* landed on the plant was recorded. Let X be the variable that controls the smell. The wasp was exposed to the three types of smell for an equal amount of time (one hour each), so we have the probabilities $p(X = \text{hv}) = p(X = \text{hz}) = p(X = \text{c}) = 1/3$, where hv stands for *H. virescence* presentation, hz means a plant bitten by *H. zea* was offered, while c stands for control: the undamaged plant. The wasp random variable Y instead records the behavior of the wasp. While in this experiment the wasp did not have a choice to land either on an hz plant, and hv plant, or the control plant (as they were not presented simultaneously), the rate at which they landed on those plants was taken as a proxy for the decision.

Here is the data from De Moraes et al., for an experiment with a tobacco plant (left panel in Fig. 2.23) and with a cotton plant (right panel). We can calculate the information that the wasp has about which herbivore bit the plant as

$$H(Y:X) = H(Y) - H(Y|X), \qquad (2.84)$$

where $H(Y)$ is the entropy of the possible wasp behaviors (choosing any of the three states hz, hv, or c). Because there are three states, the entropy is $\log_2 3 \approx 1.585$ bits. To obtain the information, we subtract the conditional entropy of selections given the state of the plant, $H(Y|Z)$, which can be calculated from the values in Figure 2.23 to be $H(Y|Z) \approx 0.75$ bits. Since the state variable X also has entropy $\log_2 3$, the Venn diagram for this information channel is given by Figure 2.23(a), and it suggests that the wasp has about 0.83 bits of information about X, the state of the plant (the cotton plant is left as Exercise **2.11**). However, this channel does not reflect the information that the wasp uses to increase its fitness, as correctly distinguishing an undamaged leaf from a leaf that was bitten by the incorrect host does not provide a benefit (unless we assume that the wasp eats its nonhost prey, which we will not).

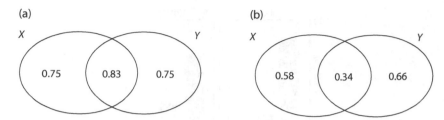

FIGURE 2.24. (a) Entropy Venn diagram showing conditional entropy and shared information when treating the tobacco data from Figure 2.23 as a three-state channel. (b) Venn diagram for the tobacco channel when the choice is between correct/incorrect host: a binary channel.

Thus, we should really treat this situation as a binary channel, in which the wasp correctly identifies its host about 83 percent of the time (164 landings out of 198 attempts). Figure 2.24(b) shows that in this case, the wasp only has about a third of a bit of information. This seems low (given that the maximum it could have is one bit), but the binary entropy function shown in Figure 2.20 shows that a third of a bit implies that indeed you will guess correctly about 83 percent of the time. Thus, there are diminishing returns in attempting to maximize the information: the cost in acquiring the necessary sensing accuracy most likely will outweigh the benefit.

This channel seems peculiar in the sense that it is not symmetric: the wasp does not signal anything to the plant, and most importantly, the plant does not change its state based on any behavior of the wasp. However, there is no doubt that the plant benefits from the signaling, as the parasitized caterpillar surely is stopped from effecting more damage. Thus, the information channel benefits both participants, and it is easy to imagine a coevolutionary process where a small difference between host/no-host probabilities selects for more disparate chemical signatures, which in turn select for more accurate sensing. In fact, there is far more potential information in the spectrum of volatiles that are being emitted by the bites of the herbivores, but all the wasp really needs of that entropy is a third of a bit.

The preceding analysis of information transmission channels in cellular and animal/plant biology demonstrates how pervasively information is used by biological systems to ensure that they survive and thrive in changing environments. Evolution tries to maximize the information that is processed (and most often succeeds in doing so), because knowing more about the environment—what it is like now, as well as how it might be in the future—translates into survival. In this sense, information is seemingly converted directly into fitness (Donaldson-Matasci et al. 2010; Rivoire and Leibler 2011). We might even say: information *is* fitness.

2.5 Summary

Shannon's theory of information is a natural description of the statistics of variation and correlations in biopolymers and allows us to quantify both the information content and the absence of information (that is, entropy) in DNA, RNA, and protein sequences. If we view biopolymers in terms of random variables, we see that evolution has changed the probability to find any particular monomer "variable" at any particular position in the sequence so as to maximize the information stored in the sequence—information about the environment within which the sequence must function. Information is always useful: it allows the holder to make predictions about the environment that are better than chance, and "better than chance" is what an organism needs to thrive in complex settings. Thus, we expect information to translate into fitness for the organism almost all of the time.

"Entropic profiles" of molecules can be used to study variability and sequence conservation, but also to predict the secondary structure of ribozymes. And while the total amount of information stored in a protein or a gene can be estimated using Shannon's theory, we can also study subtle evolutionary changes to *where* in a sequence information is stored, by tracking the entropic profile of a gene across species. Molecules that carry more information appear to be more functional than those that carry less, because this information is used to better predict the environment within which the molecule is functioning.

But molecules are not only used to store information; they are also used to *convey* information about what the state of the environment is at any particular point in time, and they can signal changes in the environment to the organism. We can imagine that evolution has managed to optimize these information transmission channels so that, given certain constraints, as much information is processed per unit time as possible. This involves adapting the cellular machinery to match how the environment changes, so as to be maximally receptive to the most likely changes. Clearly, information is useful for biological organisms; it would not deserve its name if it was not.

Information theory is not a panacea to understand the genetics of evolution and molecular function, because it is a statistical description that relies on ensembles of sequences that need to be large. Averages can wash out and obscure important genetics: the entropy at a site ignores which nucleotide or residue is preferred at that position. But the averaging introduced by information theory allows us to follow overall trends and discern macroscopic laws, such as the trend for organisms to accumulate and not lose information in the long term, a trend studied in more detail in the following chapters.

Exercises

2.1 Suppose you are given a random DNA sequence of length N and interpret it as the record of N trials where each of the four values A, C, G, T have a probability of p_A, p_C, p_G, p_T of occurring.

  (a) What is the probability of observing any particular nucleotide n times?

  (b) What is the *most likely* value for n_A/N in the limit $N \to \infty$, by maximizing the probability distribution found in (a)?

  (c) Show that the expectation value of the frequency of A nucleotides is $\langle n_A/N \rangle = p_A$.

  (d) Show that the variance is $\text{var}(n_A/N) = \frac{p_A(1-p_A)}{N}$.

  (e) For the sequence (2.2), verify the *standard error* for the four probabilities p_A, p_C, p_G, p_T given in (2.3).

2.2 Let X be a random variable with a uniform probability distribution over an ensemble $\mathcal{A}$ of size m, for example $\mathcal{A} = \{1, 2, \ldots, m\}$. Thus, we have $p(x) = \frac{1}{m}$ for $x \in \mathcal{A}$.

  (a) Determine the entropy $H(X)$. If $m = 128$, what is $H(X)$?

  (b) How many bits are necessary to describe an alphabet $\mathcal{A}$ of 128 elements without using coding?

  (c) Show that this entropy is larger than that of any other random variable Y with a probability distribution over $\mathcal{A}$. Hint: use the technique of Lagrange multipliers.

2.3 Given a random variable $X = \{1, 2, 3, 4, 5\}$ with a probability distribution $\{1/2, 1/4, 1/8, 1/16, 1/16\}$, calculate the entropy $H(X)$.

2.4 A perfect coin is tossed until it shows heads. Let the random variable X represent the number of tosses necessary to achieve this.

  (a) What is the entropy $H(X)$?

  (b) Devise a series of yes/no questions to determine the value of X. Compare the entropy of X with the mean number of questions necessary to determine the value of X.

2.5 Given two random variables X and Y with a probability distribution $p(x, y)$ given by

$p(x,y)$	0	$Y=1$	2
0	0	1/6	1/6
X= 1	1/6	0	1/6
2	1/6	1/6	0

(a) Calculate $H(X)$ and $H(Y)$.
(b) Calculate $H(X|Y)$ and $H(Y|X)$.
(c) Calculate $H(X, Y)$.
(d) Calculate $H(X:Y)$.
(e) Show the Venn diagram relating (a)–(d), and enter the numerical values.

2.6 Given two binary random variables X and Y, construct a third variable Z by performing the logical XOR (exclusive OR) operation: $Z = X \oplus Y$ as in the table below:

X	Y	Z
0	0	0
0	1	1
1	0	1
1	1	0

(a) Assuming that both X and Y take on values 0 and 1 with equal probability, draw the entropy Venn diagram between X and Y and fill in all three numbers.
(b) Draw the diagram for the remaining two pairs.
(c) Draw the diagram for all three variables, and fill in the seven entries by using the values for the seven quantities $H(X)$, $H(Y)$, $H(Z)$, $H(XY)$, $H(XZ)$, $H(YZ)$, and $H(XYZ)$.
(d) Draw the pairwise diagram of any two variables when the third one is *given*. Convince yourself that the third variable acts as a cryptographic key that, when given, reveals the information between the two other variables. (This is the information-theoretic version of the Vernam cipher.)

2.7 The specific information of a measurement can be negative in some cases. The following example will illustrate the concept of specific information by constructing a situation where measurements can take on surprising results.

Let X be a random variable whose states are the possible locations of the keys to the house. We will assume that the keys could be in eleven different places: they could be in your pocket ($X = x_0$) with a probability of $p(x_0) = P(X = x_0) = 0.9$, or they could be in one of ten places in the house, with equal probability: $p(x_i) = P(X = x_i) = 0.01$ for $i = 1 \cdots 10$.

To find the key, we can perform measurements: we can check our pocket, or we can check the ten possible locations within the house. To quantify the outcome of these measurements, we introduce a joint measurement variable $Y = Y_0 Y_1 \cdots Y_{10}$, where the states of each of the variables can be zero or one, depending on whether the key is found at any of these locations. Y is a "mutually exclusive" variable; only one of the eleven variables can take on the value $Y_i = 1$ at a time (the key can only be in one location at the time).

(a) Calculate your prior uncertainty $H(X)$ (prior to any measurement)

$$H(X) = - \sum_{i=0}^{11} p(x_i) \log_2 p(x_i). \qquad (2.85)$$

(b) Assuming that the probability to find the key in your pocket equals the prior, that is, $P(Y_0 = 1) = p(x_0)$, calculate the probability that the key is in any of the ten different places in the house given that you did *not* find the key in your pocket, i.e., calculate $p(x_i | Y_0 = 0)$.

(c) Use the conditional probability found in (b) to calculate the conditional entropy of X given that the key was not found in your pocket

$$H(X|Y_0 = 0) = - \sum_{i=0}^{11} p(x_i|Y_0 = 0) \log_2 p(x_i|Y_0 = 0). \quad (2.86)$$

(d) Calculate the specific information of the measurement result $Y_0 = 0$

$$I(X : Y_0 = 0) = H(X) - H(X|Y_0 = 0). \qquad (2.87)$$

(e) Demonstrate that the average specific information $I(X : Y_0)$ is still positive.

2.8 Using the expectation values for the mean and variance of $\hat{p}_i$ obtained in (**2.1**c, d) (where $\hat{p}_i = \frac{n_i}{N}$ is the estimated probability to find outcome i), show that the leading term (in $1/N$) in the bias correction for entropies (2.41) can be obtained via Taylor expansion of the expectation value of the entropy estimator

$$\hat{H} = H(\hat{p}_1, \ldots, \hat{p}_s) = - \sum_{i=1}^{s} \hat{p}_i \log_D \hat{p}_i \qquad (2.88)$$

to order $(\hat{p}_i - p_i)^2$.

2.9 Show that if the conditional probability for a sender and receiver pair is given by (2.67) and $p(\text{low}) = q$ and $p(\text{high}) = 1 - q$, then the conditional entropy $H(R|S)$ is independent of q and given by (2.68), by writing

$$H(R|S) = q\,H(R|S=\text{low}) + (1-q)H(R|S=\text{high}). \quad (2.89)$$

2.10 (a) Show that the Shannon entropy of a Gaussian variable X with probability distribution

$$p(x) = \frac{1}{\sqrt{2\pi\sigma^2}}e^{-\frac{(x-\bar{x})^2}{2\sigma^2}} \quad (2.90)$$

is given by

$$H(X) = \frac{1}{2}\log_2(2\pi e\sigma^2) \quad (2.91)$$

using the continuous-variable equivalent of the discrete Shannon entropy (also sometimes called *differential entropy*, see Cover and Thomas 1991)

$$H(X) = -\int_{-\infty}^{\infty} p(x)\log_2 p(x)\,dx \quad (2.92)$$

and the formula for the Gaussian integral

$$\int_{-\infty}^{\infty} e^{-a(x+b)^2}\,dx = \sqrt{\frac{\pi}{a}}. \quad (2.93)$$

(b) Show that the probability distribution of a convolution of two Gaussians $p(x)$ with variance σ_x^2 and $p(y|x)$ with variance σ_y^2 gives rise to the distribution

$$p(y) = \frac{1}{\sqrt{2\pi(\sigma_x^2 + \sigma_y^2)}}e^{-\frac{y^2}{2(\sigma_x^2+\sigma_y^2)}} \quad (2.94)$$

with entropy $H(Y) = \frac{1}{2}\log_2\left(2\pi e(\sigma_x^2 + \sigma_y^2)\right)$.

2.11 Calculate the information the parasitic wasp has about the identity of the caterpillar $H(X{:}Y)$ from the cotton data in the right panel of Figure 2.23, using $p_{\text{hv}} = 49/70$, $p_{\text{hz}} = 18/70$, and $p_c = 3/70$. Assume first a ternary channel, and then a binary channel.

3

Evolution of Information

Evolution is a light which illuminates all facts, a curve that all lines must follow.
— P. TEILHARD DE CHARDIN (1959)

Darwinian evolution is a property of the physical world (albeit of a very particular one), and as such obeys all laws of physics. In this chapter, we will look at evolution from a statistical point of view and use arguments from thermodynamics and information theory to discover a general trend in Darwinian systems: the information content of living systems has—on the whole—been increasing since the origin of life.

3.1 Evolution as a Maxwell Demon

We have seen previously that genomes are vast repositories of information about the world in which they evolved. How did this information get there? Why does it not fade and deteriorate with time? To understand this process from a physical (rather than biological) point of view, we need to study the *physics of measurement* in terms of information theory. While this exposition may seem extraneous at first glance, we will be rewarded with an unusual view of the process of evolution, so bear with me!

A measurement is the most general term for an "information acquisition event." But what actually happens in a measurement? Rather than entering a mathematically rigorous discussion of measurement (see, e.g., Krantz et al. 1971), we'll study here instead a simple example and formulate it in terms of information theory.

We will take as our example a simple length measurement. Just like in our introduction to information theory in the previous chapter, we will couch this discussion in terms of random variables. Let one random variable describe the objects whose length we would like to measure: say, sticks of irregular length.

101

Another variable describes our measurement device: say, a ruler. Now, in principle a lot of thought needs to go into the construction of a measurement device. This device is supposed to reflect as accurately as possible a particular character of the system that we are investigating, while being insensitive to other characters. In some way, our measurement device represents a *model* of the character we are interested in (Potter 2000), which implies that choosing a bad measurement device—like choosing a bad model—can result in a very accurate estimate of something very much unrelated to what we think we are measuring. In the case of sticks, however, there is no difficulty. A ruler will do.

Now let us define the states of our random variables. For both the sticks and the ruler, we clearly are dealing with *continuous* random variables, as opposed to the discrete ones we are used to. But while the mathematics of continuous random variables is not more difficult, we can easily convince ourselves that discrete variables are sufficient here too. Consider our ruler in more detail. For a length measurement, we need to attach a *number* to the property "length" in such a way that the relationship between the numbers reflects the relationship between the properties. The actual *scale* used to do this is arbitrary, of course, as long as the numbers are chosen in such a way that the objects being measured can be distinguished and possibly ordered. We can construct a ruler for example by carving markings at equidistant intervals onto a "reference stick," and assign increasing numbers to each mark. These numbers will then serve as the states of our "ruler random variable."

3.1.1 Perfectly resolving measurement

What should be the distance between markings? This distance defines your measurement *resolution*. It would be useless, for example, to space the markings so close together (if that was even physically possible) that you could not decide which marking better represents the height of the stick you are measuring. At the other extreme, neither would it be satisfying if you only had, say, two markings on the ruler, classifying sticks as "either long or not." Thus, the resolution of our ruler should be determined by what kind of a discrimination is sensible to us given the purpose of the objects being measured. Once this determination is made, the total number of states of the ruler—now a discrete variable—has been defined, and *with it the different possible states of our sticks*, since the length of the stick (while a continuous variable in principle) is now practically defined only in terms of the resolution of the ruler. In other words, the resolution of the ruler *defines* the possible "length states" of our sticks, while at the same time the resolution of the ruler is influenced by what kind of differences between sticks are interesting to us. Think of it this way:

the sticks do not actually have a length property independent of a measurement device: the concept "length" is, inherently, an expression of a correlation between two physical devices: a comparison in this case. But in reality our measurement devices will not be "perfect" (in the sense that no other device could better resolve the differences between the states) and we will deal with imperfect devices later below.

Say now that our (perfect) ruler variable R can take on $n + 1$ states with probabilities p_i (what these are will turn out to be irrelevant), while our ensemble of sticks S can take on the same states with probabilities q_j. Before a measurement, these two random variables are independent and thus uncorrelated. This means they do not share any entropy, and their joint entropy is given by the sum of their respective entropies:

$$H(R, S) = H(R) + H(S). \tag{3.1}$$

A measurement is the act of associating the state of the measurement device with that of the object. Let $\{r_0, r_1, \ldots, r_n\}$ denote the possible states of R, where we identified a special state r_0 for a ruler that does not measure anything, that is, one that points to a default value. The values $\{s_1, \ldots, s_n\}$ instead denote the sticks' possible states.

We will most often use the default state r_0 as the state before measurement (that is, we *prepare* our measurement device to take on this state). Then, an *ideal* measurement is an interaction between R and S such that

$$r_0 s_i \to r_i s_i, \tag{3.2}$$

that is, the ruler's state is determined after holding it up to the stick being measured and counting the marks up to where the stick reaches. (Alternatively, we could imagine that the ruler is in an arbitrary state r_i with probability p_i before measurement, but is forced into state r_j when interacting with a stick in state s_j.) If our ruler is perfectly resolving, then we know that the ruler's reading perfectly determines the state of the stick:

$$\text{Prob}(S = s_i | R = r_j) = p(s_i | r_j) \to \delta_{ij}, \tag{3.3}$$

where δ_{ij} is the Kronecker symbol ($\delta_{ij} = 1$ if $i = j$, and $\delta_{ij} = 0$ otherwise). If further the ruler's state given the state of the stick is fully determined, that is, that given a particular state $S = s_j$, we will find $R = r_j$ with probability one and zero otherwise, then the measurement is *noiseless*, so that in addition

$$\text{Prob}(R = r_i | S = s_j) = p(r_i | s_j) \to \delta_{ij}, \tag{3.4}$$

which implies that both the conditional entropies $H(R|S)$ and $H(S|R)$ vanish, and the only remaining entropies are correlated to each other and represent the acquired information, as depicted in Figure 3.1a.

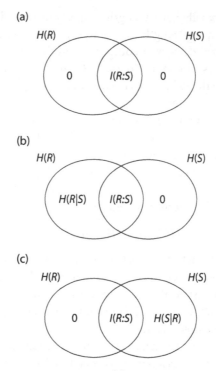

FIGURE 3.1. Entropy Venn diagram for (a) a perfectly resolving noiseless measurement, (b) a noisy perfectly resolving measurement, and (c) a noiseless nonresolving measurement.

3.1.2 Nonresolving and noisy measurements

In a more general measurement process, the measurement device does not fully resolve all the different states of the system being measured, so that several possible states of s_i correspond to one particular state r_i. A trivial example would be sticks of different colors, which obviously cannot be resolved by a length measurement. In that case, the measurement interaction forces

$$r_0 s_i \rightarrow r_i \langle s \rangle_i, \qquad (3.5)$$

with $\langle s \rangle_i = \sum_j p(s_j|r_i)s_j$, and where the sum is over all the unresolved states of S. Note, however, that this construction presupposes that a measurement device exists *in principle* that resolves these states s_j, as we otherwise would not be able to enumerate them to begin with. The crucial part of the measurement then is the "sharpening" of the conditional probability distribution $p(s_i|r_j)$, ideally up until $p(s_i|r_j) = \delta_{ij}$. All this can succinctly be expressed in terms of entropies as follows.

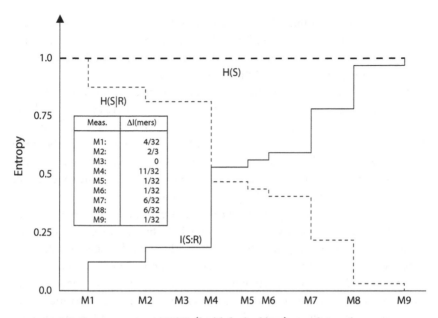

Meas.	ΔI(mers)
M1:	4/32
M2:	2/3
M3:	0
M4:	11/32
M5:	1/32
M6:	1/32
M7:	6/32
M8:	6/32
M9:	1/32

FIGURE 3.2. System entropy $H(S)$ (bold dashed line), conditional entropy $H(S|R)$ (dashed) and information (shared entropy) $I(R:S)$ (solid line) given a sequence of measurements $M1 - M9$ of S by measurement device R. Acquired information per measurement in inset.

As pointed out above, condition (3.4) imposes $H(R|S) = 0$ (the "size-pointer" is the only variable of our ruler), but a noisy measurement can still be perfectly resolving so that $H(S|R) = 0$ (see Fig. 3.2b). Fundamentally, a measurement allows us to split up what is unknown about S (that is, $H(S)$) into two terms. First, the information provided by R (that is, what R knows about S: the information $I(S:R)$), and second: that which remains, that is, that which R cannot resolve, $H(S|R)$:

$$H(S) = H(S|R) + I(S:R). \qquad (3.6)$$

This equation holds true for all possible measurements between R and S, so it is as general as can be. Before any measurement, as we noted earlier via Equation (3.1), the variables R and S are uncorrelated, which implies $H(S) = H(S|R)$, that is, R has no information about S [$I(S:R) = 0$]. Now let us imagine a process where a measurement device continuously acquires information about the system, that is, interactions between the system S and the measurement device R are going on so that more and more states of R and S become correlated. Each measurement acquires an amount of information that lies between a completely uninformative measurement ($I(R:S) = 0$, because R tells us nothing about S), and a perfectly resolving measurement, that is, $H(S|R) = 0$ as implied by (3.3). Then, because the left-hand side of Equation (3.6) is constant, we see that as the process continues, $H(S|R)$ decreases commensurately with the increase of $I(S:R)$.

Figure 3.2 shows how the entropies change in one particular sequence of measurements on a random variable whose entropy is between zero and one mer. As each measurement acquires more and more information, the conditional entropy—our (or more precisely, R's) remaining uncertainty about the system—decreases. In this particular example, we assume that once information is obtained, it is not later forgotten, so that at the end of the sequence of measurements we know everything there is to know.

For ordinary physical systems, perfect memory is a rather unusual characteristic because memory, being nothing but the record of the acquired information, is inherently unstable. Whether the record is kept as a string of letters, or as voltage patterns in the memory of a computer—or as connections and biases in the synapses of our neurons, for that matter—the second law of thermodynamics dictates that these records must deteriorate with time, unless they are continuously renewed. This is a commonplace fact for anyone who ever has dealt seriously with information preservation, and indeed our digital computers are equipped with a "refresh" system that guarantees just that. As we will see below, information acquisition via evolution is a very tedious and time-consuming process, so it would make sense to do this within a framework that ensures that any acquired information is not lost. And so it will not come as a surprise that information preservation, along with its acquisition, can be seen as the sole purpose of all life on Earth (we will discuss this idea in much more detail in section 10.1).

3.1.3 Evolution as measurement

Let us now take our information-theoretic tools and use them to characterize *evolutionary transitions* in terms of measurements. As we did previously in section 2.3, we will use a DNA sequence as our measurement device, represented by a random variable D (for DNA, as opposed to the random variable P describing proteins that we introduced earlier). The system to be measured will be *the entire external world*, also represented by a random variable E (for "environment"). If we do this, we have to be prepared that neither will our measurement device be perfectly resolving, nor that it is noiseless. But we do know that it can store information, and that it is information about the world in which the sequence evolves. Let us further assume for the moment that the world is unchanging, and that our sequence is of finite length L, that is, $D = D_1, \ldots, D_j, \ldots, D_L$. Such a sequence can maximally store L mers

(2L bits). Both restrictions can be relaxed, and we can in particular allow sequences to become longer, in case there is a need to store more than 2L bits. Let us say that (as an idealization) prior to any adaptive event our DNA sequence is random, meaning that each base takes on its four possible states with equal probability so that its entropy is maximal: $H(R) = L$ mers. (Clearly, as there is no special state r_0 for a nucleotide, this random state will be the state before measurement.)

How many possible states can the world take on? This seems like a question that is impossible to answer, but fortunately this is neither necessary, nor is it a well-defined question. Just as in the ruler/stick toy problem we considered earlier, the possible states of the world are only determined in terms of the possible states of our measurement devices. And as we are restricting ourselves here to those aspects of the world that can be reflected in DNA sequences, we can imagine a *perfect* description of the world in terms of a sequence e that represents the record of an ideal, that is noiseless and fully resolving, measurement describing the world. This sequence e represents one possible state of an ensemble E, whose maximal entropy $H(E)$ is, of course, unknown to us a priori. In short, we imagine the best description of the world e to be given in terms of a genetic sequence, that is, e is given in terms of the optimal DNA sequence in environment e. It is this description of e that evolution attempts to discover.

As we saw in the first chapter, evolution is what happens if three processes occur in an intertwined manner: inheritance, variation, and selection. Here, we look at this process under the microscope, as it were, and focus on a single adaptive event: the "fixation" of a nucleotide in a population subsequent to a mutation. The word "fixation" is apt because it describes a dynamical process wherein the probability to find a particular value (or "allele") of a gene within a population of related sequences changes in time, from random to determined. These probabilities can be obtained via alignment of the sequences in the population, just as we did in chapter 2.

Let us focus here on a single fixation, at position j described by random variable D_j. In the absence of selection, D_j takes on any of its possible four states with equal probability:

$$p(D_j) = \{p_A(j), p_C(j), p_G(j), p_T(j)\} = \{0.25, 0.25, 0.25, 0.25\}. \quad (3.7)$$

Now, let us imagine that selection becomes active.[1] What determines success in this "measurement model" of evolution is comparison with the description

---

1. This is a highly idealized scenario (because in reality we cannot just turn selection on and off) but such a "turning on selection" event is nevertheless common if a mutation elsewhere on the gene has created a situation where one of the four states suddenly becomes more beneficial than the other. In other situations, D_j initially takes on one or several particular nonbeneficial

of the environment e. If D_j takes on the value found at e_j, then the organism carrying this variation experiences a replicative advantage, and starts to out-compete variants with different values. Say, as an example, $e_j = A$. The variant with $D_j = A$ now increases in frequency in the population, to the detriment of the other variants. If we could constantly monitor the probabilities $p(D_j)$, we would see $p_A(j)$ rising while all others are decreasing. After many generations, we can imagine that all the alternate values have become extinct, and (in this scenario of asexual replication) every organism in the population is a descendant of the original mutation: fixation has taken place.

In terms of the measurement model, we can write this process simply as

$$D_jA \to AA, \tag{3.8}$$

where the second variable stands for the environment e_j. This measurement is the equivalent of the interaction (3.2) and implies a change in probabilities given by

$$p_j = \{p_A, p_C, p_G, p_T\} = \{0.25, 0.25, 0.25, 0.25\} \to \{1, 0, 0, 0\}. \tag{3.9}$$

Such an adaptive event reduces the entropy of D by one mer, while increasing the information we have about e by one:

$$H(D_j) = H(D_j|e) + I(D_j : e), \tag{3.10}$$

where now, of course, $H(D_j|e) = 0$ on account of the probabilities in (3.9). Thus, this adaptive event has turned all of D_j's entropy into information. While this is the analogue of the *noiseless* measurement paradigm we encountered above, a more general adaptive event might more likely involve a sharpening of the uniform distribution $\{0.25, 0.25, 0.25, 0.25\}$ into a more peaked one, leaving some uncertainty in D_j in the form of a nonvanishing $H(D_j|e)$.

If we imagine an evolutionary process of successive adaptive events fixing sites in the random variable $D = D_1 \ldots D_L$, each event removing some entropy from $H(D)$ and turning it into information, we seem to be faced with a curious conundrum. While $H(D)$ remains constant, the conditional entropy $H(D|e)$ is always decreasing (in the same manner as depicted in Fig. 3.2). Does this not violate one of the fundamental laws of physics, namely the second law of thermodynamics?

Before answering this in the negative, we should first take a look at this theorem, but without entering into any of the more technical aspects of thermodynamics. In essence, the second law says that if you leave a closed physical system alone after having disturbed it, it will approach its most disordered

---

values and mutates to the value conferring a selective advantage—the outcome will be the same regardless.

state, which is the one with the highest entropy, almost all of the time. A measurement is actually a very good example of a disturbance, so what this theorem says is really nothing but a statement about forgetting: no memory is permanent, in other words. The theorem does not say anything about what happens during the disturbance, because it explicitly only talks about the period after. So, clearly our measurement does not violate the second law on account of this fact alone. Furthermore, the law speaks about "closed" systems. This means precisely what we think it means: it talks about the entropy of all parts that are interacting, together. But the system here consists out of at least two parts, the variable D and the environment within which it evolves. Only the conditional entropy $H(D|e)$ decreases, while $H(D)$ remains constant.[2]

So what about the permanence of memory in our sequence of evolutionary fixation events described above? As alluded to earlier, memory must deteriorate for physical systems. Are we not violating a law now? In fact, this question was posed quite a while earlier by the Scottish physicist James Clerk Maxwell (co-creator, with the Austrian physicist Ludwig Boltzmann, of the kinetic theory of gases and thermodynamics). In his groundbreaking book *Theory of Heat* (Maxwell 1871), he postulated the existence of "a being" (dubbed later a "demon") that, by actions chosen in accordance with certain measurements, could lower the entropy of a system in violation of the second law. To better understand this idea, we should take a closer look at Maxwell's imaginary construction.

3.1.4 The Maxwell demon and the second law

In Figure 3.3, we can see a demon inhabiting one side of a room partitioned into two, connected only by a small conduit that can be opened or shut by a mechanical device controlled by the demon. The latter is also equipped with a *measuring device*, imagined to allow the demon to measure the speed of molecules within his side of the chamber. Now, the second law can be formulated in such a way that it predicts that no physical process will allow two bodies that are in contact with each other (and themselves isolated from the rest of the world) to develop a significant temperature differential. In other words, it states that you cannot expect a body to heat up at the expense of

---

2. In this case, we even have to find fault with the standard formulation of the second law, because it postulates that the entropy of a closed system should increase in a transition "from a non-equilibrium to an equilibrium state." But an information-theoretic analysis (Adami 2011) shows that the entropy of a closed system can never change, as it is simply given by the logarithm of the number of possible states the system can take on. What increases in equilibration is actually a conditional entropy instead, while memory (in the form of information) is lost.

A   B

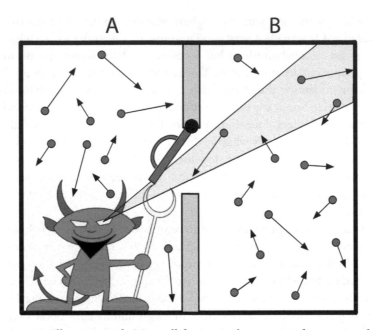

FIGURE 3.3. Illustration of a Maxwell demon in the process of measuring the speed of molecules and operating the connecting door between the two compartments A and B. Operating the door so as to only allow the fast molecules to enter A while the slow ones remain in B can create a pressure difference between the two chambers, which translates into a difference in temperature and entropy, in apparent violation of thermodynamics' second law.

another one becoming colder if these two bodies are in contact but otherwise isolated. (If an engine is connected to the two, this is of course possible: we call this a refrigerator.) But heat, as Maxwell taught us, is nothing but a manifestation of molecules moving at different speeds. If you could somehow sort molecules that have different speeds in such a manner that the fast ones end up on one side of the partition and the slow ones on the other, then we would have created a temperature differential. This is precisely what Maxwell's demon supposedly can achieve armed with his measurement device. He spots molecules moving toward the door that separates the two rooms, and for a very brief moment opens the door only for the fast molecules, while keeping it closed for the slow ones. A detailed analysis of this process reveals what Maxwell had intuited: you can indeed create a temperature differential this way. Only it does not violate the second law because of a very fundamental reason, discovered by the German-born American physicist Rolf Landauer, working at IBM. Landauer realized that the demon could only successfully do his work if he could *record* the results of his measurements, as specific actions depend on the outcome. But a recording device must be physical, and hence "information is physical," as Landauer later put it in a review article worth reading (Landauer 1991). So, the recording device itself must be regarded as a *third* system that both halves are in contact with, so that the two halves are not isolated at all (the demon itself and the measurement device can be mechanized in principle, and are "thermodynamically neutral"). A detailed calculation shows that any entropy decrease that the demon achieves in his half of the chamber must be accompanied by a commensurate increase of entropy in that part of the chamber that the demon's recording device is in contact with, because the act of recording an arbitrary variable forces the *erasure* of the state the variable previously was in.

What Landauer discovered (Landauer 1961) is that erasure creates heat, and thus creates entropy in the half of the chamber that the demon tries to keep cool. It turns out that the fixation process (3.2) of a random nucleotide represents just such an erasure, because a previously undetermined (random) nucleotide is, via the force of natural selection, taking on a specific value, irrespective of its previous value. Erasure, from the point of view of classical physics, is inherently irreversible, hence the heat generation. So, while each mutation—just like pressing the "delete" key on your computer—creates heat, the physical amount is negligible compared to the energy dissipated in other processes going on, both in computers and in life.

So the second law is not violated by the Maxwell demon after all. But the process of sequential acquisition of information and the commensurate lowering of entropy in evolution is strangely reminiscent of this process. The reason is that the self-replication of genomic information in living systems— life itself—is essentially a reincarnation of the demon (Adami 1998). Indeed, the three principles of Darwinian dynamics that we encountered in chapter 1 have precise analogues with the Maxwell demon. A type of "inheritance," for example, is ensured when the demon opens the gateway only briefly enough to let new fast molecules go through, but does not allow the fast ones in the "hot" chamber to escape. In a sense, by this mechanism, the fast molecules on one side of the partition increase in number, as if by replication. Variation is guaranteed by the different speeds of molecules in the chamber: the Maxwell distribution of molecular velocities implies that different speeds are available given any temperature. Finally, selection is enforced by the demon's ability to measure molecular speed and open and close the gate accordingly. Note that, just as in evolution, the absence of variation (for example, a set of objects all with the same speeds) prevents the demon from creating an entropy difference, because he would not be able to differentiate between molecules and make appropriate decisions.

To examine the implications of this thermodynamic view of evolution in more detail, let us go back to our discussion of the microscopic dynamics of evolution in terms of measurement. We have seen that in evolution a random mutation can become a measurement if that mutation discovers a hitherto unknown feature of the environment. The acquired information is preserved only if the mutation confers a fitness advantage to the bearer of said mutation. This fitness advantage allows it to outcompete, and ultimately drive to extinction, all those competitors sporting alternative symbols at that position, until every member of the population carries the same symbol there. Then, and only then, have the probabilities changed as in Equation (3.2).[3]

What happens if a nucleotide that is information rather than entropy—in other words, a nucleotide that is important for the success of the organism carrying the sequence—is changed by a mutational event? We must assume that this happens all the time, as mutations are not limited to those nucleotides that are entropic, that is, that do not already carry information. Such mutations are—as the reader has certainly surmised—an attempt of the laws of physics to institute the second law (that is, to establish disorder), reinforcing the idea that information cannot persist indefinitely in a noisy world. But such a mutation must, by definition, be detrimental to the organism, maybe even dooming its carrier to die without offspring. Even if such an impaired organism manages to produce a few offspring, those offspring would be impaired in the same way because they inherit the detrimental mutation, ensuring that this aberration is ultimately removed from the population. Thus, it is selection that acts as the demon, encouraging beneficial mutations by boosting their frequencies in the population (akin to opening the door for fast molecules), and eliminating detrimental ones (shutting the door for the slow ones). This "natural demon" acts like a semi-permeable membrane for our DNA sequence, allowing information to accumulate, but never to leave.

Such an ideal view of evolution as measurement seems to imply that the information content of genes can only increase in evolution. But there is sufficient evidence that this is not always the case. What can cause the demon to be "leaky," allowing information to escape once in a while? An examination of these exceptions turns out to represent a large part of the mathematical

---

3. The precise dynamics of fixation have been worked out in detail in the mathematical formulation of evolution, in a theory called "population genetics." For these details, see for example (Haldane 1927). In later work (Kimura 1983) it was discovered that even neutral (as opposed to beneficial) mutations can become fixed (see section 6.1 for more details). Modern mathematical treatments can be found in Bürger (2000) and Ewens (2004). We will discuss the consequences of neutral mutations briefly in Box 3.1, while neutral evolution will be covered more fully in chapter 6.

literature of population genetics. Rather than go into those details here, I will highlight some of the more intuitive exceptions in Box 3.1 and return to these issues in later chapters.

While the exceptions mentioned do happen (large-scale extinction events due to changing environments is a particularly important one; loss of information in the evolution of drug resistance, treated in section 3.3, another), the Maxwell demon analogy strongly suggests that, all told, evolution leads to an increase in the information content of genes. Given the intricacy of higher organisms, this realization is perhaps not all too unexpected, and we will examine the relationship of the "increase of information law" and its relation to complexity further in chapter 5.

3.2 Evolution of Information on the Line of Descent

How much can we say about the evolution of information stored in proteins, from "so simple a beginning" billions of years ago, until today? We already know that life did not start out with vanishing information: the problem in understanding the origin of life lies in part in understanding how the minimal amount of information necessary to sustain self-replication can come about (see chapter 7 for a more in-depth discussion). Of course, we also know that at the time of the origin, information cannot have been stored in proteins at all, as this would require the machinery of transcription and translation to already be in place. However, it is also clear that it does not matter what the initial carrier of information was (it could well be ribonucleic acids). What we can be certain of is that this initial amount of information was comparatively small. Through evolution, more information was added in the manner described in the previous section, by adding more molecules (mostly via the duplication of existing ones and subsequent variation), and fine-tuning those that are already there.

If we take a protein that is ubiquitous among different forms of life (that is, its homologue is present in many different branches), has its information content changed as it is used in more and more complex forms? One line of argument tells us that if the function of the protein is the same throughout evolutionary history, then its information content should be the same in each variant. We saw a hint of that in the previous chapter, when we tested the information content of the homeodomain protein and found that it was the same within rodents and primates. But we can also argue instead that because information is measured relative to the environment the protein (and thus, the organism) finds itself in, then organisms that live in very different environments can potentially have a different information content even if the sequences encoding the proteins are homologous, or even identical. Thus, we

Box 3.1. The Leaky Natural Demon

The model of measurement leading to a perfect "natural demon" that only allows information to enter the genome (our measurement device) and that retains all acquired information perfectly is, of course, a gross caricature of what happens in evolving systems. Many of the assumptions of this model are constantly violated, but mostly in interesting ways. Here I briefly summarize these violations in realistic systems and discuss their possible consequences. Many of the subjects taken up here are discussed in much more detail in subsequent chapters.

1. Environments can change in time. Because information is always specifically *about something* (in particular about the world in which the sequence has evolved), if that world changes, what once was information can just as quickly turn to entropy. As a consequence, sudden and catastrophic changes in the environment can turn many genes into useless junk and usher in massive extinctions. We will study a microcosm of information loss due to changing environments in section 3.3.

2. Realistic environments have more than a single niche. The model assumes that less fit variants are outcompeted by the fitter ones. But in an ecology, variations can persist if they represent information about separate habitats. To make matters worse, species in ecosystems co-evolve, implying that each species' environment changes in time. Yet, it can be argued that the *total* information in an ecosystem must always increase as long as the ecosystem is stable. We will take up this discussion in section 5.1.5.

3. Replication is not always asexual. The competition between variants leading to fixation assumed an asexual replication process. But since most life on Earth is sexual, we find that under some circumstances the effect of beneficial mutations can be lost simply by recombining them with deleterious ones. However, some theories about the evolution of sex posit that sexual recombination evolved for the most part to *prevent* the loss of information (Kondrashov 1988).

4. Populations are not infinite. Fixation from a uniform distribution of alleles to one where all but one probability is nonvanishing, as in Equation (3.9), technically requires an infinite population size and/or a vanishing mutation rate. In small populations and at large mutation rates, beneficial mutations can be lost from a population due to random drift, akin to the demon being sloppy in closing the door. This effect is sometimes called *Muller's ratchet* after the American geneticist Hermann J. Muller (1890–1967). Evolutionary effects stemming from high mutation rates as well as finite population size are discussed in chapter 6.

could expect differences in protein information content in organisms that are different enough that the protein is used in different ways. But it is certainly not clear whether we should observe a trend of increasing or decreasing information along the line of descent. To get a first glimpse at what these differences could be like, we will take a look at the evolution of information in two proteins that are very important in the function of most organisms, but in different ways: the homeodomain protein, whose information content we examined in the previous chapter, and the COX2 protein.

The homeodomain (also called "homeobox") protein is essential in determining the pattern of development in animals: it is crucial in directing the arrangement of cells according to a particular body plan. In other words, the homeobox determines where the head and tail go. Although it is often said that these proteins are specific to animals, some plants have homeodomain proteins that are homologous to those we study here (van der Graaff et al. 2009).

The COX2 protein (the abbreviation stands for cytochrome-c-oxidase subunit 2) is a subunit of a large protein complex that comprises fourteen subunits (Balsa et al. 2012; Zong et al. 2018), three of which (COX1, COX2, and COX3) are (in humans) encoded on the mitochondrion. While a nonfunctioning (or severely impaired) homeobox protein almost certainly leads to aborted development, an impaired COX complex has a much less drastic effect: it leads to mitochondrial myopathy due to a cytochrome oxidase deficiency (Robinson 2000) but is usually not fatal (Taanman 1997). Thus, by exploring the changes within these two proteins as evolution proceeds, we are examining proteins with very different selective pressures acting on them.

For each of the two proteins, we now have to assemble sequences that we can align to measure the information content as outlined in section 2.3.1. We can do this animal by animal, but this would not teach us anything about how information evolved, only about the differences in existing animals. To track the evolution of information along the evolutionary line of descent, we would have to examine the sequences of *extinct* animals. For example, if we wanted to study the evolution of information in the homeobox protein of humans, we would compare this information to that of the common ancestor of all the "great apes," for example, which today comprises the chimpanzees, gorillas, orangutans, and us. Of course, we do not have these sequences, but it is actually not impossible to imagine getting a glimpse of them. With modern tools of computational biology, it is possible to estimate what the genetic sequence of these ancestral proteins was like, using the sequences of the existing types. In a sense, these algorithms make "best guesses" for the ancestral sequence, given that they gave rise to the known sequences of humans, chimpanzees, and gorillas, and given what we know about the process of evolution (the rate at which substitutions occur, and what types of substitutions are most likely). This guessing game is surprisingly successful: several teams have taken such inferred sequences and synthesized those proteins. This procedure is akin to "resurrecting" the proteins of extinct animals (Thornton 2004), which can then be tested for their functional properties and compared to the existing version. We can see an example of this type of work in Box 3.2.

In the absence of information about extinct proteins, we can instead group sequences according to the *depth* that they occupy within the phylogenetic tree. Humans, for example—the species *Homo sapiens*—occupies the highest level of description within *taxonomy*, which is the field of science that classifies all forms of life. The species *H. sapiens* is part of the genus *Homo*, which besides us includes extinct types such as *Homo habilis* and *Homo neanderthalensis* (which we colloquially refer to as the Neanderthal). Incidentally, one member of the genus *Homo*, *Homo floresiensis* discovered on the island of Flores in Indonesia, may have gone extinct as little as 12,000 years ago (Brown et al. 2004). The genus *Homo* is part of the family *Hominidae* or "great apes," which comprises four non-extinct genera. So, when we measure the information content of the homeobox protein on the taxonomic level of the family, we include in there the sequences of homeobox proteins of chimpanzees, gorillas, and orangutans, along with humans. As the chimpanzee version, for example, is essentially identical with the human version, we do not expect to see any change in information content when moving from the species level to the genus level. But we can expect that by grouping the sequences on the family level (rather than the genus or species level), we move closer toward evolutionarily more ancient proteins, in particular because this group is used to reconstruct the sequence of the ancestor of that group. The Great Apes are but one family of the order of "primates," which besides the apes also contains the families of monkeys, lemurs, lorises, tarsiers, and galagos. Looking at the homeobox protein of all the primates then takes us further back in time. A simplified version of the phylogeny of animals is shown in Figure 3.5, which shows the hierarchical organization of the tree.

Sequence information covering protein families can be obtained from the "Pfam" database (Finn et al. 2010). Pfam uses a range of different taxonomic levels (anywhere from twelve to twenty-two, depending on the branch) defined by the NCBI Taxonomy Project (Federhen 2002), which we can take as a convenient proxy for taxonomic depth: ranging from the most basal taxonomic identifications (such as phylum) to the most specific ones.

Box 3.2. Reconstructing the Evolution of Coral Pigments

Misha Matz and his collaborators at the University of Texas at Austin were interested in the diversity of colors displayed by a type of fluorescent reef-building coral and how this diversity may have come about. The proteins that give rise to this fluorescence are famous in molecular biology: they are part of the family of proteins that are used in most molecular biology labs: the "green fluorescent protein" (GFP) that was originally isolated from the jellyfish *Aequorea victoria*. The great star coral *Montastraea cavernosa* has several proteins that are "GFP-like," fluorescing in the colors cyan, shortwave green, longwave green, and even red. How did all these colors evolve? And what was the color of the first GFP-like protein? Fluorescent proteins produce light via a molecule called a chromophore. Because making the red chromophore is a more complex process than making green ones, it was hypothesized that the green color is ancestral, and all the other ones are derived. But how do you prove it? In this case, by reconstructing the phylogeny of these proteins, and checking what color emerged when expressed and plated on a Petri dish!

By reconstructing the proteins in the *Montastraea cavernosa* lineage, Matz and collaborators were able to show that the ancestor of this multiple color-emitting coral was only fluorescing in green (Ugalde et al. 2004). In Figure 3.4, we can see the fluorescence color of the proteins in that lineage, expressed in bacteria that were arranged on a Petri dish in the shape of the inferred phylogeny of the chromoproteins. From this, the authors concluded that the red color proteins evolved from a green-fluorescing one in a step-by-step process.

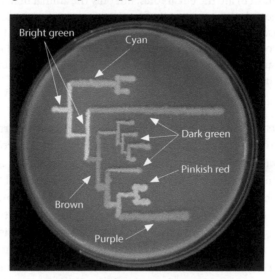

FIGURE 3.4. A Petri dish with bacteria expressing the resurrected fluorescent proteins of the lineage of *Montastraea cavernosa*, showing that the ancestor to the color diversity was a protein fluorescing only in green (original colors shown in Ugalde et al. 2004 indicated by labels). From Ugalde, J., B. Chang, and M. Matz (2004). Evolution of coral pigments recreated. *Science 305*, 1433. Reprinted with permission from AAAS.

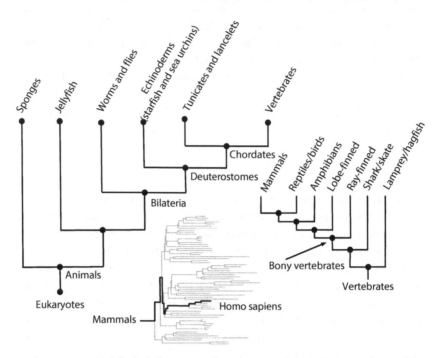

FIGURE 3.5. Simplified phylogenetic classification of animals. At the root of this tree (on the left tree) are the eukaryotes, but only the animal branch is shown here. If we follow the line of descent of humans, we move on the branch toward the vertebrates. The vertebrate clade itself is shown in the tree on the right, and the line of descent through this tree follows the branches that end in the mammals. The mammal tree, finally, is shown at the bottom, with the line ending in *Homo sapiens* indicated in bold.

For our first example, we will look at the evolution of information in the homeobox protein. I previously measured the information content of that protein in section 2.3.1, where we compared the information content of the rodent version of the protein to the primate version. We found there that the two versions both carry about twenty-five mers of information, out of fifty-seven, but that the information was encoded slightly differently. Here, we'll look at changes over much longer timescales. We can do this because the homeobox sequence is found not just in animals and plants, but in fungi and unicellular eukaryotes as well. Obviously, the associated protein is used differently in all these groups, and we should be able to see difference in the information content for that reason. We also know that the protein is used differently in vertebrates as opposed to invertebrates, and the number of copies of the gene that any group carries also differs, due to whole-genome duplication

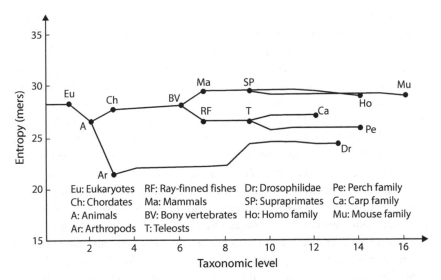

FIGURE 3.6. Entropy of homeobox protein sequences (PF00046 in the Pfam database, accessed July 20, 2006) as a function of taxonomic depth for different major groups that have at least 200 sequences in the database, connected by phylogenetic relationships. Selected groups are annotated by name. Fifty-seven core residues were used to calculate the molecular entropy. Core residues have at least 70 percent sequence in the database. Adapted from Adami (2012).

events. Thus, the homeobox protein is as good a candidate as any to test whether information theory can detect the evolution of functional changes along the line of descent.

In Figure 3.6, we can see the total sequence entropy

$$H_k(X) = \sum_{i=1}^{57} H(X_i|e_k), \tag{3.11}$$

for sequences with the NCBI taxonomic level k, as a function of the level depth. Note that sequences at level k always include all the sequences at level $k-1$. Thus, $H_1(X)$ for example, which is the entropy of all homeobox sequences at level $k=1$, includes the sequences of all eukaryotes. Of course, the taxonomic level description is not a perfect proxy for time. On the vertebrate line for example, the genus Homo occupies level $k=14$, whereas the genus Mus occupies level $k=16$. Indeed, as we saw before in section 2.3.1, the proteins in the mouse and human families do not differ much in information content. But there are dramatic differences when looking at lower levels (deeper into the phylogenetic tree).

In particular, the total sequence entropy $H_k(X)$ shown in Figure 3.6 splits along the lines of descent, indicating a split in terms of information also, as information is just $I = 57 - H$ if we measure entropy in mers. At the base of the tree, the metazoan sequences split into chordate proteins with a lower information content (higher entropy) and arthropod sequences with higher information content, possibly reflecting the different uses of the homeobox in these two groups. The chordate group itself splits into mammalian proteins and the fish homeodomain. Incidentally, there is good evidence that a major whole-genome duplication event occurred sometime before the origin of the teleosts (Taylor et al. 2003), which could be the origin of the split in homeobox gene information between the mammalian line and the ray-finned fish that we can see in Figure 3.6. Indeed, there is even a notable split in information content of the homeobox gene into two major groups within the fishes (the carp family and the perch family). This split also could be due to another homeobox duplication, but the evidence for this is less clear. A more fine-grained analysis than this one could possibly detect major changes in body-plan development, like for example the change in the arthropod body plan when the six-legged (hexapod) design evolved from the crustacean-like ancestor about 400 million years ago (Ronshaugen et al. 2002; Galant and Carroll 2002).

A similar analysis can be carried out for COX2, the subunit of the cytochrome oxidase complex of proteins that we encountered earlier. When we plot the entropy of this protein (counting only 120 residue-sites that have sufficient statistics in the database), we get a very different picture compared to what we saw in the homeodomain protein phylogeny. Except for an obvious split of the bacterial version of the protein and the eukaryotic one, the total entropy markedly *decreases* across all the lines as the taxonomic depth increases, which indicates an increase in information content. Furthermore, the arthropod COX2 is more entropic than the vertebrate one (see Fig. 3.7) as opposed to the ordering for the homeobox protein. This finding suggests that the evolution of protein information content is specific to each protein, and most likely reflects the adaptive value of the protein for each family.

3.3   Information Loss and Gain in HIV Evolution

One of the most important aspects of a measure of information is that it is *context-specific*: what is information about one environment may be useless in another. If information is useless, it has become entropy: it cannot be used to make predictions. We will consider this feature in more detail in chapter 5, but encounter here its most obvious consequence: a sudden change in a protein's environment (usually) leads to a loss of information.

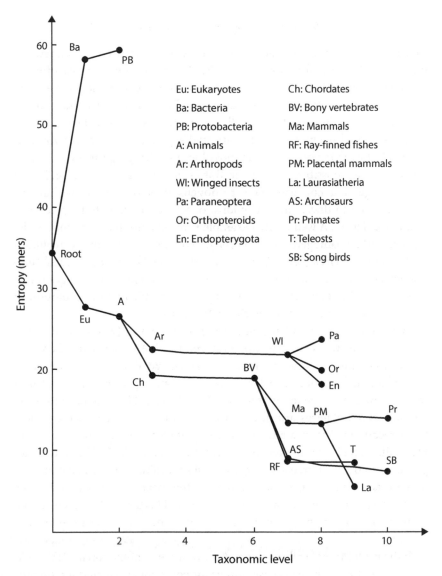

FIGURE 3.7. Entropy of COX subunit II (PF00116 in the Pfam database, accessed June 22, 2006) protein sequences as a function of taxonomic depth for selected different groups (at least 200 sequences per group), connected by phylogenetic relationships. 120 core residues were used to calculate the molecular entropy. Adapted from Adami (2012).

3.3.1 Information loss in altered environments

In nature, environments change constantly—sometimes fast, sometimes imperceptibly slowly for a long period. Yet for the most part, at least for the short period of our lifetime, we can treat environments—and hence the information content of our genes—as unchanging. But there are exceptions to this rule, and we would expect to find those exceptions in fast-evolving organisms. As we will see in chapter 4, dedicated experiments to study evolution in such fast-evolving organisms are possible, and changing the environment is just one such possible experiment. But evolution is also occurring, in real time and right under our eyes, when pathogens adapt to drugs that we humans have designed to thwart our infections, or when they mutate to evade our immune defenses. A well-known example is the evolution of antibiotic resistance in certain pathogenic bacteria (Neu 1992; Levy 1998), or the evolution of the influenza virus (see, e.g., Ferguson et al. 2003). An almost ideal "natural experiment" that allows us to measure the amount of information lost when the environment changes abruptly is exemplified by the evolution of drug resistance in the protease of the HIV-1 virus.

HIV (human immunodeficiency virus) is a retrovirus (Frankel and Young 1998)—a virus whose information is stored in RNA rather than DNA, and that carries within its viral envelope an enzyme that allows for the transcription of this information into DNA, which is subsequently integrated into the host organism's genome (see Fig. 3.8). In fact, HIV carries within its envelope *two* (not necessarily identical) copies of its genome, which are effectively recombined to create the offspring genomes. An important feature in HIV's life cycle is that the reverse-transcribed genome of HIV is, after integration into the host, translated by the host's machinery into a *polyprotein*. A polyprotein is a long chain of amino acids that needs to be cleaved into its functional pieces by an enzyme called a *protease*. This molecule recognizes the cleavage sites within the polyprotein, and cuts it into functional proteins (see Fig. 3.9).

If the protease molecule was somehow rendered unfunctional, HIV could evidently not reproduce because important components (including the protease itself) would remain inactive within the polyprotein. And indeed, one of the first highly efficient antiviral drugs was just such a protease inhibitor (Ashborn et al. 1990), a small molecule that binds to the protease's active site and prevents efficient cleavage. However, clinicians soon realized that the efficacy of drug therapy with protease inhibitors was declining with time, due to the emergence of resistance mutations (Condra et al. 1995). We can imagine the wild-type protease's genetic sequence as inhabiting a peak in a fitness landscape, that is, its sequence is optimized to maximize the virus's

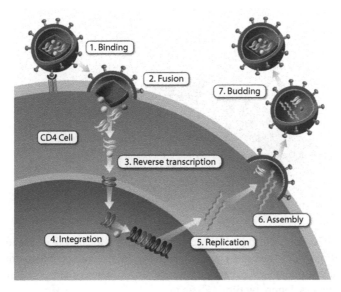

FIGURE 3.8. HIV life cycle. Image courtesy of U.S. Department of Health and Human Services.

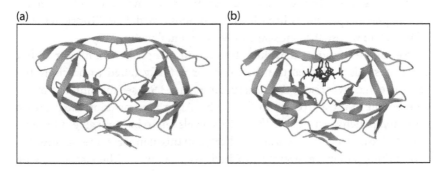

FIGURE 3.9. (a) Structure of the HIV virus's protease molecule. It is composed of two identical proteins (a *homodimer*). Note the hole in the middle that constitutes the active site of the enzyme. (b) The protease inhibitor molecule *saquinavir* (dark gray) is bound to the protease, blocking the active site (Tie et al. 2007).

replication rate, and further mutations are either neutral or detrimental, as in Figure 3.10.

Drawing fitness landscapes such as Figure 3.10 is a convenient way to visualize the constraints faced by an evolutionary process, and we will discuss them in more detail in section 5.4.2. But it is important to keep in mind that

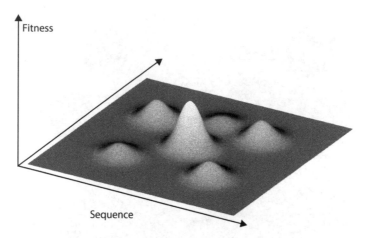

FIGURE 3.10. The wild-type HIV protease inhabits a local peak in the adaptive landscape, given by the protease's environment. Nearby peaks of lesser fitness are normally uninhabited because its denizens would be outcompeted by the fitter sequences. Figure by Bjørn Østman.

they really show only a few selected features of the overall map from genotype to phenotype, while obscuring most biophysical details (Rodrigues et al. 2016). Here we use the landscape only as a mental picture.

We can be confident that the introduction of a protease inhibitor all but eliminates this peak, because the inhibitor is finely tuned to precisely the sequence that gives rise to the structure in Figure 3.9a (see the inhibitor molecule bound to the active site in Fig. 3.9b). After administering the protease inhibitor, patients usually see their virus levels fall to become all but unmeasurable. However, the virus is not eradicated in this manner: some variants survive, although they are severely impaired (we can think of them as occupying one of the nearby peaks of lower height in Fig. 3.11).

The mutation rate of the HIV virus is high because the machinery that translates the viral RNA into the DNA that can be integrated with the host—the viral *reverse transcriptase*—is highly error prone: of every three viral copies, one on average contains a mutation (Drake and Holland 1999). And because about a billion new cells are infected each day in an average patient, every possible mutation is tried on every position (of the approximately 10,000 base-pair genome) between 10,000 and 100,000 times daily (Coffin 1995)! This means that the virus population is highly diverse, and forms a cloud of mutants around the most fit sequence—a structure known as a *quasispecies*—owing to the large number of neutral mutations that each viral gene can bear

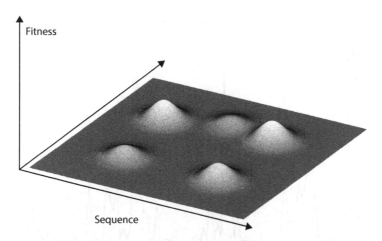

FIGURE 3.11. After administering a protease inhibitor, the landscape has changed as the initial peak is removed. However, much reduced peaks remain nearby, and can lead to higher peaks that allow the virus to function even in the presence of protease inhibitors. Mutations that lead to the new peaks are called *resistance mutations*. Figure by Bjørn Østman.

(we will discuss neutral mutations and quasispecies in much more detail in chapter 6).

Within a few weeks to a few months, mutant viruses appear that have regained some of their replicative ability. The evolution of drug resistance can be viewed in terms of the fitness landscape picture we painted above as a *peak shift*. Because the new peaks are often much lower in height than the original wild-type peak (owing to the much reduced replication rate), we might hypothesize that the virus has lost information about its environment. Indeed, what once worked does not anymore, so new tricks have to be found! But from these severely reduced peaks, new and higher peaks are reachable, allowing the virus to acquire information about this new landscape, and thus evade the virus inhibitor (see Fig. 3.11).

To characterize the information loss associated with the change in environment due to the presence of a protease inhibitor, let us try to measure the information content of the HIV protease gene using Equation (2.34):

$$I(P) = H(P) - H(P|e), \tag{3.12}$$

where $H(P)$ is the unconditional entropy of the protein, that is, its maximal entropy. Each of the molecules that compose the HIV protease dimer is a 99-amino acid polymer, thus we know that $H_{max} = 99$ mers. To calculate the

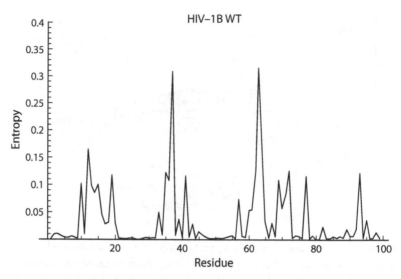

FIGURE 3.12. Entropic profile (in mers) of the HIV-1 (subtype B) wild-type protease sequence in an inhibitor-free environment (4,318 isolates, sequences downloaded from HIVdb accessed May 9, 2006).

actual entropy for the wild-type virus (and those that evolved resistance), we can make use of extensive databases that are being maintained to keep track of resistance mutations, such as Stanford University's HIV Drug Resistance Database (HIVdb; Rhee et al. 2003; Shafer 2006). For example, in May 2006 that database contained about 4,300 different sequences[4] of the wild-type virus (HIV-1 subtype B), which we can use to obtain substitution probabilities at each of the ninety-nine positions. Figure 3.12 shows the entropic profile for the wild-type (entropy at each position given the normal, meaning drug-free, environment). This profile reflects functional constraints at each position (only a select few residues can appear there), and depends very little on the sample of sequences. Figure 3.12 also documents impressively that even in the absence of drugs, the protease gene is highly diverse, with no two sequences identical. Such sets of neutral mutants are the hallmark of the "quasispecies" dynamics that we meet later in chapter 6.

Assuming for the moment that the entropy at each position is independent from all other positions (as discussed in chapter 2), we can sum up the entropies at each position to obtain the sequence entropy given the

---

4. In September 2020, this number had grown to more than 59,000 sequences of subtype B, almost 4,900 of which came from patients taking one or more protease inhibitors.

Table 3.1. Summary of information losses in drug resistance. First column denotes the environment, with the abbreviations idv ≡ indinavir, nfv ≡ nelfinavir, and sqv ≡ saquinavir. The second column shows the total entropy of the protease sequence in that environment in mers, including the bias correction (see Box 2.3) in column 3. The fourth column shows an estimate of the standard deviation of the total estimated entropy, and gives us an idea of the accuracy of the estimate. Finally, the fifth and sixth columns show the information loss in mers and bits.

Environment	H (mers)	Bias corr.	Std. dev.	ΔI (mers)	ΔI (bits)
wild-type	4.489	0.07	0.0007	—	—
indinavir	6.484	0.31	0.004	1.995	5.98
nelfinavir	6.189	0.41	0.004	1.7	5.09
saquinavir	7.862	0.77	0.009	3.37	10.1
idv + nfv	7.864	0.99	0.01	3.375	10.11
idv + sqv	9.251	2.03	0.02	4.762	14.26
idv + nfv + sqv	8.547	1.43	0.015	4.058	12.16

"wild-type environment"[5]

$$H(P|e = \text{wt}) = \sum_{i=1}^{99} H(x_i) = 4.489 \text{ mers.} \qquad (3.13)$$

According to Equation (2.34), the information content then is

$$I(P : \text{wt}) = H(P) - H(P|e = \text{wt}) = 99 - 4.489 = 94.511 \text{ mers,} \qquad (3.14)$$

that is, over 95 percent of the protease sequence is information. Let us see what happens to protease sequences exposed to the protease inhibitor *indinavir*, a small molecule that binds to the protease active site in a manner similar to the saquinavir molecule depicted in Figure 3.9b. The Stanford database (HIVdb) from May 2006 lists 986 sequences from patients undergoing a treatment with this inhibitor *only*. Note, however, that these patients have taken the drug for different amounts of times, so the genotypes in this collection may have undergone different amounts of adaptation. Nevertheless, the entropic profile for sequences that have adapted to an indinavir regimen is significantly different from the wild-type one, as Figure 3.13 shows. In particular, we can see much higher variability at almost all sites, in particular in those areas where

5. The entropy estimates in this section all include a bias correction calculated according to Box 2.3. This correction along with an estimate of the sample error appears in Table 3.1.

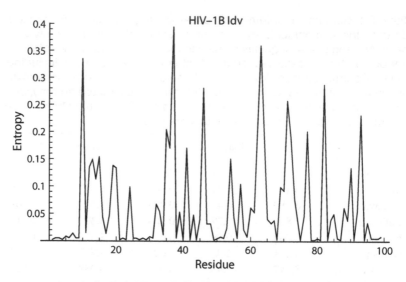

FIGURE 3.13. Entropic profile of the HIV-1 (subtype B) protease sequence adapted to an environment containing indinavir (968 isolates, sequences downloaded from HIVdb accessed May 9, 2006).

the wild-type sequence was almost perfectly conserved in the absence of the drug.

Summing up the per-site conditional entropies gives

$$H(P|e = \mathrm{idv}) = 6.484 \text{ mers}, \qquad (3.15)$$

that is, $I(P : \mathrm{idv}) = 92.516$ mers. According to this calculation, the protease in the indinavir environment—even after adaptation—has lost $\Delta I = 1.995$ mers. However, we should keep in mind that this estimate does not take into account information stored in *correlated* substitutions. Let us take a look again at the formula we use to calculate the conditional entropy $H(P|e)$ of proteins. We saw this already as Equation (2.44), where we wrote the entropy of the protein $P = P_1 P_2 \cdots P_L$ as

$$H(P) = \sum_{i=1}^{L} H(P_i) - \sum_{i<j} I(P_i : P_j) + H_{\mathrm{corr}}. \qquad (3.16)$$

Thus, the information content (3.15) does not take into account the second and third term in (3.16). Such correction terms may drastically alter our conclusions, as we note further below when we analyze how the information stored in the HIV-1 protease changes over time, in section 3.3.2. In the meantime, let us compare the information loss for different drugs. Turning to the Stanford database again, they list 739 sequences of patients taking the

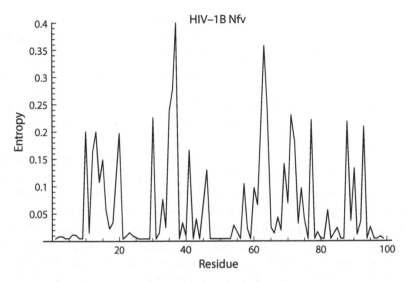

FIGURE 3.14. Entropic profile of the HIV-1 (subtype B) protease sequence adapted to an environment containing nelfinavir (739 isolates, sequences downloaded from HIVdb accessed May 9, 2006).

protease inhibitor *nelfinavir*. Repeating the analysis above, we obtain the entropic profile of nelfinavir-adapted sequences. While the general structure of the profiles is the same, there are subtle differences that can easily be detected by eye (see Fig. 3.14). The sum of conditional entropies in this case gives $H(P|e = \text{nfv}) = 6.189$ mers, for an information content of $I(P : \text{nfv}) = 92.811$ mers, and an information *loss* (with respect to the wild type) of $\Delta I = 1.7$ mers.

While this information loss is similar to the loss suffered by molecules adapting to indinavir, it is not clear whether the *same* or different information was lost and regained by the molecules adapting to different environments. We can obtain part of this answer by investigating the information loss of molecules that adapted to the presence of *both* indinavir and nelfinavir. Indeed, the quick emergence of resistance mutations has forced clinicians to combine drugs in an effort to keep the viral load in patients low. Often, these combination therapies (called HAART, for Highly Active Antiretroviral Therapy) contain two types of protease inhibitor, and an inhibitor for the reverse transcriptase of HIV. (Note that many patients in the "combined environment" ensembles took the drugs consecutively, not concurrently.) In 2006 the Stanford database listed 305 sequences adapting to the combination indinavir + nelfinavir only, giving $H(P|e = \text{idv} + \text{nfv}) = 7.864$ mers, and thus an information loss (again with respect to the wild type) of $\Delta I = 3.375$ mers, somewhat less than the sum of the individual information losses. Thus, we can conclude from this analysis that the two drugs nelfinavir and indinavir act quite differently, leading to almost independent information loss. This analysis can be extended to a third protease inhibitor: saquinavir (actually the first protease inhibitor on the market, introduced in 1995). The statistics of information loss is summarized in Table 3.1, where we also note the bias and error of the entropy estimate. For the smallest data sets (the indinavir + nelfinavir set has 305 sequences, the indinavir + saquinavir set only 147), the total bias correction becomes sizable.

From the point of view of clinical success, it is satisfying to see that the information losses for the pairs idv and nfv, as well as idv and sqv, are almost additive. However, using all three inhibitors together (209 isolates at Stanford HIVdb, accessed May 10, 2006) actually produces *less* information loss than just indinavir and saquinavir together (but more than indinavir and nelfinavir together). In principle, an information-theoretic analysis using a set of well-annotated sequences (with as much patient information as possible) should allow us to predict the best drug regimen that minimizes drug resistance and thus prolong the efficacy of drugs.

3.3.2   Evolution of drug resistance as information gain

We have seen that a changed landscape leads to decreased fitness because the nearby peaks that are still available to the virus are lower (if they had not been, the population would have inhabited that peak instead). A picture such as Figure 3.11 suggests that there are no other higher peaks available, but this is just a limitation of the fitness landscape picture. Sequence space is vast, and other peaks, possibly as high as the initial one (now removed by the drug) are likely close by. Of course, the virus is not adapted to any alternate peaks at the moment the drug is introduced, because on those alternate peaks the function (here, the cleavage of the polyprotein) is achieved in a somewhat different way. But mutations can restore function by altering the protein. The evolution of drug resistance can then be seen as exploring the landscape in this "alternate universe," climbing those alternate peaks to restore viral fitness to previous levels.

Because the Stanford HIV database stores sequences obtained from patients with time stamps from several decades now, the database allows us to monitor the evolution of drug resistance year by year, to witness the virus's climb back up. How does the virus store the information it acquires during that climb? And is the virus able to regain previous levels of fitness, even when more and more drugs are administered at the same time in ever more complex drug "cocktails" (combination of drugs)?

To study this question, let us focus on the HIV protease that we discussed in the previous section because it has been one of the first targets for antiviral drugs, and because the protein sequence is relatively short. We have seen the entropic profile of the wild-type protease sequence already in Figure 3.12, but because there we were not concerned with how the sequence changes over time, we lumped all the sequences of patients that had never received any antiviral drugs in one "pot" as it were (for sequences obtained up to the year 2006). To see how the information content changes over time, we need to calculate the information content separately for each year. Let us first do this for drug-naive patients. Because HIV evolves quickly, we expect that the protease sequence in a constant environment is optimal and does not change over time. As a consequence, the information content should stay the same, as information content and fitness must correspond to each other (for the simple reason that it is the information that ultimately produces the fitness).

We saw earlier in section 2.3.1 that the entropy of a protein can be calculated from the marginal and joint substitution probabilities estimated from a multiple-sequence alignment using the formula

$$H(P) = \sum_{i=1}^{L} H(P_i) - \sum_{i<j} H(P_i : P_j) + H_{\text{corr}}, \qquad (3.17)$$

where H_{corr} represents information stored in correlated substitutions between three or more sites. In the previous section, we ignored the second term in (3.17) to get a rough estimate of the information loss in the HIV-1 protease due to exposure to different protease inhibitors. To track how information changes over time, we will be more careful and include those correlations, because it turns out that they are encoding most of the information that the protease reacquires (Gupta and Adami 2016). According to (3.17), we write the information content $I(P) = L - H(P)$ as

$$I(P) = I_1(P) + \Delta I_2(P) - H_{\text{corr}}, \qquad (3.18)$$

where $I_1(P) = L - \sum_{i=1}^{L} H(P_i)$ is the "first-order" estimate of the information content that does not take into account correlations, while $\Delta I_2(P) = \sum_{i<j} H(P_i : P_j)$ is the information stored in pair-wise correlations between residues. Note that the contribution from $\Delta I_2(P)$ to the total information $I(P)$ is always positive, while the contribution from the third-order and higher terms, H_{corr}, can be either positive or negative. Unfortunately (as we'll discuss below), the data we have at our disposal are not sufficient to estimate H_{corr} with any reliability, and we therefore have to hope that this contribution is small.

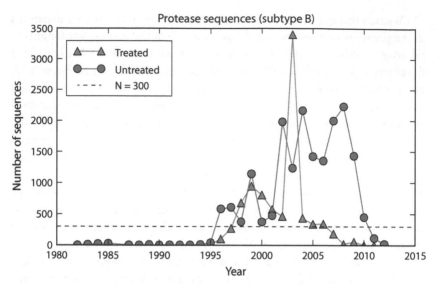

FIGURE 3.15. Number of HIV-1B protease sequences downloaded from HIV Stanford database on September 17, 2013.

As discussed in Box 2.2, estimating probabilities from finite samples creates a bias in entropy estimates, which is particularly important when estimating entropies shared between multiple sites. To have sufficient accuracy to estimate $\Delta I_2(P)$, we decided that we should have at least 300 sequences per year, which limited the analysis to the years 1998–2006 (see Fig. 3.15). Estimating the contribution to $I(P)$ from the shared entropy between three sites $H(P_i{:}P_j{:}P_k)$ would require at least twenty times as much data, which is not currently available.

Let us first check what the entropic profile looks like in the year 1998 (the first year of the analysis), for untreated individuals. The entropic profile shown in Figure 3.16(a) suggests that the profile is consistent with the averaged profile from the years before 2006 shown in Figure 3.12, but the profile looks quite different from that of patients on at least one protease inhibitor (shown in Fig. 3.16[b] for the year 2006, the last year in our analysis).

The total entropy (summed over sites) is significantly higher in 2006 for treated subjects, in particular at sites where the wild-type protein shows very little variation. If we look at the difference in entropy between treated and untreated subjects in Figure 3.17, we notice immediately that the majority of entropy changes are associated with residue positions with known *resistance mutations* or *compensatory* (accessory) mutations.

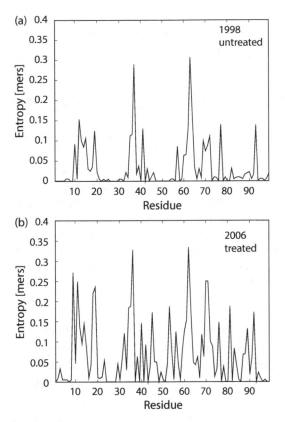

FIGURE 3.16. (a) Entropic profile of the HIV-1B protease constructed from 376 sequences from patients not exposed to protease inhibitors (drug-naive) in 1998. (b) Entropic profile constructed from 341 sequences from patients exposed to at least one protease inhibitor (year 2006).

Resistance mutations are substitutions at particular sites in the protein that confer resistance to one or more protease inhibitors. For example, Figure 3.17 shows a strong increase in entropy at position 82, which is due mostly to the known drug resistance mutation V82A (a substitution of the amino acid alanine where valine is usually found), which confers resistance to the protease inhibitor indinavir. The increase at position 10, on the other hand, is mostly due to the "accessory" mutation L10F (substitution of phenylalanine where leucine is usually found). A mutation is termed accessory when it reduces susceptibility of the virus to an inhibitor if it occurs in conjunction with another resistance mutation.

Let us now look at how I_1 has changed from 1998 to 2006, both in the drug-naive population and in those exposed to antiviral drugs. Figure 3.18

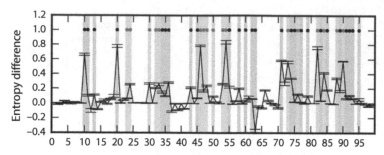

FIGURE 3.17. Changes in protease entropy due to treatment. The entropy difference at each site is obtained by subtracting the entropy of the untreated data from that of the treated data. Average values are obtained by sampling sequence data from all years (1998–2006, 10 subsamples/year of 300 sequences each). Error bars represent ±1 SE. Gray dots represent positions known to be primary drug resistance loci, while black dots mark positions of compensatory or accessory mutations (Shafer and Schapiro 2008). From Gupta and Adami (2016). CC BY 4.0.

shows that the information stored in single sites has largely remained constant for the drug-naive population, while it appears to have *decreased* for those patients taking protease inhibitors. We also note that the year 2003 appears to be an outlier. Indeed, as we can see in Figure 3.15, that year had an extraordinary number of sequences deposited in the database compared to other years. These can largely be traced back to two clinical trials of the drug tipranavir (2,900 out of the 3,399 sequences), which is a second-line drug given mostly to patients who had already failed a first-line drug. Patients in these trials (for which virus can be detected) necessarily will have on average many more mutations in the protease, which explains the low value of I_1 for this data point.

We can understand why we see no significant change in information for the untreated group: the environment is unchanged, and therefore there is no need for a change in the protease's function. While a slight decrease in I_1 can be detected for the untreated case, this decrease is not statistically significant. But it is possible that some increased variation can be due to the transmission of drug-resistant types to individuals who have never been treated, and are thus tagged as "untreated" in the database. Indeed, about 8 percent of new cases had transmitted drug resistance in a study concluded January 2007 (Ross et al. 2007), sufficient to explain the slight decrease of I_1 in Figure 3.18.

On the contrary, there is a marked decline in I_1 for the treated group. If I_1 represented the total information in the protein accurately, this would be surprising as during the years 1998–1996, several new protease inhibitors came on the market, which means that to become resistant to them, there

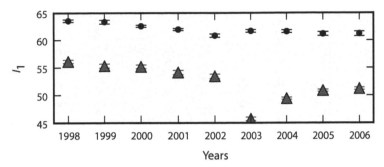

FIGURE 3.18. Information stored in single sites $I_1 = 66 - \sum_{i=15}^{90}$ (in mers) for untreated (circles) and treated (triangles) subjects. To ensure statistical accuracy, the beginning fourteen residues as well as the last nine residues were omitted in this analysis, as those have missing sequence data that drops the sample size for those regions below the threshold of 300. Error bars are one standard deviation. From Gupta and Adami (2016). CC BY 4.0.

should be an increase in information, not a decrease. Therefore, let us take a look at the piece ΔI_2 defined above, which is the information stored in pairwise correlations between residues. We discussed pair-correlation corrections earlier in the context of DNA sequences in section 2.3.3, where we in particular pointed out the dangers of overestimating this correction if sample sizes are low. This danger is even more pronounced for amino acid sequences (as opposed to DNA sequences) because the sum in $\Delta I_2 = \sum_{i<j}^{99} H(P_i{:}P_j)$ has 4,851 terms. Because entropy estimation is biased, even a small bias in each of the $H(P_i{:}P_j)$ can quickly add up. It is for this reason that we insisted on a minimum of 300 sequences per year, but we also used a more sophisticated bias correction method as the one discussed in Box 2.3. In fact, computational experiments with both real and simulated protein sequences showed that the "NSB method" (Nemenman et al. 2002) we used in the analysis accurately estimates mutual entropies for sample sizes as small as one hundred sequences (see Supplementary Text S6 in Gupta and Adami 2016.)

Let us see then how ΔI_2, the information stored in correlations between sites, changes in treated as well as untreated subjects in Figure 3.19. Two things stand out immediately: the information stored in correlations is markedly higher in sequences from treated subjects, and it is increasing over time. On the contrary, ΔI_2 is essentially constant in untreated subjects, which again squares with the idea that no evolution is taking place in a constant environment. If we add the terms I_1 and ΔI_2 we obtain our estimate for the total information stored in the protease molecule (neglecting third-order and higher correlations) shown in Figure 3.20. The trends we see are clear: no

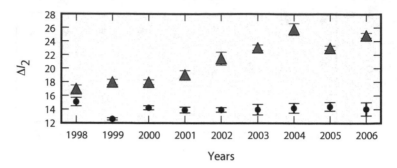

FIGURE 3.19. Information stored in correlations ΔI_2 (in mers) for sequences from untreated subjects (circles) and treated individuals (triangles). Error bars are one standard deviation. From Gupta and Adami (2016). CC BY 4.0.

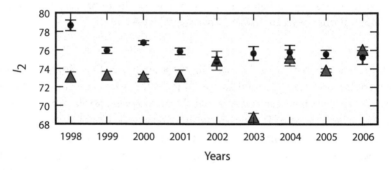

FIGURE 3.20. Total information $I_2 = I_1 + \Delta I_2$ (in mers) for sequences from untreated subjects (black dots) and treated individuals (dark-gray triangles). Error bars are one standard deviation. From Gupta and Adami (2016). CC BY 4.0.

change in information for the untreated group (this is especially clear if we disregard the data point at 1998), and a small but steady increase of information in the treated group, which is adapting to a variety of different protease inhibitors (disregarding the 2003 value for this purpose).

The difference in the amount of information stored in correlations between the untreated and the treated group is striking. When we identify which pairs of residues are correlated, we can see (in Fig. 3.21) that many of the correlated pairs involve residues close to the active site that are correlated with residues far away from it.

One of the most famous correlated pairs of residues is (82,54), with residue 82 at the edge of the active site, and residue 54 in the "flap" (the "top" part of the molecule), which plays an important part in substrate binding and moves as much as 7Å when the protease is bound to the target (Miller et al.

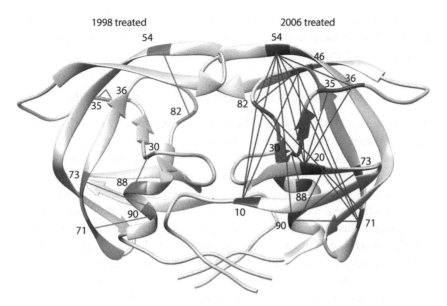

FIGURE 3.21. Correlated residues mapped onto the protease structure. Epistatic interactions in the protease sequences in treated data from the year 1998 (medium gray on the left chain) and 2006 (dark gray, right chain). The inter-acting residues are numbered. Only those interactions are shown where the information between any pair of residues exceeds 0.1 bits, indicating strong epistasis (Gupta and Adami 2016). CC BY 4.0.

1989). The correlation is present even in the absence of drug resistance, but as Figure 3.21 shows, the flap mutation I54V is correlated to many other muta-tions on the structure, many of them far away of any residue thought to be involved in substrate catalysis.

The reason why the protease accumulates all these long-distance correla-tions to evade protease inhibitors is still a matter of ongoing discussion. One theory posits that because single substitutions will on average destabilize a protein fold by about 1 kcal/mol (Bava et al. 2004), most resistance mutations must be accompanied by compensatory mutations that restabilize the fold by a commensurate amount (Bloom et al. 2006; Bloom and Arnold 2009). And because there are more residues that are far from the active site than residues that are close, those compensatory mutations are likely further away, since any residue on the sequence could potentially restabilize a fold via consensus sta-bilization, that is, replacement of the residue by the consensus residue in a multiple-sequence alignment (Amin et al. 2004). Because multiple pair-wise correlations are likely to impair evolvability, we should expect that further evo-lution in a constant environment might release some of these long-distance correlations and replace them with shorter-distance ones with fewer constraints, thus reestablishing evolvability. However, none of these hypotheses have so far been tested.

3.4   Evolution of Information in DNA Binding Sites

The previous sections illustrated the idea that evolution increases the amount of information stored in proteins, or more precisely, in the open reading frames of the genome. But it is well known, of course, that genomes contain much more than open reading frames that code for proteins. In humans, for example, only about 2 percent of the genomic sequence is translated into proteins (International Human Genome Consortium 2004). While much of the remaining sequence consists of repeating elements that are unlikely to be functional, a good part of the untranscribed code is used for the *regulation* of the expression of genes. The complexity of gene regulation depends very much on the organism. In general, regulation is quite simple in prokaryotes, where genes are usually turned on or off by a single protein called a *transcription factor* (recall the example of the lac operon in section 2.4). In eukaryotes, and in particular in the metazoans (animals), gene regulation is often extremely complex, so much so that the sequence that regulates the expression of a gene can be several times longer than the sequence for the gene that it controls. It is not a stretch to hypothesize from this observation alone that most of the complexity we observe in metazoans is due to the complexity of gene regulation patterns, rather than the complexity of the genes themselves. Indeed, while humans share about 60 percent of their genes with bacteria (Britten and Davidson 1971), it is unlikely that our complexity advantage is simply due to the 40 percent extra genes that bacteria do not have. Instead, it seems we use most of the same tools (proteins) as simple organisms, but we use them in a far more complex manner.

As an example, consider the regulatory network that controls expression of the gene *endo16* during development of the California sea urchin shown in Figure 3.22. As is common in vertebrate development, the activation of the gene must be precisely timed and localized, and is conditional on the successful completion of other developmental events. The regulatory region consists of about 2,600 base-pairs upstream of the transcription start site, and is organized into "modules" (labeled A–G in Fig. 3.22) that each consist of multiple binding sites for four to eight different transcription factors.

Given the complexity of this regulatory machinery, it is not unreasonable to speculate that most of the evolution in the last, say, two billion years, affected gene regulation through the evolution of binding sites and modules as well as their respective transcription factors, rather than the refinement of existing

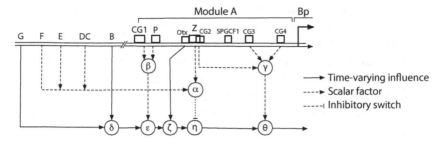

FIGURE 3.22. Logical "wiring" diagram that depicts the influence of the different regulatory modules on the transcription of *endo16*. Circles represent computational interactions between modules, conditional on the state of the system. Arrows indicate influences of transcription factor binding sites. Modified from Yuh et al. (1998).

proteins or the evolution of novel ones (Britten and Davidson 1969; Britten and Davidson 1971; Valentine 2000), and that the complexity of multicellular life is mainly due to changes in these noncoding stretches of DNA (and the associated regulatory proteins). However, decoding the complexity of such complicated regulatory systems is one of the hardest tasks in molecular biology today (Davidson 2001), and we will not be able to contribute to this endeavor here. Instead, in this section, we will limit ourselves to the simplest aspect of gene regulation, namely the binding of a regulatory protein to DNA, and study the information content and evolution of DNA *binding sites*, that is, the stretches of nucleotides to which proteins bind (called the *cis-regulatory* system). Because complex cis-regulatory systems are built out of many of these binding sites, it is conceivable that the techniques outlined below may one day help in unraveling the bigger picture of the regulation of metazoan genes.

3.4.1 Thermodynamics and information theory of DNA binding

The purpose of this section is to introduce a formalism (the Berg-von Hippel theory) that allows us to connect thermodynamic quantities (like the free energy of binding of a transcription factor to its binding site) to information-theoretic ones. This connection will enable us to view the evolution of the functional properties of a binding site in terms of the evolution of information in that site. Readers that are not interested in the technical aspects of this connection can skip to the end of this section, where the salient results are summarized.

DNA binding sites (or "motifs") are sequences of nucleotides that are recognized by a specific binding protein (the transcription factor), and that show a much higher affinity to their particular binding protein than a random sequence would. The binding of a protein to its binding site sets into motion a cascade of events that depends on the gene and regulatory system at hand, but in general the binding interaction is a *signal* that ultimately affects how (and to what extent) the protein coding region "downstream" of the regulatory site is transcribed. Thus, the specificity of a binding site has crucial implications for the fitness of the organism, and is therefore optimized by evolution.

The kinetic reaction of a protein P binding to a particular site X_s (specified by the sequence s of length ℓ) can be described as

$$P + X_s \leftrightarrow P \circledast X_s,\qquad\qquad(3.19)$$

where $P \circledast X_s$ represents the protein-DNA complex. The *binding constant $K(s)$* of this reaction depends on the particular sequence s, and is given by the normalized concentration of the complex in equilibrium:

$$K(s) = \frac{[P \circledast X_s]}{[P][X_s]}.\qquad\qquad(3.20)$$

For a good binding site s (one that binds a particular protein with high specificity), $K(s)$ should be several orders of magnitude larger than the binding constant of a random site of that length, and the sequence s itself should be rare enough that in a genome of length L it does not occur too often by chance alone, because such sites would detract the protein from the site that it should be binding to. If we ask that not more than two sequences s appear in a sequence of length L by chance alone, then the minimum size of the binding site needs to be $\ell_{\min} = \log_4(2L)$ mers. For *E. coli* bacteria (whose binding sites we will study later), this gives about $\ell_{\min} \approx 13$.

In the following, I outline a theory that allows us to connect the binding affinity $K(s)$ of a sequence s to the *binding energy ΔG* of that sequence and its information content. To do this, we will have to delve a little bit into the thermodynamics of reactions, without introducing any of the underlying concepts in detail. But we will be rewarded with a surprisingly simple view of DNA binding, and an expression that relates function directly to information content.

Berg-von Hippel theory

The statistical theory that describes binding entirely in terms of the contribution of individual nucleotides to the specificity of a particular sequence for a particular protein is due to Berg and von Hippel (1987, 1988). Here I will only summarize the principal elements of this theory, and also change their notation to suit our discussion. While this theory is valid only in the approximation that the concentration of the transcription factor is low (Djordevic et al. 2003), it can be extended to take into account arbitrary chemical potentials (that is, molecular densities). An information-theoretic analysis under such conditions (high molecular densities where transcription factors are bound to sites almost all the time) may require more sophisticated methods (Djordevic and Sengupta 2006).

A reaction such as described in Equation (3.19) can proceed in both directions, as the double arrow implies, depending on whether it takes energy to produce the complex $P \circledast X_s$, or if energy is released. In thermodynamics, the energy that is either consumed or released in a reaction is called "free energy," and the quantity that determines the direction of (3.19) is called the change in free energy, ΔG. A reaction proceeds forward if $\Delta G < 0$, that is, if the compound has lower free energy than the free protein and binding site. According to standard chemical reaction kinetics, the *reaction constant* $K(s)$ is related to the free energy change by (see, e.g., Jackson 2006)

$$\Delta G = -kT \ln [K(s)/K(r)], \tag{3.21}$$

where k is Boltzmann's constant, T the temperature of the system, and $K(r)$ is the binding coefficient of a *random* site (with sequence r). The general idea here is that random sequences do not bind specifically, so that ΔG vanishes and the reaction does not proceed forward or backward with any particular bias. In the future, we will measure free energies "per kT" (giving units kcal per mole to ΔG), so that the relation becomes

$$\Delta G = -\ln K(s) + \ln K(r). \tag{3.22}$$

We can now introduce an energy variable $E(s)$ that characterizes each binding site s so that the difference of those energies is the free energy difference ΔG:

$$\Delta G = E(s) - E(r). \tag{3.23}$$

This variable is chosen such that the energy of a site that shows high affinity to the protein is low, whereas the energy of a random site is high. These energies are only defined up to a constant (only energy differences are physically relevant, after all) and we will see later that these energies are "renormalized" any time a beneficial mutation occurs.

Using (3.22) and (3.23), we can now write the reaction constant of a site in terms of the energy of that site and the reaction constant of a random site:

$$K(s) = K(r)e^{-E(s)+E(r)}. \tag{3.24}$$

Now, let us introduce the sequence s_0 that binds the protein the best, among all sequences in an ensemble of sequences. This ensemble can be either all possible sequences of length ℓ (in which case s_0 would be unique given the transcription factor that it binds to), or it could be all sequences of length ℓ that are present in a genome of size L (there are about L of those). Note that for ℓ of the order 20, the number of possible different sequences (4^{20}) is significantly larger than the number of sequences in a genome of size L, no matter what the genome is. Also note that in the case that our ensemble of sites is given by those that are present in a genome at one time, evolutionary changes over time can change the identity of s_0. In practice, s_0 is the *consensus sequence* (Davidson et al. 1983) of the ensemble of sites, which can be obtained from an alignment of binding sites by picking the most prevalent nucleotide at each position, as we will see shortly. The consensus s_0 turns out to be the sequence that binds the strongest, both in theory (by definition) and it seems in practice (see, e.g., Mauhin et al. 1993; Fields et al. 1997).

As discussed earlier, because only energy differences matter, we have the freedom to normalize our energies $E(s)$. We will choose $E(s_0) = 0$, i.e., the sequence that binds the protein the best has zero energy. Then, from (3.24) we have $K(s_0) \equiv K_0 = K(r)e^{E(r)}$, which we can plug back into Equation (3.24) to obtain

$$K(s) = K_0 e^{-E(s)}. \tag{3.25}$$

According to Equation (3.25), the binding coefficient $K(s)$ of a sequence s is reduced from the best possible coefficient K_0 by a Boltzmann factor with the energy $E(s)$ of that site (remember we set $kT = 1$). This is the starting point for Berg and von Hippel. Their central assumption is that the binding free energy is determined by the nucleotides in the binding site *individually*, that is, that one nucleotide's contribution to the energy $E(s)$ does not depend on the contribution of another site. In other words, they assume that the energy $E(s)$ is an independent sum of the contributions of each of the ℓ nucleotides that make up s:

$$E(s) = \sum_{n=1}^{\ell} E_n(s), \tag{3.26}$$

where we just defined implicitly the contribution $E_n(s)$ of site n to the total binding energy $E(s)$. Note that this is a typical assumption in statistical physics, where each nucleotide plays the role of a statistical subsystem, and the total energy of the joint system is given by the sum above. This assumption is crucial and needs to be tested. We will see later that while the nucleotide values within the binding site are rarely totally independent, correlations between nucleotides are usually quite weak so that the independence assumption looks reasonable. In cases where correlations are strong, the theory can be modified to take this into account (see the Appendix in Berg and von Hippel 1987).

Let us then take a look at a single position n in the sequence s, and the probability to find any of the four nucleotides at that position, $p_i^{(n)}$, where

$i = 0, 1, 2, 3$. Here I introduced a convenient way of labeling the four values A, C, G, and T: we will always denote with "0" the nucleotide that has the highest probability of occurring in the sample of nucleotides at position n in the multiple-sequence alignment, and we'll denote it as s_0. Given this definition, $p_0^{(n)} \geq p_i^{(n)}$ always. We now would like to construct an "energy functional" in such a manner that the most likely nucleotide at position n (the one that occurs with probability $p_0^{(n)}$) should contribute $E_n = \epsilon_0^{(n)} = 0$ to $E(s)$, while the other alternatives $i = 1, 2$, and 3 would contribute an amount $\epsilon_i^{(n)}$ for $i = 1, 2, 3$. These $\epsilon_i^{(n)}$ are called "discrimination energies" because they contribute to the binding site's ability to discriminate between binding proteins. In analogy with the occupation probability of single-particle energy levels in statistical physics, Berg and von Hippel postulate that

$$p_i^{(n)} \propto e^{-\lambda \epsilon_i^{(n)}}, \tag{3.27}$$

where λ is a constant that ensures we can make contact to the energy $E(s)$ introduced above.

In statistical physics, λ is the inverse temperature of the system, but not here because the nucleotides are treated like physical statistical systems only by analogy. The agent that causes transitions between the "single-particle states" A, C, G, and T is not temperature, but rather the Darwinian process of mutation and selection. In a sense, then, λ represents the strength of selection in the system.

If we normalize (3.27), we obtain

$$p_i^{(n)} = \frac{e^{-\lambda \epsilon_i^{(n)}}}{\sum_{i=0}^{3} e^{-\lambda \epsilon_i^{(n)}}} = p_0^{(n)} e^{-\lambda \epsilon_i^{(n)}} \tag{3.28}$$

because $\epsilon_0^{(n)} = 0$ (the mostly likely nucleotide contributes nothing to the binding energy) implies that $p_0^{(n)} = 1 / \sum_i e^{-\lambda \epsilon_i^{(n)}}$.

We should note that Equation (3.28) does not have to be postulated. It follows by simply requiring that the entropy at each position, $H[p_i^{(n)}]$, is maximal while the average energy is constant (see Exercise **3.1**). This calculation is, in fact, the same one that yields the Boltzmann distribution in thermodynamics.

Equation (3.28) allows us to write the discrimination energies in terms of "occupation probabilities" as:

$$\lambda \epsilon_i^{(n)} = \ln \frac{p_0^{(n)}}{p_i^{(n)}}. \tag{3.29}$$

This form is extremely useful, because it allows us to estimate the $\lambda \epsilon_i$ from the observed probabilities p_i only, which can be obtained from an alignment of functional binding sites, as we will do below. The $\epsilon_i^{(n)}$ form a $4 \times \ell$ dimensional matrix, often called "weight matrix" in the literature. The contribution $E_n(s)$ to the sum $E(s)$ can then be written in terms of the weight matrix $\epsilon_i^{(n)}$ and a "sequence matrix" $S_i^{(n)}$, which carries a "1" at sites n that have nucleotide i in sequence s, and a zero otherwise, as

$$E_n(s) = \sum_i \epsilon_i^{(n)} S_i^{(n)}. \tag{3.30}$$

The total energy for sequence s is then

$$E(s) = \mathrm{Tr}(\epsilon \cdot S^T), \tag{3.31}$$

where Tr denotes the matrix trace.

We can also introduce an "information matrix" that allows us to attach an "information score" to each sequence.[6] This matrix is given by Schneider (1997); Stormo and Fields (1998):

$$R_i^{(n)} = \ln \frac{p_i^{(n)}}{1/4}, \tag{3.32}$$

and is, in a sense, the "opposite" of Equation (3.29). Instead of the log-likelihood to find the most probable nucleotide with respect to the alternatives, this information matrix reflects the log-likelihood of each nucleotide with respect to the random estimate ($p = 1/4$).

The definition (3.32) allows us to write the information score as

$$I(s) = \mathrm{Tr}(R \cdot S^T) \tag{3.33}$$

and the relationship between the weight matrix (3.29) and the information matrix (3.32) allows us to write

$$I(s) = L - \lambda E(s) + \lambda F. \tag{3.34}$$

In Equation (3.34) we introduced the notation $L = \ell \ln(4)$ (the binding site's length in units "nats" rather than the usual "mers"), and the "Helmholtz free

---

6. Keep in mind that this information score is not technically information in the sense introduced in chapter 2, because the determination of entropy (and hence information) requires statistical ensembles from which probabilities can be obtained. We will study the information score's relation to Shannon information when we explicitly examine the statistics of binding sites in the next section.

energy" of the ensemble of sites

$$F = \frac{1}{\lambda} \sum_{n=1}^{\ell} \ln p_0^{(n)}.$$ (3.35)

The reason that the combination $\ell \ln(4)$ appears rather than the sequence length proper is the choice of natural logarithms in the energy and information weight matrices. Of course, all the preceding and following analysis can be carried through in any base. As Equation (3.35) shows, the probability λ plays the role of the inverse temperature.[7]

We are now ready to link this description of the binding site in terms of its information score $I(s)$ to our starting point Equation (3.22). Earlier we found that $K(s)$ can be related to the binding constant of the consensus sequence K_0 and the energy of the site $E(s)$ via

$$K(s) = K_0 e^{-E(s)}.$$ (3.36)

At the same time we can write $e^{-E(s)} = e^{-\sum_n \epsilon_i^{(n)}}$, where we sum over all the $\epsilon_i^{(n)}$ given by the particular nucleotides i in sequences s. Invoking the independence of the contribution of nucleotides to the overall binding strength again, we also write ($s^{(n)}$ denotes the n-th position in the sequence s)

$$K(s) = \Pi_{n=1}^{\ell} K(s^{(n)}),$$ (3.37)

and the same decomposition holds for the consensus site

$$K_0 = \Pi_{n=1} K(s_0^{(n)})$$ (3.38)

so that altogether

$$K(s_i^{(n)}) = K(s_0^{(n)}) e^{-\epsilon_i^{(n)}}.$$ (3.39)

We can now average the left-hand side of (3.39) over the four possible nucleotides, which by definition gives us the binding constant of a *random* nucleotide:

$$K(r^{(n)}) = \frac{1}{4} \sum_{i=0}^{3} K(s_i^{(n)}) = \frac{1}{4} K(s_0^{(n)}) \sum_i e^{-\epsilon_i^{(n)}},$$ (3.40)

7. If the concepts from statistical physics such as the Helmholtz free energy and partition function are not familiar to the reader, no need to worry. They do not add anything fundamentally new to the discussion, but they do allow us to cast the formulas into a form that provides us with interesting interpretations. Still, keep in mind that the Helmholtz free energy F, because it is information-theoretic in nature, is unrelated to the thermodynamic free energy difference of binding ΔG introduced earlier.

so that finally

$$\frac{K(s_i^{(n)})}{K(r^{(n)})} = \frac{e^{-\epsilon_i^{(n)}}}{1/4 \sum_i e^{-\epsilon_i^{(n)}}}. \tag{3.41}$$

Summing over the ℓ sites we now see that

$$\sum_{n=1}^{\ell} \ln\left(\frac{K(s_i^{(n)})}{K(r^{(n)})}\right) = \ln \frac{K(s)}{K(r)} = L - E(s) - F, \tag{3.42}$$

which would be the right-hand side of Equation (3.34) if $\lambda = 1$. But it is possible to expand Equation (3.39) around $\lambda = 1$ to relate Equation (3.34) to ΔG. Such an expansion allows us to write (see Exercise **3.2**)

$$\Delta G = -I(s) + (1 - \lambda)(E(s) - \langle E \rangle), \tag{3.43}$$

where $\langle E \rangle = \sum_{n=1}^{\ell} \sum_{i=0}^{3} p_i^{(n)} \epsilon_i^{(n)}$ is the average energy of the ensemble of sequences used to construct the weight matrices (the "seed" ensemble). Because λ is usually close to 1, we find that

$$\Delta G(s) \approx -I(s), \tag{3.44}$$

that is, the binding energy of a site s in kcal/mole is given precisely by its information score. This holds in particular for sequences s whose energy $E(s) \approx \langle E \rangle$, that is, for sequences s whose binding energy is close to the average energy of the alignment of sites that gave rise to the weight matrix in the first place. This is the relationship between the thermodynamic quantity ΔG and an information-theoretic quantity that we were looking for at the beginning of this section. We find that, at least approximately, higher binding affinity implies higher information score, proving once more that information is power indeed! (We will discuss the relationship between the information score and Shannon information in the following section.)

A few comments are in order. While Shannon information can never be negative, the information score can be, so that reactions such as (3.19) only proceed forward at a pace exceeding that of random collisions if $I(s) > 0$, and that large free energy differences require a commensurately large information score. Naturally, reactions with $\Delta G > 0$ can occur if this energy is provided to the molecules. Furthermore, this equation is valid only for those genomes for which random sequences have an entropy of one mer per nucleotide, that is, those genomes that do not have an appreciable base composition bias (Stormo 2000). The formulas can be generalized by replacing the probability "1/4" in the information weight matrix (3.32) by the base composition distribution. While the base composition bias encodes information (as noted in chapter 2, footnote 1), it is not information about the transcription factor, and therefore does not contribute to the determination of ΔG.

We can also ask about the distribution of energy $E(s)$ in an adapting population of sequences with binding sites s, and how often the consensus sequence s_0 (with $E(s_0) = 0$) actually occurs. The distribution of individual binding energies in an alignment from an ensemble with average $E(s)$ is approximately Gaussian (Schneider 1997), with the functional sequences in the left tail, usually without encompassing the value $E = 0$. In fact, the consensus sequence rarely (if ever) appears in the set of aligned sequences (Schneider 1997) (or, for that matter, in the genome as a whole). In other words, the "target of optimization" s_0 is never achieved. There are a number of hypotheses to explain this (as pointed out by Berg and von Hippel 1987), perhaps chief among them mutational drift, and the fact that less-specific sites are far more numerous than specific ones. Furthermore, there may be an advantage to binding sites that are not absolutely specific, for example because transcription factors need to be released efficiently after transducing the signal, or more simply there may be a threshold binding energy above which all binding sites are functional (Kotelnikova et al. 2005). Another explanation that suggests that a binding site carries only as much information as is necessary to find the target site in the genome is discussed below in section 3.4.3.

Summary

In this subsection I introduced a formalism (Berg-von Hippel theory) that allows us to connect the functional characteristics of a binding site s (its binding free energy $\Delta G(s)$) to an information-theoretic construction, the information score $I(s)$. The functional value $\Delta G(s)$ can be measured in experiments, while the information score $I(s)$ comes purely from sequence data. The relationship is approximately

$$\Delta G(s) \approx -I(s), \qquad (3.45)$$

that is, the binding free energy (which is negative for a functional site, measured in kcal/mole) is minus the information score (measured in nats).

3.4.2 Information in CRP binding sites

It is high time we applied some of the theory above to real sequence data, in order to check the assumptions that went into it and study the relation between information and binding in detail. Many transcription factor binding sites have been analyzed in terms of the theory outlined above (see Stormo 2000, for a review), but we will focus here on an illustrative example: the DNA binding site for the cyclic AMP (cAMP) receptor binding protein (CRP) in

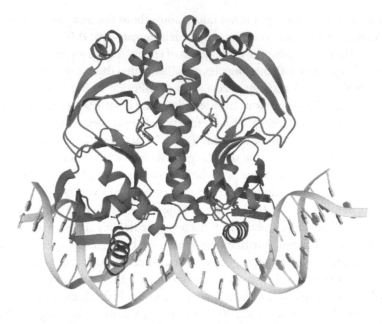

FIGURE 3.23. The transcription factor CRP (dark gray dimer, with two smaller cAMP molecules bound to each monomer), shown attached to its DNA binding motif (light gray). Note the bend in the DNA brought about by CRP, which is its main function. For this bend to occur reliably, the distance between the two motifs that make up the CRP binding site must be exactly six nucleotides.

E. coli. We encountered CRP before: when discussing the lac operon information transmission channel (Fig. 2.19), CRP went under its other name, CAP (catabolite activator protein).

CRP is a transcription factor that initiates the expression of a number of different genes in *E. coli* (as opposed to other factors that are much more specific, that is, regulate only a few or only one gene). CRP is a homodimer that must be bound to the smaller cAMP protein in order to be active (see Fig. 3.23 and Fig. 2.19). Note that CRP binds to two similar (but not identical) stretches of DNA, and *bends* the DNA at a point between them, by 90 degrees or more.

Our first task is to build a weight matrix (3.29) from known CRP binding sites. The database DPInteract (Robison et al. 1998) lists forty-eight such sites (of length twenty-two nucleotides each),[8] and we will use that alignment to estimate the ratio $p_0^{(n)}/p_i^{(n)}$ from those. Because CRP is a symmetric dimer (see Fig. 3.23), it should bind equally well to the top or to the bottom of the DNA strand. Thus, for every binding site, the site's *reverse complement*

8. Retrieved July 2006. These sites are also available form the Supplementary Information in Brown and Callan Jr. (2004).

FIGURE 3.24. The consensus binding site for CRP is palindromic, meaning that the reverse complement of the sequence on the "bottom" strand of the DNA is equal to the forward sequence on the "top" strand. (We omit here the six-nucleotide spacer, which is also palindromic in the consensus.) In principle, because CRP is a homodimer, CRP should bind equally well to the top or bottom strand. However, there is some evidence that CRP *does* discriminate between the two halves of its binding site, see Berg and von Hippel (1987).

should also be a binding site. (The reverse complement of TGTGA, for example, is TCACA.) Figure 3.24 shows a palindromic CRP binding site where the sequence and its reverse complement are equal (I omitted the random six nucleotide spacer). To estimate the weight matrix from sequence data, we count the number of occurrences N_i of each base for each position, and call the frequency of the most abundant base N_0. The most unbiased predictor for the weight matrix can then be taken as

$$\lambda \epsilon_i^{(n)} \approx \ln \frac{N_0^{(n)} + 1}{N_i^{(n)} + 1}. \tag{3.46}$$

The "+1" in the numerator and denominator of (3.46) are known as "pseudo-counts," and ensure that the fraction is meaningful even if a particular base is not found in the alignment (this is also sometimes called "Laplace's rule of succession," and is standard in estimation theory). This procedure yields the matrix shown below in Table 3.2.

Because of Equation (3.46), the most common base in the alignment is assigned a weight 0.0 in Table 3.2, while less common bases have correspondingly higher weights. It is clear that the weight matrix constructed using only a selected set of binding sites is unlikely to be very specific. Ideally, the seed set should consist of many sites with the same binding strength, and cover all functional sites in the genome.

This weight matrix can now be used to assign a *score* to sequences of length 22 that are candidate binding sites. Given a particular threshold energy, those sequences that have scores below the threshold are predicted to be functional binding sites, while those with scores above the threshold are predicted not to bind CRP. Clearly, it would be useful if the sites that are predicted to be functional could be verified experimentally. If we are given a 22 base-pair sequence from *E. coli*, preferably from an intergenic region (where we suspect

Table 3.2. Weight matrix (3.46) for the first eleven nucleotides of the CRP binding site based on forty-eight sequences in DPInteract.

n	A	C	G	T
1	0.00	2.48	1.57	0.13
2	0.00	2.48	1.23	0.23
3	0.00	1.34	0.64	0.05
4	3.66	1.47	2.56	0.00
5	2.59	3.00	0.00	1.74
6	3.00	1.49	3.69	0.00
7	2.10	3.71	0.00	2.10
8	0.00	2.37	2.37	3.76
9	1.23	0.78	0.88	0.00
10	0.59	0.00	0.59	0.25
11	0.69	0.22	1.20	0.00

true binding sites reside), we can use the weight matrix to assign an *energy score* $\lambda E(s)$ to the sequence by summing up its particular contributions in the weight matrix. Let us denote a sequence by a $4 \times \ell$ matrix containing "1"s at the base in that position, and a zero elsewhere. For example, the 5-mer s=GTGAC would be represented by the matrix

$$S(s) = \begin{matrix} & \begin{matrix} A & C & G & T \end{matrix} \\ \begin{matrix} 1 \\ 2 \\ 3 \\ 4 \\ 5 \end{matrix} & \begin{pmatrix} 0 & 0 & 1 & 0 \\ 0 & 0 & 0 & 1 \\ 0 & 0 & 1 & 0 \\ 1 & 0 & 0 & 0 \\ 0 & 1 & 0 & 0 \end{pmatrix} \end{matrix}. \tag{3.47}$$

As shown earlier, the energy score can be calculated as

$$\lambda E(s) = \text{Tr}(W \cdot S^T), \tag{3.48}$$

where $W_{ni} = \ln(p_0^{(n)}/p_i^{(n)})$ is the weight matrix introduced in Equation (3.29). Based on this, the energy $\lambda E(s)$ can be calculated for any sequence s. For example, the site TTCTGTGATTGGTATCACATTT has $\lambda E = 0.13 + 0.23 + 1.34 + 0.00 + 0.00 + 0.00 + 0.00 + 0.00 + 0.00 + 0.25 + 1.20 + \cdots = 3.8$, a very good score. The zero-energy consensus sequence, the strongest binder according to this theory, is predicted to be AAATGTGATCTAGATCA-CATTT (see Fig. 3.24), but this sequence is nowhere to be found in the entire

Table 3.3. Statistics of sites from the full *E. coli* genome scored with the weight matrix (3.46) that was generated with the forty-eight experimentally verified sequences from DPInteract and their reverse complements (except the last column). First column: energy cutoff, second column: number of sites with energy less than cutoff, third column: fraction of binding sites in coding region, fourth column: fraction of sites in intergenic region, fifth column: fraction of input sites below cutoff from weight matrix constructed with reverse complements, last column: without reverse complements (adapted from Brown and Callan Jr., 2004).

Cutoff λE	No. of sites	Coding %	Intergenic %	Known (96)	Known (48)
5.00	31	3	97	4/48	4/48
7.00	105	9	91	10/48	18/48
9.00	375	26	74	27/48	34/48
11.00	1,495	53	47	39/48	45/48
15.00	26,873	72	28	48/48	48/48

E. coli genome. Indeed, the lowest score in that genome (assessed with the weight matrix above) is $\lambda E = 1.96$.

As any 22-mer from *E. coli* can be assigned a score in this manner, the algorithm predicts that sites that receive a low score $\lambda E(s)$ should be functional. Let us take a look at what happens if we scan *all* possible stretches of twenty-two nucleotides in a genome. Does the algorithm return just the sites that were put in (the sequences from DPInteract), or does it return a few more viable candidates (along with the input set) that we could check experimentally? Are most of the putative sites in the intergenic region (which comprises only 15 percent of the *E. coli* genome)?

A search of all 22-mers of the 4,639,221 base-pair genome of *E. coli* K-12 (NCBI accession No. NC004431) in fact yields many more binding sites, depending on the threshold (Brown and Callan Jr. 2004), and does not necessarily return the input sites! Table 3.3 (reproduced from Brown and Callan Jr. 2004) shows that a low cutoff (using only sequences with $E \leq 5$) yields thirty-one sites, with only four of them in the input set. Increasing the threshold to $E \leq 9$ yields 375 candidates while still missing twenty-one of the experimentally known sites used to create the weight matrix in the first place. In order to pick up all forty-eight input sites, the threshold has to be increased to an unlikely $\lambda E \leq 15$, yielding a total of 26,873 candidates, only a fraction of whom can possibly be genuine CRP sites. Note, however, that this weakness of the weight matrix can be partly blamed on using the reverse complements of each of the forty-eight sequences in the construction of the weight matrix. If

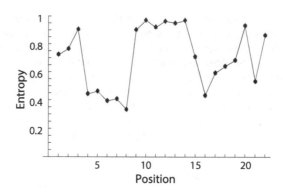

FIGURE 3.25. Entropic profile of the twenty-two base-pairs of the CRP binding site, obtained from forty-eight experimentally verified sequences in DPInteract, with logarithms taken to base 4.

instead the reverse complements are excluded, the algorithm does much better (last column in Table 3.3), suggesting that there is an asymmetry in the binding of CRP that is biophysical in nature.

We can get a better idea of the variability at each site by calculating the entropy per-site as we did in section 2.2. Again using the forty-eight sequences from DPInteract yields the entropic profile in Figure 3.25. We can recognize very clearly the three components of the binding motif, namely the five base-pair contact areas with low entropy and the six base-pair intermediate sequence that is almost random. Using our formula for the information content of DNA sequences Eq. (2.50), we can estimate the information content by subtracting the sum of entropies from the sequence length (we are ignoring correlations between positions here, as we did before in the Berg-von Hippel theory)

$$\langle I \rangle = \ell - H[p_i] = \ell + \sum_{n=1}^{22} \sum_{i=0}^{3} p_i^{(n)} \log_4 p_i^{(n)} = 6.5 \pm 0.2 \text{ mers.} \quad (3.49)$$

This application of information theory to binding sites was pioneered by Tom Schneider, an investigator at the NIH's National Cancer Institute (Schneider et al. 1986), and used by Berg and von Hippel in their statistical theory.

We can now test the relationship between thermodynamic quantities like the binding free energy to information-theoretic constructs. We can rewrite the Helmholtz free energy Equation (3.35) in terms of the average energy $\langle E \rangle$ and the Shannon entropy to obtain

$$F = \langle E \rangle - \frac{1}{\lambda} H. \quad (3.50)$$

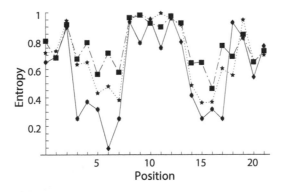

FIGURE 3.26. Entropic profile of sets of candidate CRP binding domains (twenty-two mers with approximately constant energy) from the entire genome of *E. coli*. Diamonds: $\lambda E = 5.9$, stars; $\lambda E = 8.2$, squares: $\lambda E = 10.3$.

Comparing with Equation (3.34) (and paying close attention to the base of logarithms) gives the relation between the information and energy scores and the respective average values within the ensemble:

$$I(s) = \langle I \rangle - \lambda(E(s) - \langle E \rangle). \tag{3.51}$$

This relation, in turn, yields for us the energy score of a random site, $E(r)$, by comparing Equation (3.51) with Equation (3.23)

$$E(r) = \langle I \rangle + \langle E \rangle, \tag{3.52}$$

a relation we can use in Equation (3.24) to deduce the binding constant of a random site in terms of that of the best (i.e., consensus) site and the parameters $\langle I \rangle$ and $\langle E \rangle$ that we can glean from the ensemble (Berg and von Hippel 1987):

$$K(r) = K_0 e^{-\langle I \rangle - \langle E \rangle}. \tag{3.53}$$

To test these predictions, we can use the whole *E. coli* genome again to calculate energy and information scores. First, let us check the entropic profile of the binding site for ensembles with different energy[9] scores λE. By scanning all the 22-mers that appear in the entire *E. coli* genome and assigining scores to them based on the position weight matrix, we can create three groups of sequences: predicted strong binders (with $\lambda E = 5.9 \pm 0.5$), sequences predicted to have intermediate binding strength (with $\lambda E = 8.2 \pm 0.5$), and weak binders (a set with $\lambda E = 10.4 \pm 0.5$). The entropic profile for each of these sets is shown in Figure 3.26. Clearly, the lower the energy, the more

9. Because λ is generally unknown, the discrimination energies are given in units "$1/\lambda$."

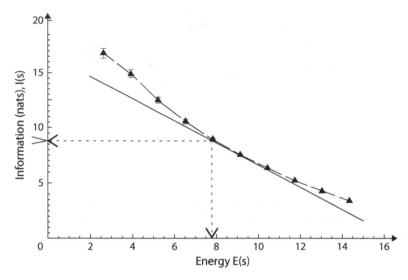

FIGURE 3.27. Information score $I(s)$ (solid line) and Shannon information (triangles and dashed line, in units "nats") of *E. coli* twenty-two mers in energy intervals with constant $\lambda E(s)$, for putative binding sites with $\lambda E \leq 15$ obtained with the weight matrix in Table 3.2 [see also Eq. (3.34)]. The dotted line (downward arrow) points to the average energy score and the corresponding information score (left-pointing arrow) of the reference ensemble (the sequences in the position weight matrix). The Shannon information of that ensemble (right pointing arrow, at 9.01 nats corresponding to 6.5 mers) is only imperceptibly larger than their mean information score.

pronounced the entropy differences, and therefore the higher the Shannon information of the ensemble.

Because we know that a lower energy score also implies a higher information score [from Equation (3.51)], we can surmise that the Shannon information content of an ensemble with fixed λE (call this the set S_E) and the information score of sequences with score λE will be related. The Shannon information of sequences with fixed E is

$$I_E = \ell - H[S_E], \tag{3.54}$$

where $H[S_E]$ is the sum of the per-site entropies of the set S_E of aligned sequences with energy λE in the *E. coli* genome. The information score in turn, is given by Equation (3.34). In Figure 3.27, we can see both the linear information-energy relation using all sequences that have an energy score less than $\lambda E = 15$ in the *E. coli* genome (solid line) and the Shannon information of ensembles with fixed energy (triangles and dotted lines) calculated using

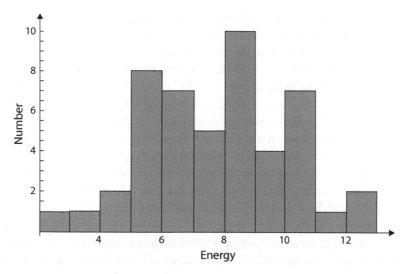

FIGURE 3.28. Distribution of energy scores λE in a seed set of forty-eight sequences from DPInteract. The mean energy of these sequences is $\lambda E = 7.8$.

Equation (3.54). These lines almost coincide at the mean energy score of the seed ensemble (the distribution of energy scores can be seen in Fig. 3.28) that was used to obtain the weight matrix. This means that the information score is closest to the Shannon information for ensembles with energy close to that of the seed ensemble. Furthermore, the Shannon information of all binding site sequences with an energy score equal to that of the average energy of the seed ensemble is $I \approx 6.35$ mers. This is approximately equal to the Shannon information of the binding sites in the seed ensemble itself $[I \approx 6.5 \pm 0.2$ mers, see Equation (3.49)]. It is also possible to check Equations (3.44) or (3.43) directly, by experimentally measuring the free energy of binding of a set of sites. Clearly, this is a much more complex undertaking than the numerical test I performed here, but preliminary data collected by Berg and von Hippel (1988) as well as more extensive work on the *Mnt* repressor site (Fields et al. 1997) seem to bear out the predicted linear relationship.

3.4.3 Evolution of binding sites

If the strength of transcription factor binding to the site is directly proportional to the information content of the site, then a simple model of the evolution of binding sites (where the binding energy of the site is being optimized) would be sufficient to show that the information content of the binding site increases during this adaptation. But as mentioned at the end of section

3.4.1, there is no evidence that evolution is trying to increase the binding energy past some threshold, that is, that there is a limit in the amount of information in binding sites that is useful to the organism. Schneider et al. (1986) propose (see also Schneider 2000) that binding sites are selected to carry only enough information to be reliably located by their respective transcription factor (but not more). We can calculate this necessary information as follows.

The number of possible binding sites of size ℓ in a genome of size N is about N, and as a consequence, the uncertainty about the location of the site (without knowing *which* site we are looking for) is $H_{max} = \log_4(N)$ mers. Now, given a *particular* binding site, imagine that there are ν such sites that are necessary to achieve a particular function in the genome. Then, all else being equal (that is, each of these sites is equally accessible and transcription factors have the same concentration everywhere), the conditional uncertainty about the site's location (given we are looking for one of the particular ν sites) is simply $H_{cond} = \log_4 \nu$. The amount of information needed to locate any of those ν sites is then

$$I = H_{max} - H_{cond} = \log_4 \frac{N}{\nu} \text{ mers.} \qquad (3.55)$$

Schneider's hypothesis is that the information content of any particular site, measured as the length of the site minus the sum of the per-site entropies [that is, Eq. (3.49)], must be equal to the information [Eq. (3.55)].

At first sight, such a hypothesis might seem preposterous: what does the information content of the binding site (which reflects the affinity to a particular transcription factor) have to do with the information it takes to locate any particular such site in the genome? It turns out that the biophysics of transcriptional regulation gives us the answer: transcription factors "search" for their target site by first binding unspecifically to DNA and then sliding along the strand in order to locate the site it binds best to (Phillips et al. 2008). It is thus entirely possible that the information content of a binding site is indeed information about where to find it.

Let us first check what this prediction implies for the CRP sites we investigated earlier. The information stored in the CRP site is about 6.5 mers as per Equation (3.49), if we use the forty-eight experimentally verified sites for our ensemble. Instead, we could use a group of sites with approximately the same energy score as the set of sites in DPInteract to estimate the information content of CRP binding sites. The average energy of the seed ensemble is $\lambda E = 7.8$, translating to an information score (see Fig. 3.27) of 8.8 nats, or about 6.35 mers. If we take the latter estimate, Equation (3.55) predicts that there are about $\nu = Ne^{-I \ln 4} \approx 700$ CRP binding sites in the *Escherichia coli* genome, far more than are known experimentally (for I measured in mers). In fact, this is precisely the conclusion reached by Brown and Callan Jr. (2004), who also checked that orthologs of the putative CRP binding sites (discovered by checking for sites with low E) are conserved in *Salmonella typhimurium*, which uses a CRP transcription factor that is essentially identical to that of *E. coli* and has a similar genome length. Thus, we can conclude that perhaps some of the 700 or so predicted sites are functional, but to resolve this question we need more specific position weight matrices that are built from more sites than the ones in this analysis.

Schneider (2000) proposed an evolutionary model to test whether the requirement of correctly binding to all existing binding sites of a particular type in a sequence (thus specifying $\log(N/\nu)$) gives rise to a Shannon information content equal to this information. In this model, the evolution of transcription factor binding was simulated by evolving a position weight matrix that is used to recognize binding sites in genomes that are initially random. In these simulations, the position weight matrix co-evolves with the sequence specifying the transcription factors, and when the information content of the binding sites was evaluated after all the sites had been successfully recognized, it is indeed close to the prediction $\log(N/\nu)$. In other words, the information stored within the binding site represents information on the site's location and identity: just what the transcription factor needs to find the site successfully. While this evolutionary model necessarily abstracts the relation between transcription factor and binding sites, a more careful mathematical analysis essentially corroborates these results (Kim et al. 2003).

3.4.4   Information among DNA binding sites

According to the preceding section, DNA binding sites encode information about the transcription factor it is to bind (and where to find it), as well as the strength of that interaction. But it turns out that sometimes these binding sites encode other information, for example, information about other binding sites. A particularly striking case is made by the regulatory machinery that determines one of the principal axes of development in the *Drosophila* embryo: the dorsal-ventral axis (the axis from "back-to-belly") as opposed to the orthogonal anterior-posterior, and medial (left-right) axes. The dorsal-ventral axis is specified by the maternal concentration of the Dorsal protein, which enters the nucleus of the cells in the syncytial blastoderm at different rates, depending on the maternal concentration. When inside the nucleus, Dorsal activates (or inhibits) the expression of specific genes in specific tissues, so that development of these tissues can begin (see Fig. 3.29).

The Dorsal transcription factor controls the expression of many genes, perhaps as many as one hundred proteins either directly or indirectly (Biemar

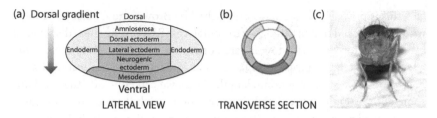

FIGURE 3.29. (a) Dorsal expression regulates tissue differentiation in the dorso-ventral axis (lateral view). (b) Transverse view of tissue differentiation due to Dorsal activity. The mesoderm forms at the fly's ventral side. (c) Frontal view of a fully developed *Drosophila melanogaster*, with ventral side down and dorsal side up. Photo credit: André Karwath, Wikimedia Commons (CC-BY-SA-2.5).

et al. 2006). For that reason, Dorsal is termed a *morphogen*, that is, a substance that guides the pattern of tissue development. While the expression of different proteins under Dorsal control is concentration-dependent, it is clear that other factors must be important so that Dorsal can control multiple different genes with high specificity. One of the transcription factors that Dorsal cooperates with is the transcription factor "Twist," whose binding sites are often found close to Dorsal sites, in particular in the genes active in the neurogenic ectoderm, namely *rho, vn, vnd, brk*, and *sog* (see Table 3.4 for gene names and where they are predominantly expressed). In fact, the concentration-dependence that activates these genes differentially appears to be encoded in the length of the spacer between the Dorsal and Twist binding sites (Crocker et al. 2010). But how can the Dorsal transcription factor know about the presence of Twist, anywhere between just a few base-pairs (bps) and 30 bps away (or even further)? We can hypothesize that a Dorsal binding site that is in the vicinity of a Twist binding site is actually *different* from a Dorsal binding site by itself, and that this difference encodes information about the proximity of Twist into the DNA binding motif itself.

We can test this hypothesis directly, by aligning cooperative (meaning those with a proximal Twist site) and noncooperative (with a distal Twist site) Dorsal sites independently and asking whether these binding motifs are different; whether the cooperative sites predict the presence of Twist sites, and whether a "Dorsal detector" using separate conditional (conditioned on the presence of Twist) and unconditional position weight matrices does better than a Dorsal detector using the combined position weight matrix. In the following, we'll see that the answer to all of these questions is yes.

To perform this test, we need to first create multiple-sequence alignments for the Dorsal transcription factor, either in proximity to a Twist site

Table 3.4. Genes (and their abbreviation) under Dorsal control and expressed in dorsal-ventral (DV) polarity patterning. The region of expression is approximate: most genes are also expressed in adjacent regions.

Gene name	Abbreviation	Expression
twist	twi	mesoderm
rhomboid	rho	neural ectoderm
vein	vn	neural ectoderm
ventral neurons defective	vnd	neural ectoderm
brinker	brk	ventro-lateral
short gastrulation	sog	ventro-lateral
decapentaplegic	dpp	dorsal/lateral ectoderm
tolloid	tld	dorsal ectoderm
zerknüllt	zen	amnioserosa

(a proximal site) or not (a distal site). Here we take proximal to mean "within a window of 30bp in either 5' or 3' direction" of the Dorsal site, that is, 30 bp upstream or downstream (Clifford and Adami 2015). First, we downloaded and aligned Dorsal sites that activate nine genes in the DV (dorsal-ventral) patterning system: the aforementioned *rho, vn, vnd, brk,* and *sog* (active at high to medium Dorsal concentrations), as well as *twi, zen, dpp,* and *tld* (see Table 3.4). We did this not just for the binding sites found in *Drosophila melanogaster*, but also for their ortholog sites in eleven related *Drosophila* species. Specific care must be taken to have a flexible alignment of the conditional and unconditional sites separately, and also to allow for variations in the Twist binding site. These details are described in Clifford and Adami (2015), but we skip them here. Figure 3.30 shows the DNA sequence logos of length 9 bps that we obtained for the conditional (that is, cooperative with Twist), unconditional, and combined PWMs.[10] The conditional and unconditional sites (Fig. 3.30[a] and [b], respectively) are markedly different visually already, and a statistical test confirms it. The unconditional and combined PWMs do look similar, however, due to the fact that the combined ensemble has more unconditional sites in it. The sites also differ by the total information content (obtained by adding the height of all the letters in the logos). The conditional sites have a total information content of 13.5 bits, the unconditional sites contain 9.1 bits, while the combined set harbors 9.6 bits of information. This

10. The effect of extending binding sites to include flanking regions up to 3 bps on either side can be studied (Clifford and Adami 2015). Doing so does not affect the conclusions.

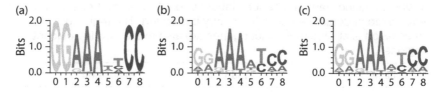

FIGURE 3.30. Sequence logos for position weight matrices (PWMs) obtained using (a) the conditional set of binding sites, (b) the unconditional set, and (c) the combined set. In these sequence logos, the height of a letter indicates how much of the total information at that site is comprised of that nucleotide. Sites with small letters (such as site 5) carry little information and mostly entropy. From Clifford and Adami (2015).

already confirms to us that the conditional sites carry additional information, but what is this information about?

We can test this by calculating the information shared between sequence and class label of a site. The class label of a site is determined by scanning the genome for the presence of a Twist site within a 30-bp window. For each Dorsal sequence $S = s_i$, if a Dorsal sequence is within the 30-bp window, then the class random variable is set to $C = p$ (proximal), otherwise it is set to $C = d$ (distal). We can then calculate the information

$$I(S:C) = \sum_{s_i} \sum_{c,d} p(S = s_i|C)p(C) \log_2 \frac{p(S = s_i|C)}{p(s_i)}. \qquad (3.56)$$

Here, $p(S = s_i|C)$ are the conditional PWMs, namely the conditional $p(S = s_i|C = c)$ and the unconditional $p(S = s_i|C = d)$. We can do this same test for a window 60 bp away and 90 bp away. We expect very little information about a potential Twist site being encoded as far away as 90 bp, and indeed this is what we find. Table 3.5 confirms that the conditional Dorsal targets encode about 0.5 bits of information about Twist sites within 30 bps, about 0.3 bits about Twist sites between 30 bps and 60 bps away, and virtually no information about such sites between 60 bps and 90 bps away. So indeed: Dorsal sites close to Twist sites know about Twist's presence! Now let us try to answer the final question. We have previously argued that PWMs are often notoriously unreliable predictors of functional binding sites because they predict many more active sites than can reasonably be expected, and many predicted sites turn out not to be functional. We have also speculated that one of the reasons behind such poor predictability could be that several sites that are actually distinct are mixed into the same PWM, thus reducing the specificity (this was certainly clear when the reverse complement of experimentally verified

Table 3.5. Information between functional Dorsal binding site sequences and putative Twist sites that match the Twist motif 5′-CAYATG (Y stands for "C or T") using a sliding spacer window scheme. From Clifford and Adami (2015).

Spacer	[0,30] bp	[31,60] bp	[61,90] bp
Information (bits)	0.49	0.29	0.04

binding sites was added to the PWM of CRP proteins; see Table 3.3). Since this appears to be precisely the case here (mixing the conditional and unconditional sites together into a combined data set), we can test whether a "Dorsal detector" based on *separate* sets (constructed from conditional or unconditional sites alone) performs better in predicting Dorsal sites than one based on the combined set. In particular, we can make the "condition-specific" detector such that it will predict a Dorsal site if either the conditional PWM or the unconditional PWM predicts a Dorsal site at that particular cutoff energy E_c (recall that, as in the prediction of CRP binding sites in *E. coli* [see Table 3.3], PWMs make binding site predictions given a particular cutoff energy). We call this Dorsal predictor the "OR-gate" predictor. We can then compare this prediction to one made by the "combined" (CB) predictor that uses the combined data set.

We can measure predictor performance in the following manner. Every detector is a channel that has a 9-bp sequence $S = s_i$ as input, and an output of a predictor variable P that takes the value $P = 1$ if a Dorsal site is predicted, and $P = 0$ otherwise. The performance of the detector is given by the shared entropy (information) between the true identity of a site ($T = 1$ for an annotated Dorsal site, $T = 0$ for a non-Dorsal site) and the predictor variable. To measure predictor performance, we test the predictor with a set of candidate 9-mers, and ask it to classify them as Dorsal or not, given the energy cutoff. In that case, a probability $p(P = 1|T = 1)$ is the *true positive rate* (*TPR*), while $p(P = 0, T = 0)$ is the *true negative rate* (*TNR*), and so on. If p is the likelihood of a sequence being a Dorsal site (this is the Bayesian prior and depends on the input data set) while q is the likelihood that the predictor fires ("decides the site is Dorsal"), then the information $I(T{:}P)$ that predicts whether the Dorsal detector "fires" for a particular site given whether the site is actually a Dorsal binding site can be written as (see Exercise **3.3**)

$$I(T : P) = H[p] - qH[TPR] - (1 - q)H[FPR]. \tag{3.57}$$

This information is shown in Figure 3.31(a) for the "OR-gate" detector in black and the CB (combined) detector in gray, with a Bayesian prior $p = 1/2$.

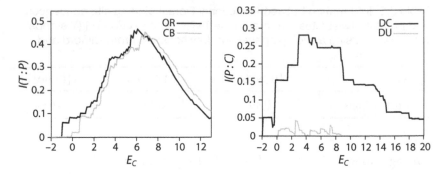

FIGURE 3.31. (a) Information about the true identity (random variable T) in the predictor (random variable P) of a 9-bp DNA binding site, for two different predictors: the "OR-gate" detector and the "combined" (CB) predictor. The OR-gate predictor has more information about the true identity for specific sites (small E_c), while the reverse is true for unspecific sites. (b): Information that a predictor has about whether a Twist site is nearby. A predictor based on the conditional PWM (DC, black line) has up to 0.3 bits of information about Twist proximity, while the DU PWM (light gray), as expected, carries almost no information about the class variable (since it is constructed from sites that have both close and distal Twist sites). Clifford and Adami (2015).

It is clear that for small cutoff energies, the OR-gate detector does better than the combined detector, that is, the detector based on conditional PWMs predicts Dorsal sites with higher accuracy than the average (combined) detector.

We can also ask whether this better performance is due to the information that the conditional PWM has about the Twist site, by calculating the information that the respective PWM has about the class variable C, $I(P:C)$ (again, $C = p$ for proximal Twist sites, and $C = d$ for distal Twist sites, defined using a 30-bp window). Figure 3.31(b) shows that if the cutoff energy is too low or too high, neither detector can distinguish those classes of sites very well. But for an intermediate energy cutoff, the conditional PWM (black line) has up to 0.3 bits of information about the proximity of Twist, while the unconditional PWM (light gray line) has virtually none.

This analysis of transcription factor binding sites that cooperate in regulating transcription (the Dorsal and Twist sites) demonstrates how important a good understanding of the biophysics of transcription is when constructing position weight matrices. Because the Dorsal and Twist transcription factors interact when both binding to sites that are within 30 bps of each other, the binding site of Dorsal depends on whether or not a Twist site is nearby. As we have seen, the difference between those sites is *predictive* about the proximity of a Twist site. It is likely that the Twist site also carries information about the proximity of the Dorsal site, but we did not investigate this question in this study. Indeed, we can argue that if we were to distinguish proximal and distal Twist sites, the information in Dorsal sites about proximal Twist sites would increase, as the PWM for Twist becomes more specific by excluding Twist sites that are not proximal.

3.5 Summary

While the principal laws of evolution—inheritance, variation, and selection—are simple enough, the interaction of these laws with a world that itself becomes more and more complicated with each adaptive event creates an intricate system that seems to defy analysis. A model of adaptation in terms of a measurement process, although a caricature of the complicated and messy dynamics that are actually going on, can give us some guidance about general trends that we can expect in evolution, such as an increase (on average) of the information content of the biosphere. In this picture, evolution is seen as a "natural Maxwell demon" that allows information to enter the genome but not to leave. This demon is not perfect, and sometimes fails altogether, but only for short periods of time. All in all, he appears to do a fabulous job! But to follow the amount of accumulated information through evolutionary history even for a single gene is a difficult job, and we find that many factors can impede an accurate estimate of the information content. Still, changes in information content at major branches in evolutionary history may be measurable. If we do so, we can see both losses and gains of information as taxa are adapting to different niches, as the law of increasing information only holds within any particular niche.

The law of increasing information is also violated when environments can change, and this effect can be documented in detail in the evolution of drug resistance in HIV-1 antiviral therapy with protease inhibitors. Introducing the drug changes the virus's world dramatically, and the virus struggles to regain some of the information it lost. But often enough it learns about its new environment quickly via chance mutations, as the virus's mutation rate is very high. These resistance mutations, as beneficial as they are to the virus, are not at all welcome to the host, who would prefer if the virus had no "memory" at all.

Finally, much of the information is not stored in our classical genes defined as open reading frames, but rather in regulatory regions that are untranscribed and untranslated, but crucial to the functioning of the organism. The information content of DNA binding sites can, however, be measured by the same technique, and moreover the information about the transcription factor is directly proportional to the binding strength (in terms of the free energy change) of the sequence. In turn, this information can be seen as information that is necessary to locate the binding site within the genome. Thus, we see again that information is a symmetric quantity: what the transcription factor knows about the binding site is equal to what the binding site knows about the transcription factor. Besides information about the transcription factor itself, a binding site can also store information about the presence of other binding sites that the particular transcription factor may interact cooperatively with. The example of the Dorsal and Twist binding site interaction in early development of the fly taught us another fundamental and important lesson in information biology: sometimes sequences contain information that is difficult for us to read or extract, or even to recognize as such. Indeed, in the absence of the clue that a Dorsal binding site may be different depending on the presence of the Twist site, it would have been computationally intractable to divide the combined set of Dorsal binding sites into the conditional and unconditional set, which made the inherent information in the conditional site apparent. In a sense, this information is encrypted in the site, and only knowing the key (the presence or absence of Twist) allowed us to unlock that information. We can expect that a significant amount of information is encrypted in this manner in biology, and we may not always be so lucky that a bioinformatic analysis (and biophysical intuition) hands us the key.

Exercises

3.1 Given an ensemble of sequences s_i that appear with probabilities p_i and energies ϵ_i, show that the constraint

$$\langle E \rangle = \sum_i p_i \epsilon_i, \tag{3.58}$$

together with the requirement that the entropy $H[p] = -\sum_i p_i \log p_i$ is maximal, leads to the probability distribution (3.28). Hint: introduce Lagrange multipliers λ and μ in a Lagrange function

$$L = H[p] + \lambda \left(\langle E \rangle - \sum_i p_i \epsilon_i \right) + \mu \left(\sum_i p_i - 1 \right). \tag{3.59}$$

3.2 The contribution to the binding strength of a sequence s of length ℓ from the nth base $s_i^{(n)}$ is

$$K(s_i^{(n)}) = K(s_0^{(n)})e^{-\epsilon_i^{(n)}} = K(s_0^{(n)}) \left(\frac{p_i^{(n)}}{p_0^{(n)}} \right)^{1/\lambda}. \tag{3.60}$$

Show by using the Taylor expansion

$$\ln K(s^{(n)}, \lambda = 1) = \ln K(s^{(n)}, \lambda) + (\lambda - 1)\frac{\frac{d}{d\lambda}K(s^{(n)}, \lambda)}{K(s^{(n)}, \lambda)} + \mathcal{O}(\lambda - 1)^2$$

(3.61)

that to linear order in $\lambda - 1$

$$\sum_n \ln K(s^{(n)}, \lambda = 1) \approx \ln K(s_0) + (1 - 2\lambda)E(s) \qquad (3.62)$$

and

$$\sum_n \sum_i \ln K(s^{(n)}, \lambda = 1) \approx \ln K(s_0) - \sum_n \ln p_0^{(n)} + (1 - \lambda)\langle E \rangle.$$

(3.63)

Show further that these equations imply Equation (3.43).

3.3 We can define *true positive rates* (TPR) and *false positive rates* (FPR) in terms of a conditional probability for the prediction variable P and the "true" variable as follows

$$p(P = 1 | T = 1) = TPR,$$
$$p(P = 0 | T = 1) = FNR = 1 - TPR,$$
$$p(P = 1 | T = 0) = FPR,$$
$$p(P = 0 | T = 0) = TNR = 1 - FPR. \qquad (3.64)$$

Show that the information $I(T : P)$ can be written as

$$I(T{:}P) = H[p] - qH[TPR] - (1 - q)H[FPR], \qquad (3.65)$$

where $H[x]$ is the binary logarithm function

$$H[x] = -x \log_2 x - (1 - x) \log_2 (1 - x) \qquad (3.66)$$

and p is the probability that any site in the ensemble is a true Dorsal site, while q is the probability that the Dorsal detector fires, that is, $H(T) = H[p]$ and $H(P) = H[q]$.

4

Experiments in Evolution

The observations I refer to were made with a view to discovering whether it was possible by change of environment (...) to superinduce changes of an adaptive character, if the observations extended over a sufficiently long period.

—REV. DR. WILLIAM H. DALLINGER, F.R.S. (1887)

Most scientists are well aware of the classical paradigm of science: the interplay of experimental evidence and theoretical hypothesis that paves the path from mystery toward knowledge. We also know that this path is not always a straight one, but rather leads us via sometimes lengthy detours stumbling—more than running—toward a greater understanding of the physical phenomena around us. This paradigm of scientific investigation is most familiar from physics, where increasing experimental technology and mathematical sophistication have built an unparalleled edifice. In biology, however, this interaction between theory and experiment is largely absent, because it seems unlikely that we will ever have a coherent structure or framework that we could call a "theory of the cell or organism" (but see Bialek 2018 for an optimistic view). The reason for this is that a large fraction of the features of biology are the products of a historical development (namely evolution) whereas in physics the role of history is absent except in some rare cases. But it turns out that the very process that prevents a theoretical description of biology is *itself* amenable to a theoretical treatment, allowing for the classical paradigm to apply to the study of evolution, as long as experiments can be designed to test such a theory. But what is an "experiment in evolution"? Should we not be able to learn about evolution just from observing it unfold before our eyes?

Of course, this is exactly how Darwin learned about evolution: by observing. But if you are limited only to observation, it is difficult to test specific hypotheses, which is generally the way science advances. When we test hypotheses (say, "trait X can evolve only when the environment contains element A"), we need to construct an environment that has element A present, as well as one where element A is absent, and then let evolution proceed forward in both. We then have to run the evolution experiment for long enough that we have a reasonable expectation that trait X can evolve if A is present. Naturally, some progress can be made about such hypotheses without a dedicated experiment, if we can identify environments on Earth that are quite similar but only differ in the presence of element A, and indeed Darwin used observations he collected precisely in this manner. But clearly, some luck must be involved in this, and because of that the experimental approach is preferable by far.

Speaking of luck, sometimes experiments in evolution start inadvertently. In 1958 and 1959, a team of scientists intended to study whether one species of lizard that was resident on a small island in the Adriatic Sea would be able to outcompete another similar species that was resident on an adjacent island. "Who can replace who?" is a question often asked in biogeography, and the principle being tested is often called the competitive exclusion principle, which states that in a fixed niche, only a single species can reside. But it is not clear what decides which species will emerge victorious when two fairly similar species are made to compete in the same niche. This is what a team of ecologists sought to test by transplanting an invading species on islands where a different species was resident (they selected three such islands). When revisiting those islands in 1971, they found that in two of the islands, the invading species had completely disappeared, while on the third island the invading species appeared to be replacing the native form (Nevo et al. 1972). This, it turns out, was the planned part of the experiment. The unplanned part came next. Naturally, the team asked themselves what would happen if they took the resident species that resisted invasion and transplanted it to the island of the invader; the "reciprocal" invasion experiment, so to speak. In 1971, the team transported five adult breeding pairs of the Italian wall lizard *Podarcis sicula* to the tiny island of Pod Mrčaru (0.03 km^2) from the neighboring island Pod Kopište (0.09 km^2) where *Podarcis sicula* was native (see Fig. 4.1).

At the time, the species resident on Pod Mrčaru was the related lizard species *Podarcis melisellensis*. While the scientists likely planned to return about a dozen years later to study whether the invaders were extinct or not, the Croatian War of Independence erupted and thwarted those plans. Unwittingly, the lizards on Pod Mrčaru were left to their own devices. After the war ceased, another team of scientists visited the island (first in 2004 and several times after) and went looking for the resident *Podarcis melisellensis*, but it was nowhere to be found. Instead, a different species of lizard was everywhere, but it did not look like the species that was placed there thirty-three years earlier. For one, the head was bigger and the jaws were stronger. Moreover, the

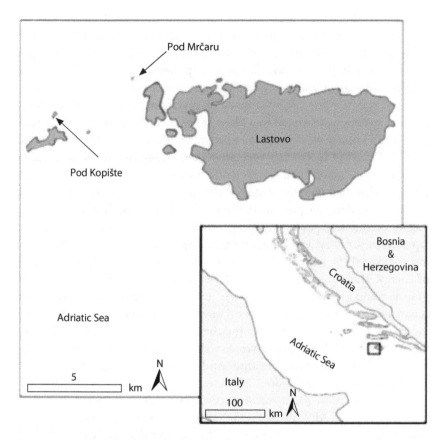

FIGURE 4.1. The islands Pod Mrčaru and Pod Kopište in the Adriatic Sea. Adapted from Wehrle et al. (2020).

new resident species (there were over 5,000 individuals) was found to have an anatomical difference with the introduced *Podarcis sicula*: it had a cecal valve. Cecal valves are muscles that separate the intestine into compartments, and are particularly useful for digesting plant matter that must stay in the intestines for extended amounts of time. The introduced *Podarcis sicula* did not have one, but the lizard they found on Pod Mrčaru did. The team turned to what any team would have turned to in the age of DNA sequencing: they obtained the species' DNA and compared it to the DNA of the types on Kopište where *Podarcis sicula* was still thriving. To their astonishment, they found an exact match (Herrel et al. 2008), suggesting that the species they encountered on Mrčaru was indeed the one that was introduced thirty-three years before, and had evolved rapidly into the form they found—including the evolution of a brand-new organ.

But what drove these adaptations? And if the introduced species was not well adapted to the island's conditions, how were they able to displace the resident type? It turns out that the introduced *Podarcis sicula* was more aggressive than the resident *Podarcis melisellensis*, perhaps because it was used to a diet of insects, and was constantly engaged in territorial defense. Pod Mrčaru, however, had very few insects for the invading species to dine on, and indeed the resident species was adapted to a plant-based diet. Once the aggressor took over (and of course we do not have a record of how precisely this happened), it needed to adapt to a mostly vegetarian diet. Strong jaws to chew on leaves are one of the "musts" in this new world, but the other problem, for "meat-eaters," is that their digestive system is not suited to such a cellulose-rich diet. To digest cellulose, the plant matter has to be exposed to a broth that usually includes cellulose-digesting microorganisms. But this process takes time, so a rapid conveyor ferrying food matter from input to output might do well for insect-based diets, but not when grasses and leaves provide the only source of energy. Many animals in the past have, of course, adapted to a cellulose-only diet, and inevitably this consists of evolving an expanded digestive tract, separated into compartments by cecal valves. Which is exactly how the invading *P. sicula* responded, along with the changes to head morphology allowing for more efficient chewing of leaves, as well as changes in social behavior that resulted in a less aggressive posture. And all of these changes happened in about thirty years of evolution.

It is not surprising that an organism can adapt to local conditions to increase their "fit." After all, this is the tale that Darwin found everywhere he visited, once he knew what to look for. What is astonishing is the pace at which these changes happened. Darwin interpreted his observations to imply that the changes that shaped species are slow and imperceptible. And indeed this is probably true in most cases, but it cannot be elevated to a rule. When conditions are dire, change can happen quickly, as it did on Pod Mrčaru, and as it may have happened on countless other, less well documented, occasions. Even significant changes in anatomy can occur in a few decades, which in this particular case represents perhaps about thirty generations. It is therefore *this* insight that makes the case for experimental evolution: evolution is *not* intrinsically slow. It is mostly slow because in many niches the resident organism is close to perfectly adapted, and most changes are detrimental. But when conditions change, adaptation will be rapid and observable.

Looking back, the case for experimental evolution is an obvious one, but resistance to the idea that evolution can be observed experimentally was commonplace not too long ago (Garland, Jr. and Rose 2009). Given such a hostile environment for the concept of experimental evolution, it is perhaps shocking that not only was the idea of experimental evolution entertained during

Darwin's times, but the first long-term experiment testing evolution was conducted while Darwin was still alive, and Darwin even corresponded with the experimenter!

4.1   The Dallinger Experiment

Reverend William Dallinger (1839–1909 Fig. 4.2) was a Methodist preacher and microscopist, who specialized in documenting the life history of a group of microorganisms (amoebae) that he had isolated from rotting food (Haas Jr. 2000b). Around the time Dallinger perfected his skills in microscopy, the concept of the "spontaneous generation" of life from nonlife was a fairly commonly held position (see, e.g., Bastian 1872), but Dallinger instead sought to document that what appeared to be spontaneous generation was really just the hatching of microorganisms from very hardy spores. Together with his colleague, the physician John Drysdale (1817–1892), he observed specific groups of flagellated algae (amoebae now commonly known as *flagellates*) under the microscope, painstakingly, through their life cycle of fission, fusion, and spore production. To make sure that they did not miss any development

FIGURE 4.2. Portrait of William Henry Dallinger. Oil painting by Edgar Thomas, ca. 1884. Wellcome Collection (CC BY 4.0).

in the amoeba's life cycle, Dallinger and Drysdale manned the microscope for twenty-four hours a day, taking turns, sometimes up to two weeks straight. The series of papers they produced was noticed by Darwin himself, who in correspondence encouraged Dallinger, writing that he had "read all your and Dr. Drysdale's papers, & they seem to me to possess higher value than anything that has been published on such subjects."[1] Dallinger, in turn, was inspired by Darwin's statement that the "lowest and least visible organisms could be used to demonstrate the manner in which living creatures adapt to changed circumstances and produce what are called new species" (Haas Jr. 2000a).

Dallinger then set out to prove Darwin right by using precisely those organisms that he had studied for so long, by showing that they evolved to adapt to changed circumstance. His plan was to expose them to an increased temperature that they were ordinarily unable to withstand, by making slow and gradual changes in their environment. To do this, he needed to create an environment where the populations (he planned to study three independent populations in parallel), as well as temperature, could be precisely controlled. Furthermore, the experiment had to be set up in such a way that the microbes could be examined and studied without interfering with the experiment. After considerable planning and experimentation, Dallinger designed an incubator that would allow him to study microbial populations in controlled conditions for an extended period of time.

The incubator depicted in Figure 4.3 was, according to his own report (Dallinger 1887), delivered early in 1879. Dallinger actually wrote Darwin about his preliminary experimental results at the end of June 1878 (thus before he even took possession of the device that was to house his long-term experiment), suggesting to Darwin that his experiment would "palpably demonstrate your great doctrine."[2] Darwin responded a few days later, writing back

> I did not know that you were attending to the mutation of the lower organisms under changed conditions of life; and your results, I have no doubt, will be extremely curious and valuable. The fact which you mention about their being adapted to certain temperatures, but becoming gradually accustomed to much higher ones, is very remarkable. It explains the existence of algae in hot springs.[3]

The exact nature of the preparatory experiments Dallinger carried out are unknown to us, but in the end he focused on three flagellates that he had

1. Letter to W. H. Dallinger, after January 10, 1876.
2. Letter to C. Darwin, June 28, 1878.
3. Letter to W. H. Dallinger, July 2, 1878.

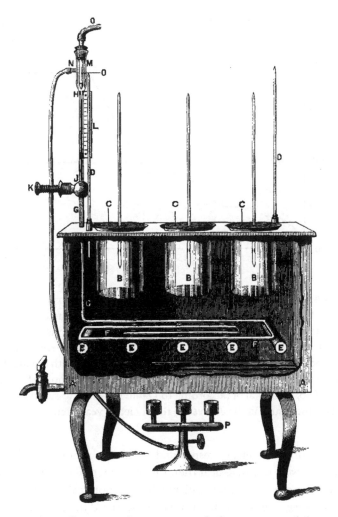

FIGURE 4.3. The Dallinger incubator consisted of a copper vessel that contained water at a controlled temperature and three completely isolated chambers holding the nutrient media as well as the amoeba. Thermometers are housed in the chambers as well as the vessel itself. A burner is integrated with the system that allows for the accurate (to half a degree) homeostatic regulation of temperature via a mercury column (a thermostat). From Dallinger (1887).

studied previously in detail: *Tetramitus rostratus, Monas dallingeri,* and *Dallingeria drysdali.* Knowing every aspect of their physiology and life cycle in detail, he was perfectly poised to detect any changes if they appeared. He began the experiment at a temperature of 60° Fahrenheit (F) (about 15° Celsius), and raised the temperature slowly to 70° F over a period of four months.

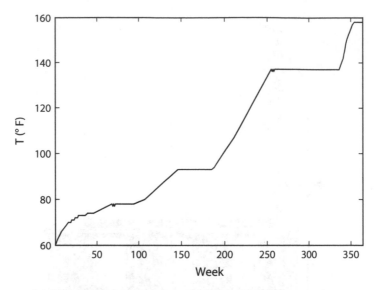

FIGURE 4.4. Estimated temperature profile of the Dallinger experiment as a function of time (in weeks). The temperature profile is extracted from Dallinger's narrative in (Dallinger 1887) and is approximative as Dallinger mixed days, weeks, and months as time intervals. The experiment likely began when the incubator was delivered (early 1879), and ended with the incubator's destruction shortly before Dallinger's address to the Microscopical Society in February 1886, for a total of (according to Dallinger) close to seven years. He reported on the results in the address to the society a year later.

Noting that all his flagellates responded by dividing more quickly, he realized that they were nowhere near their maximum temperature, and therefore increased the temperature more quickly. But he soon realized that there were boundaries. Over the course of seven years, he raised the temperature gradually but found that at certain points no increase was possible without killing the entire stock (see Fig. 4.4 for the temperature profile of the experiment gleaned from a reading of Dallinger's 1887 narrative). For example, for periods up to 18 months, he was unable to raise the temperature but found that suddenly they would adapt and then allow a fairly fast increase, only to find a new upper limit. During the whole time, he would examine the organisms to detect any changes in their life cycle or their physiology. Twice during the experiment, Dallinger alternated between temperatures (possibly daily), indicated in Figure 4.4 by wiggly lines. This treatment occurred when any advance in temperature seemed impossible.

While Dallinger saw some significant changes in the amoeba's cell bodies, in particular the formation of small and large vacuoles (cavities) within the cell body when the organisms were particularly stressed, these changes reversed themselves when the organisms adapted to the raised temperature. Remarkably, all three species appeared to react to the temperature challenge in a very similar manner: when one type suffered, so did the other two. When one recovered and its vacuoles disappeared, the same happened to the others. Because of this synchrony, in hindsight it is not clear whether all the adaptation observed in the experiment was genetic in nature. But according to Haas (2000b), the three species Dallinger selected had life cycles that averaged four minutes per generation (360 generations per day), allowing for well over 500,000 generations over the period of the entire experiment, giving ample time for genetic changes to establish themselves. However, as flagellates produce many spores per generation, this number is misleading when comparing to the doubling time of a bacterium, say, which is typically higher than twenty minutes even under ideal conditions, and significantly longer than that in the wild.

Dallinger also tested whether his adapted forms could survive in their native environments of 60°F, but found that they instantly perished. The physiological changes he observed thus were very profound and likely had at least some genetic basis. We can surmise that Dallinger intended to reach much higher temperatures, such as those that the microorganisms in the hot springs mentioned by Darwin tolerate (they are known to withstand temperatures in excess of 185°F). Unfortunately, Dallinger's experiment was interrupted by an accident that resulted in the complete destruction of the incubator, which ended the experiment at 158°F. Dallinger was severely shaken by the untimely end of his experiment, writing with typical British understatement (Dallinger 1887):

> Here, with such pain as I presume is natural, I have to close the story. The accident happened, destroying the use of the instrument, and causing the whole to collapse.

Even though he wrote in his report to the Microscopical Society that he had restarted his experiment, nothing would be heard from it again, and the report from 1887 turned out to be his last scientific publication on evolution.

4.1.1   Design requirements for experiments in evolution

The Dallinger experiment was unique in many ways: it clearly took place long before it could be appreciated, and even Darwin did not (it appears) fully recognize its importance. But it is also unique in that when Dallinger designed the experiment, he considered many of the variables that more modern instantiations of evolutionary experiments have also considered very carefully.

1. **Use a well-described organism.** For his experiment, Dallinger chose organisms that were so well studied that he would notice any changes that occurred in them. Indeed, two of the organisms were discovered and described by him and Drysdale. For the same reason, evolution experiments are usually carried out on "model organisms," which are organisms that are widely studied in many laboratories, and for which a large number of well-established results exist. The thinking behind focusing on well-studied organisms is not difficult to discern: when changes occur in response to adaptation to a novel environment, these changes must be evaluated in the light of the unchanged organism, and the more is known about this "wild-type," the easier and the more accurate the characterization of changes. In modern times it is essential that the organism has a well-annotated genetic sequence, and that the genomes of evolved specimens can be compared to the ancestral sequence. Naturally, this was not a concern for Dallinger.

2. **Use an organism that replicates fast.** The faster an organism replicates, the more generations can be studied in a given time interval. Obviously, longer experiments that accumulate more generations and therefore have the capacity to display more evolutionary changes are preferable to shorter experiments. For this reason, microbes are the organism of choice for the majority of evolution experiments today (but there are exceptions). Dallinger studied flagellates, which are not model organisms today, but as Dallinger was by all accounts the world expert on this organism at this time, it was an obvious and appropriate choice. Flagellates can grow very rapidly, so that as many generations as possible could be studied in the duration of the experiment. The 360 generations per day that is usually quoted (Haas Jr. 2000a), however, is not the number of generations that evolutionary biologists would count, as it should be divided by the number of offspring generated at each burst. In practice, the growth rate of an organism is limited by the rate in which nutrients (the "growth medium") can be provided, and what carbon source is available in the medium. As Dallinger had been growing his flagellates for decades, he was able to choose a medium that led to rapid growth, and his incubator was designed so that medium could easily flow into the chambers, while waste products were flowing out.

3. **Use a stable and controllable environment.** Evolution experiments usually test the mode and manner in which the organism adapts to its environment, so this environment should be well understood, stable, and controllable. For microorganisms, this environment usually includes the growth medium, the manner in which the organism interacts with the medium (liquid broth, solid surface, or structured environment), as well as external parameters such

as temperature. For higher organisms such as flies, keeping a constant and controllable environment is a much more complex task. As the main degree of freedom for Dallinger was temperature, accurate temperature control in the incubator was essential. Using an ingenious feedback control system, Dallinger was able to control temperature accurately to the level of half a degree Fahrenheit, over the entire length of the experiment and across the entire range of temperatures he tested. Evidently, however, the system was less stable than he had anticipated, as it underwent "rapid unscheduled disassembly" after seven years.

4. **Ensure continuous observation of the organisms without interfering with the experiment.** To monitor the progress of the evolution experiment, specimens must be observable without interfering with the experiment. This design requirement is perhaps less obvious than some of the others at the outset, but it is no less important. In the case of the Dallinger experiment, this requirement was particularly significant, as the organisms were being adapted to high temperature. Because after several years of evolution the adapted organisms could no longer survive at room temperature, Dallinger designed a heated microscopy stage to observe the flagellates at their current temperature. At the same time, the system was designed in such a way that organisms could be extracted without affecting the temperature of the culture.

5. **Run replicate experiments.** The fate of evolving organisms is impacted by three elements: adaptation, chance, and history (Travisano et al. 1995). To unravel the relative contribution of each of these elements, it is necessary to run several (ideally many) replicate populations in parallel. While this requirement follows from basic statistical principles that were not known to Dallinger at the time, he nevertheless intuitively designed his experiment in triplicate. The three species he chose to investigate were closely related, and thus he could expect that they would respond in a similar manner. He may also have realized that it would be unwise to use three populations of the same species, as it would be impossible to guarantee that the three populations evolve independently. Indeed, this is another design requirement for running replicates: ensure that the replicate lines can be identified unambiguously to guarantee that one population does not contaminate another. By choosing three different species that were identifiable visually under the microscope, Dallinger would always be able to detect any cross-contamination between his replicates.

6. **Keep a "fossil record" of the line of descent.** Darwin pieced together the story of evolution by observing life as it is today, noting common descent and adaptations wherever he went to study it. He was also acutely aware that it would have been much easier to unravel the story of evolution if an unbroken line of descent was preserved somehow, allowing us to peruse adaptation

following adaptation, from a simple ancestor to a complex adapted form. Because of the haphazard nature of the fossil record, such unbroken lines are not available to us (but there are some fairly complete reconstructions of lines of descent, such as the line covering fifty million years of horse evolution, see e.g., MacFadden 1992). If you conduct an evolution experiment, however, it would behoove you to preserve this line to the fullest extent possible, by storing away intermediates as often as possible. Of course, this is not possible for all model organisms (yet even fruit fly embryos can now be preserved in the freezer; Mazur et al. 1992), but some microorganisms are very hardy and can be maintained at minus 80°C indefinitely. Dallinger would probably have given away all of his possessions to be able to continue his experiment from a stored frozen intermediate after the explosion, but alas, preserving a fossil record of his experiment had not occurred to him. Future experimental evolutionists, however, would heed Darwin's call.

4.2 The Lenski Experiment

On February 24, 1988, a young assistant professor of evolutionary biology at the University of California at Irvine filled twelve Erlenmeyer flasks with a sugar-rich medium and inoculated each with *E. coli* bacteria, the kind of bacteria that reside in people's guts (and other mammalian guts across the planet). Dr. Richard Lenski hoped that living from here on out in the flask rather than in guts would teach those bacteria some new tricks, and that he could watch as they adapted to this new lifestyle. In other words, he started a long-term evolution experiment. This experiment is still ongoing today, and after (as of 2022) over 75,000 generations of evolution and more than thirty-four years have passed, it is the longest-running microbial evolution experiment in history, over four times longer than Dallinger's experiment.[4]

Lenski, like Dallinger before him, thought long and hard about experimental design (but was unaware of the Dallinger experiment when he designed it).[5] Like Dallinger, Lenski performed preliminary experiments—including one aborted nine days earlier—to test various aspects of the daily routine of keeping an experiment like this ongoing for an extended period of

4. It is not the longest running evolution experiment: that honor goes to the Illinois Long-Term Selection Experiment, in which researchers are selecting corn with either high or low grain protein content, as well as high or low oil content (Moose et al. 1996), and plant the best/worst performers. But because corn produces only one generation per year, progress is much slower even though the experiment began in 1896 and has now accumulated over 120 generations. Because this experiment is subject to changing weather as well as farming technology, it is not possible to keep the environment variable constant.

5. R. E. Lenski, personal communication.

time. In designing what is now known as the Long-Term Evolution Experiment (LTEE) (or simply the "Lenski experiment"), Lenski checked off the list of design requirements described above in the following manner:

1. **Use a well-described organism.** The commensal bacterium *Escherichia coli* is one of the most well-studied organisms of all time. Not only is it ubiquitous because its natural habitat is the mammalian gut, it is also the workhorse of countless molecular biology studies. Its genome is extremely well known, and so is its life cycle. The strain that Lenski used as the ancestral type was a particular laboratory strain derived from the "B strain," which itself has a long and illustrious heritage (Daegelen et al. 2009). Lenski selected it not just because of its well-known lineage, but because it did not harbor any *plasmids* (short circular DNA sequences that code for genes that can be exchanged between bacteria), nor any viruses that can similarly move DNA between cells. This choice ensured that genetic information could only be transferred "vertically" (meaning from ancestor to daughter), as opposed to "horizontally," meaning between individuals by sharing plasmids. The reason for this precaution is that horizontal gene transfer makes evolution more complicated and messy, and Lenski wanted to observe evolution in a system that is as simple as possible. The strain Lenski used to start the experiment (the strain was designated "REL606") had other "built-in" advantages that made it an ideal progenitor, and we will discuss some of these. Because of the importance it acquired as the parent strain of the LTEE, the provenance and genetic makeup of REL606 has been ascertained and documented as perhaps no other *E. coli* strain before (Studier et al. 2009; Daegelen et al. 2009; Jeong et al. 2009).

2. **Use an organism that replicates fast.** Under ideal conditions, *E. coli* divides in about twenty minutes, but this alone does not predict how many generations can be achieved per day. If an organism doubles in twenty minutes, there will be eight in an hour (three doubling periods, giving rise to $2 \times 2 \times 2 = 2^3$ cells). In ten hours there are thus thirty doubling periods, leading to a total of 2^{30} cells. That is just over a billion, and there simply is not enough food in one of the Erlenmeyer flasks to feed that many bacteria. Thus, the number of bacteria will increase exponentially within the first few hours, and then level off. In fact, this is how microbial populations behave in general when faced with a limited food supply (see Fig. 4.5). When exposed to food they can consume, the bacteria first "get ready" (synthesizing DNA, enzymes, and other molecules that they will need to double). This is the "lag phase" in Figure 4.5. With all the ducks in a row, the doubling begins, leading to exponential growth.[6] Lenski chose to house each of his twelve populations

---

6. The mathematically inclined reader fully realizes that a doubling process giving rise to 2^n copies after n doublings is synonymous with an exponential increase giving rise to $e^{\ln(2) \times n}$ copies.

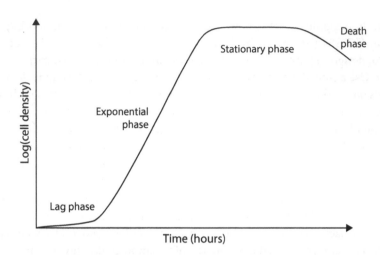

FIGURE 4.5. Growth chart of a typical bacterial population exposed to a finite supply of nutrients (note log scale). During the lag phase, the bacteria are gearing up for reproduction, then engage in full exponential growth until the food supply is exhausted. After that, the cells stop dividing and enter the "stationary phase," followed by cell death when food remains absent.

in 50 ml Erlenmeyer flasks, but barely filled them. Each flask would be filled with 10 ml (milliliters)[7] of solution (we will discuss the composition of the bacterial diet shortly), and 1 percent of the previous day's starving population (mostly bacteria, as all food was consumed shortly before stationary phase was reached) would be transferred to those flasks. Because a milliliter of nutrients supports about fifty million cells, about five million bacteria make it into the next day's flask, down from about 500 million when they reached stationary phase the previous day. Talk about thinning the herd! Those five million that were lucky enough to live another day have quite a day, of course. They grow in number by a factor of 100, going from 5 to 500 million because that is all the room afforded to them on Erlenmeyer Street. After they reach that number, the food in the nutrient medium (technically known as DM25)[8] is exhausted and the cells settle into the stationary phase shown in Figure 4.5. How many doublings can happen during those heydays when food is aplenty? From the number transferred, they increase one-hundred-fold. What number, in the exponent of two, gives one hundred? The answer is $\log_2(100) \approx 6.644$.

7. A milliliter is a measure of volume equal to a cubic centimeter. A milliliter of water weighs precisely 1 gram.

8. DM25 is a basic medium known as "Davis Mingioli Broth," supplemented with the sugar glucose (the bacteria's main source of food) at 25 mg per liter.

FIGURE 4.6. A close-up of some of the Erlenmeyer flasks holding media and bacteria. The flask in the center that appears "cloudy" houses the population known as Ara-3, which we will discuss in section 4.4.2. Photo courtesy Brian Baer and Neerja Hajela, Michigan State University (CC-BY-SA 3.0).

So there will be about six and two-thirds generations per day if we transfer 1 percent of the population each day into 10 ml of food. That looks like a poor use of the bacterium's potential, but this is, in the end, the reality of life on limited resources: even if per day those bacteria *could* grow to 10^{21} from a single cell (seventy-two doublings in principle, in a 24-hour day), there just is not enough food to allow this. Yes, you could get seventy-two generations per day if you had about 800 Olympic-size swimming pools filled with DM25. Eight hundred pools for each population, each day. I would not want to be on that cleaning crew. Alternatively (if swimming pools full of solution are out) you could transfer the cells two or even three times per day, which creates other trade-offs on the demand of the lab running the experiment (see Exercise **4.1** for ideas to increase the number of generations per day in a long-term experiment).

3. **Use a stable and controllable environment.** We have already discussed the living conditions for Lenki's bacteria: a well-shaken flask with "minimal medium" supplied with a controlled amount of glucose. But the food environment is not the only thing an organism could adapt to. Dallinger taught us that temperature (for example) is another important variable, and indeed Lenski and collaborators studied adaptation to temperature in an evolution experiment as early as 1990 (Bennett et al. 1990), in a "short-term" evolution experiment covering 200 generations of evolution at 42°C. The

temperature environment for the LTEE is fairly well controlled at 37°C, which as it happens is the "native" environment for *E. coli* (the temperature of our intestine). On the other hand, the strain that the ancestors of this experiment stem from has been "in the lab" since as early as 1918 (Daegelen et al. 2009), and thus 37°C may come as a surprise to the LTEE denizens. What is important for the experiment is that the conditions are kept almost perfectly constant for the duration. After all, the purpose of the experiment is to understand the dynamics of adaptation and its repeatability (Lenski 2017), as well as whether the changes in speed of evolution that are commonly seen from the fossil record are due to environmental changes, or instead arise from the stochastic nature of the evolutionary process. To investigate such a question, keeping the environment strictly constant is essential.

　　4. Ensure continuous observation of the organisms without interfering with the experiment. There is a trope, which is often traced back to the uncertainty principle in quantum physics: observation disturbs that which is being observed. While this may be true in quantum physics, it is not common in the classical domain. But clearly, there are dangers in an observation affecting the system being watched, in particular when the system's future can depend on small changes which, of course, is particularly true for evolution. Thus, in experimental evolution we must ensure that observation of that which evolves does not affect the future evolution. In Dallinger's experiment, this was secured by observing microbes without changing the constancy of the temperature. In Lenski's experiment this is particularly easy as 99 percent of the population is "removed from further consideration" every day anyway. Of those 99 percent, a fraction is kept for a while in case an error occurred during the daily transfers (more on this below), and periodically another fraction is frozen permanently to keep a "fossil record" (see design requirement 6, below). Thus, if experimental samples can be adequately preserved, the observation of organisms without interfering with the experiment is straightforward, as it is sufficient to observe the fraction of the population that did not win the lottery to live for another day, simply because there is essentially no difference between the winners and losers from one day to the next.

　　5. Run replicate experiments. Because evolution strives to adapt but is beholden to chance, a single evolutionary trajectory (the path taken by a single population observed for some time) only has limited explanatory potential. If a particular trait evolved (or was lost) at a particular point in time, we cannot infer whether it evolved or was lost due to adaptation, or due to drift (that is, chance). If a trait does *not* evolve during a fixed period of time, we will not know whether the trait is simply not beneficial, or whether the mutation giving rise to that trait just never occurred, or perhaps if it did occur, it had not spread to everyone in the population. To test hypotheses concerning the evolution of traits, we must gather statistics. And to obtain statistics about evolutionary trajectories, we need to run replicates of the same experiment, ideally with identical starting conditions. Dallinger designed his experiment with three replicates (albeit not perfectly identical for the reasons we discussed), and this gave him some confidence that the results he observed reflected some generalities and would be repeatable. Lenski started with *twelve* populations, but those, too, were not all identical. As we saw earlier, it is essential that an experiment with replicates is protected from the danger of cross-contamination, where a member of one population somehow ends up in another, potentially eradicating the resident type. Evidently, such an event would ruin the independence of the replicate lines, and such an accident becomes ever more likely if (as in the Lenski experiment) daily transfers are required. To safeguard the independence of the lines, Lenski created a variant of REL606 that had a mutation in the gene *araA*. This strain, called REL607, was used as the ancestor of six of the twelve lines of the LTEE (REL606 founded the other six, of course). To appreciate the importance of the *araA* mutation, we have to spend a little time understanding how bacteria gain their energy.

Like all forms of life, bacteria need fuel to power their cellular machinery, and the prime source of fuel is sugars (they can also use other sources of carbon). Different sugars, however, require somewhat different cellular "machinery" to break them down, and some bacteria have the tools to break down (and thus consume) one type of sugar, but not another. In general, bacteria can grow on a variety of different sugars, and the simplest of those (the monosaccharides) are glucose, fructose, and sucrose. Of those, *E. coli* much prefers glucose, and indeed this is the sugar provided for the cells in the growth medium DM25. However, *E. coli* can also grow on more complex sugars (such as the disaccharide maltose), and in particular on the simple sugar arabinose. To perform this feat the gene *araA* encodes one of the required proteins, but in the REL606 strain that gene was mutated and thus defective, so it is designated as "defective in Ara utilization," or simply: Ara⁻. As a consequence, REL606 cannot grow on a medium that has arabinose as its sole carbon source. The mutation in the *araA* gene that defines REL607 restores the capacity to grow on arabinose, and it is thus denoted as Ara⁺. Because the day-to-day growth medium DM25 does not contain arabinose, both REL606 and REL607 grow exactly the same way within the flask (this has proven to be true many times over); thus their evolutionary path should not be affected by the arabinose utilization marker at all. However, the difference between the strains provides for the cross-contamination protection in an ingenious way.

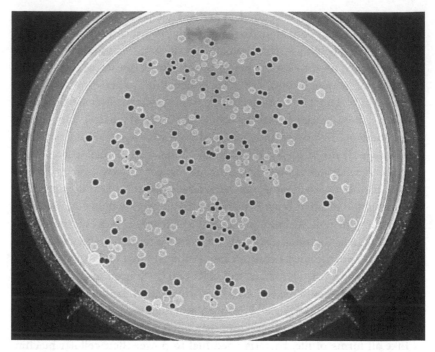

FIGURE 4.7. Petri plate containing a mix of Ara$^-$ and Ara$^+$ colonies. Ara$^-$ colonies are colored deep red (here, dark gray) while Ara$^+$ appear pink (here, light gray). Richard Lenski, Michigan State University.

When REL606 is grown on a standard Petri dish that contains the indicator tetrazolium and the sugar arabinose, REL606 forms colonies that are deep red in color (this is visible to the eye, see Figure 4.7 where the red color appears dark gray). Not so for REL607, however. Breaking down arabinose turns the color of REL607 colonies from red to white (or pinkish), so that REL607 colonies can be easily distinguished from those that REL606 makes (but only on those tetrazolium-arabinose, or TA plates). The different colonies are clearly visible in the example plate in Figure 4.7. How does this distinction help to protect from cross-contamination, since neither arabinose nor tetrazolium are present in the day-to-day life of the evolving bacteria?

Let us imagine in our minds how the daily transfer protocol proceeds. You are faced with twelve labeled flasks filled with starving bacteria, and twelve new flasks filled with medium, but no bacteria. Six are labeled Ara$^-$ (1 to 6), and six are labeled Ara$^+$. Your task is to remove 0.1ml from yesterday's flask (using a pipette or syringe) and inject it into the corresponding fresh flask, prepared with 9.9 ml DM25. Then you move to the next flask. Lenski figured that the most likely contamination will be that bacteria from the last flask-transfer somehow make it into the next, even when taking precautions to avoid cross-contamination. To ensure that such a mistake can be detected if it happens, the protocol calls for alternating transfers between Ara⁻ and Ara⁺ lines. In this way, a spot check (by plating the colony on TA plates) will reveal any intruders. If a mixture of red and white colonies is detected, the culture is thrown away and the experiment is restarted from an earlier time point that was saved in the freezer (more on this below), after checking that the last saved transfer was free of contamination.

It turns out that this procedure has saved the LTEE many times over, as undetected cross-contamination events could potentially end any of the twelve independent lines. How well this procedure has worked to prevent accidental cross-contamination can only be ascertained for sure by sequencing the entire genome of representatives of the twelve populations. Such a test was performed on the genomes of all twelve lines after 50,000 generations had passed. Sequencing hundreds of genomes from all twelve lines (Tenaillon et al. 2016) showed that they were indeed still independent after 50,000 generations. Given that many thousands of transfers had taken place to get to this point, the proof that all twelve lines were uncontaminated underscores the discipline, perseverance, and ingenuity of Team Lenski.

Cross-contamination among competing lines is not the only possible source of contamination, however. The human beings that perform the daily transfers are themselves hosts of countless E. coli bacteria, and it might seem impossible to prevent them from entering the flasks and taking over the local population. While such a contamination is unlikely due to standard precautions, Lenski chose his ancestral strain in such a way that a contamination could readily be detected. Here, another excursion is in order, this one to understand how E. coli interact with the viruses that infect them.

One of the reasons that E. coli is one of the best-studied model organisms today is that it became the organism of choice of a group of scientists that unraveled the fundamentals of bacterial genetics: the "phage group" around the former physicist Max Delbrück, starting around 1940 (Cairns et al. 1966). This group used viruses that infect bacteria—the bacteriophages (or phages, for short)—to understand how information is encoded in molecular sequences. The hosts to these phages were E. coli bacteria. There are many different documented types of bacteriophages that infect E. coli, denoted by letter codes such as T4, T5, and T6. Many E. coli strains are resistant (or otherwise immune) to both T5 and T6, but REL606—being an offspring of a particular lineage of the "B strain" (Daegelen et al. 2009)—is sensitive to T5, while resistant to T6. To test whether the bacterium in the flask is truly a descendant of REL606 rather than a newer contaminant, bacteria from a previous day's flask are cultured on a plate along with the T5 virus (kept refrigerated in the lab for this purpose). If the virus can kill the bacterium (an interaction called a lysis), this will be readily visible on the plate. If it cannot, the virus's host is likely an intruder, and a contamination with bacteria has probably taken place.

We can now appreciate a fundamental difference between Dallinger's experiment and Lenski's. Because Dallinger used eukaryotes and not bacteria, he was able to test for contamination and establish the integrity of his lines simply by observing them under a microscope. However, as amoebae are much more complex than bacteria and are propagated under recombination, they are less suited to study the fundamental principles of evolution. Bacteria are much more conducive to such an endeavor, but observation under a microscope will not protect from accidents, as any differences are very unlikely to be revealed under a light microscope. To use E. coli for a long-term experiment, much more sophisticated means must be deployed to safeguard the integrity of the lines. But as accidents will happen, a procedure to identify and correct such errors must be in place.

While those safeguards are essential to the long-term experiment, the feature where Lenski's experiment truly blazed a trail is the preservation of the fossil record in the freezer.

6. **Keep a "fossil record" of the evolving populations.** Every organism on Earth has at least one parent, who has at least one parent, and so forth. Following the line backward in this manner, we ultimately will arrive at the last common ancestor of all life on Earth. All evolutionary changes that make us different from that last common ancestor are encoded on that line, which therefore tells the story of evolution. Of course, such a line is not available to us, but if we conduct an evolutionary experiment, we ought to attempt to preserve as much of that line as possible. Even in an experiment as meticulous as the LTEE, we still cannot save every generation, for the obvious reason that almost seven generations pass in the flasks each single day.

To safeguard against mishaps (such as flasks that have cracks, leading to the loss of the population of that day), the previous day's population is kept in a refrigerator overnight. But every seventy-five days (this corresponds to about 500 generations)[9] the entire population is saved by freezing a sample of it at −80°C. While a sample every 500 generations is hardly a perfect line of descent, the series of frozen samples serves multiple purposes. First, it provides for a fail-safe in case a contamination is detected that compromises the independence of lines. In fact, this fail-safe was triggered several times, which forced the team to discard the population and restart the line using the frozen sample, effectively setting back the clock by 500 generations on the affected line. For this reason, the twelve lines do not all share the same

---

9. If we use the exact number of $\log_2(100)$ generations per transfer, 75 transfers equals about 498.3 generations.

generation number. Second, the frozen stocks represent an opportunity to revive populations from bygone days.

In describing the experiment after 10,000 generations had elapsed (Lenski and Travisano 1994), Lenski asked the reader to entertain a fantastical scenario in which Darwin's dream of an unbroken line of descent is not only realized, but where the preserved "fossil beds" had the magical quality of being able to spawn a resurrection of the populations preserved in them:

> Imagine that you could resurrect these organisms (not merely bits of fossil DNA but the entire living organisms) and reconstruct their environment exactly as it was during the thousands of generations preserved in the fossil bed. You could measure not only the organism's morphology, but also its functional capacities and genetic composition. You could even place derived and ancestral forms in competition to determine their relative fitness in the "fossil" environment.

This story sounds fantastical, but Lenski assures us: (Lenski and Travisano 1994)

> Yet this fantasy is not fiction; it is fact. We have many such "fossil beds" preserved, and we have "traveled in time" to manipulate populations with respect to their history and environment. The fossil beds are preserved in a freezer and contain populations of the bacterium *Escherichia coli*. Our time travel thus far extends over 5 years, representing $> 10,000$ generations in this system, and we have manipulated many populations each comprising millions of individual organisms. In essence, our approach might be called experimental paleontology.

At the time this book went to print, the team had frozen over 75,000 generations, making "time travel" that extends to thirty-four years possible. An end to the long-term experiment is not in sight, and with any luck the intrepid evolving bacteria have many more surprises in store for us. We will discuss some of the surprises the experiment has already delivered in section 4.4, but before we savor these we should visit one other model organism that is used extensively in experimental evolution. This organism, however, does not thrive on rotting food, nor does it populate your intestines. Instead, it lives inside of computers.

4.3   Digital Life: A Brief History

The idea that the basic processes of life could be implemented within nonbiological materials is an old one and goes back at least to the nineteenth century, with its fascination with machines. To wit, on June 13, 1863, the British writer Samuel Butler wrote a letter to the newspaper *The Press* entitled "Darwin

Among the Machines." In this letter, Butler argued that machines had some sort of "mechanical life," and that machines would some day evolve to be more complex than people, and ultimately drive humanity to extinction:

> Day by day, however, the machines are gaining ground upon us; day by day we are becoming more subservient to them; more men are daily bound down as slaves to tend them, more men are daily devoting the energies of their whole lives to the development of mechanical life. The upshot is simply a question of time, but that the time will come when the machines will hold the real supremacy over the world and its inhabitants is what no person of a truly philosophic mind can for a moment question.
>
> SAMUEL BUTLER, 1863

It is difficult, reading such dire predictions from over 150 years ago, to avoid being reminded of today's ever louder doomsday prophecies involving the imminent takeover by intelligent machines. We'll discuss the prospects of artificial intelligence a few chapters hence, and in particular weigh the chances that people (not just men, as Butler would have us imagine) will be "bound down as slaves" to tend to the machines (section 9.4). But the general idea that life is, in its essence, mechanical (as opposed to, say, magical) goes back further than Butler. In 1651, Thomas Hobbes in *Leviathan* asked (Hobbes 1651)

> Why may we not say, that all "automata" (engines that move themselves by springs and wheels as doth a watch) have an artificial life?

While Hobbes was really more concerned with the image of society as a whole being viewed like an organism, it should not be much of a surprise that such ruminations were unusual in 1651 or 1863. But in the twentieth century, with technological developments beyond the imagination of the centuries that preceded it, dreams of artificial life forms multiplied. Yet, because of the enormous biochemical complexity of even the simplest natural organism, all attempts to create artificial (biochemical or mechanical) organisms so far have predictably failed.

Still, researchers pressed on. For example, the idea that all life is based on information (see chapter 2) suggests that perhaps it is possible to create a form of life that is noncorporeal, consisting *only* of information. Armed with this idea alone, it is not a stretch to attempt to implement the concept of self-replication within an information-processing machine, as was suggested by the mathematician (and pioneer computer scientist) John von Neumann a few years before the discovery of DNA's structure (as we'll see shortly).

In retrospect, we learned this lesson in the early days of the computer age, when we were forced to deal with ever more cunning computer viruses that (beginning in the 1980s, coinciding with the introduction of the personal computer) infected the machines we work with. These man-made programs for the most part had only one overarching goal, namely to copy themselves into as many hosts as possible (nowadays, computer *malware* is significantly more sophisticated). But these computer viruses differ in one crucial aspect from the forms of life that we are used to: they do not evolve autonomously, but rather rely on the actions of programmers to adapt them to the environmental changes conceived by computer security experts (see Fig. 4.8). And indeed these viruses are noncorporeal: they are pure information!

4.3.1 The early years

When the history of digital life is written one day, the era of "Digital Artificial Life" (as opposed to "Biochemical Artificial Life") will likely begin with the aforementioned John von Neumann, the ultimate polymath who had his hand in, it sometimes seems, just about every branch of modern science (Macrae 1992). John von Neumann was a mathematician, a physicist, and computer scientist, born in 1903 into a new century as the son of a fairly well-to-do family in Budapest, Hungary. His father was elevated into nobility when John (then called János) was ten years old, and János Neumann became János Neumann von Margitta. When he became a professor at the University of Berlin in 1928 (he was only twenty-five years old at the time) he changed his name simply to Johann Neumann.[10]

In the 1940s von Neumann became involved (as many of his Hungarian expatriates at the time) in the war effort, and his calculational skills and intuition became crucial for the success of the Manhattan project because he was the only one (besides the nuclear physicist Hans Bethe) who had mastered the mathematics of shock waves. It was there at Los Alamos that von Neumann became exposed to the first electronic computers (the IBM tabulating machines), and he started to imagine what truly programmable computers could do. Indeed, the concept and design of every modern computer today has its roots in von Neumann's design (outlined in the classic Burks et al., 1946), a design that is now commonly referred to as the "von Neumann" architecture.[11]

After the war, von Neumann continued to be fascinated by computers, and in particular by automata theory. He was already familiar with Alan Turing's

10. Von Neumann anglicized his name in 1933 when he was a professor at Princeton, and became a US citizen in 1937.

11. In all fairness, von Neumann's design of the architecture of the EDVAC (Electronic Discrete Variable Automatic Computer) was based on J. Presper Eckert and John Mauchly's design, who invented ENIAC: the Electronic Numerical Integrator And Computer.

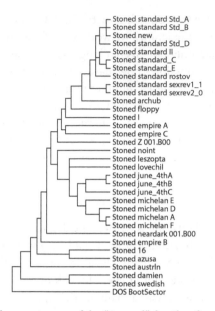

FIGURE 4.8. A phylogenetic tree of the "Stoned" family of computer viruses that infected the boot sector (on floppy disks and hard drives) of IBM PCs in the late 1980s. The "mutations" of the virus are alterations performed by hackers to counteract the changes made in the IBM operating system meant to protect it. In a sense then, the phylogeny represents the result of a co-evolutionary arms race between virus and host. Reconstruction and image courtesy of D. H. Hull.

work that introduced the concept of "universal machines" (Turing 1936). The concept of a universal machine greatly influenced von Neumann's thinking (as he readily acknowledged) when designing the programmable computer's architecture. Turing (the "other genius," so to speak) also was a mathematician by trade, who had decided to tackle the "Entscheidungsproblem" (decision problem) of mathematics (see Box 4.1), which led him to formalize what it means to perform a computation.

In particular, Turing realized that the process of computation could be thought of quite generally as a series of mechanical operations. This is different from creating a mechanical computing device, something that was already achieved in antiquity.[12] Such devices can "compute" the answer to a very

12. Perhaps the most famous mechanical "computer" of antiquity is the "Antikythera mechanism," an artifact that was recovered in 1901 from a shipwreck off the Greek island of Antikythera. The device (which dates back to 100–150 BCE) is a complex clockwork mechanism used to predict eclipses and other astronomical occurrences.

Box 4.1. Mathematics, Logic, and Computers

Mathematics in the early twentieth century became dominated by considerations of mathematical logic, as opposed to the more concrete advances in number theory common to nineteenth-century mathematics. One of the proponents of a logical approach to mathematics was David Hilbert who, in 1900 (as a thirty-eight-year-old professor at the University of Göttingen, the world's center of mathematics at the time), gave a talk at the International Congress of Mathematicians in Paris in which he outlined mathematics' unsolved problems (Hilbert 1902). Among them were classical problems such as the (still unsolved) Riemann hypothesis about the real part of the zeros of the Riemann zeta function (Connes 2016). That problem is quintessentially a nineteenth-century problem, and the metamorphosis of mathematics in the twentieth century is perhaps best illustrated by Hilbert's changing focus toward the "Entscheidungsproblem" (German for "decision problem"). Hilbert issued this challenge in 1928: "Can you construct an algorithm that can decide whether any given mathematical statement is provable using the rules of logic?" It is called the decision problem because the algorithm must decide whether any particular statement is true or false. Three years later, the mathematician Kurt Gödel proved his famous (and eponymous) Incompleteness Theorem (see Goldstein 2005 for an introduction to the man and the theorem, or Hofstadter 1979 for a trip down the rabbit hole).

Gödel's theorem established that there cannot be a consistent set of axioms that can be used to prove all correct statements about natural numbers (the positive and negative integers, including zero). This means that there are true statements about these numbers that will remain unprovable within the axiomatic system. From there, it was just a few steps to answering Hilbert's challenge with a resounding "No!": Alonzo Church, a mathematician at Princeton University, used Gödel's theory to show that the Entscheidungsproblem was unsolvable. At the same time (and independently), Alan Turing used Gödel's insights to show that it is impossible to construct a universal computer (now known as a "Turing machine") that will determine whether an arbitrary program will ever "halt" (that is, issue a decision). While both Church and Turing are equally credited with answering Hilbert's challenge, Turing's solution turned out to be far more consequential, as his mathematical construction of the Turing machine effectively ushered in the computer age.

specific set of questions only, that is, they are limited to the purpose they were designed to. Instead, Turing was able to show that *all* computers (past, present, and future) could be described in terms of just a few mechanical operations. Machines of this sort (later called "Turing machines") are useful because their abstract formulation makes it possible to prove *general* results about what computers can (and cannot) do. It is *this* insight that led von Neumann to design the computer architecture we still use today, but it also led von Neumann to think more broadly about arguably the two most fundamental problems of biology: the origin and evolution of complex life, and the nature of intelligence.

The idea of computing viewed as a mechanical process influenced von Neumann's thinking about the relation between computing and brain function (von Neumann 1958). But before he thought about using computers to create artificial intelligence, von Neumann was thinking about using computers to create *artificial life*. The influence of Turing's thoughts are unmistakable. After all, one of the essential aspects of the universality of Turing machines was that you could make one machine emulate another. It is this powerful property that makes it possible to construct proofs that do not depend on the particular way in which a computer operates: they are all interchangeable because they can all be made to act like each other. The way Turing achieved this universality was by showing that you could give a description of one machine as a *program* to another machine, which could use that description to act just as the machine described. Von Neumann took this idea to the next level: What if you give a machine not the description of how it operates, but how to *build* it? Executing those instructions, the machine will then construct another such machine, which, when given its own description, can construct another, and so forth. With this, von Neumann had hit upon the central idea of self-replication of information in biology, years before the discovery of DNA. Indeed, the South African biologist and Nobel Laureate Sydney Brenner (who once shared an office with Watson and Crick) later wrote:

> And it is one of the ironies of this entire field that were you to write a history of ideas in the whole of DNA (. . .) you would certainly say that Watson and Crick depended upon von Neumann, because von Neumann essentially tells you how it's done. But of course no one knew anything about the other.
>
> (BRENNER 2001)

You could imagine, given that von Neumann was a mathematician, that he would have been satisfied to simply point out this relationship between computation and life. Instead, he attempted to use this insight to literally construct life inside of a computer: to *create* artificial life. How do you manufacture a machine that contains the instructions to construct itself? Von Neumann imagined that it would be possible to build a "universal constructor" that would be able to construct *any* machine if a blueprint was provided, while making a copy of the instructions at the same time. If a universal constructor were given a blueprint of itself, you would automatically have a self-replicating machine. We can immediately see here the analogy to Turing's idea, to give a "halting-program evaluator" its own program to examine.

Mechanically speaking, making a universal constructor was impossible then, just as it is now. But given that computers can be thought of "mechanically" (according to Turing's construction), von Neumann wondered whether it is possible to implement (or simulate) such a machine inside of computers. In discussions with his friend Stanisław Ulam, von Neumann came up with the concept of a *cellular automaton* (abbreviated simply as CA). CAs are simple "machines" that have a finite number of states (just as Turing machines), but they can in principle do more than just read and write bits from a tape (they could also do less). And while in Turing's construction a single machine roams a one-dimensional data tape, von Neumann imagined a grid in which each machine has four neighbors in its cardinal directions. The state of the machine is updated depending on its own state and that of each of its four neighbors, just as in John Conway's famous "Game of Life" CA that most readers will be familiar with. Except von Neumann's CA predates Conway's by about thirty years, and is far more complex.

To design the self-replicating machine, von Neumann imagined a machine with twenty-nine states that can be used to construct a pattern of cells that can read the instructions that are written on a "tape" of sorts (really just a pattern of zeros and one), and construct the pattern described by the instructions at another location on the grid, using a "construction arm." The sequence of instructions itself is not copied by the universal constructor, but instead must be copied by a separate "copy machine." While von Neumann completed the entire rule set to implement his universal constructor in the 1940s, he was never able to test it. He passed away in 1957 (at only fifty-three years of age) due to complications of cancer, quite possibly a consequence of conducting experiments with radioactive nuclides without proper shielding during the war effort. How the loss of this great mind affected the future of humankind we will never know.

Von Neumann's design of the universal self-replicator was published posthumously in the book *Theory of Self-Reproducing Automata* (von Neumann and Burks 1966), but it took another thirty years to test whether the design actually worked. It turns out it did not work, which is not too surprising given the complexity of the code and the fact that von Neumann was never able to run it. However, two Italian scientists were able to "fix" von Neumann's design so that it would self-replicate (Fig. 4.9; Pesavento 1995; Nobili and Pesavento 1996), but it was still not a "universal constructor" (it could not duplicate the tape). But that implementation also showed how far advanced von Neumann's ideas were: not only did it anticipate the division between construction machinery, copying machinery, and program that we find in molecular life, the design also required a tape ("genome") of 145,315 grid cells, and takes sixty-three billion steps to replicate. Even in 1995 when Nobili and Pesavento published the revised version of von Neumann's design, they were unable to run it on any computer available at that time. The program was just too complex. In 2008, finally, the algorithm could be run due to significant increases in computing power (and advanced algorithmic implementations), and took a few minutes to produce a single offspring. This accomplishment is a fitting tribute to von Neumann's genius: it took over sixty-five years, and the development of the most powerful computing machines known to man, to realize a process that von Neumann had imagined in his incomparable brain.

4.3.2   Chris Langton and the artificial life revival

The story of von Neumann's creation of the field of Artificial Life is a sober reminder of how far advanced his thinking was compared to his contemporaries. Not only was he able to reduce life down to its abstract principles (anticipating the idea of genetic material as a program) long before these principles were established in molecular biology, but he was also able to go so far as to write a program that created a new kind of life inside a computer, except that the computers required to run and thus test such a design would only become available about sixty-five years later. Perhaps this explains why the field of artificial life became dormant after von Neumann's efforts. Besides a few unrelated attempts that were only discovered much later (for example, Barricelli 1962), the field essentially ceased to exist.

However, von Neumann's work kept inspiring researchers periodically, and one of the first to revive that work with impact was Chris Langton, a computer scientist and complexity researcher with a rebellious streak, fresh out of graduate school. Langton revived the fledgling field in the late 1980s by organizing a series of workshops that brought together researchers who had been working on the sidelines (some may call it fringes) of more established fields of science, to investigate whether the principles of life could be understood by implementing or simulating them in a computational medium. The interaction of Langton with the group at the Santa Fe Institute (a complexity research institute not far from Los Alamos), in particular Doyne Farmer and Norman Packard, is well described in Steven Levy's book *Artificial Life* (Levy 1992) that chronicles these early days of the revival.

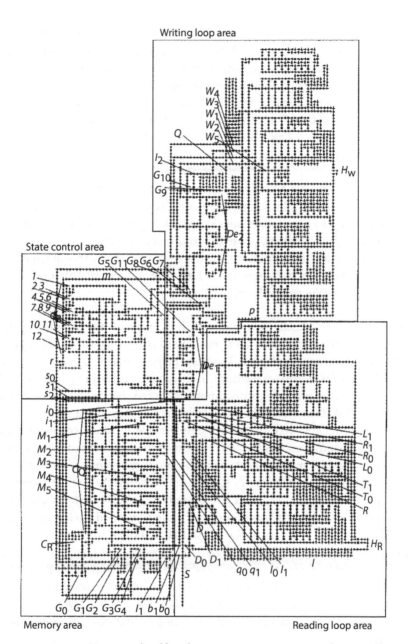

FIGURE 4.9. von Neumann's self-replicating automaton was implemented by Pesavento (1995) using 6,329 cells in a 97 × 170 configuration. This implementation was not fully functional (as it could not replicate the information tape). Further, Pesavento also had to introduce three more states (in addition to the twenty-nine in von Neumann's original construction).

Langton's artificial life workshops (the first of which was held at the Los Alamos National Laboratory in 1987) electrified a community that was, by their own admission, working in the shadows of acceptable computer science. One of the researchers toiling at the interfaces of physics, computer science, and biology was the Danish physicist Steen Rasmussen, also employed at the time at the Los Alamos National Laboratory (the same place where von Neumann had worked to help the US war effort). The artificial life workshop thrilled Rasmussen. He recalled the atmosphere as euphoric:[13]

> The workshop was a bit like a bar scene from a Star Wars movie with very excited, and a bit comical creatures (us scientists), young and old from across the world, computer scientists, engineers, roboticists, molecular biologists, chemists, physicists, and more, all moving around and in between multiple workstations and displays with demos assembled in the Oppenheimer Study Center hall at Los Alamos National Laboratory, the birth place of the atomic bomb.

One of the demonstrations at the workshop involved the computer game "Core War" that was popular in the mid-1980s (Dewdney 1984). This game was unlike your standard game in which players control fighters that battle each other for control of territory. In this game, the territory was a virtual computer, and the players wrote computer programs that battle each other for control of that computer. Experienced players had discovered that the most successful programs were those that simply made identical copies of their code (like a computer virus) and overwhelmed the opponent's memory core by proliferating all over it. These winning programs were, in essence, self-replicators in a simulated world. Rasmussen immediately saw the potential to study life inside of a computer and started a project where he could study the properties of these self-replicators in a more controlled environment: the VENUS simulator (Fig. 4.10; Rasmussen et al. 1990). In this virtual Petri dish, self-replicating computer programs—domesticated computer viruses, really—would be able to thrive, and maybe evolve to become something altogether different. While Rasmussen was mostly interested in using the VENUS simulator to gain insights into the possible *origins* of life (which we discuss in more detail in chapter 7), the resulting evolutionary dynamics were disappointing.

Rather than evolving, the programs instead were bent on Armageddon. Rasmussen witnessed the programs over-writing each other and effectively destroying each other's codes (Rasmussen et al. 1990). It was a digital carnage:

13. S. Rasmussen, personal communication

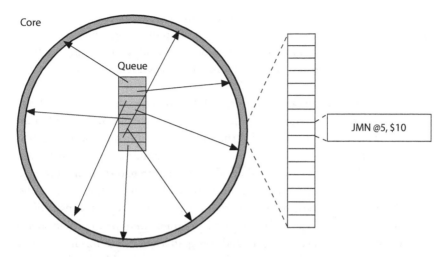

FIGURE 4.10. Coreworld programs are arranged along a circular "core" in the VENUS core world. The core consists of 3,584 addresses that each contain one "word." Each word is composed of an instruction and up to two operands (addresses of instructions). The queue of execution pointers determines the order of execution of each program. Adapted from Rasmussen et al. (1990).

only the smallest snippets of code managed to escape being over-written, but they were effectively rendered too short to self-replicate. Life on VENUS (VENUS stood for Virtual Evolution in a Non-Deterministic Universal Simulator) was doomed.

While ultimately unsuccessful, Rasmussen's work attracted the attention of another researcher who would leave an indelible mark on digital life research. At the time, Tom Ray was an assistant professor in the Department of Biology at the University of Delaware, a department that had just denied him tenure (Levy 1992). As a tropical ecologist, Ray studied how ecosystems evolve and adapt, but when not in the field he had become fascinated with the idea of implementing those kinds of dynamics within a computer. Lacking any training in programming, he nevertheless was fascinated by the idea that life could be thought of as a computational process, and sought out communities of researchers that were thinking along the same lines. Ultimately he found Langton's Proceedings volume documenting the first artificial life workshop. Eager to learn more, he visited the Los Alamos group, and finally came face-to-face with Rasmussen's VENUS. As retold in Levy's book (Levy 1992), it was at the Santa Fe Institute where Ray realized (after the resident Doyne Farmer and Norman Packard gently nudged) that making self-replicating programs on real computers was not necessary (Ray had imagined computers in steel cages removed from any possibility to interact with other computers).

Instead, Farmer and Packard suggested creating a *simulated* computer to avoid having to quarantine *actual* self-replicating computer programs and prevent them from escaping to other computers on a network and creating a cyber-meltdown of global proportions. Pretty much in the same way Rasmussen had done in his Coreworld.

Indeed, what Rasmussen's research with VENUS had shown is that a meltdown can happen in a virtual computer as well, albeit a meltdown of a different sort. How do you prevent the destruction that Rasmussen's programs had unleashed on each other? Watching the carnage unfold at a demonstration at the Santa Fe Institute, Ray had an idea. In fact, he had two: one for each of the two main impediments for evolution in VENUS. First, he had to prevent the unruly over-writing of code that Rasmussen's programs engaged in. Second, he had to change the language the programs were written in, in such a manner that modified (that is, mutated) programs had a fighting chance to be functional themselves. The first problem was easily solved by implementing *write-protection* to the program's code. Indeed, a clever system that allocates read-, write- and execute privileges to different owners of code is at the heart of every modern computer system. By preventing one CPU from over-writing the code of another, the destructive forces were stopped. The second problem, however, was more sophisticated.

The Core War programs were written in a language called "Redcode" that only had sixteen instructions. But a complex addressing system made it so that effectively there were 7,168 different instructions in Redcode, each with a potentially different action. As a consequence, most mutated programs were nonfunctional: the language was not robust. [Indeed, a test carried out ten years later (Ofria et al. 2002) revealed that on average barely three out of 1,000 mutated programs written in Redcode were still able to still self-replicate.]

Ray's idea was to change the Redcode addressing system, one in which an instruction would point to a particular address in the memory block, in such a way that addresses are determined using *patterns in the code* instead, reminiscent of how proteins can find specific locations on the DNA to bind to. In this manner, changes elsewhere on the code would not affect the program flow, while the patterns themselves could change when necessary. It turned out that these two ideas were sufficient to create a complex ecosystem of evolving programs, and Ray promptly presented the first results of these experiments at the second of the artificial life workshops (Ray 1992). Ray called the program that implemented the virtual computer within which digital life unfolded *Tierra*, the Spanish word for Earth.[14]

---

14. As a conservation ecologist, Ray also ran a biological station in Costa Rica to study rainforest conservation.

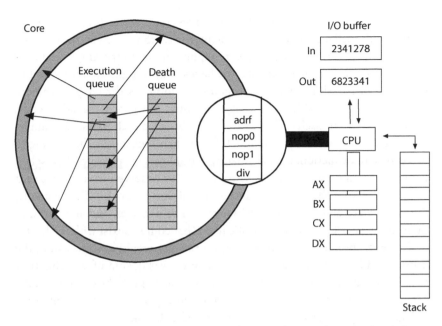

FIGURE 4.11. Sketch of the Tierra system showing the circular core, divided up into programs that each have their own "identity," and whose order of execution is determined by the "execution queue." When a program replicates, another program must be removed, in the order determined by the "death queue." The central processing unit (CPU) of the Tierra virtual computer consists of four registers (labeled AX-DX) and a stack. The I/O buffer was added later in order to enable computations on externally provided numbers.

In some ways, Tom Ray's Tierra was an iteration of Rasmussen's VENUS, and in important ways, it was not. Tierra retained the circular core structure of VENUS, along with the queuing system that regulated the order of execution of programs (see Fig. 4.11). Reflecting the simplified addressing system, the virtual computer executing the tierran code was simple: just four registers, and a single stack that could be used to store intermediate numbers that needed to be recalled later. The order of execution of tierran programs was determined by an execution queue that served as a fitness determinant: by moving a program up or down in the queue, you could affect the expected number of offspring a program can produce. When a program split off a daughter program, that program was entered at the bottom of the execution queue, and the program residing there was removed ("killed"). The order of death was determined by a separate queue, called the "reaper" by Ray (1992).

Evolution in Tierra was rapid, but proceeded only in one direction: toward smaller and smaller programs. In hindsight, the reason was clear: the environment within which the programs were replicating was very simple and only rewarded replication speed. In fact, the fastest way to replicate turned out to be a form of parasitism where a program would hijack the copy-machinery of a host program. (While the programs could not over-write other code, the execution of nonself code was allowed.) Because such parasites could be very small, they could replicate much faster than their hosts, threatening to drive them to extinction. But of course, without hosts the parasites themselves would be doomed so that the dynamics that Ray observed were reminiscent of the periodic infections seen in the biological world (Ray 1992). Eventually, the hosts were able to evolve resistance to the parasites, and the parasites themselves evolved into a mutualistic form that could coexist with the host.

However, besides this coevolution, no novelty would emerge within the Tierra world because the world they inhabited was devoid of anything they could take advantage of: it was too simple. In that respect, the outcome of the experiment was similar to the classic evolution experiment carried out in Sol Spiegelman's lab (Mills et al. 1967), described in Box 4.2.

The loss of complexity in both the tierran virtual world and Spiegelman's flask were not in contradiction to the increase in complexity seen (on average) in the biosphere. Rather, it was the predictable outcome when a complicated organism is transplanted into a simple world. To observe growth in complexity, it is necessary to adapt to a world with new opportunities, a world in which there is "something to do."

This is the point where my own contributions to artificial life begin. In 1992 I arrived as a newly baked postdoctoral "division fellow" in the laboratory of Steve Koonin, at the Kellogg Radiation Laboratory at Caltech. Koonin's lab specialized in computational nuclear physics, but his group was famously interdisciplinary and tackled any problem they thought they might have a good shot at answering. A few weeks after my arrival, I needed to go back to Stony Brook University (where I had obtained my Ph.D. in 1991, and subsequently spent one postdoctoral year) for some paperwork. As Koonin and I had not yet decided on a topic that I might work on, I asked him for some papers to read on the plane. Among the stack he handed me, I found a set of papers on $1+1$-dimensional light-cone QCD (that, frankly, bored me to tears) and one paper that was unlike all the others: Tom Ray's paper on Tierra (Ray 1992). When I returned, Steve asked me whether I found anything interesting in the papers he sent along, and when I fished out Ray's paper from the stack and dangled it by the corner, he smiled mischievously and asked, "Do you want to work on it?" After I enthusiastically agreed, Steve shared his thoughts with me. His vision was that perhaps one day we would not have to write computer programs, but have evolution create them instead. He thought

Box 4.2. Spiegelman's Monster

The bacteriophage Qβ is a common virus that infects *E. coli* bacteria (we encountered the phages called T4, T5, and T6 earlier). Qβ is a double-stranded RNA virus with a fairly short genome (4,215 nucleotides, or less than half of HIV-1 that we encountered in section 3.3) that codes for only four proteins. The key protein for the virus is the Qβ replicase, which is an RNA-dependent RNA polymerase, meaning that it takes an RNA strand and makes a copy of it. Spiegelman wondered what would happen if you evolved the Qβ genome in an artificial world: a flask in which the replication machinery (in the form of the Qβ replicase enzyme) as well as the necessary raw material (in the form of nucleotides) were provided. In a serial transfer experiment somewhat similar to the Lenski experiment we discussed earlier, Spiegelman incubated the virus genome in this mixture, and transferred a fraction (usually about a tenth) of the resulting broth into a next-generation flask, with fresh replicase and nucleotides. Over time, Spiegelman reduced the incubation time to select for faster and faster replication. Specifically, Spiegelman asked: "What will happen to the RNA molecules if the only demand made on them is the Biblical injunction multiply!, with the biological proviso that they do so as rapidly as possible?"(Mills et al. 1967). The answer turns out to be: "Jettison all code that is not strictly necessary to survive in this world," which in the case of Qβ-in-a-flask implied an extraordinary reduction in sequence length within seventy-five transfers, to only 17 percent of the original sequence, according to the original publication. A subsequent re-analysis (Kacian et al. 1972) revealed that the RNA sequence isolated after those seventy-five transfers was really only 218 bases long, a reduction of almost 95 percent. Of course, we need to keep in mind that because the replicase enzyme was provided in the test tube, the amount of information necessary to replicate was automatically minimal: it was a sequence that did not need to store any other information than being a template for replication: a virus within a virus, so to speak. Viewed information-theoretically, Spiegelman's "monster" was not a monster at all: instead it was merely the contour cast by the replicase, chased by the relentless pressure to rid itself of everything but a shadow of its former self.

Tom Ray's Tierra was a first step, but (as alluded to above) those programs ended up not doing anything interesting besides replicating. We decided that as a first step, perhaps evolution should teach those programs how to add numbers.

A day later, I had downloaded Tom Ray's Tierra code and realized it was written in plain C (the programming language). As I had only programmed in the Fortran language during my entire career to this point, I went to see Koonin and mentioned this problem. His answer to my predicament was

"So?," which sent me directly to the university bookstore to acquire "Introduction to C," and to start learning the language. I soon became familiar with the Tierra code and started my own modifications of it. I imagined that the calculations that we wanted the tierrans to do were the equivalent of a simple metabolism, as the code necessary to perform these calculations would provide energy to the tierrans that they could use to replicate faster. In short, I figured that tierrans should work for a living, just like everyone else.

I quickly learned that just rewarding such calculations with extra energy was not sufficient for these functions to evolve, because the code to trigger the rewards was too rare to simply occur via random mutations. I first had to reward other aspects that I imagined would be helpful in generating the necessary code. First, I gave a small reward for simply having read and write instructions in the code. These were new instructions that I provided to tierrans that would read numbers from the environment into the newly created input buffer (shown as I/O buffer in Fig. 4.11). Given those rewards, evolution would fairly quickly fill tierrans with these get and put instructions, but still code for adding or subtracting would not evolve. I figured I should perhaps reward programs for "clearing the channel," meaning that they would obtain some energy from writing to the output whatever they had read in the input. I called this task "echo." At the same time, I rigged the Hewlett Packard 700 Series workstation (on which I ran the Tierra program) to emit a beeping sound if any of the critters had figured out how to add. The next morning, I opened the door to my office to find the workstation emitting an almost continual scream of beeps.

When examining the programs (after stopping the run and the annoying alarm), I found that the tierrans had figured out how to add all right. Examining them closer, I discovered that once the programs had evolved the capacity to echo what they saw in the input, they were also performing all kinds of mathematical operations on numbers. Via mutations to the code that could echo, they would add or subtract a constant number, or multiply a number by two, all purely by chance. Among those random changes to the code, a simple add function would also appear. But unlike all the other ones, the add function would be rewarded with a burst of extra energy, which allowed the program that performed the task to run much longer, and thus make *two* offspring. As on average all other programs would only make one daughter program before being removed, these rewarded programs would quickly dominate the population, and from then on out they would attempt to make as many add operations as they could fit into their genome. Hence the screeching workstation in the morning.

After I submitted this work (Adami 1995a) for publication, one of the reviewers asked whether all tierrans had evolved the same algorithm to perform the add operation. I thought this was an interesting question and proceeded to dig through tierran code to classify the evolved algorithms (I had run many parallel experiments). It turned out that there were two main types of algorithm that tierrans would discover over and over again. But then I also found one that seemed complicated and unwieldy. In fact, it took me quite a while to understand how it was able to perform the function at all. On the face of it, the program appeared extremely unlikely, indeed utterly "insane" (see its description in the Appendix of Adami 1995a). That such an intricate and seemingly impossible program (it used a staggering twenty-four instructions to achieve something that can be done in five) was able to evolve at all indicated the potential of digital evolution to create far more complex programs, and hinted that the awesome creative power of evolution could be unleashed inside a computer as well.

4.3.3   Avida and the advent of digital genetics

After doing some more work with Tierra (Adami 1995b), I decided that I needed a system that was more flexible, and in particular one that allowed me to study evolution proceeding on a physical surface, like bacteria growing in a Petri dish, for example (see Fig. 4.5 as an example). Fortunately, two undergraduate students visited me at the Kellogg lab at the time: Titus Brown from Reed College and Charles Ofria, who was an undergrad at SUNY Stony Brook (Charles and Titus knew each other from high school on Long Island). They both declared that they would be happy to help me out in this matter, as both had been programming since they were in middle school. While I had asked Titus to just modify Tierra, he ended up writing his own code, and with that the first version of Avida was created in 1993. The name "Avida" was a play on the Spanish rendition of "A-Life" (short for artificial life), echoing the Spanish "Tierra" for Earth that Tom Ray had coined.

Besides many improvements in the "guts" of the digital life system, the main change in Avida at that time was that avidians (as we called the denizens of the Avida world) could live on a grid where each program had eight immediate neighbors. By introducing a spatial structure, we could investigate how self-replicating entities grow in physically structured environments like, say, a Petri dish. At the same time, we also introduced a "well-mixed" mode, where each avidian could potentially interact with any other organism in the population, as you would find in a well-mixed beaker (like for example those in Fig. 4.6). The "virtual Petri dish" allowed us to visualize the competition between different strains (see Fig. 4.12[b]) and made it possible to examine the impact of spatial structure on every aspect of the evolutionary process.

In 1994 Charles Ofria returned to Caltech as a graduate student and a year later joined my lab (I had moved from postdoc to faculty) as my first graduate student. I asked Charles to continue developing the version of Avida that Titus Brown had written (who at this point had returned to Reed College to continue his studies there). Naturally, Charles instead proceeded to write his own version of Avida, and made a number of fundamental changes. In particular, he developed a redesigned simulated CPU that worked together with a completely new artificial metabolism, and a new instruction set that was adapted to both (see Table 4.1). In particular, the language he crafted created a new way to define matching patterns that went beyond the paradigm introduced by Tom Ray. In the tierran language, there were two instructions that had no function on their own: the instructions nop0 and nop1. Patterns created by a concatenation of those instructions could be searched for in a complementary manner, for example the pattern nop1-nop0 is complementary to nop0-nop1. The complementarity was a necessary choice so that a call

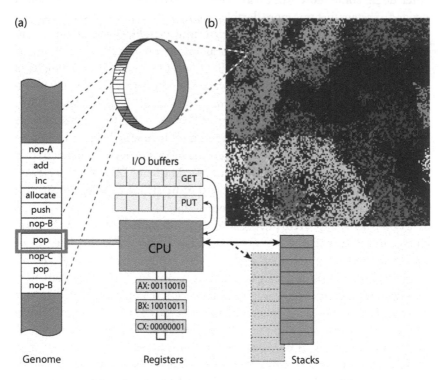

FIGURE 4.12. (a) A sketch of the Avida CPU architecture, and (b) an avidian population on a regular grid. The debt to the tierran CPU structure (Fig. 4.11) is obvious. The CPU shown here is the one designed by Charles Ofria.

Table 4.1. Standard instruction set of the avidian programming language. The notation ?BX? implies that the command operates on a register specified by the subsequent nop instruction (for example, nop-A specifies the AX register, and so forth). If no nop instruction follows, use the register BX as a default. The notation ?$\overline{BX}$? refers to the complement register (see text for definition of complement). The ASCII symbol used to abbreviate the instruction is shown in the column at right.* More details about this instruction set can be found in Ofria et al. (2009).

Instruction	Description	Symbols*	
nop-A	no operation (type A)	a	a
nop-B	no operation (type B)	b	b
nop-C	no operation (type C)	c	c
if-n-equ	Execute next instr. only if ?BX? $\neq$?$\overline{BX}$?	d	d
if-less	Execute next instr. only if ?BX? $\leq$?$\overline{BX}$?	e	e
pop	Remove number from current stack and place in ?BX?	f	p
push	Copy value of ?BX? onto top of current stack	g	o
swap-stk	Toggle the active stack	h	q
swap	Swap the contents of ?BX? with ?$\overline{BX}$?	i	r
shift-r	Shift all the bits in ?BX? one to the right	j	k
shift-l	Shift all the bits in ?BX? one to the left	k	l
inc	Increment ?BX?	l	m
dec	Decrement ?BX?	m	n
add	Calculate sum of BX and CX; put result in ?BX?	n	s
sub	Calculate BX minus CX; put result in ?BX?	o	t
nand	Bitwise NAND on BX and CX; result $\rightarrow$?BX?	p	u
IO	Output value ?BX? and replace with new input	q	y
h-alloc	Allocate memory for offspring, size of offspring in AX	r	w
h-divide	Divide off code between read-head and write-head	s	x
h-copy	Copy instr. from read-head to write-head, advance both	t	v
h-search	Find complement template and place flow-head after it	u	z
mov-head	Move instruction pointer to flow-head position	v	g
jmp-head	Move instruction pointer by value in register CX	w	h
get-head	Write position of instruction pointer into register CX	x	i
if-label	Execute next instr. only if template complement copied	y	f
set-flow	Move flow-head to position specified by ?CX?	z	j

*The first column of symbols refers to the ASCII assignment before 2009 (for example in Lenski et al. 2003), while the second column shows the current assignment (post-2009).

to search for the pattern nop1-nop0, for example, does not automatically return the call itself.

To expand the number of patterns that could be recognized (without making the patterns too long), Ofria introduced three nop instructions, which also could be used to modify other instructions so that they could use one of the three registers AX, BX, or CX. At the same time, the complementarity scheme was designed to be circular, so that the complement of AX is BX (notation: $\overline{AX}$=BX), the complement of BX is CX, while CX's complement is AX. For example, the complement of the pattern ABC is BCA. Another radical change affected how the copying of information worked in the new CPU. Rather than having a single instruction pointer that executes the code, there were now four "machines" that roamed the code: the instruction pointer, a read-head, a write-head, and a flow-head (see Fig. 4.13). Just as the molecular machines (typically, proteins) that attach themselves to biological sequences (typically DNA and RNA), the computational machines attach to the avidian code. The instruction pointer reads the current instruction and instructs the CPU to execute that instruction. The read-head is part of the copying machinery: during the copy process it is placed at the to-be-copied instruction. The write-head is usually placed outside of the executed code, namely at the position in memory where the instruction is to be copied into (this is why the write-head does not appear in Fig. 4.13). The flow-head is used to mark particular locations in the code (in a manner somewhat reminiscent of methylation in biochemical code). Typically, it marks the beginning of the copy loop (as in the example Fig. 4.13), but it is also used to mark an execution branching point, as in instructions 53–54 of the code shown in Table 4.5 (Exercise **4.1c**).

In Tierra (and in Brown's version of Avida), the computational tasks that were rewarded with extra CPU time were mainly addition and subtraction. With Ofria's Avida, digital organisms could evolve much more complex tasks that were based on Boolean logic. In particular, it was now possible to construct complex logic expressions from simpler logic elements. With that change, avidians would now benefit from being proficient at two equally important tasks: replicate the genetic code, and gain the energy necessary to perform that replication. Only a single instruction was provided in the set shown in Table 4.1 to perform Boolean logic: the nand instruction. On a pair of bits, NAND returns a 1 (which stands for the logical TRUE) if any of the two inputs is 0 (standing in for the logical FALSE), and a 0 if both inputs are 1. It is known that NAND is a logical *primitive*, which means that any logical statement can be built with this logical element alone.[15]

---

15. The NOT operator can be implemented with the NAND operator by providing a TRUE along with the variable to be inverted to the NAND gate, see Exercise **4.1**(a).

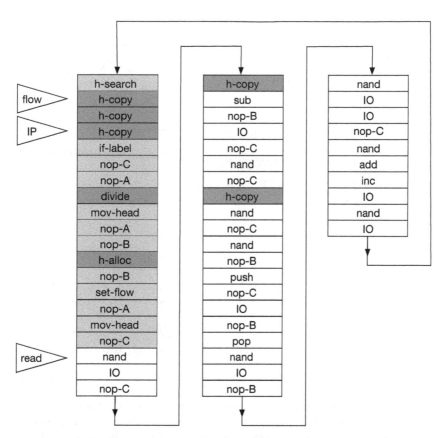

FIGURE 4.13. A typical evolved avidian of fifty instructions, using the standard instruction set described in Table 4.1. The code is organized in a circular manner (like the genetic code of most bacteria), with instructions shaded in terms of the three main functional categories. Instructions in dark gray handle replication of information, while instructions in light gray are responsible for regulating the computational machinery (the various pointers). Unshaded instructions are "metabolic," that is, they are used to compose the logical functions that are used to gain energy. The triangles attached to the code represent the machines that interpret the code. A fourth machine (the write-head) is not attached to the progenitor code because it moves inside the offspring code.

With the logical primitives in place, all the experimenter has to do is to set up a reward structure that determines how much extra CPU time is awarded to an organism that correctly performs a logic operation. Let us say we wanted to reward the performance of the EQU operation. This is a logic gate acting on two inputs, and it returns TRUE if both inputs are either both TRUE or both FALSE. Avidians can read in numbers into the code by executing the IO instruction. Of course, they do not read in just a single bit,

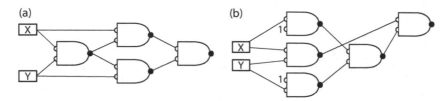

FIGURE 4.14. Building complex logic using the NAND primitive. (a) To perform the EQU operation, a minimum of four NAND gates have to be combined in the manner shown. (b) Performing the XOR operation is even more complex, requiring five NAND gates. In the implementation shown, two NAND gates are used as NOT gates.

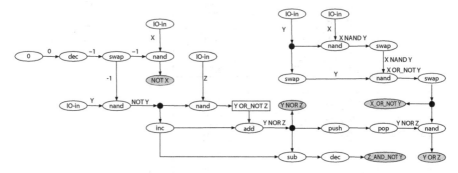

FIGURE 4.15. Computational-metabolic pathway of a digital organism that produces five rewarded computations (gray ovals). The white ovals represent instructions executed by the organism, and numbers or variables represent the information that is being processed. From Adami (2006a).

as a random output would be correct half the time. Instead, the numbers are actually 32-bit binaries, which makes obtaining the reward by random guessing extremely unlikely (the probability of correctly predicting the result of a coin flip thirty-two times in a row).

It turns out that it is not at all easy to create the EQU operation armed only with the NAND operator, because it takes a minimum of four NAND gates to achieve that (see Fig. 4.14[a]). The XOR operation is even more complex, as it requires one additional gate (as the example in Fig. 4.14[b] shows). Since it is known that we can build all computational tasks out of the nand instruction primitive, providing rewards for intermediate functions will allow for the evolution of fairly complex logical functions, including all possible two-input logical operations, and all possible three-input operations.

I will close this brief tour of Avida by showing the flow of information in the metabolism of an avidian that has mastered five different logical operations, combining the three inputs X, Y, and Z. This particular sequence manages to clinch rewards for NOT X ($\neg X$), Y NOR Z ($\neg X \vee Y$), $Z \wedge \neg Y$, and $Z \vee \neg Y$, as well as Y OR Z. To accomplish all these, the organism only uses six nand instructions in total (see Fig. 4.15), using complex ways to operate on partial results. Note that only the results in gray ovals are rewarded, other intermediates are only used as inputs to obtain more complex results. This reliance on building blocks, with some intermediate results used directly while others are only used as ingredients to create even more complex structures, is very reminiscent of how glycolysis (for example) works in bacteria. In essence, these digital critters eat numbers for lunch, and use them to cook up even fancier numbers for profit.

4.4 Promises and Rewards of Experimental Evolution

Experimental evolution in the biochemical and in the digital realm has opened up new ways to explore evolutionary dynamics and to test the validity of diverse hypotheses (Kawecki et al. 2012; Barrick and Lenski 2013). It goes without saying that in any sufficiently complex system, we can come up with many competing ways in which certain observed phenomena can be interpreted. But without the capacity to manipulate a system so as to falsify one hypothesis at a time, progress toward a better understanding of the phenomena would be excruciatingly slow, waiting for the right initial conditions to occur by chance (as, for example, in the accidental experiment on the Croatian islands). While today there are many different experimental evolution systems (including long-running experiments with fruit flies to understand the genetics of aging, Rose et al. 2004) the majority of long-term experiments use microorganisms with short replication time (see, for example, Dunham 2010 for an overview of experiments with yeast). The Lenski long-term experiment discussed in section 4.2 in particular has turned out to be an extraordinarily fertile system, even though the initial conditions (once set in 1988) could not be changed anymore. However, many experiments with different conditions and new hypotheses have been spun off of Lenski's experiment, making it even more fertile than just the original experiment (Fox and Lenski 2015).

The Avida digital life system is a very different tool compared to the Lenski experiment, and has given rise to as many publications, albeit from several different groups. In Avida, the initial conditions can be set almost arbitrarily and varied tremendously, but the focus is not on long-term evolution at all. However, both systems share a number of similarities: an emphasis on simplicity without giving up essential complexity, the consciousness of the power of experimental replication, and the ability to measure what is observable and to preserve history.

It is not possible in this book to review all the insights that experimental evolution in *E. coli* and digital organisms, respectively, have generated. Instead, we will focus here on two examples that have become touchstones of the discussion of the evolution of complexity over time: an experiment to study the evolution of complex features using the Avida system, and the evolution of a new complex trait in one of the lines of the Lenski experiment. Both studies aim to understand how complex traits can evolve from existing traits, but the approach is different. In the Lenski experiment, the complex trait (a new way for the bacterium to obtain energy) was unexpected, and the analysis was largely retrospective: even though the novel trait occurred in only one of the twelve lines, the discovery also led to new prospective experiments (Blount et al. 2008; Blount et al. 2020). In the Avida experiment, on the contrary, the complex trait was expected (because the experimenters could define the different ways in which energy could be obtained beforehand) and the emergence of the complex trait could be observed in multiple replicates. Despite the many differences—in time scale, in trait complexity, and in evolutionary parameters (e.g., mutation rate and population size)—the two experiments provided similar lessons about the unfolding of complexity via the evolutionary process. We cannot help but surmise that the reason for this congruence is that the substrate of evolution is less important than the process that shapes those carriers of information, whether they be nucleotides or bits.

We begin with the tale of how complexity evolves in avidians, followed by the dizzying account of how a bacterium can acquire a whole new bag of tricks. The reader might be wondering how it is possible that Richard Lenski was the lead on both of these projects. In fact, I met Lenski in 1994 when giving a seminar on digital evolution at Michigan State University, which started a long-term collaboration that is still ongoing today. The history of our entanglement via digital and biochemical experimental evolution is retold in Lenski (2020).

4.4.1   The rise of digital "equality consumers"

The fitness landscape of avidians is rigidly defined: you need to replicate as fast as possible, which you can do by gaining as much energy as possible, and use this energy as efficiently as possible in producing offspring. Without a doubt there will be trade-offs involved in managing these different requirements (as obtaining energy takes time), but the evolutionary marching orders are clear. In Avida, energy can be obtained in different ways as we discussed in section 4.3.3, but they all involve computations, and the experimenter sets the relevant demographic, genetic, and environmental parameters. In the digital experiment, Lenski et al. (2003) simplified the fitness landscape such that only nine different logic operations were accessible (also known as the Logic-9 landscape),[16] and the "target" complex feature was the EQU ("equality") logic operation. Indeed, we encountered this operation before in Figure 4.14: it is one of the two most complex $2 \rightarrow 1$ logic operations, and must be built out of simpler components, since only the nand instruction is available as a logic primitive. The study we will discuss aimed at quantifying the mode and manner in which the EQU function evolved. How does evolution build up a complex trait? How does the evolutionary path to the complex trait depend on the presence of other traits that could promote, or interfere with, the eventual emergence of the target trait?

To answer this question, we ran a series of replicate experiments to understand how likely a complex function could evolve from scratch. Now, it should be clear that in biological evolution, nothing evolves "from scratch." Even the simplest organism (that lived at least three billion years ago, maybe even longer) was probably complex enough that there was a significant amount of "material" (meaning genes, or gene-like sequences) that could be reused, recycled, or repurposed. So, the experiment carried out here with avidians investigates the worst-case scenario, in a way. What if you had almost no existing material to aid you in constructing a complex trait?

First, we wanted to know how important a fitness landscape that rewarded steps *toward* the goal was for the emergence of the goal function. In the "Logic-9" landscape, there are nine rewarded functions, and we assume that on the path toward the complex trait EQU, there are other traits that are rewarded as well, and that turn out to be stepping stones and components of the EQU function.

A very instructive example of a complex trait that has gradually evolved from components is the evolution of the vertebrate eye (see Box 4.3), through a sequence starting with nondirectional photoreception, over low-resolution vision to high-resolution vision (Nilsson 2013). At any stage, the existing system is functional in some way, and additions are made by the reuse and repurposing of other existing genes and their encoded products to add to the complexity of the emerging trait.

To wit, people are often puzzled about how it is possible that a protein such as *crystallin* can evolve. Crystallins are the main structural component in the vertebrate lens, but as we can see in Box 4.3, there was a light-sensitive

---

16. In principle, it is possible to write down sixteen different $2 \rightarrow 1$ logic operations, but many of them are equivalent. By default the Logic-9 set includes the $1 \rightarrow 1$ NOT (flip all input bits) but not ECHO (do not change the input) as this operation does not require a NAND. The landscape that rewards all possible $3 \rightarrow 1$ logic operation, is called "Logic-77," and was used in creating Figure 4.15.

Box 4.3. The Evolution of the Vertebrate Eye

How an organ as complex as the vertebrate eye could evolve via a gradual process of refinement has been the subject of discussions ever since Darwin's *Origins*, since he brought it up as something that needs an explanation. Since those times, we have learned a great deal about the evolutionary history of how the eye formed, in particular that eyes (or eye-like structures) have evolved many times, in many different lineages, because of the obvious survival advantages they confer (Nilsson 2009). However, this does not mean that all these different eyes have an independent evolutionary history: the development of the camera eyes of vertebrates and squids, for example, as well as the compound eyes of flies, is controlled by a single gene called Pax-6.

Eyes appeared on the evolutionary scene just when the Cambrian explosion got under way (about 540 million years ago), which makes sense as this period saw a proliferation of hard-shelled large predators that hunted smaller more squishy prey. In fact, it is likely that the earliest precursors to eyes evolved long before that time, in the form of light-sensitive patches, or "eyespots." These light-sensitive cells would gradually evolve to become photoreceptors that are connected to the brain via nerve fibers (see Figure 4.16[a]). Folding these photoreceptors in (Fig. 4.16[b]) allows for more directionality, while shaping the fold into a pinhole increases acuity, as seen in Figure 4.16(c). Replacing the water in the fold by a liquid with higher refractive index (the "humor," see Fig. 4.16[d]) improves vision even more by sharpening the image. Finally, the evolution of the iris and lens (the latter shaped out of the crystallin protein, see Figs. 4.16[e–f]) makes a fully focused image possible, albeit upside down on the retina (which, however, the neural system corrects for by interpreting the image in the correct orientation).

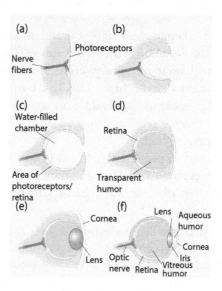

FIGURE 4.16. Major stages in the evolution of the vertebrate eye. Modified from Wikimedia (CC-BY-SA 3.0).

structure before it was filled with crystallin proteins. How could evolution "know" to evolve a crystallin protein, when the beneficial effect of the protein can only be apparent when it is fully evolved and functional?

The answer is that the crystallin proteins have many functions, and proteins of the alpha-crystallin family (there is also a sister group of crystallin proteins called betagamma-crystallins, see for example Andley 2007) in particular are also chaperones, which are proteins that help other proteins fold in time of stress.[17] What this means is that evolution was not simply "waiting" for the crystallin proteins to become available until they could be used in the eye. They were already present and functional, helping out other proteins respond to stressful conditions, and when a mutation arose that revealed that the protein could *also* be used to make a lens (maybe because it stuck to other proteins to make a stable whole) the organism that carried that mutation had somewhat better eyesight, which allowed it to become dominant so that further mutations in the protein could improve the new use as a lens structural protein.

In his *Origin of Species*, Darwin showed that he was fully aware that understanding the evolution of complex traits posed a serious challenge to his theory. He wrote about the vertebrate eye in particular:

> To suppose that the eye with all its inimitable contrivances for adjusting the focus to different distances, for admitting different amounts of light, and for the correction of spherical and chromatic aberration, could have been formed by natural selection, seems, I freely confess, absurd in the highest degree.
>
> (DARWIN 1859, 186)

Those that have a vested interest in discrediting evolution as the generative process of biocomplexity have pounded on that sentence, while ignoring Darwin's far-sighted resolution coming immediately after:

> Yet reason tells me, that if numerous gradations from a perfect and complex eye to one very imperfect and simple, each grade being useful to its possessor, can be shown to exist; if further, the eye does vary ever so slightly, and the variations be inherited, which is certainly the case; and if any variation or modification in the organ be ever useful to an animal under changing conditions of life, then the difficulty of believing that a perfect and

---

17. The co-option of crystallin-like proteins apparently happened more than once: the lens-protein in some octopus species are similar to the crystallin in vertebrates (Zinovieva et al. 1999), even though cephalopod mollusks and vertebrates last shared an ancestor over 560 million years ago.

Table 4.2. Rewards (multipliers of execution "speed") for performing nine one- and two-input logic functions on inputs A and B. The symbol ¬ is the logical negation.

Function name	Logic operation	Computational multiplier
NOT	$\neg A; \neg B$	2
NAND	$\neg(A \text{ and } B)$	2
AND	A and B	4
OR_NOT	$(A \text{ or } \neg B); (\neg A \text{ or } B)$	4
OR	A or B	8
AND_NOT	$(A \text{ and } \neg B); (\neg A \text{ and } B)$	8
NOR	$\neg A \text{ and } \neg B$	16
XOR	$(A \text{ and } \neg B) \text{ or } (\neg A \text{ and } B)$	16
EQU	$(A \text{ and } B) \text{ or } (\neg A \text{ and } \neg B)$	32

complex eye could be formed by natural selection, though insuperable by our imagination, can hardly be considered real.

(DARWIN 1859, P. 186–187)

This story of reusing, recycling, and repurposing existing proteins is found again and again when researchers are able to unravel the evolutionary history of a complex trait. It is also exactly (albeit in the computational domain) what we found in the evolution of the EQU trait, as we'll now see.

The study of how the evolution of the EQU function unfolded has two parts. In the first part, we tested statistically how important building blocks are for the emergence of the fully formed trait. This allowed us to test how important a modular structure of the fitness landscape (a landscape where complex function is achieved by combining building blocks) is for the evolution of complex functions. In the second part, we analyzed a particular evolutionary path in detail, to be able to document the emergence of a new trait mutation by mutation, in unprecedented detail.

To test the importance of building blocks, we first tested how likely it was that the EQU function would emerge if evolutionary parameters such as population size and mutation rate were kept constant, while systematically varying which functions are rewarded. Keep in mind that EQU is only one of nine potentially rewarded logic functions; the other eight are shown in Table 4.2. From Table 4.2 we can immediately see how complex logic operations can be built from simpler ones. The XOR function, for example, can be cobbled together using two AND_NOT functions, which themselves require the NOT and the AND functions. We can also see this in the computational "metabolic pathway" in Figure 4.14.

When all of the nine functions in Table 4.2 were rewarded, the EQU function evolved in almost half of the experiments we ran (twenty-three out of fifty). We then asked, "What if one or two of the remaining eight possible building blocks are not rewarded?" There are thirty-six such distinct alternate fitness landscapes (eight in which one task was unrewarded, and $\binom{8}{2} = 28$ where any combination of two out of the eight potential building blocks did not reap a benefit). In each of those landscapes, EQU evolved at least once out of ten tries, showing that no single building block, nor even any pair of building blocks, is essential to the evolution of EQU. Indeed, EQU evolved 124 times in these 360 experiments, only slightly fewer than the 23/50 we found in the fully rewarded landscape (Lenski et al. 2003).

But what if we *only* rewarded the EQU function? We found that in that case EQU evolved in *none* of fifty experiments, even though these populations typically evolved smaller genomes, and as a consequence were able to "test" more genotypes (about twenty million, on average) during the experiment[18] than the populations exposed to the full landscape (those experiments tried out about ten million different genotypes). We can conclude from these observations that having some building blocks is indeed essential for the evolution of the EQU trait. This really is an essential lesson of evolution: complex things are built from other (perhaps somewhat less complex) things. The evolutionary process is opportunistic, as if thinking: "What do we have here that we can take advantage of? We need something to make a lens! How about we use this protein that tends to clump, but also happens to be transparent? What if we modify that just a bit?" Of course, evolution does not think. The message here is that evolution tends to tinker like a worker in the shop does, who must make do with what is already there but is also equipped with the capacity to make small modifications (Jacob 1977). We should also note that none of these populations will ever explore even the tiniest fraction of all *possible* avidians, since for sequences of length 50 (the length of the ancestral sequence used to start all the experiments here) there are about 5×10^{70} possible digital critters, a truly astronomical number.

How then does the evolutionary process take advantage of building blocks in order to build a complex function? One way to find out is to investigate the *line of descent* of a population that evolved the function. In these asexually reproducing populations, the line of descent can be reconstructed by picking

---

18. All experiments were ended after 100,000 updates had elapsed. The population is "updated" if each organism in the population has executed thirty instructions on average. As these experiments had 3,600 organisms each, an update corresponds to $30 \times 3,600$ instruction executions. This means that if it takes each organism 210 instruction executions to replicate (say), a generation would take seven updates.

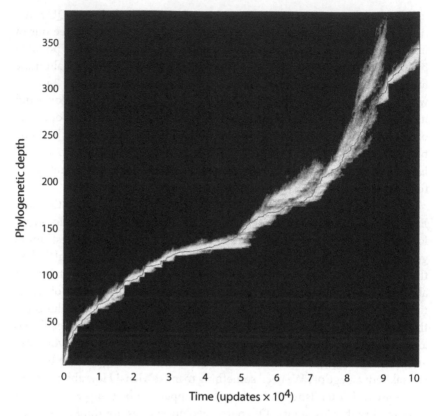

FIGURE 4.17. Phylogenetic depth of an example population that evolved the EQU function, over 100,000 updates. Different shades of gray indicate the relative abundance of genotypes at a given depth, with brighter shades indicating higher abundance than the darker ones. The actual line of descent anchored at an arbitrary genotype at update 10^5 is superimposed in gray. From Lenski et al. (2003).

an arbitrary organism from the population at the end of the experiment, and then following the line backward from offspring to parent, until one arrives at the organism used to begin the experiment.

Figure 4.17 shows a graph of the *phylogenetic depth* of a population over 100,000 updates, with the line of descent superimposed as a gray line. The phylogenetic depth can be calculated for every organism in the population by measuring how many mutational changes have occurred when reconstructing that organism's particular line of descent back to the ancestor. Because many organisms in the population do not belong to the final line of descent (because ultimately their descendant lineage became extinct), the phylogenetic depth is, at each moment in time, a distribution of distances. In Figure 4.17, this distribution is rendered using a gray scale where a brighter shade indicates more organisms in the population at that distance. We can see from this figure that during an experiment there can be groups of organisms that appear to form a separate "species" that diverges from the main line (for example, the lobe at the top right of Fig. 4.17). However, we know from the competitive exclusion principle (Hardin 1960) that in a single-niche environment (such as the one we constructed here) only one ecotype will eventually prevail, and therefore ultimately the line of descent is an unbroken line that only branches at the very end of the experiment, because the fate of each of the branches has not yet been decided. The actual line of descent thus coalesces quickly when going back in time (it is indicated in gray in Fig. 4.17).

Let us take a closer look at the line of descent in the evolution of EQU consumers, now moving forward in time instead of the backward thinking that allowed us to identify that lineage. The line begins with the ancestral avidian of course, whose genome consists of fifty instructions that contain a handful of instructions that allow the avidian to replicate, and otherwise consists of only the inert nop-C instruction that does nothing when devoid of any context (see Table 4.1). Thus, that ancestor cannot compute any logic functions: all it can do is replicate. Figure 4.18 shows two excerpts of this line of descent (LOD): the beginning and the crucial part that ends with the first EQU-capable sequence. The first sequence on that line is of course the ancestor, and the letter c is the ASCII abbreviation of the nop-C command (all the abbreviations are listed in Table 4.1).

The next sequence on the LOD has a single point mutation at the thirty-fifth instruction (a nop-C), which was turned into a swap-stk, that is, a mutation c35h using the ASCII abbreviation of instructions. We know this level of detail because in an Avida experiment the LOD can be recorded in such a way that not only is the sequence stored, but also its fitness, along with when the mutation occurred (see Fig. 4.18). Our ability to reconstruct this LOD in such exquisite detail validates the investment in creating this digital life system: the amount of information we can gather is really unprecedented in evolution (experimental or otherwise). We see that this c35h mutation was neutral (it had no effect on fitness), whereas the next mutation (an insertion of a p just after instruction 30, which is possible because genome size was not constant in this experiment) actually increased fitness by 1 percent.

The Avida system allows us to examine each sequence entirely by itself (in a test environment) where we could also, in principle, analyze the functional consequences of that p insertion, that is, how and why it resulted in a 1 percent speed-up in replication. Interestingly, p stands for the nand instruction, but

PD	Born	Functions									Fit	Genome Sequence
0	0	0	0	0	0	0	0	0	0	0	—	rucavccccccccccccccccccccccccccccccccccccutycasvab
1	32	0	0	0	0	0	0	0	0	0	1.00	rucavcccccccccccccccccccccccccccchccccccutycasvab
2	93	0	0	0	0	0	0	0	0	0	1.01	rucavccccccccccccccccccccccccccpcccchccccccutycasvab
3	143	0	0	0	0	0	0	0	0	0	1.36	rucavcccccccccccccccccccccccccccp_ccchcccccccutycastvab
4	225	0	0	0	0	0	0	0	0	0	1.01	rucavccccccccccccccccccccccccccpcnchcccccccutycastvab
5	284	0	0	0	0	0	0	0	0	0	1.01	rucavcczcccccccccccccccccccccccpcnchccccqcutycastvab
6	352	0	0	0	0	0	0	0	0	0	1.00	rucavcczccccccccccccccccccccccqcnchccccqcutycastvab
7	361	0	0	0	0	0	0	0	0	0	1.00	rucavcozccccccccccccccccccccccqcnchccccqcutycastvab
8	379	0	0	0	0	0	0	0	0	0	1.01	rucavcozccccccccccccccccccccxccqcnchccccqcutycastvab
9	411	0	0	0	0	0	0	0	0	0	1.00	rucavcozcccccccccccccccccccacxccqcnchccccqcutycastvab
10	567	0	0	0	0	0	0	0	0	0	1.00	rucavcozcccccccccccccccccccacxecqcnchccccqcutycastvab
11	734	0	0	0	0	0	0	0	0	0	1.00	rucavcozccccccccccccccccccamxecqcnchccccqcutycastvab
12	775	0	0	0	0	0	0	0	0	0	1.00	rucavcozcccscccccccccccccamxecqcnchccccqcutycastvab
13	793	0	0	0	0	0	0	0	0	0	0.99	rucavcozcccscicccccccccccamxecqcnchccccqcutycastvab
14	835	0	0	0	0	0	0	0	0	0	1.00	rucavcozcccscicccccccccccamxelqcnchccccqcutycastvab
15	949	0	0	0	0	0	0	0	0	0	1.00	rucavcozjccscicccccccccccamxelqcnchccccqcutycastvab
16	963	0	0	0	0	0	0	0	0	0	0.99	rucavcozjccscicccccccccccamxelqcnqhccccqcutycastvab
17	1118	0	1	0	0	0	0	0	0	0	2.01	rucavcozjccscicccccccccccamxelqcnqhccpcqcutycastvab
18	1194	0	1	0	0	0	0	0	0	0	1.01	rucavcotzjccscicccccccccccamxelqcnqhccpcqcutycastvab
19	1250	0	1	0	0	0	0	0	0	0	1.01	rucavcotzjcisccicccccccccccamxelqcnqhccpcqcutycastvab
20	1252	0	0	0	1	0	0	0	0	0	2.01	rucavcotzjciscicccccccccccamxelqcnqhpcpcqcutycastvab
⋮	⋮										⋮	
100	22404	1	1	1	1	1	1	1	0	0	2.90	rmzavcgtzciqptqpqcpctltncogctbnamqdtqcptipqfpqqcutycuastttvab
101	22412	1	1	1	1	1	1	1	0	0	1.00	rmzavcgtzciqptqpqcpctltncogctbnamqdtqcptipqfpqqcutycuastttva_
102	22487	1	1	1	1	1	1	1	0	0	1.00	rmzavcgtmciqptqpqcpctltncogctbnamqdtqcptipqfpqqcutycuastttva
103	22586	1	1	1	1	1	1	1	0	0	0.99	rmzavcg_mciqptqpqcpctltncogctbnamqdtqcptipqfpqqcutycuastttva
104	22629	1	1	1	1	1	1	1	0	0	1.05	rmzavcgmciqptqpqcpctletncogctbnamqdtqcptipqfpqqcutycuastttva
105	22864	1	1	1	1	1	1	1	0	0	1.00	rmzavcgmciqptqpqcpctletncogctbkamqdtqcptipqfpqqcutycuastttva
106	22886	1	1	1	1	1	1	1	0	0	0.99	rmzavcgmciqptqpqcpctletncogc_bkamqdtqcptipqfpqqcutycuastttva
107	23002	1	1	1	1	1	1	1	0	0	1.02	rmzavcgmciqqptqpqcpctletncogcbkamqdtqcptipqfpqqcutycuastttva
108	25881	1	1	1	1	1	1	1	0	0	1.00	rmzavcgmciqqptqpqcpctletncogcbqamqdtqcptipqfpqqcutycuastttva
109	26343	1	1	1	1	1	1	1	0	0	1.01	rmzavcgmciqqptqpqcpctletncogcbeamqdtqcptipqfpqqcutycuastttva
110	27437	1	0	1	1	1	1	1	0	0	0.49	rmzavcgmciqqptqpqcpctletncogcbeamqdtqcptipqfpqqxutycuastttva
111	27450	1	0	0	1	1	1	1	0	1	8.00	rmzavcgmciqqptqpqcpctletncogcbeamqdtqcptipqfpgqxutycuastttva

FIGURE 4.18. Part of a line of descent (LOD) of the case-study experiment (shown also in Fig. 4.17). This LOD has 112 sequences on it (counting the ancestor), and the first 20 and the last 11 before emergence of EQU are shown here. (The full LOD can be perused in the Supplementary Information of Lenski et al. 2003.) The first column gives the position on the LOD, followed by the update at which that mutant was born, as well as an array that indicates which of the nine logical functions the sequence can perform, in the order shown in Table 4.2 from least complex to most complex. This is followed by the *relative* fitness of the sequence compared to the sequence that immediately preceded it on the LOD (except for the ancestor, whose fitness is defined as 1.0) and finally the sequence itself, written in terms of the ASCII instruction code (see Table 4.1).

it does not trigger a reward here because at this point in time, no numbers were being read or written out. Here, the insertion of nand creates a fitness advantage because an increase in length (by design) carries some benefit so as to offset the increased time to replicate the longer sequence. Thus, this advantage was entirely unrelated to the actual function of the inserted instruction.

The next mutation on the LOD (genotype 3) was a deletion, and it occurred right after the instruction that was just inserted. This deletion was enormously beneficial (to the tune of 36 percent), and we note immediately that it would have been better to just mutate instruction c31p rather than inserting a p and then deleting the following c. It is events like these that remind us that in the evolutionary process there is no requirement, dictate, or imperative for parsimony. Evolution takes what happened and works with it.

The first mutation that creates a logic function is genotype 17, which carries the mutation c38p. That mutation, along with others that occurred before, confers the capacity of carrying out NAND, and as a consequence genotype 17 experiences a replicatory speed-up of a factor 2. In fact, Figure 4.18 shows a 2.01-fold increase, but this might be because the organism just preceding it on the LOD actually carried a mutation that was deleterious (by 1 percent) and the substitution *reversed* the deleterious effect of that mutation at the same time. This means that the substitution c38p occurred on the background of a genotype that was inferior to the current wild-type. In other words, we recorded a *valley-crossing* event, where a deleterious mutation is rescued by an advantageous mutation that soon follows.

How is it possible that the NAND function was activated via the single substitution c38p, when the NAND function requires at least five instructions? A typical sequence that achieves NAND would be qncqpq, which translates into "read the first number into the BX register, add it to the CX register (which at this point is empty, thus moving the number into the CX register), read in another number into BX, perform NAND on the pair, then write it out." The actual code in genotype 17 that achieves the function is more complicated. Leaving out nonfunctional instructions, the sequence that performs NAND is qcnqpcqc. And indeed, in the absence of the c38p mutation, the sequence is not functional. In this particular sequence, the computation is actually performed in the CX register rather than the default BX register, which makes sense in hindsight as the c instruction that modifies the input/output (and other instructions) is the nonfunctional "filler" in the ancestral sequence. But it is clear that without the presence of the two qs that read in the two numbers, as well as the q that writes the result into the output, the c38p substitution would have no effect. Indeed, we can see that the first q occurred as the fifth substitution, and it had a 1 percent advantage at the time. But it was part of a double substitution (z was also substituted at position 8), so again without testing in detail it is not clear what this advantage is due to. The second q was introduced as mutation 8, and was completely neutral. Thus, we see that the instructions necessary for the final c38p to work were completely unrelated to the NAND function and emerged earlier, and were either neutral or influenced replication speed in a completely independent manner.

It is hard not to see the parallels to other evolutionary innovations, for example the co-option of the existing crystallin protein in the evolution of the eye, or the co-option of an existing transporter protein, which we will encounter in the bacterial example that we will discuss in the following section 4.4.2. In all of these cases, single mutations (whether substitutions or more complex events, as in the next section) take advantage of existing variation that is present in the genome either by chance or because it serves a different function.

Let us now skip ahead and look at how the EQU function was finally obtained. The mutation that triggers EQU looks innocent enough: q46g, a substitution that turned an IO (which writes a result to the output and at the same time reads in a new number) into a push instruction, which in this case pushes the number in the BX register onto the stack. A look at the fitness gained in Figure 4.18 shows that while this mutation (number 111, the last one in the list) finally produced EQU, the sequence did not gain the multiplier 32 that performing EQU should reap (see Table 4.2). Instead, it gained a factor of 8 only. A closer examination of the functionality of this genotype (and its immediate progenitor) reveals (see Exercise **4.1[c]**) that this reduced benefit is due to the mutation q46g also *destroying* the rewarded function AND at the same time. Indeed, q is an input/output (IO) operation, and with hindsight we can see that this output previously triggered the bonus for AND, resulting in a four-fold speed increase. By converting this into a push, the AND function and its resulting bonus was lost, but at the same time the machinery that was used to trigger AND was integrated into a grander scheme that now triggered the even more valuable EQU. But the line of descent has more surprises to offer.

The fortuitous q46g occurred on a sequence that carried the highly deleterious mutation c48x, which abolished the NAND function that had evolved much earlier. Was it just happenstance that the beneficial EQU-triggering mutation occurred on a damaged sequence (one that replicated at half the speed than most others in the population) or was that flaw an integral part of the "grander scheme"? In digital evolution, we can easily answer such questions directly, by manually creating the genotype in which the deleterious c48x has been reverted, and testing the fitness of this construct (this can also be done with much more labor in experimental evolution;

see for example Khan et al. 2011). What we found is that q46g is *not* functional without the prior c48x. In other words, c48x was only dele-terious when looking backward, but not when looking forward, that is, in conjunction with the future mutation q46g (much like the case discussed in Box 4.4).

The q46g mutation (along with the mutation c48x that preceded it) is a beautiful example of how evolution operates: when mutations create novelty, an existing function may be disrupted or destroyed at the same time, for the simple reason that the code that is modified is likely already functional. At the same time, mutations that appear to be detrimental when they occur can turn out to be crucial in the future.

This type of interaction between mutations was in fact statistically com-mon in the digital experiment: of the twenty-three "pivotal" mutations that triggered the EQU bonus in the study, five occurred on a background that was a deleterious mutation (a valley-crossing event). When the deleterious background mutation was reverted in these five cases, the EQU function was eliminated in three of those cases (Lenski et al. 2003).

Either we were lucky to capture this lesson in the same LOD that taught us another lesson in the evolution of complexity (namely the co-option of existing variation into a new trait) or these mechanisms are so common in evolution that they would show up frequently anytime you analyzed the line leading to a complex trait in detail. Examples in the biological world abound. For example, in the evolution of resistance to the influenza drug oseltamivir, the mutation that (by itself) confers this resistance to the influenza virus actually compromises the virus's fitness. However, a pair of mutations that occurred just before the emergence of the resistance mutation in the wild counteracted that fitness decrease and thus "permitted" the evolution of resis-tance (Bloom et al. 2010). A similar example has been documented in the evolution of antibiotic resistance (Wang et al. 2002). For another example, this time in the evolution of intracellular communication, see the molecular analysis of steroid receptor evolution on the LOD in Box 4.4.

4.4.2   The rise of bacterial citrate consumers

When scientists want to characterize bacteria, they often turn to the ency-clopedic *Bergey's Manual of Systematic Bacteriology*, a voluminous tome first published in 1923 (now in five parts). If one were to look up *E. coli* in *Bergey's*, one of its defining characteristics has been that it cannot grow on citrate (also known as citric acid, a substance that occurs naturally in citrus fruits) when oxygen is present. This is particularly surprising given that many other bacteria in the same family are perfectly capable of growing on citrate. Indeed, it is

Box 4.4. The Evolution of a New Steroid Receptor

Much of what goes on in a cell is communication. Cells need to know what is going on in neighboring cells (or other parts of the organ or body); and within a cell signals need to be transmitted from one organelle to another, and sometimes from one protein to another. A typical example of communication within the cell is the steroid communication system. Steroids are hormones that are used extensively for signaling throughout the tree of life, which suggests that their origin is ancient. There are many different steroid molecules, so to be useful for signaling each steroid should have a particular receptor (a molecule that can detect the presence of the steroid, and which is often embedded in a membrane). We can think of a steroid/receptor pair as a lock-and-key system: when the right key binds to its lock, the signal is recognized by the receptor, which in turn leads to further events within the cell. This observation has caused people to wonder: how can a lock-and-key system evolve, given that the key can only be useful if a lock to fit it is already present, and why would a lock evolve before there is a corresponding key?

By mechanisms similar to those we discuss in the evolution of digital EQU consumers and bacterial citrate consumers, steroid communication systems can evolve because there are locks that can fit several different keys, and there are keys that can fit several different locks. Bridgham et al. (2006) sought to understand how two different lock-and-key systems could evolve: the glucocorticoid receptor that is activated by cortisol, and the mineralocorticoid receptor that is "unlocked" by the hormone aldosterone. Because the glucocorticoid receptor regulates metabolism, inflammation, and immunity while the mineralocorticoid receptor controls electrolyte homeostasis, it is important that neither of the hormones activate the wrong receptor. It is known from phylogeny that the two receptors derive from the same ancestral sequence, having evolved from an ancient gene-duplication event over 450 million years ago. However, we also know that aldosterone evolved long after that duplication. How could the mineralocorticoid receptor lock have evolved if there was no steroid key to activate it? To answer this question, Bridgham et al. reconstructed the ancient receptor using methods that makes it possible to "resurrect" ancient proteins from their predicted sequences. They found that the ancient receptor was able to bind both cortisol and aldosterone (even though aldosterone was not present at the time), and that two mutations, L111Q and S106P, cooperated to create the glucocorticoid receptor by losing the aldosterone specificity. It is interesting that the two mutations strongly depend on each other: L111Q makes the receptor insensitive to any hormone, while S106P by itself destroys cortisol binding, but retains weak binding to a third steroid. Together, the two mutations were critical to creating the glucocorticoid receptor. While Bridgham et al. argued that this tells us that S106P has preceded L111Q on the line of descent, we can argue that the opposite order is also allowed, as long as S106P follows L111Q quickly enough (Adami 2006b), just as in the last two steps of the line of descent that produced EQU.

Table 4.3. Ingredients of Davis-Mingioli
(DM) Broth with glucose

Ingredients	grams/liter
Glucose	0.025
Ammonium sulphate	1.0
Dipotassium phosphate	7.0
Monopotassium phosphate	2.0
Sodium citrate	0.5
Magnesium sulphate	0.1

likely *because* of this peculiarity that microbiologists chose to characterize *E. coli* in this manner. Being an intermediary in the *citric acid cycle* (also known as the TCA cycle), citrate is an important component of the metabolism of all aerobic organisms.

We briefly looked at the growth medium Richard Lenski used when designing the LTEE in section 4.2: the "Davis Broth," supplemented with glucose. If you were to read the list of ingredients, it looks innocent enough: If it was not for the glucose, there is nothing for *E. coli* to eat in this mixture. It is basically a bunch of minerals, plus sodium citrate as a chelating agent (and a small amount of thiamine, also known as vitamin B1, not shown in Table 4.3). The citrate cannot interfere with growth because the inability to digest citrate results from the fact that *E. coli* cannot take up the citrate from the medium (at least not when oxygen is present). But keep in mind that the DM25 medium is Davis-Mingioli Broth supplemented with 25 milligrams of glucose per liter, while the citrate concentration (in the form of sodium citrate) is *twenty times as high* (see Table 4.3). In other words, there is an ecological opportunity hidden in this medium that is the equivalent of a triple burger with double fries, washed down with plenty of soda, sitting right next to a spartan meal of a few crumbs of cereal. For *E. coli* bacteria, a bonanza so close, but so far at the same time. And for the first 30,000 generations or so, that jackpot remained out of reach. However, at generation 33,127 (about fifteen years after the experiment was started) one of the twelve populations showed very disturbing signs: the normally clear mixture of medium and bacteria turned alarmingly cloudy (Blount et al. 2008).

Usually this excess cloudiness is a sign of contamination, like a foreign bacterium accidentally colonizing the flask—one that actually *can* grow on citrate—and becoming so abundant that the mixture becomes increasingly turbid. This had happened a few times before, so once the team noticed the cloudy flask, they went back to the previous day's flask, to try the transfer again. But when the turbidity emerged several times in the same manner, the team began to suspect that something much more interesting had happened to the bacteria. First they made sure that this was really the *E. coli* bacteria they expected, and not some interloper. The strain was indeed negative for growth on the sugar arabinose (the flask in question was the "Ara-3" population, see the picture in Fig. 4.6), and it was sensitive to the bacteriophage T5, while resistant to T6: exactly like the ancestor REL606. So while the careful design of the experiment paid off once more, the question still was: What had happened to the population in Ara-3?

The answer is that the Ara-3 population evolved to be able to grow on citrate, and the story of how exactly this happened is as breathtaking as it is complex. The bigger story that emerges (as we shall see) is that the evolution of a novel function is likely to be complicated because in any existing organism that has long made its living in one environment, the existing genes all are pretty much adapted to each other in an optimal way, because if they were not there would be a selective pressure to optimize them. As a consequence, a change will likely be detrimental and will have to be compensated by changes somewhere else. According to this view, the emergence of novel functions occurs via a three-step process (Blount et al. 2012): potentiation, actualization, and refinement. We will take a look at how these three processes played a role in the emergence of these citrate-utilizing *E. coli* bacteria, arguably a new species of bacteria (Turner et al. 2015; Blount and Lenski 2013).

It was clear from the very beginning that the evolution of citrate utilization was not a single-point mutation, because by generation 30,000 every population must have "tried" all those mutations multiple times (Blount et al. 2008). Furthermore, the reason why *E. coli* cannot grow aerobically on citrate was known. It is not that *E. coli* lacks the genes to utilize citrate (a transporter to bring citrate into the cell, and enzymes to use it for energy). It does in fact have such genes, except that the transporter, in particular, is contained in an operon that is only expressed in the absence of oxygen. When oxygen is present, those genes are as silent as if they did not exist.[19]

Could that transporter have been activated somehow? Sequencing of the genome revealed that just such an activation had happened, via a tandem duplication of the operon that includes the citrate transporter. The duplication (of a 2,933 base pair genomic segment) put the citrate transporter operon under the control of a different promoter: one that was activated even in

---

19. In fact, the LTEE ancestral strain REL606 has an additional limitation to using citrate. At some point in its history (prior to the LTEE), a transposable element called IS1 hopped into the citrate operon (Schneider et al. 2000), quite likely disrupting the cell's ability to grow on citrate even in the absence of oxygen.

the presence of oxygen, just as the glucose was exhausted. This duplication-mutation created an active citrate transporter by recombining an existing gene-regulatory region with an existing protein-coding gene—the *actualization* of the novel function. However, a careful analysis of types isolated from earlier time points showed that the duplication was actually present quite a bit earlier, by around 31,500 generations. The clone isolated from that time point (remember clones are stored every 500 generations) could utilize citrate, but it grew only very slowly on citrate—so slowly, in fact, that it did not affect the cell density and therefore did not cloud the Ara-3 culture. What had happened between generations 31,500 and 33,000 that strengthened the actualization that occurred earlier?

Further careful analysis revealed that two processes were important for this change. To understand the genetics of it, we have to briefly delve into the biochemistry of membrane transporters (like the citrate transporter). This transporter is actually an *antiporter*: as it imports citrate from outside the cell, it must simultaneously *export* succinate out of the cell. There is a good reason for this quid pro quo: when citrate is metabolized, it is broken down sequentially into succinate, and needs to leave the cell, lest it accumulates there. But if there is no succinate in the cell to pump out, the antiporter cannot operate. It is either give-and-take or nothing.

This is precisely the kind of situation that we discussed earlier: an innovation that looks good by itself, but that does not work in the context of the other genes. So what would be needed? For example, if there was a way to independently import succinate into the cell, that might work. And it so happens that between the origination of the citrate transporter around generation 31,500 and the rise of the efficient citrate utilizers at generation 33,127, just such a mutation occurred. By analyzing all the mutations that the team (Quandt et al. 2014) found on the line of descent between 31,500 and 33,127 in isolation, they found one that conferred that ability: a mutation in the promoter region of a gene called *dctA*: a succinate importer! Other changes that made citrate utilization more powerful were easier to understand, such as having more copies of the active antiporter (which evolution produced via gene duplication) so that more citrate can be imported into the cell. However, other questions remained. Why did it take so long for this new ability to arise? And why have none of the other eleven lines evolved this way of life?

To answer this question, Zachary Blount (then a postdoc in Lenski's lab) performed extensive "re-evolution" experiments, to learn whether the likelihood to evolve the citrate-utilization phenotype ("Cit$^+$," for short, indicating an active citrate transporter—as opposed to the "Cit$^-$" phenotype in which the transporter is inactive) depends on time, in particular on other genetic changes that had happened in the Ara-3 population *before* the actualization event. The results were unambiguous: the emergence of the Cit$^+$ phenotype was more likely to take place for clones sampled after generation 30,000, so some mutations most have occurred during this time period that made the evolution of citrate usage much more likely (Blount et al. 2008). That would be the *potentiation* event (or events): some mutations must have potentiated the evolution of the Cit$^+$ phenotype, but which mutations? After all, there are seventy-seven mutations that separate the ancestor REL606 and the earliest Cit$^+$ clone (Leon et al. 2018)!

While investigating this question, the team found something that seemed even more puzzling. When they inserted the functioning citrate machinery into clones taken from various earlier time points (a "knock-in"), they found a surprise. Because this knock-in was lacking the "refining" mutations that came later, it was expected that the fitness effect of the knock-in would be only slightly beneficial. This was certainly the case in many clones (even in the ancestral strain). However, for several time points between 20,000 and 30,000 generations, the presence of the citrate transport system was actually detrimental, meaning that during that time period the population was not potentiated for the evolution of the Cit$^+$ phenotype, it was in fact *antipotentiated* (Leon et al. 2018). This means that between roughly 20,000 and 30,000 generations into the experiment, even if the activating mutation had occurred, it would never have survived (gone to fixation) as it was actually detrimental to the cell! How is it possible that the background of mutations can change so dramatically?

There appear to be multiple reasons why the Cit$^+$ phenotype did not emerge during the first 10,000 generations, even though during that time citrate utilization was marginally beneficial. For one thing, the actualizing mutation—the specific tandem duplication seen in the Ara-3 population— occurs at a very low rate (much lower than a typical point mutation), even in the potentiated genetic background (Blount et al. 2008). Moreover, beneficial mutations are by no means guaranteed to survive. In fact, most of them— especially those that are only slightly advantageous—will be lost before they become numerous, owing to the random process of drift during the 100-fold dilutions that occur each day (see Box 6.1). On top of that, *E. coli* was not "used to" (had not evolved in) the particular laboratory environment in which they grow for a few hours, only to find themselves starving for the rest of the day (and with no predators to defend against, no competing species to contend with, etc.). As a consequence of the novelty of these experimental conditions, there were many obvious improvements that evolution could make more easily, with much larger fitness benefits (Barrick et al. 2009). During this initial period, even if the Cit$^+$ mutation had occurred, it could not compete with those other beneficial mutations because the fitness benefit of the Cit^+ trait (without that advantage of the refinement) was so slight. Other changes in the glucose metabolism (in particular the evolution of cross-feeding on acetate) may also have interfered with establishment of a Cit^+ mutation during the period between 10,000 and 30,000 generations (Quandt et al. 2015; Leon et al. 2018).

The story of an evolutionary innovation, as told by this tale of citrate utilization, is not a simple one. But the message is unambiguous: trying something new in the background of established structures is difficult. The new "idea" does not fit in well with how things have been done in the past. At times, other recent changes may even work against these new ideas. But with luck, more changes will mitigate those obstacles, until the day comes that the innovation is—even if reluctantly at first—given a chance to show its potential for good, and shine supreme. If only this was a metaphor that worked in other endeavors!

4.5 Summary

Evolution is often dismissively characterized as "just a theory," in an attempt to confound the word "theory" with "speculation." However, a scientific theory is much more than a mere guess or conjecture: it is a consistent framework of ideas that makes it possible to formulate qualitative and even quantitative predictions that can be tested by anyone. In this sense evolution is indeed a theory, and Darwin was able to make predictions armed with this theory right away. For example, he encountered an orchid that had an astoundingly long nectary that would make it impossible for any of the known pollinators to reach it. But given that the orchid was there, he was able to predict the existence of the pollinator and wrote: "In Madagascar there *must* be moths with proboscises capable of extension to a length of ten to twelve inches" (Darwin 1862). This prediction was ridiculed by some entomologists, but not by Alfred Russel Wallace who agreed, writing:

> That such a moth exists in Madagascar may be safely predicted; and naturalists who visit that island should search for it with as much confidence as astronomers searched for the planet Neptune—and they will be equally successful!
>
> A. R. WALLACE (1867)

In particular, Wallace made an even more detailed prediction, namely that the pollinator is likely a sphinx moth. In 1903 (twenty-one years after Darwin died) scientists discovered a sphinx moth in Madagascar that fit the bill (so to speak) and promptly named it *Xanthopan morganii praedicta*, after Wallace's prediction.

The element of Darwinian theory that allowed both men to make this prediction is the law of co-evolution. But evolutionary theory can make predictions not just about the presence or absence of particular species; it can also make predictions about the future, in many ways similar to the laws of physics. However, unlike the laws of physics that can predict the future of any particular system as long as the initial conditions are known with sufficient accuracy, the theory of evolution cannot predict particular futures because chance is such an integral part of the theory. Instead, the theory makes predictions about statistical averages (we will encounter a specific prediction concerning the form of fitness trajectories in the next chapter). Predictions about the future cannot be tested armed only with observations; instead, dedicated experiments must be carried out. Dallinger's experiment, Lenski's long-term experiment, and experiments with digital life are the tools that allow us to test the theory of evolution at various levels of detail, and up to this point the theory has withstood all these tests with flying colors. Indeed, as I argue in chapter 6, experiments with digital life have even suggested extensions of the theory, something that we should expect whenever theory and experiment combine in furthering our understanding of nature.

Exercises

4.1 In Avida, performing bit-wise logic operations on inputs (as in Fig. 4.15) speeds up replication and thus leads to increased fitness. In this exercise we will try to understand the different ways in which certain logic operations can be achieved. As there is no requirement for parsimony (or even intelligibility) for evolution, unraveling the path from simplicity to complexity can be tedious, but ultimately rewarding.

(a) Verify (by applying to the four bit patterns of x and y) the logic identities ($\vee$ and $\wedge$ are the logical OR and AND operator, respectively, and $\neg$ is the logic negation)

$$\neg x = \text{nand}(-1, x), \tag{4.1}$$

$$\neg x = -x - 1. \tag{4.2}$$

$$\neg x \vee y - y = \neg x \wedge \neg y \tag{4.3}$$

using the "two's complement" encoding of negative numbers in binary.

(b) Verify the fundamental identities of Boolean logic

$$\neg(x \vee y) = \neg x \wedge \neg y, \tag{4.4}$$

$$\neg(x \wedge y) = \neg x \vee \neg y, \tag{4.5}$$

$$x \wedge (y \vee z) = (x \wedge y) \vee (x \wedge z), \tag{4.6}$$

$$x \vee (y \wedge z) = (x \vee y) \wedge (x \vee z), \tag{4.7}$$

$$x \wedge (x \vee y) = x, \tag{4.8}$$

$$x \vee (x \wedge y) = x. \tag{4.9}$$

(c) ★ Tables 4.4–4.6 show the trace of execution of two organisms on the line of descent shown in Figure 4.18, namely organisms 109 (Tables 4.4–4.5) and organism 111 (Table 4.6). The trace of the first forty-nine instructions of organism 111 is the same as that of 109, as the two organisms only differ in mutations at instructions 46 and 48. In those tables, instructions and register values are in italics if an instruction is conditionally *not executed*. Instructions and register values that differ when executing organisms 109 and 111 are shown in bold. When an instruction is modified by a pattern that follows, then the instruction plus the modifying pattern are listed together. Inputs to the logic operations are denoted by "a," "b," and "c." Only three inputs are provided, so that after input "c" is read, the next input will be "a" again. The divide command fails if the resulting daughter program differs from the progenitor by more than two instructions.

Using the instruction set shown in Table 4.1 and the trace of execution flows in Tables 4.4–4.6, as well as the logic identities above and those in Table 4.2, construct a narrative that explains how the two mutations q46g and c48x create the new logic function EQU while destroying the functions AND and NAND at the same time. Explain why the deleterious mutation c48x is *necessary* for q46g to be functional. In particular, explain the importance of the jump out of the copy loop just before division (instructions 53–54).

4.2 The strain REL606 used as the ancestor in the Lenski experiment doubles about once per hour when grown in the minimal medium DM25, and when diluted at a ratio 1:100 once per day, can evolve for $\log_2(100)$ or about six and two-thirds generations per day. If instead a rich medium is used, *E. coli* can double every twenty minutes, and can in principle deliver seventy-two generations per day if enough growth medium were available

Table 4.4. Trace of the first forty-five instruction-executions of organism 109 on the LOD in Figure 4.18. Here, we use $\bar{x}$ to denote $\neg x$.

n	symbol	instr	BX	CX	stack	out	logic
1	r	allocate	0	0	0	0	
2	m	dec BX	−1	0	0	0	
3–4	za	set-flow to AX	−1	0	0	0	
5–6	vc	move write-head to flow	−1	0	0	0	
7	g	push BX → stack	0	0	−1	0	
8–9	mc	dec CX	−1	−1	−1	0	
10	i	swap(BX,CX)	−1	−1	−1	0	
11	q	i/o BX	a	−1	−1	−1	
12	q	i/o BX	b	−1	−1	i1	
13	p	nand(BX,CX)→ BX	$\bar{b}$	−1	−1	0	
13	t	copy	$\bar{b}$	−1	−1	0	
15	q	i/o BX	c	−1	−1	$\bar{b}$	NOT
16	p	nand(BX,CX)→ BX	$\bar{c}$	−1	−1	0	
17–18	qc	i/o CX	$\bar{c}$	a	−1	−1	
19–20	pc	nand(BX,CX)→ CX	$\bar{c}$	$\bar{a}\vee c$	−1	0	
21	t	copy	$\bar{a}$	$\bar{a}\vee c$	−1	0	
22	l	inc BX	$\bar{c}+1$	$\bar{a}\vee c$	−1	0	
23	e	if BX<CX	$\bar{c}+1$	$\bar{a}\vee c$	−1	0	
24	t	copy	$\bar{c}+1$	$\bar{a}\vee c$	−1	0	
25–26	nc	add(BX,CX)→CX	$\bar{c}+1$	$\bar{c}+1+\bar{a}\vee c$	−1	0	
27	o	sub(BX,CX)→BX	$-\bar{a}\vee c$	$\bar{a}\wedge\bar{c}$	−1	0	
28–29	gc	push CX → stack	$-\bar{a}\vee c$	$\bar{a}\wedge\bar{c}$	$\bar{a}\wedge\bar{c}$	0	
30	b	nop-marker	$-\bar{a}\vee c$	$\bar{a}\wedge\bar{c}$	$\bar{a}\wedge\bar{c}$	0	
31–32	ea	if-less (AX,BX)	$-\bar{a}\vee c$	$\bar{a}\wedge\bar{c}$	$\bar{a}\wedge\bar{c}$	0	
33	m	dec BX	$-(\bar{a}\vee c)-1$	$\bar{a}\wedge\bar{c}$	$\bar{a}\wedge\bar{c}$	0	
34	q	i/o BX	b	$\bar{a}\wedge\bar{c}$	$\bar{a}\wedge\bar{c}$	$a\wedge\bar{c}$	AND-NOT
35	d	if-n-eq(BX,CX)	b	$\bar{a}\wedge\bar{c}$	$\bar{a}\wedge\bar{c}$	0	
36	t	copy	b	$\bar{a}\wedge\bar{c}$	$\bar{a}\wedge\bar{c}$	0	
37–38	qc	i/o CX	b	c	$\bar{a}\wedge\bar{c}$	$\bar{a}\wedge\bar{c}$	NOR
39	p	nand(BX,CX)→BX	$\bar{b}\vee\bar{c}$	c	$\bar{a}\wedge\bar{c}$	0	
40	t	copy	$\bar{b}\vee\bar{c}$	c	$\bar{a}\wedge\bar{c}$	0	
41	i	swap(BX,CX)	c	$\bar{b}\vee\bar{c}$	$\bar{a}\wedge\bar{c}$	0	
42	p	nand(BX,CX)→BX	$b\vee\bar{c}$	$\bar{b}\vee\bar{c}$	$\bar{a}\wedge\bar{c}$	0	
43	q	i/o BX	a	$\bar{b}\vee\bar{c}$	$\bar{a}\wedge\bar{c}$	$b\vee\bar{c}$	OR-NOT
44	f	pop stack → BX	$\bar{a}\wedge\bar{c}$	$\bar{b}\vee\bar{c}$	−1	0	
45	p	nand(BX,CX)	$a\vee c$	$\bar{b}\vee\bar{c}$	−1	0	

Table 4.5. Trace of the instruction executions 46 and beyond of organism 109 on the LOD in Figure 4.18.

n	symbol	instr	BX	CX	stack	out	logic
46	q	i/o BX	b	$\bar{b}\vee\bar{c}$	−1	$a\vee c$	OR
47,48	qc	i/o CX	b	c	−1	$\bar{b}\vee\bar{c}$	NAND
49	u	set flow next instr.	0	0	−1	0	
50	t	copy	0	0	−1	0	
51–52	yc	if last-copied = 'a'	0	0	−1	0	
53–54	ua	*set flow to 'b'*	−24	*1*	−1	*0*	
55	s	divide	0	0	−1	0	
56–58	ttt	copy 3 instr.	0	0	−1	0	
59–60	va	move IP to flow	0	0	−1	0	
31–32	ea	if-less (AX,BX)	−24	1	−1	0	
33	m	*dec BX*	−25	*1*	−1	*0*	
34	q	i/o BX	a	1	−1	−24	
35	d	if-n-eq(BX,CX)	a	1	−1	0	
36	t	copy	a	1	−1	0	
37-38	qc	i/o CX	a	b	−1	1	
39	p	nand(BX,CX)→BX	$\bar{a}\vee\bar{b}$	b	−1	0	
40	t	copy	$\bar{a}\vee\bar{b}$	b	−1	0	
41	i	swap(BX,CX)	b	$\bar{a}\vee\bar{b}$	−1	0	
42	p	nand(BX,CX)→BX	$a\vee\bar{b}$	$\bar{a}\vee\bar{b}$	−1	0	
43	q	i/o BX	c	$\bar{a}\vee\bar{b}$	−1	$a\vee\bar{b}$	OR-NOT
44	f	pop stack → BX	−1	$\bar{a}\vee\bar{b}$	$\bar{a}\wedge\bar{c}$	0	
45	p	nand(BX,CX)	$a\wedge b$	$\bar{a}\vee\bar{b}$	$\bar{a}\wedge\bar{c}$	0	
46	q	i/o BX	a	$\bar{a}\vee\bar{b}$	$\bar{a}\wedge\bar{c}$	$a\wedge b$	AND
47,48	qc	i/o CX	a	b	−1	$\bar{a}\vee\bar{b}$	NAND
49	u	set flow next instr.	0	0	−1	0	
50	t	copy	0	0	−1	0	
51–52	yc	if last-copied ='a'	0	0	−1	0	
53–54	ua	set flow to 'b'	−24	1	−1	0	
55	s	divide	−24	1	−1	0	

(there is not). But how many generations could we achieve if we diluted the cells more, and transferred them more often per day?

(a) Calculate the number of generations per transfer if the cells are diluted at a ratio 1:1000.

(b) How many generations can you achieve per day if you used the 1:1000 dilution, rich medium, and transferred three times per day?

(c) Why is diluting by larger ratios at every transfer ultimately not going to increase the number of generations per day?

Table 4.6. Trace of the instruction executions of 46 and beyond of organism 111 on the LOD in Figure 4.18.

n	symbol	instr	BX	CX	stack	out	logic
46	g	**push BX→ stack**	a ∨ c	$\bar{b}$ ∨ $\bar{c}$	a ∨ c	**0**	
47	q	**i/o BX**	b	$\bar{b}$ ∨ $\bar{c}$	a ∨ c	a ∨ c	**OR**
48	x	**get-head**	b	$\bar{b}$ ∨ $\bar{c}$	a ∨ c	0	
49	u	set flow next instr.	0	0	a ∨ c	0	
50	t	copy	0	0	a ∨ c	0	
51–52	yc	if last-copied = 'a'	0	0	a ∨ c	0	
53–54	ua	*set flow to 'b'*	−24	*1*	a ∨ c	*0*	
55	s	divide	0	0	a ∨ c	0	
56–58	ttt	copy 3 instr.	0	0	a ∨ c	0	
59–60	va	move IP to flow	0	0	a ∨ c	0	
31–32	ea	if-less (AX,BX)	−24	1	a ∨ c	0	
33	m	*dec BX*	−25	*1*	a ∨ c	*0*	
34	q	i/o BX	c	1	a ∨ c	−24	
35	d	if-n-eq(BX,CX)	c	1	a ∨ c	0	
36	t	copy	c	1	a ∨ c	0	
37–38	qc	i/o CX	c	a	a ∨ c	1	
39	p	nand(BX,CX)→BX	$\bar{a}$ ∨ $\bar{c}$	a	a ∨ c	0	
40	t	copy	$\bar{a}$ ∨ $\bar{c}$	a	a ∨ c	0	
41	i	swap(BX,CX)	a	$\bar{a}$ ∨ $\bar{c}$	a ∨ c	0	
42	p	nand(BX,CX)→BX	$\bar{a}$ ∨ c	$\bar{a}$ ∨ $\bar{c}$	a ∨ c	0	
43	q	i/o BX	b	$\bar{a}$ ∨ $\bar{c}$	a ∨ c	$\bar{a}$ ∨ c	OR-NOT
44	f	pop stack → BX	a ∨ c	$\bar{a}$ ∨ $\bar{c}$	−1	0	
45	p	nand(BX,CX)	a ≡ c	$\bar{a}$ ∨ $\bar{c}$	−1	0	
46	g	**push BX→ stack**	a ≡ c	$\bar{a}$ ∨ $\bar{c}$	a ≡ c	0	
47	q	**i/o BX**	c	$\bar{a}$ ∨ $\bar{c}$	a ≡ c	a ≡ c	**EQU**
48	x	**get-head**	c	48	a ≡ c	0	
49	u	set flow next instr.	0	0	a ≡ c	0	
50	t	copy	0	0	a ≡ c	0	
51–52	yc	if last-copied = 'a'	0	0	a ≡ c	0	
53–54	ua	set flow to 'b'	−24	1	a ≡ c	0	
55	s	divide	0	0	a ≡ c	0	

5

Evolution of Complexity

And as natural selection works solely by and for the good of each being, all
corporeal and mental endowments will tend to progress towards perfection.

—CHARLES DARWIN (1859)

Everybody seems to know that complexity increases in evolution.

—DANIEL MCSHEA (1991)

Life on Earth is, by any and all ways in which we could possibly express it,
extraordinarily complex. The sheer variety of crown-group organisms (the
animals, plants, and fungi) is balanced by an extraordinary (and largely unexplored) diversity of unicellular eukaryotes which, even though they do not
display the type of complexity that is visible to the naked eye, nevertheless
stun by a rich palette of intricate mechanisms geared toward preserving their
genetic information. At the same time, the simpler bacteria and archaea that
carry the echoes of past lineages are still around and thriving. Thus, new and
old are coexisting on the same planet, each "making their living" in their own
special and complex way.

The relationship between those forms of life is not, of course, given by the
great chain of being that we encountered earlier in section 1.2 (see Fig. 1.5).
Rather, life has continually branched so that all forms have a common ancestry (see Fig. 5.1) where remnants of the earliest forms (the branch of bacteria
and archaea) coexist with later modifications (everything else). If we imagine, for a moment, that the organism at the very root of this tree is indicative
of the earliest form of life and follow its line all the way up to any particular
crown-group organism, would we be able to see a trend in a particular character? Are animals getting bigger on average, or faster? While organism size
has been increasing to some extent over evolutionary time (we will discuss
this observation, sometimes called "Cope's rule," in more detail below) did
complexity increase over time? What is complexity, anyway?

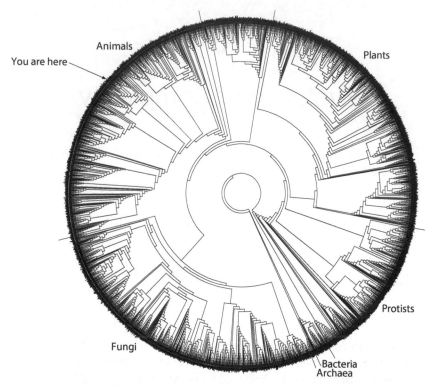

FIGURE 5.1. The Tree of Life. Source: D. M. Hillis, D. Zwickl, and R. Gutell, University of Texas.

The word "complexity" conjures up images of intricate and complicated systems and structures. It is used in everyday speech to describe objects or systems that are difficult to analyze, comprehend, or predict. Most of the time, we apply the term "complex" to systems with many different and interacting parts. But what makes the concept of complexity interesting from a scientific point of view is that, in almost all cases, the complexity disappears at some point during the process of breaking the system into its parts. And, of course, we usually do not know how to put those parts back together again to make a functioning complex system—because if we did, we would be less likely to think of the system as complex in the first place. Thus, complexity is a property of systems that have many parts that are put together in a nonrandom fashion. Also, complexity seems to vary in the eye of the beholder (what is complex to one may be simple to another) and even worse, depends on the level of description. But in any case, even if we are entirely ignorant of the nature of the complex object or system—for example whether it is useful for something or not—the complexity itself seems to require an explanation.

Even though we will be mainly interested in *biological* complexity in this chapter, it is worth pondering whether or not we can expect there to be a *universal* measure of complexity, applicable to evolved and designed systems, natural and artificial, equally. Such a measure would be useful for a host of reasons, not the least because it would allow us to *compare* the complexity of systems from different realms (for example, natural and artificial, evolved and designed). But before entering this fray, we should remind ourselves that complexity—as commonly understood—is not an objective category. Rather, as alluded to above, we tend to classify things as complex when they are difficult for us to understand, build, or replicate. As a consequence, our idea of what is complex is bound to change over time as we learn more, and thus this intuitive concept is useless for science because its value—should we be able to assign one—would not be observer- or even time-independent. Moreover, it seems that complexity should be context-dependent: one object may be complex in one environment (performing a set of intricate functions, for example) but simple and inert in another. So while perhaps a search for a universal measure of complexity is ultimately foolish, it is worth bearing in mind that the complexity concepts that are specific to the systems they are applied to (some of them we will encounter in this chapter) have value on their own, as they allow us to quantify differences within the system. And it would certainly be useful if we encountered a measure that quantifies the complexity of biological organisms.

I have argued throughout this book that the existence of all this biological complexity is a consequence of the process of evolution. According to those arguments, evolution naturally leads to an increase of information, which encodes and specifies the complexity we see. But does evolution really produce a trend in *complexity*, so that an inexorable march toward the world we see around us today was unavoidable? Or is the complexity we see around us the consequence of a trend that is merely a statistical by-product of the accumulation of variation? But first, we should ask the following.

5.1 What Is Complexity?

People appear to have an inherent need to classify the world that surrounds them. This quest is perhaps nowhere more apparent than in the historical zeal to attach to all living organisms a tag that reveals their relationship to ourselves. More recently, biologists and laypeople alike have, sometimes unconsciously, given in to the same need by classifying organisms according to their perceived complexity. In that vein, viruses are often perceived as the least complex organisms, followed by bacteria, unicellular eukaryotes, fungi, plants, inverte-brates, vertebrates, mammals, and, ultimately, ourselves. Such a hierarchical view of biological complexity with *Homo sapiens* at its top has repeatedly been decried (see, e.g., Gould 1996 and more recently Nee 2005) because it smacks of human self-importance, and a lack of respect (or understanding) for other forms of life. In particular, the success and diversity of prokaryotic life are often used as a standard for the superiority of this class over all other forms. How-ever, success and diversity are hardly good proxies for complexity, and in the absence of a *quantitative* measure of biological complexity, neither a position in favor nor against a hierarchical ordering of species can authoritatively be maintained. Indeed, even though it has become less and less accepted to view *Homo* as the "crown jewel" of evolution, there is no reason a priori why this cannot in fact be true.

Many different measures for biological complexity have been introduced in the literature, some of them with obvious biological roots (such as number of cells, number of different tissue types, genome length, etc.), some of them inspired by the physics of dynamical systems. None of these measures has been put forward as a candidate for a universal measure of complexity, nor has anybody claimed that they represent the best proxy for a biological complex-ity measure. Indeed, most have very obvious shortcomings. As we learn more about the genome and proteome of different organisms, it might appear as if we are well on our way to accumulating a sufficient amount of detailed knowl-edge concerning the structure and function of organisms that a universal measure of organism complexity is at last imaginable. For example, we might imagine that the set of all expressed proteins and their interactions (including their regulation and post-translational modification) could conceivably enter into an equation whose result is a number: the biological (functional) com-plexity of an organism that allows for comparisons across the tree of life. And while the result would not be a ladder, attaching numbers to the nodes of the evolutionary tree that reflect the complexity of the taxon could perhaps pin-point lineages in the tree that managed to evolve more complexity than others.

This dream, however, is almost certainly misguided unless we can enu-merate, along with the list of proteins and their interactions, the possible *environments* within which an organism can make a living, that is, their *niche*. After all, as I alluded to earlier, it is clear that the complexity of an organ-ism must depend crucially on the environment within which it functions and is therefore not just a single number. Consider for example a 150-pound rock and an average human of the same weight, transplanted to the surface of the moon. Functionally, the two are very comparable in that environ-ment because most of the characteristics that make humans interesting (and complex compared to a rock) are absent on the moon, for the simple reason that humans cannot survive there (unless they bring their own environment with them, in the form of a spacesuit). Likewise, a human is much less complex than even an algal bloom if forced underwater for an extended period of time. Given that we are unable to characterize even a single environment for most bacteria (environmental microbiologists estimate that less than 1 percent of all bacteria can be cultured in the laboratory; Amann et al. 1995) it is highly unlikely that the full functional complexity of most organisms can be ascertained.

Still, for organisms that live in comparable environments, a classification in terms of their biological complexity given their bioinformatic and ecological pedigree is not unreasonable, and the concepts and issues discussed below are steps toward such a goal. In the following sections, I review standard types of biological complexity measures, how they compare to each other, and how they fail to capture the essence of biological complexity. I first discuss measures that attempt to quantify the *structural* complexity of an organism. I continue with first an overview and then an in-depth look at measures seeking to quantify the complexity of the *sequence* that gives rise to that structure. There, we will see that the amount of information about the environment that an organism uses to survive appears as a possible candidate for a universal complexity measure. After discussing the complexity of networks and graphs, I will discuss whether we can see long-term (millions of generations) trends in the evolution of groups of organisms, using measures that could be taken as proxies for complexity. Finally, we'll take a look at possible trends over the short term (thousands to tens of thousands of generations) in a single lineage of organisms. The distinction between short-term and long-term trends, as well as the distinction between trends visible in average lines compared to a single lineage, is important because it is possible to observe opposite trends, as the case may be.

5.1.1   Structural complexity

In biology, complexity is staring us in the face every day. Biological organisms are complicated structures, and their behavior is difficult to predict. But we can also see clear differences. An amoeba, we intuit, cannot possibly be as complex as, say, you or me, simply because it consists of a single cell while plants and metazoans (multicellular animals) are made of very many cells, from the roughly 1,000 cells of the nematode *C. elegans* to the hundred trillion cells in humans. Furthermore, the physiological functions—and interactions with the environment—of unicellular organisms are vastly different compared to metazoans, but the latter reflect the *functional* complexity of the organism, not the structural one. As a consequence, when discussing structural complexity, most people intuitively think of multicellular animals, ignoring most of the tree of life (and plants, for that matter) in the process. This appears justified from the structural point of view because prokaryotes and unicellular eukaryotes are by definition structurally simple: on the level of cells they only have one part. A quick glance at the functional complexity of unicellular life forms (see, e.g., Margulis 1988) informs us, however, that structure on the cellular level is just one way in which complexity of function can be achieved. There is much structure on the subcellular level that is ignored with this approach (cf. the astonishing complexity of the unicellular eukaryotic parasite *Trypanosoma brucei* that we meet in section 6.4), which can however be investigated using either functional or network complexity approaches.

Trying to quantify the structural complication of an organism is perhaps the most obvious way to assess organism complexity. Clearly, structural complexity is not limited to biological organisms alone, and a good measure of this kind might allow us to compare biological organisms to human feats of engineering. Two problems are immediately apparent for any measure of structural complexity. First, generating a scalar (that is, a single number) that will rank all possible structures appears to be a formidable task mostly because there appears to be no good way to create an exhaustive list of all possible structures. Second, there is no guarantee that structural complexity is always a good predictor for an organism's success in the biosphere. On the one hand, something that looks like a complicated contraption could, in principle, be an evolutionary artifact: a nonadaptive feature that is either a necessity for or a consequence of another feature (such a feature was called a "spandrel" by Gould and Lewontin 1979). On the other hand, a complicated device could conceivably only be functional in an environment that differs in crucial ways from the one we use to study the organism's functionality, perhaps an environment in which the organism can no longer survive. In other words, complex structure is not necessarily predictive of complex function, although we expect this to be true in the majority of cases. This observation is related to the earlier observation that a good estimate of functional complexity requires a good understanding of the ecological niche of the organism.

While it may be difficult to accurately describe an organism's ecological niche, quantifying structural complexity may be even harder. A common theoretical measure of structural complexity is "the shortest complete description of the structure" (Löfgren 1977; Papentin 1980; Papentin 1982). This measure is clearly inspired by the concept of Kolmogorov complexity ("shortest complete program that will create a particular symbolic sequence") that we shall encounter later. The idea here is that complex structures require complex blueprints, and conversely that anything simple can also be described simply. While appealing, this measure has the same problem as the Kolmogorov complexity, namely that it is not in principle possible to prove that a particular

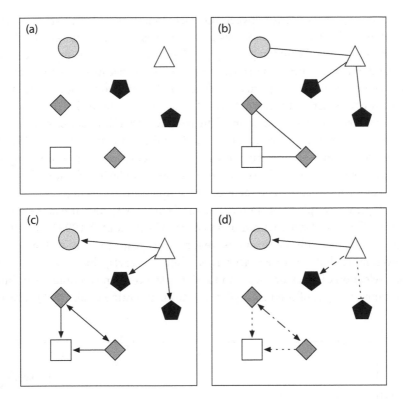

FIGURE 5.2. Representation of structural complexity of metazoans in terms of parts and their interactions. (a) Different parts only, (b) parts and interactions, (c) parts and directed interactions, (d) parts and directed interactions of different kinds.

description is in fact the shortest: it is always possible that a much shorter, much more elegant description exists (and would thus reveal that the object so described is not complex at all). But the measure suffers from another problem: the alphabet used to encode blueprints is not specified. In principle, it is always possible to imagine different encodings that differ drastically in the length of the description of the structure, but actually describe the same structure. As a consequence, these "minimum description length" measures are not practical.

A more workable measure of structural complexity attempts to count the number of (different) parts and their connections. For example, we could decompose the organism into parts at a given level, and apply a measure of "kind" or "type" that allows us to distinguish the parts. If parts are all we are interested in, then a representation of the organism would look perhaps like Figure 5.2 in panel (a). Often, the way the parts interact is more representative of structural complexity than the parts are themselves. In a graphical form, we can draw connections between the parts to indicate which part interacts with what other part (or sometimes parts can interact with themselves). Such interactions are indicated in panel (b) of Figure 5.2. A further sophistication in the characterization of the *process* complexity (the nature of the interaction between the parts) is to specify the *direction* of interaction (panel c) or even the *type* of interaction (different types of arrows), as in Figure 5.2(d). Once we have decided on a representation of the parts and their interactions, we must forge a complexity measure out of the graph. In the simplest case, we could just count the number of different parts and their interactions, but more complicated measures are possible. We will discuss the complexity of graphs in more detail in the last section, and focus on the simplest measures here.

A typical example for a simple structural complexity measure is the number of different cell types within an organism (Valentine et al. 1994). Clearly, it may be difficult to obtain this measure for some organisms, and certainly this is even more difficult for extinct forms of life. However, this data does exist for a selection of metazoan taxa, and a plot of the cell-type complexity versus the time of origin of the taxon (Fig. 5.3, from Valentine et al. 1994) reveals

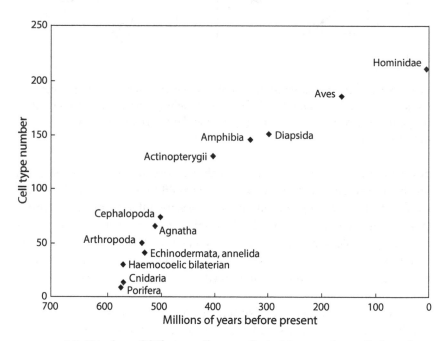

FIGURE 5.3. Number of different cell types of primitive members of selected metazoan taxa (estimated from counts of cells in extant organisms), using those taxa that are believed to be near the upper bound in cell-type number at the time of origin of the taxon (from Valentine et al. 1994).

an increase in complexity from the origin of multicellular life until today. We should keep in mind, however, that the increase depicted in this figure does not necessarily reflect an increase in complexity throughout evolutionary time because there are many different cell types per taxon (each taxon represents a number of different species with different cell type numbers), and the trend is only apparent in the *maximum* number of different cell types per taxon. Indeed, Valentine et al. (1994) offer a simulation of a stochastic process without selection that includes cell-type addition and subtraction that reproduces the general features seen in Figure 5.3. If a stochastic process can reproduce an observed trend, we can rule out that evolution via natural selection is necessary for this trend. But of course, we cannot deny that selection can also produce such a trend. If any process (driven by selection or not) can produce a trend in a particular candidate for complexity, then perhaps we should conclude that such a measure is not useful in our quest to quantify organismal complexity. We will be discussing trends in the evolution of complexity, whether they are driven or passive, in more detail in section 5.3.

Bell and Mooers analyzed cell-type data but without confining themselves to the maximum diversity in each taxon (Bell and Mooers 1997), and studied cell-type number as a function of organism *size*. They found that larger organisms tend to have more differentiated cell types, and base this trend on the general principle that larger systems need to operate on the cooperative "division-of-labor" principle in order to be effective. While cell-type number robustly classifies animals as more complex than plants, and plants as more complex than algae, the measure remains a very crude estimate of complexity, unable to shed light on finer gradations of the tree of life.

A more sophisticated analysis of animal structural complexity describes animal skeletons in terms of a vocabulary that appears to cover the "space" of all existing and extinct structures: the "skeleton space" of Thomas and Reif (1991, 1993). These authors describe skeletons in terms of general properties, such as internal or external skeleton, multiplicity of parts, materials, shapes, growth modes, assembly, and interconnections, giving a list of seven categories that can take on twenty-one states. In principle, fairly complex structures can be described in this alphabet. For example, the skeleton of the human finger would be described as "internal, multiple parts, rigid, jointed, rods, that are remodeled, as they grow in place" (Thomas et al. 2000). Because the design space spanned by these multiple combinations of attributes is large, Thomas and Reif instead study the space of skeletal possibilities by counting how many times any combination of only two of these properties exists in organisms either extant or extinct (some combinations are actually mechanically infeasible or implausible). For example, they study how often endoskeletons can be found that are made of flexible material, or exoskeletons that molt (molting of endoskeletons being technically infeasible). Of the 182 possible pairs of properties, over half are abundantly used by organisms in different phyla, and two-thirds of the pairings are quite common (Thomas and Reif 1993). A full 146 combinations were already in use by Burgess Shale animals, indicating that the matrix of combinations was filled fairly rapidly during the Cambrian explosion (Thomas et al. 2000). An analysis of the skeletal complexity of organisms in terms of the theoretical morphospace (where the complexity score of various phyla is given by the aforementioned use of different combinations by animals in those phyla) seems to imply that the most adapted organisms are not necessarily those that are morphologically the most complex (Thomas and Reif 1993), reiterating that structural complexity is not always a reliable predictor of functional complexity.

The structural complexity measures we just reviewed take into account parts and connections at one specific level of description, but ignores the organization of the system into hierarchies of structures or modules. Such a hierarchy, however, is readily apparent for biological organisms: parts on one level form strongly interconnected groups that themselves can be viewed as parts on a higher level. This is obviously the case for skeletons, for example, but hierarchical modularity appears to be a common theme on the molecular level too. Generally speaking, a measure of hierarchical complexity seeks to quantify the number of levels needed to build a biological system, for example as the minimum amount of hierarchical structuring needed to build an understanding of the system (Nehaniv and Rhodes 2000). The problem with a hierarchical scale of complexity for biological systems is that there are only four clear hierarchies: the prokaryotic cell, the eukaryotic cell viewed as a symbiotic assembly of prokaryotic cells, the multicellular organism, and colonial individuals or integrated societies (McShea 2001). However, it is possible to introduce a higher-resolution scale by decomposing each hierarchy into levels and sublevels, for example by differentiating a monomorphic aggregate of elements of the lower hierarchy from a differentiated aggregate and an aggregate of nested differentiated elements (McShea 2001). Tissues and organs are examples of aggregates on an intermediate level. Even though a hierarchical measure of complexity necessarily represents a fairly coarse scale, it is one of only a few measures of structural complexity that shows an unambiguous increase in complexity throughout the fossil record.

5.1.2 Sequence complexity

Given that all forms of life on Earth contain a genetic code that is responsible for generating their form and function, we might naively assume that the amount of haploid DNA, measured either in picograms (pg) as was done

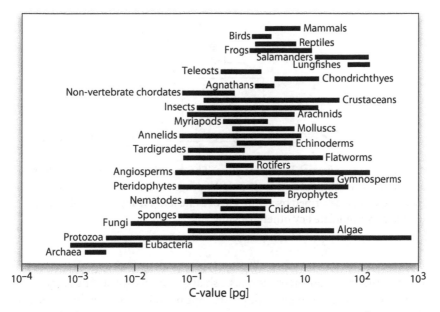

FIGURE 5.4. Ranges in haploid genome size measured in picograms (the C-value) in different organisms (after Gregory 2004).

before the advent of whole genome sequencing, or in millions of base pairs (mbp) as is more common today, would reflect—even if only roughly— the complexity of the organism. This hope was quashed relatively early on: Britten and Davidson (1971) showed conclusively that no correlation between genome size and perceived complexity exists. This disconnect has been termed the "C-value paradox" (Cavalier-Smith 1985; reviewed in Gregory 2004) and is perhaps best exemplified by the giant free living amoeba *Amoeba dubia*, whose total DNA content was estimated at 700 pg (which would correspond to about 675,000 mbp if it was all haploid). This is equivalent to about 200 times the C-value of humans. However, the haploidy of the *A. dubia* genome is now in doubt (Gregory 2005). The variation in genome size and the absence of a correlation to a complexity scale such as that given by the classical chain of being is clearly apparent in Figure 5.4. Furthermore, the genome of many eukaryotes is actually quite variable, either during the life cycle of the organism or even within a population (Parfrey et al. 2008). Because sequence length proper cannot be used as a measure, the focus for measures of sequence complexity instead has been on mathematical measures. The literature on mathematical sequence complexity, or more specifically, the complexity of symbolic strings, is far richer than that concerning functional or structural and morphological complexity. Among sequence

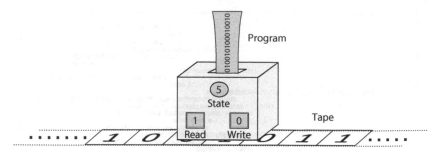

FIGURE 5.5. A universal Turing machine. The machine reads and writes from an infinite tape using rules provided on the program. Because every Turing machine can be described by a program of finite length, any universal Turing machine can simulate any other machine.

complexities there are many different types, such as Kolmogorov, compositional, or information-theoretic ones. I shall give brief expositions of examples of each of these types here, without attempting to be even nearly exhaustive. For a good review of sequence complexities used in physics, see Badii and Politi (1997).

The most well-known sequence complexity is that introduced by the Russian mathematician Andrey Kolmogorov. He proposed to assign to each symbolic sequence a scalar that represents the *regularity* of the sequence (Kolmogorov 1965). So, for example, a string consisting of the repetition of a symbol or pattern is classified as regular, whereas a string with no discernable pattern would be classified as irregular, and therefore complex according to this notion. Note, however, that this algorithm does not classify a string as complex just because no pattern is readily identifiable. The Kolmogorov measure assumes that *all possible* computer programs that run on a *universal Turing machine* are tested so as to find the shortest one that produces the sequence in question. Turing machines are very simple abstract computing machines that, due to their simplicity, can be described mathematically. Generally speaking, all existing computers can be simulated with a universal Turing machine by providing to the machine a description of the computer to be simulated, on the input tape that is read by the Turing machine (see Fig. 5.5).

Mathematically, the Kolmogorov complexity of a string s is given by the length of the shortest program p (denoted as $|p|$) that produces s when executed on a universal Turing machine T:

$$K(s) = \min_p \left\{ |p| : s = C_T(p) \right\}, \tag{5.1}$$

where $C_T(p)$ denotes the result of running program p on Turing machine T. So, for example, the binary equivalent of the irrational number π (the sequence s_π) is random prima facie; however, a concise algorithm (a short program p_π) exists to produce it, leading to a fairly low complexity for s_π. We can think of the randomness of a string (the flip side of regularity, so to speak) as its *compressibility*: random strings have zero compressibility, and perfectly regular strings can be compressed to zero length. There is a vast amount of mathematical and information-theoretical literature concerning the Kolmogorov complexity (see, e.g., Li and Vitanyi 1997; Calude 2002), but a thorough analysis of the measure in terms of the foundations of computing reveals that the measure has three significant flaws.

First, the procedure to generate the Kolmogorov complexity is uncomputable because the search for the smallest program may never end (the computation may not halt, see Box 4.1). Second, the measure is impractical because the value $K(s)$ is only defined in the limit of infinitely long strings. This is because the value given by Equation (5.1) depends on the type of Turing machine used: different computers can use programs of different length to obtain the same sequence. Because of this limitation, the value $K(s)$ is defined only up to a program of finite length that specifies one computer to another (this program is independent of the sequence being analyzed). So, if the complexity of string s when analyzed on computer C_1 is $K_1(s)$ but yields $K_2(s)$ on computer C_2, it is possible to prove that

$$|K_1(s) - K_2(s)| \leq c, \qquad (5.2)$$

where c is a constant given by the "simulation penalty." Only in the limit of infinite-length sequences is this constant negligible.

Third, truly random strings, that is, those that cannot be generated by any computation, are assigned maximal complexity in this definition. But in physics (and in particular in biology!) truly random sequences are meaningless, and therefore should not be assigned a large complexity. In fact, the Kolmogorov complexity is even logically inconsistent. Remember that the procedure to calculate the Kolmogorov complexity is one where an automaton returns the smallest program that produces the sequence in question. For a random sequence, the algorithm is to return the sequence itself (and thus the complexity estimate of a random sequence becomes the length of that sequence, the largest possible result). But, logically, a random sequence can never be the result of a computation, because a computation is by definition a deterministic process. These flaws can be partly fixed by focusing instead on conditional and mutual Kolmogorov complexities, as I outline further below. In summary, while the Kolmogorov complexity is a mathematically sound measure of *mathematical* sequence regularity, it is not a measure of the complexity of a sequence that describes a physical object; in other words, it is not a physical complexity.

I suggest that the key concept that will allow us to quantify the complexity of objects in the physical—rather than mathematical—world is the *conditional Kolmogorov complexity* [first introduced by Kolmogorov himself (1965)] of a string *s given* another string *t*, defined as the length of the smallest program $|p|$ that produces *s* using that program and an external string *t*:

$$K(s|t) = \min_{p} \left\{ |p| : s = C_T(p|t) \right\},\tag{5.3}$$

with the notation $(p|t)$ pronounced as "*p* given *t*." Now, *t* can be a sequence that actually exists in the physical world, or it might describe an object in the physical world. In either case, the sequence *t* can be accessed by the Turing machine in its quest to produce sequence *s*. The program *p* is small if the sequence *s* can be obtained from *t* using a simple computation. Using this construction, the conditional complexity of a random sequence *r* can be defined rigorously, because the shortest program to produce a random sequence *r* involves both the sequence *r* as input to the machine and the vanishingly small (in the limit of infinite strings) program $p =$ "print." In other words, the conditional Kolmogorov complexity of random strings *vanishes* in the limit of long sequences (rather than being maximal):

$$K(r|r) = \min_{p} \left\{ |p| : r = C_T(p|r) \right\} = 0\tag{5.4}$$

because when the string *r* exists in the world, then the smallest program to produce it (from that sequence) is essentially zero (in the limit of long sequences).

Given what we have just learned, to define the physical complexity of a symbolic string we can imagine a sequence *e* that represents everything that can be measured in that world. We now ask of our Turing machine to compute string *s* given everything that is knowable about the physical world, that is, given *e*. In that case, the conditional complexity $K(s|e)$ represents everything that *cannot* be obtained from knowing the physical world. In other words, it represents the "remaining randomness": the unmeasurable. Naturally, the physical complexity of the string is then just the unconditional complexity *minus* the remaining randomness (Adami and Cerf 2000):

$$C_P(s) = K_0(s) - K(s|e).\tag{5.5}$$

Here I defined the unconditional complexity $K_0(s)$ as the Kolmogorov complexity in the absence of both a "world" string *e and* the rules of mathematics, that is, $K_0(s)$ is the complexity of the string without any means of compression. This is of course just the length of s: $K_0(s) = |s|$.[1] This notation is chosen in anticipation of results from information theory introduced in the following section. In particular, it allows us to see that $C_P(s)$ is simply a *mutual Kolmogorov complexity* (Kolmogorov 1965), between string s and the world e. Put in another way, the physical complexity of a sequence, as defined by Equation (5.5), is that part of the sequence that can be obtained from the world string e using a concise (and therefore short in the limit of very long sequences) computation using program p. Only *those* sequences can be obtained by computation from the world e that *mean* something in world e, or refer to something there. Note that by its construction, the physical complexity (5.5) represents a special case of the "effective complexity" measure introduced earlier by Gell-Mann and Lloyd (1996).

As an example, consider a world of machines in which blueprints are stored in such a way that they can be represented as sequences of symbols. In this world, e represents all the blueprints of all the possible machines that exist there. A sequence s is complex in this world if a part—or all—of the sequence can be obtained by manipulating, or translating, the world tape e. (It is conceivable that s contains part of a blueprint from e in encrypted form, in which case the program p must try to compare e to encrypted forms of s.) Of course, it would be an accident bordering on the unreasonable to find that a string that is complex in e is also complex mathematically. Instead, from a mathematical point of view, such a sequence most likely would be classified as random, or rather, the search for a shortest program would not halt. Similarly, it is extremely unlikely that sequence s would be classified as complex in a world in which e represents, say, all the literature produced on Earth (unless there are a few books on the blueprints of certain machines!). Thus, the complexity of a sequence s, by this construction, is never absolute (like in mathematics) but always conditional with respect to the world within which the sequence is to be *interpreted*.

This is precisely what we need to quantify the complexity of biological sequences, because it is immediately clear that a biological sequence only means something in a very specific environment, given very specific rules of chemistry. So, according to this argument, the sequence describing a particular organism is complex only with respect to the environment within which that organism "makes its living," that is, its niche. Take the organism out of its

---

1. A definition of the physical complexity using the standard Kolmogorov complexity $K(s)$ instead of $K_0(s)$ would be practically identical because the fraction of compressible programs is exponentially small (Cover and Thomas 1991; Calude 2002). In other words, $K(s) \approx |s|$ for almost all sequences anyway.

niche and it is unlikely to function as well as in its native niche; some of its structural complexity may turn into a useless appendage or, worse, a liability.

In the following section, we will see under what circumstances this physical complexity can be understood in terms of information theory. In particular, we will be able to deduce that the physical complexity of a sequence will be related to the information content of that sequence, discussed previously in section 2.3.

5.1.3 Sequence complexity and information

It is tempting to speculate that the information content of a biomolecule is related to the complexity of an organism. After all, possessing information about an ensemble enables the *prediction* of the possible states of an ensemble with accuracy better than chance, something that is highly valuable for a biological organism whose "ensemble" is an uncertain environment. Let us recall the measure of sequence complexity ("physical complexity") introduced in the previous section:

$$C_P(s) = K_0(s) - K(s|e), \tag{5.6}$$

for any sequence s, given an environment sequence e. If we take the average of this quantity over an infinite ensemble of sequences s_i (drawn from an ensemble S), we obtain

$$\langle C_P(s) \rangle_S = \sum_{s_i} p(s_i) \left(K_0(s_i) - K(s_i|e) \right), \quad s_i \in S. \tag{5.7}$$

It is easy to prove (see Exercise **5.1**) the following inequality between average Kolmogorov complexities and the Shannon entropy (Zurek 1990):

$$\sum_{s_i} p(s_i) K(s_i) \geq \sum_{s_i} p(s_i) \log \frac{1}{p(s_i)} = H(S). \tag{5.8}$$

The inequality in Equation (5.8) reflects the possibility that the complexities $K(s_i)$, which can be viewed as compressed *encodings* of the sequences s_i, do not necessarily form a *perfect code* that saturates the Kraft inequality (Cover and Thomas 1991) (see also Exercise **5.1**). However, because the $K(s_i)$ do represent the *smallest* program encoding s_i, it is reasonable to assume that the average Kolmogorov complexity is given by the Shannon entropy of the ensemble up to an additive constant, which represents the length of a program that tells one computer how to simulate another, that is, the simulation penalty constant c we encountered earlier (see, e.g., Cover and Thomas 1991):

$$\langle K(s_i) \rangle_S \approx H(S) + c. \tag{5.9}$$

In that case (and assuming that the overall constant cancels from the difference), the average physical complexity becomes

$$\langle C_P(s) \rangle_S \approx H(S) - H(S|e), \tag{5.10}$$

where $H(S)$ is the unconditional Shannon entropy of the ensemble of sequences. If the ensemble S consists of sequences of fixed length L, then the unconditional entropy is $H(S) = L$. $H(S|e)$, in turn, is the conditional entropy of the sequences as in Equation (2.34). (Note that, technically, the Kolmogorov complexity for fixed length sequences $K(s|L)$ is related to the arbitrary length complexity $K(s)$ via $K(s) \leq K(s|L) + 2 \log L + c$, where c is again the simulation penalty.)

We should also point out that this difference of entropies should *not* be viewed as the "redundancy" of the sequence, as implied for example by a construction of Barlow (1989), who gives this name to the difference between the maximal information capacity of a sequence, and the actual entropy. Barlow refers to this difference as redundancy because (like Kolmogorov before him) he intuits that an unpredictable signal (a random sequence, with maximum entropy) must have zero redundancy, and that a signal that has less entropy therefore must have some redundancy, in the sense that the sequence is more predictable. But the prediction that biological sequences are making are not about what other characters appear in the sequence (which you could easily do in English words where part of the word is obscured, for example) but instead what kind of a world the carrier of these sequences is inhabiting. Thus, Barlow calls this difference redundancy because he mistakenly thinks that the importance of a sequence is in how well you can perform *horizontal* prediction, that is, a prediction of symbols *along* the sequence. Instead, we know from chapter 2 that the importance of symbols on a sequence lies in the capacity to do *vertical* prediction, that is, a prediction of the identity of a symbol at a particular position in a multiple-sequence alignment, because that is a measure of how well-conserved the sequence is. Indeed, Table 2.1 makes this distinction very clear: we can make vertical predictions armed with only a subset of these sequences, but we certainly cannot make horizontal predictions. In a sense, the concept of redundancy is only important when we are concerned with the *encoding* of information. The same information can be encoded in a redundant manner (for example in a block code), or in a nonredundant manner, for example in a maximum entropy or encrypted code. To summarize, the average physical complexity is (assuming perfect coding) equal to the Shannon information that the ensemble has about the environment, that is, a sequence's information content as defined earlier in section 2.3.1.

This interpretation of complexity is particularly satisfying from an evolutionary point of view. The value of information lies in the ability of the observer who is in possession of it to make *predictions* about the system that the information is *about* (see also Rivoire and Leibler 2011). Organisms, armed with the functionality bestowed upon them by their genetically encoded information, do precisely that to survive. An organism's metabolism is a chemical machine making predictions about the availability and concentrations of the surrounding chemicals. A cell's surface proteins make predictions about the type of cells it might interact with, and so on. Viewed in this way, informational complexity should be a near perfect proxy for functional complexity, because information must be used for function: if it were not so used, a sequence would represent entropy, not information. In fact, an investigation of the informational complexity of evolving computer programs [an instance of "digital life" (Adami 1998; Adami 2006c), described in much more detail in section 4.3.3] has shown that this complexity indeed increases in evolution (Adami et al. 1999) and correlates well with the functional complexity of the programs (Ofria et al. 2007).

5.1.4   Application to DNA, RNA aptamers, and proteins

A good example for how the equivalence of function and information is achieved in biochemistry is the evolution of functionality in ribozymes by in vitro evolution, achieved by Jack Szostak's group at Massachusetts General Hospital. This group evolved short GTP-binding RNAs (aptamers) in vitro, and found eleven distinct structures with different binding affinities (Carothers et al. 2004), each evolved in separate experiments. Because within each pool where the structure was found there were also genetic variants of that same structure, the team was able to create a multiple-sequence alignment to determine the substitution probabilities $p_n(x)$ (of finding nucleotide x at position n), and thus to measure the information content of the sequences as outlined here. Using this information content, the team could show that increased functional activity went hand in hand with increased information content, so much so that they were able to derive a simple law that predicts, within this GTP-binding set of ribozymes, that a ten-fold higher binding affinity is achieved by about ten bits of extra information. (This work was described earlier in section 2.3.1.) In other words, the informational complexity is linearly proportional to the functional complexity. Even better, the structural complexity, as measured by the number of different stems (ladders) within the secondary structure of the enzyme, also seemed to increase with functional activity.

Inspired by the observed relationship between structure, function, and information in these ribozymes, Szostak proposed a new measure of functional complexity (Szostak 2003; Hazen et al. 2007) that is based both on function and on information. For a particular function x, let E_x represent the *degree* to which this function is achieved by a system (a typical degree of function would be the level at which a reaction is catalyzed, as measured by the catalytic constant of the reaction). Then, the *functional information* is defined as (Hazen et al. 2007)

$$I(E_x) = -\log(F(E_x)), \tag{5.11}$$

where $F(E_x)$ is the fraction of all possible configurations of the system that possess a degree larger or equal to E_x. For sequences, the function could represent a binding affinity, or the number of ATPs produced by a pathway within which the enzyme is the bottleneck factor, or any other real-valued attribute that characterizes the performance of the sequence. This measure introduces a clear link between information and function, but fundamentally turns out to be a coarse-grained version of the information content (2.34), as we will now see.

Suppose we are interested in measuring the information content of a sequence s that performs function x to the degree E_x. We can obtain the functional information of s by creating all possible mutants of s and measuring the fraction $F(E_x)$ of sequences that have the same function as s, given by $v(s)/N$, where $v(s)$ is the number of neutral (or better than neutral) mutants of s within the ensemble S, and N is the total number of possible sequences. Thus,

$$I(E_x) = \log N - \log v(s). \tag{5.12}$$

The conditional probability to find a sequence s_i given environment e in an evolving population of sequences of the type s is given by $p(s_i|e)$. If e specifies the function x at level E_x for s_i, then $p(s_i|e) = 1$ if s_i performs the function at the required level, and zero otherwise (coarse-graining of the entropy). There are $v(s)$ such sequences in the ensemble, and thus

$$H(S|e) = -\sum_i p(s_i|e) \log p(s_i|e) = \sum_{v(s)} \frac{1}{v(s)} \log \frac{1}{v(s)} = \log v(s). \tag{5.13}$$

As $\log N = \log D^L = L$, Equation (5.12) recovers Equation (2.34), that is, functional information is a coarse-grained version of the Shannon information content of a sequence. We will encounter functional information again in chapter 7 when we discuss how much information a sequence must have in order to *replicate* that same sequence.

5.1.5 Complexity of ecosystems

An obvious shortcoming of informational complexity is that it refers to a particular niche only and furthermore cannot quantify the complexity of genes that are adapted to *varying* or multiple environments. So, for example, at this point we cannot use the informational complexity to estimate the complexity of an ecosystem, nor of an organism that spends part of its life cycle in one environment, and another part in a completely different one (simple forms of life such as arboviruses and the more complex trypanosomes are notorious for such a cycle). However, a natural extension of the informational complexity exists that may cover multiple environments both in time and space. Recall that the informational complexity of an ensemble of sequences S refers to a single environment description $E = e$:

$$I(S:e) = L - H(S|e), \qquad (5.14)$$

where $H(S|e)$ is the ensemble entropy of the sequences: it is a *specific* information. We can generalize this expression by promoting the environment to a true random variable E that can take on states e_i with probability $p(e_i)$. This formalism can describe environments that are composed of different (spatially separated) niches e_i, as well as environments that take on the different states e_i periodically or even randomly, in time. The informational complexity then becomes

$$I(S:e) \rightarrow \sum_e p(e)I(S:e) = I(S:E) = L - H(S|E). \qquad (5.15)$$

Here, $H(S|E)$ is the average conditional entropy of the sequence ensemble S given the environment E. Whether this construction will turn out to be useful for characterizing the complexity of ecosystems or variable environments remains to be seen, as the practical obstacles are only amplified by having to measure the informational complexity in multiple environments. But at the very least this construction addresses one of the fundamental problems in assessing functional complexity that we encountered in the introduction, namely that for organisms that have adapted to be functional in a variety of environments, we can find genes that appear to show no phenotype upon knockout in the laboratory. Such genes, however, may very well show a phenotype in a particular environment that the organism encounters in the wild, and this functional capacity of an organism needs to be taken into account when assessing functional complexity. If the random variable E accounts for a multitude of environments with the correct probabilities $p(e)$, then the full functional complexity of the organism may be characterized using Equation (5.15).

But to apply this measure, more efforts need to be expended toward understanding the modes in which an organism functions in its native environment(s) (as opposed to the unnatural laboratory conditions that are the norm today). If such an effort is made (surely we can expect an exponential increase in sequence data in the coming years), the prospects for a general understanding of biological complexity in terms of sequence complexity are good.

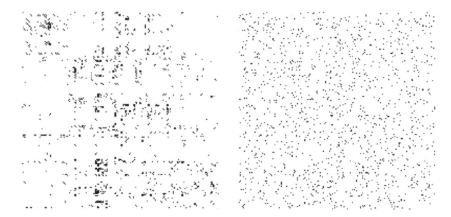

FIGURE 5.6. Adjacency matrix of 179 of the 302-neuron neural network of a
C. elegans brain (left), and a random network of the same size and connectivity
(right). In an adjacency matrix, a dot indicates a connection between the neuron
numbers listed in the *x*- and *y*-axes. From Adami (2009).

   In the absence of this much detailed data, it is possible to quantify the complexity of interacting systems in a different way, by abstracting away most (if
not all) of the complexity of any single entity, and instead focusing on the *connection* between individuals or groups. This is the focus of *network biology*,
or the study of complex systems in terms of graphs and their components:
modules and motifs.

5.2   Complexity of Networks, Modules, and Motifs

What is the complexity of a network? Can this complexity be deduced from
the stucture (i.e., the topology) of the network alone? The difficulty in assessing network complexity from network topology is clear to anyone who has
studied the multitudes of networks arising in engineering and biology. Biological networks usually have thousands of nodes and several thousand edges
and often appear to be unstructured. For example, the network summarizing
the connectivity of neurons in the brain of the nematode *C. elegans* shows very
little modularity or structure at first sight, but is markedly different from a
random network (Reigl et al. 2004) (see Fig. 5.6). Because the functionality
of a network is not necessarily reflected in its topological structure, the best
hope for assessing the complexity of a network is to measure the complexity
of the *set of rules* used to construct it. In biology, this set of rules is encoded
in the genome, so a first-order estimate of the complexity of a network should
be given by the complexity of the genome that produced it. Of course, this is

difficult for all the reasons given in the previous section, but even more difficult in this case because a network of proteins, for example, is specified not just by the open reading frames coding for the proteins, but also all the untranscribed regulatory regions, as well as the transcription factors affecting them.

We can test the evolution of network complexity in an artificial cell model (ACM) where a genome represents the functionality of a cellular network (Hintze and Adami 2008). In this model, all of the chemistry and genetics is encoded in a simple linear (circular) code based on the quaternary alphabet 0, 1, 2, 3, where enzymatic proteins with variable specificity act on fifty-three precursor molecules. These molecules are built from three different kinds of monomers (called 1, 2, and 3, not to be confused with the code letters 0, 1, 2, 3) to form up to 555 metabolites, indicated by the linear molecules with monomers in different shades of gray in Figure 5.7. The figure also shows import and export proteins with affinities to the molecules they import or export, as well as catalytic proteins that drive an artificial metabolism that creates long-chain molecules from simple precursors or shorter chains, defined by a particular artificial chemistry.

The rules of the artificial chemistry are simple: linear molecules are formed such that each monomer-type must be connected to as many other monomers as the number indicates. The simplest molecule then is 1-1 while the second-simplest is 1-2-1. Longer chains can be formed (such as the chain 1-2-3=3-2-2-1) via enzymatic reactions catalyzed by proteins that have an affinity toward the component molecules. For example, the reaction 1-2-2-3=3-3=2 + 1-2-3=3-2-2-3=2 shown in Figure 5.8 is catalyzed by a computational enzyme that has affinities to the four fragments of the molecules, which it recombines. This affinity is indicated in Figure 5.7 graphically by symbols that have particular shapes indicating the affinity.

The metabolic reactions involving transport and enzymatic proteins are obtained by a translation of the genetic code into reactions (see Fig. 5.9), and implementing chemostat physics and reaction kinetics. Evolution proceeds from a simple ancestral genome with only three genes (one for importing the molecule 1-2-2-2-2-1 as well as two molecular reactions), toward large and complex metabolic networks of thousands of nodes and several thousand edges, in a completely asexual Wright-Fisher process[2] acting on two chromosomes.[3]

---

2. In the Wright-Fisher model of evolution, organisms are placed into the next generation with a probability given by their relative fitness (see, for example, Ewens 2004). Generations do not overlap.

3. Because of the absence of sexual recombination, one of the two chromosomes consistently degenerates in evolution; see Hintze and Adami (2008).

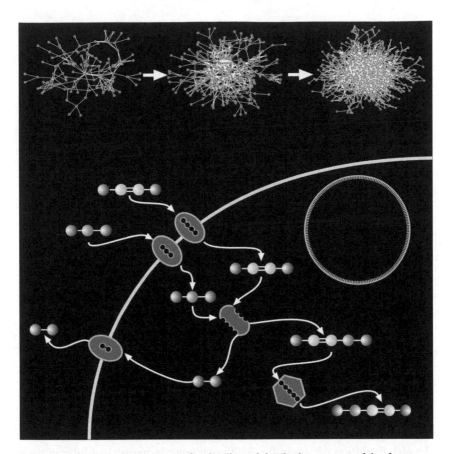

FIGURE 5.7. Evolution in an artificial cell model. The lower part of the figure shows a cell wall with a circular chromosome that contains all the genetic information of the artificial cell. Inserted into the cell wall are import and export artificial proteins that recognize specific precursors or metabolites, and enzymes catalyze the formation of longer-chained molecules from precursors or shorter-chain sequences. The top of the figure shows the evolution of the metabolic network powering the artificial cell, over evolutionary time.

$$\boxed{1-2-2-3=3-3=2} \; + \; \boxed{1-2-3=3-2-2-3=2} \; \rightarrow$$

$$\boxed{1-2-2-3=3-2-2-3=2} \; + \; \boxed{1-2-3=3-3=2}$$

FIGURE 5.8. Schematic of a reaction that involves the rearrangement of the fragments of two chemicals.

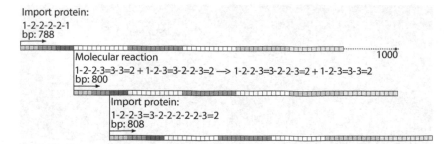

Import protein:
1-2-2-2-2-1
bp: 788

Molecular reaction
1-2-2-3=3-3=2 + 1-2-3=3-2-2-3=2 —> 1-2-2-3=3-2-2-3=2 + 1-2-3=3-3=2
bp: 800

Import protein:
1-2-2-3=3-2-2-2-2-3=2
bp: 808

1000

FIGURE 5.9. Genetic encoding of cellular dynamics in the ACM. The figure shows a segment of the chromosome that encodes two import proteins and an enzyme in overlapping reading frames. Each artificial protein is encoded by an open reading frame indicated by a start codon (light gray), followed by segments that indicated the type of protein, a putative regulatory region (dark gray), the protein's expression level (white), and then followed by the domains that specify the affinity of the protein. Here, the beginning of the pattern encoding the expression level of the import protein at "base pair" (bp) 800 reads like a start codon, giving rise to the overlapping frames. Adapted from Hintze and Adami (2008).

To be considered fit, an artificial cell has to import precursor molecules that are available outside the cell walls and convert them into metabolites. The fitness of an organism is determined by calculating the produced biomass of metabolites, where longer chains count more than shorter chains (Fig. 5.10; the total chain length is limited to twelve; see Hintze and Adami 2008 for more details). Evolution occurs in three different environments that differ in their predictability. In the simplest environment, the location of precursor sources and their abundance does not change during evolution (the "static" environment), while in the quasistatic environment one randomly selected precursor source location is changed per update. In the dynamic environment, the source location of *all* precursors changes randomly, and 25 percent of all precursors are made unavailable, giving rise to a highly unpredictable environment.

We can measure the information content of the evolved genomes just as outlined above, that is

$$I = L - H(s), \tag{5.16}$$

where L is the total length of the sequence and $H(s)$ is the sum of per-site entropies

$$H(s) = \sum_{x=1}^{L} H(x) \tag{5.17}$$

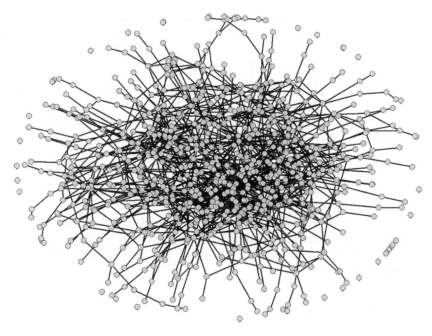

FIGURE 5.10. Evolved metabolic network with 969 nodes and 1,698 edges (from Hintze and Adami 2008), rendered with PAJEK (Batagelj and Mrvar 2003).

and the per-site entropy $H(x)$ is obtained by summing over the substitution probabilities p_i at that site:

$$H(x) = -\sum_{i=0}^{3} p_i \log_4 p_i. \tag{5.18}$$

Because we only have four possible symbols per site, taking the logarithm to base 4 again ensures that the per-site entropy lies between 0 and 1.

As the evolutionary mechanism allows for insertions and deletion of entire genes or genetic regions along with point mutations, genomes can change length during evolution. As a consequence, an alignment of genomes in a population to ascertain substitution probabilities is problematic. Instead, we can use an approach that determines the substitution probabilities p_i from the fitness effect of the substitution on organism fitness, along with an application of population genetics theory. If a substitution of allele i has fitness effect w_i, then the probability to find this allele in an equilibrated population evolving at mutation rate μ can be determined by iterating the discrete replicator-mutator

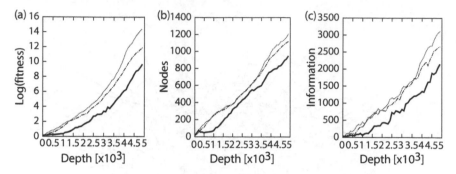

FIGURE 5.11. Evolution of complexity in artificial metabolic networks. (a) Log fitness for networks evolving in a static (dotted line), quasistatic (dash-dotted), and dynamic environment (solid line). (b) Evolution of the number of nodes (number of edges follows a similar trend). (c) Evolution of informational complexity, lines as in (a). From Hintze and Adami (2008); Adami (2009).

equation (Huang et al. 2004)

$$p_i^{(t+1)} = \frac{p_i^{(t)} w_i}{\bar{w}}(1 - \mu) + \frac{\mu}{4}, \tag{5.19}$$

where $\bar{w} = \sum_{i=0}^{3} p_i^{(t)} w_i$ is the mean fitness of the four possible alleles at that position. The fixed point as $t \to \infty$ gives the equilibrium p_i. Figure 5.11 shows the evolution of network fitness, size, and complexity (measured in units "mer," where one mer equals two bits) for the three different environments discussed above, as a function of the phylogenetic depth of the organism. The phylogenetic depth is the position of the genome on the line of descent of the particular evolutionary run: The genome with the highest fitness at the end of the run is used to reconstruct the line of descent by following the sequence's direct lineal ancestry and increasing the depth counter whenever a cell's genome differs from that of its direct parent. When arriving at the initial cell, this counter is set to zero, so that the phylogenetic depth counter increases up until the last cell on the line. Because on average we find about one new cell on the line of descent per generation in these runs, the phylogenetic depth is a good proxy for evolutionary time, even though in principle many generations could pass without an advance on the line of descent.

Because the fitness of a cell is multiplicative in the biomass (discovering how to produce a new metabolite multiplies the previous fitness by a number greater than one), the log of the fitness grows about linearly for all three environments (Fig. 5.11[a]). The fitness grows fastest for the static environment that is the easiest to predict (dotted line), while it takes longer for complexity to emerge in the dynamic environment (solid line). The same trend is reflected in the growth of the number of nodes and edges in these environments (Fig. 5.16[b]). Finally, the information content as calculated by Equation (5.16) using the substitution probabilities (5.19) follows the same trend: the informational complexity grows the fastest for the static and quasistatic environments and lags behind for evolution in a dynamic environment. The reason for the slower growth of complexity for networks evolving in dynamic environments is interesting: because the availability of precursors necessary for the production of complex metabolites cannot be relied upon in such environments, the cells end up manufacturing the precursor molecules *within* the cells (rather than importing them from the outside). This machinery is complex in itself but takes time to evolve. Ultimately, we expect networks evolving in dynamic environments to be more complex than those evolving in static environments because of the added flexibility of producing precursor molecules within the cells. However, such networks lag behind slightly during the time this complexity is generated.

Note that the informational complexity of the networks used to seed these evolutionary experiments is rather low: the network with three genes is specified with a genome of informational complexity of just thirty-six mers (seventy-two bits), even though the starting genome has 1,000 "nucleotide" positions in each of the two chromosomes. The noncoding part of these initial genomes thus does not contribute to the informational complexity, because changing any of these positions to any other allele cannot change the fitness of the organism (we do not take beneficial mutations into account in the fitness tests). For these noncoding nucleotides, the p_i calculated by Equation (2.29) all are exactly equal ($p_i = 1/4$), guaranteeing that they do not contribute to I in Equation (5.16), as is easily checked. However, the informational complexity grows rather quickly once more and more metabolic reactions are discovered and optimized, at a pace of about 0.5 mers (1 bit) per depth step (roughly one bit per generation).

The artificial cell model confirms a valuable hypothesis, namely that it is *genomic complexity* (measured in terms of information) that codes for the structural complexity of metabolic networks. Many other insights can be gained from performing evolution experiments using the ACM. For example, it is possible to study how modules emerge, and how the modularity of the networks depends on the predictability of the environment (Hintze and Adami 2008). In fact, modules—groups of nodes that are more connected to each other than other such groups—play an important part in systems and network biology. In the next section, we will explore a method to identify such modules in a network using methods from information theory.

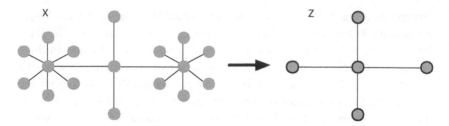

FIGURE 5.12. Collapse of a topology described by the nodes X to a more succinct one described by Z in which clusters are replaced by a cluster assignment variable Z. From Adami (2009).

5.2.1 Modules from information theory

The path toward an understanding of the functional organization of a network in the absence of genomic information usually involves the *decomplexification* of the network, either by clustering nodes that are related in function (Watts and Strogatz 1998), removing those nodes that are immaterial to (or redundant in) function, or else analyzing the subgraph decomposition (Milo et al. 2002) as discussed further below. A particularly insightful method to decompose networks into modules—both overlapping and nonoverlapping—uses information theory to estimate how much information about the original network is present in the abstract (that is, decomplexified) version, while maximizing a variable that measures the *relevance* of the abstraction. This method, sometimes called the "network information-bottleneck" approach (Tishby et al. 1999), was applied to biological and engineering networks by Ziv et al. (2005).

The *Network Information Bottleneck* (NIB) approach attempts to replace a complex network by a simpler one while still retaining the essential aspects of the network. For example, a highly connected star topology could be replaced by a single node that represents the modular function of the star, as in Figure 5.12. The main idea of the method is that while there are many different ways in which one can collapse a topology, the optimal mapping is one where the new topology retains as much information as possible about the original one, while maintaining as much *relevance* of the description as possible.

Say, for example, that a random variable X describes nodes $x \in X$ in a network that occur with probability $p(x)$, and total number of states $|X| = N$, where N is the size of the network. A *model* of this network can then be made using a random variable Z with *fewer* states $|Z| < N$, and where a cluster assignment is given by a set of probabilities $p(z|x)$: the probability that z is assigned to a particular cluster given the input node x. Ideally, we would like to maximize the mutual entropy (information) between the random variables

X and Z, but we shall do this with a constraint given by the relevance variable mentioned earlier. This relevance variable will distinguish different ways in which clusters are assigned. In this application of the NIB to networks, the relevance variable involves diffusion on the network: those nodes that are close to each other are preferentially visited by a diffusion algorithm and are more likely to be clustered together. To implement this algorithm, the relevance is represented by a random variable Y that is defined such that a diffusive process determines the probability to arrive at node y. The relation to the network variable X is given by the joint probability $p(x, y) = p(y|x)p(x)$, the probability to arrive at node y via a diffusive process given the process started at node x, times the probability that we started at node x. The latter probability is always assumed to be uniform, that is, $p(x) = 1/N$.

The NIB algorithm makes it possible that several different nodes z are assigned to any given input node x. This flexibility allows for the possibility of *overlapping* clusters or modules (soft clustering), but in the following we will only describe the algorithm for *hard clustering*, so that for any choice of x and z, $p(z|x)$ is either one or zero. One version of the algorithm (agglomerative clustering) begins with a random variable Z that has exactly one fewer nodes than X, and attempts to find the optimal pair of x nodes to join to produce a model of the network with one fewer nodes. For each possible cluster assignment $p(z|x)$ we can execute a diffusion process to determine the matrix $p(y|z)$, that is, the probability to arrive at node y given a node z as starting point. We choose to merge those nodes that maximize $I(Y:Z)$, and the algorithm repeats with a set Z smaller by one node until all nodes have been merged. At each step, we can calculate the normalized variables $0 < I(Z:X)/H(X) < 1$ and $0 < I(Z:Y)/I(X:Y) < 1$ and plot them against each other, giving rise to the *information curve* (Ziv et al. 2005), as in Figure 5.13. A completely random network gives rise to the diagonal in Figure 5.13 and represents the least modular network.

We can define a modularity score, the *network modularity*, as the area under the information curve. Perfectly modular networks then have a maximal modularity score of 1, whereas random networks have a score of $1/2$. Ziv et al. (2005) analyzed several different networks using this modularity measure, such as the network of co-authors for papers presented at a meeting of the American Physical Society. This network has 5,604 nodes and 19,761 edges, and yielded a modularity core of 0.9775. The regulatory network of the bacterium *E. coli* (328 nodes and 456 edges), also analyzed by these authors, yielded a score of 0.9709. Thus, both of these networks are highly modular according to this measure. Interestingly, the network of connections of the *C. elegans* brain (see Fig. 5.6, we will study this brain in more detail below) has a network modularity score of 0.9027, whereas a randomized version retaining the same number of nodes and edges scores 0.4984 on average, as

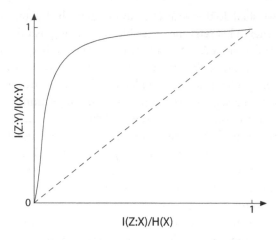

FIGURE 5.13. The information curve for a modular network (solid line), obtained by starting with a model network Z of the same size as X (maximal information $I(Z:X)/H(X) = 1$ and $I(Z:Y)/I(X:Y) = 1$, upper right corner), and merging nodes while maximizing $I(Z:Y)$. This process generates the information curve from the upper right corner all the way down to the lower left corner, where $|Z| = 1$ and the mutual entropy vanishes. The dashed line represents the information curve for a random network. The modularity score is given by the area under the information curve. From Adami (2009).

predicted for a random network. The evolved metabolic networks discussed in section 5.2 also score high on this scale. For the largest connected component of a 453-node network, we find a modularity score of 0.8486 for the 5,000th organism on the line of descent (about one organism per generation).

While modularity is an important concept in network biology, it is not clear what aspect of complexity modularity really reflects, if any. A modular organization of function is important in engineering design (Suh 1990) because it allows the engineer better control over the components of the structure or device. But since biological networks are not designed, such considerations are probably not applicable. The artificial neural networks (ANNs) often used in computer science and machine learning, for example, are strongly connected networks that lack modularity. It is often argued that a modular structure makes a network more evolvable (Clune et al. 2013) and in particular renders it more robust to change. While there certainly is evidence for this phenomenon, we cannot argue that a more modular network is either more complex (or less complex) than a non-modular one. If complexity cannot be found in modularity, can it perhaps be found on a smaller scale in network *motifs*?

5.2.2 *Information in motifs*

We have seen that networks that are functional are built from modules, or at least can be understood in terms of strongly connected sets of nodes that are only weakly connected to other such clusters. This clustering—carried out using information theory in the previous section—can also be performed on the basis of topology alone. For example, every network can be analyzed in terms of its *subgraph composition* (Milo et al. 2002), that is, the frequency with which particular subgraphs or motifs appear within the entire network. The degree with which certain motifs are overutilized—and some others underutilized—with respect to a uniform distribution or that obtained from a random network—reflects the *local structure* of the network and can be used to classify networks from very different realms into similar categories (Milo et al. 2004).

Subgraph abundances can also be used to study the modular composition of networks, in analogy to the modular composition of sentences in written language. Meaning can be conveyed in text only because the utilization frequency of letters in words, and words in sentences, is different from uniform (see also the discussion in section 7.2). For example, the letters e, t, a, i, o, and n appear in decreasing frequency in an average text written in English, while the rank-abundance distribution of the words follows a scale-free distribution known as Zipf's law (Zipf 1935; Shannon 1951). If we assume that a random sequence of letters contains no information, then the deviation from the uniform distribution could be used to distinguish and perhaps classify functional (that is, meaningful) text from gibberish. In the same vein, it is possible that functional networks differ significantly from random networks in the subgraph utilization, and we can study this difference by estimating the *subgraph information content* as follows.

Suppose we compute the probability to find any of the two possible motifs that can be made of three nodes (see Fig. 5.14). For simplicity, we are considering here only undirected graphs and do not allow self-interactions; that is, nodes that link to themselves. We can then compare these empirical probabilities to the probabilities with which these subgraphs appear in a random network (also known as an Erdös-Rényi network, see Box. 5.1).

A priori, we might think that any of the two motifs of size $n = 3$ should appear with equal probability in a random network, giving rise to a motif entropy that is maximal:

$$H_3(\frac{1}{2}, \frac{1}{2}) = -\frac{1}{2}\log_2\frac{1}{2} - \frac{1}{2}\log\frac{1}{2} = 1. \qquad (5.20)$$

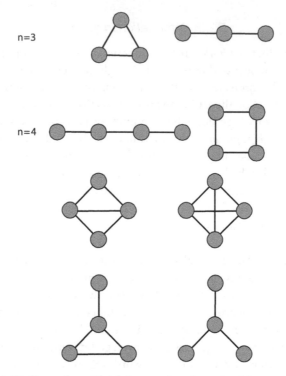

FIGURE 5.14. Undirected motifs of size $n = 3$ and $n = 4$, without self-interaction.

Here, I defined the size-n motif entropy (Adami 2009)

$$H_n(p_1, \cdots, p_m) = -\sum_{i=1}^{m} p_i \log_m p_i, \qquad (5.21)$$

where m is the number of possible connected motifs of size n, and the p_i are the probabilities to find the ith motif in the network. (Because the base of the logarithm is also m, this entropy is normalized to be between zero and one.) The information stored within $n = 3$-motifs would then be (the superscript (u) refers to the uniform baseline distribution)

$$I_3^{(u)} = H_3(\frac{1}{2}, \frac{1}{2}) - H_3(p_1, p_2) = 1 - H_3(p_1, p_2), \qquad (5.22)$$

while the information stored in n-motifs is naturally

$$I_n^{(u)} = H_n(\frac{1}{m}, \cdots, \frac{1}{m}) - H_n(p_1, \cdots, p_m) = 1 - H_n(p_1, \cdots, p_m). \quad (5.23)$$

Box 5.1. Random Graphs

Random graphs (sometimes called random networks or Erdös-Rényi networks) are graphs in which the probability that an edge connects two randomly chosen nodes is drawn from a binomial distribution.

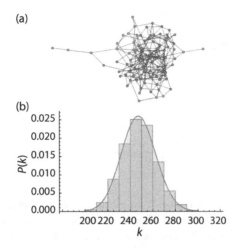

(a)

(b)

FIGURE 5.15. (a) An example random graph with $N = 100$ and $p = 0.05$. (b) Probability density function of random graphs with $N = 100$ and $p = 0.05$ (created from 10^4 realizations of such graphs), with the binomial distribution—obtained with the same parameters—superimposed in light gray.

They are used most often to create a "null" model in which the edges do not represent any special relationship between the nodes, but they also play a role in pure Math (Erdös and Rényi 1960; Bollobás 1985). Random graphs can be created in different ways, but in the simplest method (due to Gilbert 1959), you first create a number of nodes n, and then place an edge between any pair of nodes (there are $\frac{N(N-1)}{2}$ such pairs) with probability p. Figure 5.15(a) shows an example network with $N = 100$ and $p = 0.05$, which on average has $p\frac{n(n-1)}{2}$=247.5 edges. The probability density function $P(k)$ of edges (the probability that a particular realization of a graph has k edges) is a binomial distribution

$$P(k) = \binom{N-1}{k}p^k(1-p)^{N-1-k}. \tag{5.24}$$

We see this distribution in Figure 5.15(b) (gray histogram), along with the (theoretical) binomial distribution with $p = 0.05$ and $N = 100$. We can use random networks to estimate how often we would expect to find any particular motif of size n in a large network, to test whether a particular distribution of motifs reflects the function of the network.

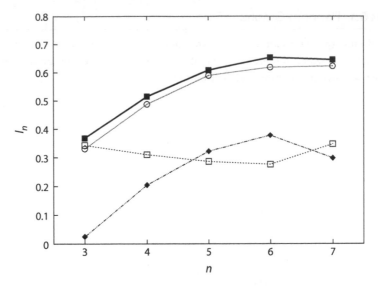

FIGURE 5.16. Motif information I_n for motifs of size n for evolved metabolic networks with a uniform distribution as a baseline (solid line, filled squares), an Erdös-Rényi network of the same size and number of edges and connectivity $p \approx 0.005$ (dashed line, open squares), and a randomized network of the same size with a scale-free edge distribution (dotted line, open circles). The motif information using the probability distribution of a random network as the baseline ($I_n^{(r)}$) is the difference between the solid and dashed lines (dash-dotted line, filled diamonds). From Adami (2009).

However, even random networks do not have a uniform distribution of motifs, and we can instead consider the information stored in motifs compared to a random network (see Box 5.1) as baseline, so (for $n = 3$)

$$I_3^{(r)} = H_3(p_1^{(r)}, p_2^{(r)}) - H_3(p_1, p_2) . \tag{5.25}$$

where the $p_i^{(r)}$ refer to the probability of finding motif i in a random network.

In Figure 5.16, we can see the information $I_n^{(u)}$ stored in motifs of size n for $n = 3 - 7$ (filled squares, solid line), for a single, evolved, functional, simulated metabolic network of 598 nodes in the ACM, discussed in section 5.2. The network information increases as more complex motifs are used for encoding, but appears to stabilize. This behavior mirrors the statistics of n-gram entropies in English text, as noted early on by Shannon (1951). Note that because the shortest path between any two nodes is on average about four to five in these artificial metabolic networks (Hintze and Adami 2008), motifs of size 7 or larger are not well sampled.

We can study whether the network information is dictated by functionality, edge distribution (or both) by constructing analogous networks that have the functionality removed. This can be done by randomizing connections but keeping the scale-free edge distribution, and by randomizing the network but destroying also the edge distribution. If, for example, we randomize the connections in our functional evolved network while keeping the scale-free edge distribution, we find that the network information is only slightly lowered (open circles and dotted line in Fig. 5.16). On the other hand, if we randomize the network in such a way that the degree distribution is that of a random graph (but still keeping the same number of nodes and edges), the dependence of the network information as a function of the subgraph size is markedly different (open squares, dashed line in Fig. 5.16). This suggests that the network information is significantly dictated by the biological scale-free distribution of edges per node, and only weakly by the actual function of the network.

Because random graphs do not harbor subgraphs with equal probability (as witnessed by the nonzero network information in Erdös-Rényi networks), it is more appropriate to use the random network probabilities as the baseline to calculate the network information. If we do this, we see the network information $I_n^{(r)}$ increase from small values at $n = 3$ up to $n = 6$ (dash-dotted line, filled diamonds in Fig. 5.16). The decrease noted for $n = 7$ is due to incomplete sampling of size $n = 7$ networks for this small graph (598 nodes), and is not significant.

In summary, network information as measured by subgraph "n-gram" entropies behaves similarly to the dependence of n-gram entropies in written language and can be used to distinguish functional networks from random ones. However, the network information appears to be controlled mostly by the form of the edge distribution. Insight into the modular structure of networks will likely depend on understanding how the subgraphs of networks are assembled into modules, much like how letters are assembled into words in written text.

5.2.3 Colored motifs

In the previous section, I considered the complexity of network motifs as a possible proxy for the complexity of the cell (or organism) it represents. It is clear, however, that the structural complexity of motifs only carries part of the functional complexity. In particular, the nodes of the network are all assumed to be the same, while we know that in realistic networks, nodes have different attributes that may influence what other nodes they connect to. In the network that describes the interaction of proteins in a cell, for example, most proteins are functionally different, yet the motifs determined by the usual network description ignores this difference. What if we could take these functional differences into account when defining network motifs? We could do this, for example, by coloring nodes of a network that are functionally similar by the same color. In a social network, for instance, we could define colors by age groups and find that there are many more connections between same colors than different colors. The exception to this rule would likely be connections between family members, so that an analysis of *colored motifs* could be used to detect family subgroups within large social networks.

A quick calculation shows that using colored motifs to assess both the structural *and* functional complexity is only possible if the list of possible node types—that is, colors—is small. While for uncolored networks there are six possible motifs of size 4 (if the edges are undirected), it is clear that each of the motifs could come in many different colorations. For example, there are 50 different motifs of size 4 if nodes come in two different colors, but this rises to 201 if a node can take on three colors. If edges can be both undirected and directed (meaning that the edge could be an arrow), there are then 199 motifs of size 4 if nodes are uncolored, but 13,770 different motifs when three colors are considered (Qian et al. 2011). Clearly, because of this combinatorial explosion in the number of possible motifs, we must be very careful when deciding on a coloring scheme.

For proteins as nodes in a protein-protein interaction network, one possible coloring is the effect that a knockout mutation has on a protein, giving rise to three colors: say, green for "neutral" (the knockout does not affect the cell or organism's fitness), red for "lethal" (the knockout results in a dead organism), and yellow for "detrimental" (knocking out the gene that codes for this protein results in an impaired cell or organism). When this is done for the network of proteins of baker's (or budding) yeast (*Saccharomyces cerevisiae*), the resulting network (a detail of the network is shown[4] in Fig. 5.17) can be used to understand how function is encoded in robust pathways (Jeong et al. 2001).

Another network where colored motifs can throw light on the functional organization of the network is the "connectome" of the worm *C. elegans*, displayed as an adjacency matrix in Figure 5.6. This worm is about a millimeter long (shown as a render in Fig. 5.18[a], with the neuronal network revealed in Fig. 5.18[b]) and is one of the best understood animals on the planet: it only has about a thousand cells, and there are two sexes: a male with 1,031 cells in total, and a hermaphrodite that has only 959 cells. The hermaphrodite can propagate by self-fertilization as the name suggests, while the male has to find

---

4. In the original figure colors are used instead of grayscale. The translation of the colors in the original is indicated in the legend of the figure.

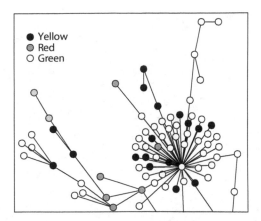

FIGURE 5.17. Detail of the largest connected cluster of the protein-protein interaction network of yeast (adapted from Jeong et al. 2001). Nodes colored white (green in the original) do not affect the fitness of the organism, notes colored in gray are lethal to the yeast, while nodes colored in black represent proteins whose knockout is detrimental to growth, but not lethal.

another hermaphrodite to further its lineage. The brain of this animal is also exceedingly well understood, thanks to the pioneering work of John White and his colleagues (White et al. 1986). Using this worm as a model animal was the idea (and life work) of Nobel laureate Sydney Brenner (see, e.g., Brenner 1974).

The brain of the hermaphrodite worm (almost all worms in natural populations are actually hermaphrodite: males are very rare) has exactly 302 neurons, 279 of which have been described—along with their connection to other neurons—in detail (see, for example, the most up-to-date compendium by Varshney et al. 2011). Twenty of the remaining neurons can be found in the pharyngeal system that form a ring of nerves around the pharynx of the animal and are completely independent from the much larger somatic nervous system comprising the 279 neurons. There are three neurons left over: they apparently do not connect to anything, and their function is therefore somewhat obscure. Nothing happens to the animal if their function is interfered with.

Given the wiring diagram of the worm brain (shown as Fig. 2 in Varshney et al. 2011, with a different coloring scheme for nodes), we can now look for motifs. This has been done (without taking into account neuron type) by Reigl et al. (2004) and then again by Varshney et al. (2011), with the aim of discovering the computational motifs that the *C. elegans* brain is using in its decisions.

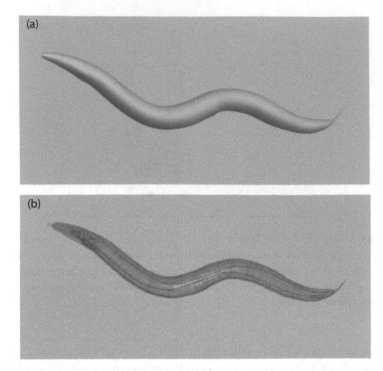

FIGURE 5.18. (a) A rendering of the worm *C. elegans* in the OpenWorm project (Szigeti et al. 2014). (b) The worm's nervous system superimposed on its body.

The motifs in the *C. elegans* neuronal networks are actually quite a bit richer than the motifs we have encountered earlier because there are two types of connections between neurons in the worm: synaptic connections that have a directionality (and are indicated by an arrow with a single tip), and "gap junctions," which are undirected and represented with arrows on both ends of the connection. In the network described below, we used 3,606 edges (the majority of which are synaptic), but about as many connections have been added to the network since (Varshney et al. 2011). In Figure 5.19, we can see a variety of motifs of two and three neurons using the two different kind of edges (fifteen altogether). But looking for "over-represented" motifs (motifs that occur much more often than expected by chance) in the network proves to be somewhat disappointing: the search only turns up four fairly common motifs (highlighted in Fig. 5.19).

Why does a search for over-represented motifs not lead us to a better understanding of the *C. elegans* brain, if evolution shapes the connectivity of the network to better carry out its function? The most plausible reason (as we

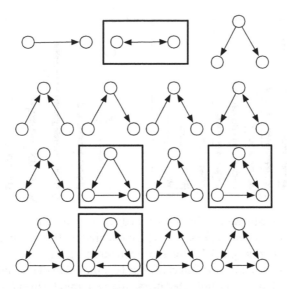

FIGURE 5.19. Motifs of size two and three in the *C. elegans* neuronal network. Only four of those motifs are consistently over-represented, indicated by a box around the motif. Adapted from Reigl et al. 2004.

(a)   (b)

FIGURE 5.20. (a) A three-node motif in the uncolored *C. elegans* network that is unremarkable. (b) The motif that connects a sensory neuron (white) to an interneuron (light gray) and then finally to a motor neuron (dark gray) is highly over-represented in the network (Qian et al. 2011).

had already guessed) is that network structure alone is unlikely to reflect function (Ingram et al. 2006), simply because in real networks, different nodes take on different functions. In particular, the neurons of the worm can be classified broadly into three types: sensory neuron, interneuron, and motor neuron (Varshney et al. 2011). If we color nodes according to this function (for example, green for a sensory neuron, red for an interneuron, and blue for a motor neuron,[5] a simple motif that was *not* over-represented in the study by Reigl et al. (2004) becomes dramatically over-represented for a very particular color assignment, namely the one seen in Figure 5.20. Even among

---

5. Because this book is printed in black-and-white, green is replaced by white, red by light gray, and blue by dark gray.

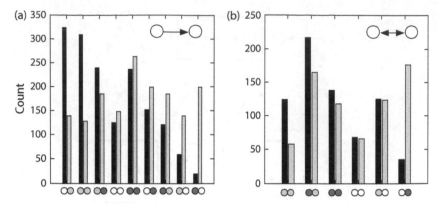

FIGURE 5.21. Histogram of abundances of directed two-neuron motifs with particular coloration in *C. elegans* (black), compared to the color randomization of the same network (gray). White: sensory neuron, light gray: interneuron, dark gray: motor neuron. (a) Directed pairs (the direction of information flow is left-to-right). (b) Undirected pairs. Adapted from Qian et al. (2011).

two-neuron motifs (shown in Fig. 5.21) there are obvious colored motifs that "make sense" in a functioning brain, and many others that do not.

To check which motifs make sense, we can create a "randomized" worm brain by shuffling colors while keeping the connectivity unchanged. Those motifs for which the count in the real worm brain (black bars in Fig. 5.21) is significantly higher than the gray bars (color-shuffled worm brain) are the ones that are functionally relevant. Among three-neuron motifs, the one shown in Figure 5.20(b) actually stands out the most, followed by the same structural motif but with two interneurons feeding into the motor neuron (Qian et al. 2011). Testing four-node colored motifs for significantly over-represented motifs becomes quite a bit more difficult, as there are 13,770 different possible colored motifs, but only 8,310 of those actually occur in the *C. elegans* network. In such a case, it is imperative that significance (of the overcount or undercount, with respect to the count expected by chance) is estimated by correcting for multiple-hypothesis testing.[6]

A few examples of four-node colored motifs are shown in Figure 5.22(a). It is immediately apparent that colored motifs that appear more often than chance in a network also play important functional roles when we dissect the part of the *C. elegans* brain that is responsible for forward locomotion, shown in Figure 5.22(b). The function of this particular part of the network can be simulated (Karbowski et al. 2008), and that simulation shows that the network

---

6. For this study, we corrected for multiple hypotheses by using the single-step min-P procedure for multiple-hypothesis adjustment (Dudoit et al. 2003; Westfall and Young 1993).

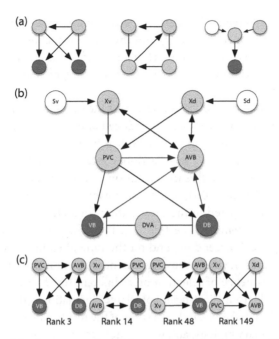

FIGURE 5.22. (a) Three examples of four-neuron motifs. The middle one is uncommon in the *C. elegans* brain. (b) The core of the forward locomotion network of the worm, after Karbowski et al. (2008). Nodes are shaded as in Figure 5.20. Arrows with single points are excitatory connections via a chemical synapse, while edges ending in a bar denote inhibition via a chemical synapse. Edges with two arrow heads represent gap junctions. Neurons are "named" according to the standard nomenclature. Figure 9.23(b) shows the entire forward locomotion module. (c) A selection of four-node motifs that occur much more often than chance in the network (rank number out of 8,310 motifs that appear in the network listed) that also appear in the locomotion network shown in (b). Adapted from Qian et al. (2011).

drives the motor neurons (in dark gray) in an oscillatory manner, propelling the worm forward. Within this small network fragment we can see some of the motifs that are highly over-represented in the *C. elegans* brain, for example the third-ranked bi-fan motif, plus other variations on the theme.

It is also possible to study how much information is encoded in these motifs in the same manner as we did in Figure 5.16, except now with colors (Adami et al. 2011). We find that information is now encoded both in structure *and* in function (coloring), and that the information per motif increases with the size of the motif, as we would have expected.

It is clear from the examples studied here that some of the complexity of a network is reflected in the motifs of the network, in the sense that these motifs carry information that is maintained by selection. However, there are significant differences between the structural complexity of the motifs (information in connectivity only) and the functional complexity (information stored in the color assignment). The motif-based complexity measure cannot be universal, however, because there is no "right" way to assign colors to particular nodes.

We have now discussed several different measures of complexity (structural, functional, hierarchical) and introduced different measures of sequence complexity (Kolmogorov, physical, as well as information-content). In some examples (the evolution of information content of digital life sequences in Adami et al. 1999, as well as the evolution of artificial metabolic networks, see Hintze and Adami 2008), we saw an unambiguous trend of increase in complexity, as anticipated in the information-theoretic discussion of evolution as a Maxwell demon in section 3.1. But how general is such a trend, and how might such a trend depend on the type of complexity measure we use? Are we witnessing there the consequences of a universal trend, or is this increase a consequence of selection on a different trait? How can we distinguish true trends from short-term correlated fluctuations (the equivalent of the "hot hand" in basketball)? This is one of the most fundamental questions in all of evolutionary biology, and we will now consider it in some depth.

5.3   Long-Term Trends in Evolution

This question of long-term trends in complexity has been discussed at length in the biological literature; in particular it was the subject of a well-documented disagreement between two giants of evolutionary biology: Stephen J. Gould and Richard Dawkins. Gould, at the time a paleontologist at the American Museum on Natural History, had already established himself as one of the leading popularizers of evolution in the United States. Dawkins, an evolutionary biologist at New Oxford College, was his *pendant* in the United Kingdom: also an accomplished author and frequent lecturer. Their disagreement over the existence of a trend in complexity culminated in dueling books (Gould 1996; Dawkins 1996), and dueling book reviews (Gould 1997; Dawkins 1997) where they each reviewed the other's book, respectively.

Gould's main point was that an "adaptationist" point of view falsely paints evolution as "progressive" to satisfy a "human chauvinist" desire to put *Homo sapiens* at the top of a ranking of sorts. Dawkins, who does not actually subscribe to such a tendency, instead argues that within any particular lineage the evidence of progress is incontrovertible. The arguments that Gould uses for "denying that progress characterizes the history of life as a whole" (Gould 1996, p. 3) rely on the research of Daniel McShea, a paleontologist at Duke University. McShea argued that generally speaking it is important to make

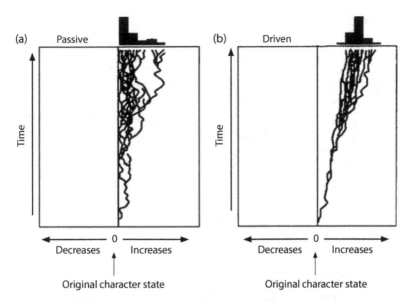

FIGURE 5.23. Illustration of passive and driven trends in dynamical processes. (a) Branching random walks subject to a boundary (the character trait cannot be negative). As a consequence, the variance as well as the maximum value of traits will tend to increase even though there is no bias in trait changes. (b) A branching random walk where changes toward increasing trait value are more likely than those that lead to a decrease, giving rise to a driven trend. Adapted from Gregory (2008).

a distinction between active (driven), and passive trends (McShea 1991; McShea 1996).

A passive trend can occur if a variable changes randomly up-and-down, but is bounded on one side (for example, a trait value that must be positive, as in Fig. 5.23[a]). If the random walk branches, there will be many branches with different trait values each as time goes on, and in particular we would be able to see the variance of traits increase with time, as well as the value of the mean and of the maximum trait value in the group. In comparison, if a random walk starts at trait value zero but is not constrained to be positive, we would also see an increase in the variance, but the mean would stay constant. In a driven trend, on the contrary, each particular random walk is assumed to be biased, so that (for the example we discuss here) the likelihood of an increase in trait values is higher than the likelihood of a decrease. Such a driven trend, in which also the *minimum* increases, is apparent in Figure 5.23(b).

It is possible to test the nature of trends using data meticulously obtained from fossils. For example, there is good evidence that Cope's rule, which posits that animal size has been increasing on average along a lineage, does hold. This

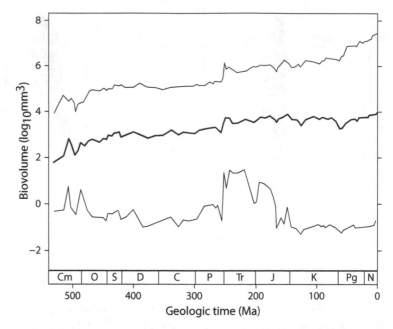

FIGURE 5.24. Body size evolution of marine animals across the past 542 million years. Shown is the distribution of fossil marine animal biovolumes since the Cambrian explosion. The thick black line indicates the stage-level mean body size. The thin black lines demarcate the fifth and ninety-fifth percentiles. Cm: Cambrian; O: Ordovician; S: Silurian; D: Devonian; C: Carboniferous; P: Permian; Tr: Triassic; J: Jurassic; K: Cretaceous; Pg: Paleogene; N: Neogene. Adapted from Heim et al. (2015).

rule, named after the American paleontologist Edward Cope (1840–1897), is assumed to hold because often fitness is correlated with size. But is this increase due to a passive or a driven trend? Heim et al. (2015) compiled adult body size measurements for over 17,000 genera of marine animals (more than 74 percent of animal diversity in the fossil record) and compared the trend shown in Figure 5.24 to simulations of trajectories based either on an unbiased random walk or a size-biased walk in which size explicitly provides for a selective advantage. They concluded that the size-biased model provides for a much better fit to the data than a neutral model: for the mean, the maximum, as well as the minimum size (Heim et al. 2015). Thus, in the case of body size, we can conclude that the overall trend is indeed due to selection.

Trends can be observed in many other traits as well. We can expect evolution to give rise to a driven trend as long as selection tends to result in more changes in one direction than the other. Of course, evolution acts on the level of DNA, so we might also look for observable long-term trends in DNA sequence composition or genome length over time. As we will see later, while there is an overall trend toward increasing genome length, the "DNA content" of an organism is a poor predictor of its complexity (the "C-value paradox" that we discussed earlier). What about sequence composition? In section 2.1 we defined the "GC content" of a sequence as the sum of the probabilities to observe the nucleotide guanine and cytosine, that is, $GC = p_G + p_C$. This value varies tremendously among species, ranging from 20 to 80 percent. It is also variable across genes in a single species. In humans, for example, the mean $GC = 46.1$ percent, but varies from 35 to 60 percent in 100 kb (kilobase) fragments (Romiguier et al. 2010). What is driving these differences?

It turns out that changes in GC content may be adaptive in some cases. For example, increasing the GC content at the beginning of an mRNA sequence (the 5' end) leads to a less stable secondary structure of the mRNA, which can be shown to lead to more efficient recognition of the start codon (Gu et al. 2010). However, it seems that the majority of changes in GC content are driven by the rate at which the genome undergoes recombination and thus is not adaptive in nature. The correlation with recombination is explained by base mismatches that are created during homologous recombination events. Those mismatches are often fixed by mismatch repair in a biased manner, creating more GC pairs than AT pairs on average (this is called GC-biased gene conversion, see Galtier et al. 2001). And even though bacteria do not undergo homologous recombination, intragenic recombination is frequent, leading to patterns similar to those observed in eukaryotes (Lassalle et al. 2015).

If selection-driven long-term trends cannot be seen (on the whole) on the level of DNA evolution, what about proteins? The functional elements of proteins are called "domains": they are conserved parts of the protein that exist independently of the other domains in a multidomain protein. As is described in Box 5.2 below, proteins evolve via duplication, diversification, and domain shuffling. Can these processes alone give rise to a bias in the evolution of domain sizes? It turns out that this is possible, but while within a particular lineage a long-term trend can indeed be observed, it is not the same trend across lineages. Wolf et al. (2007) find increasing domain size among the lisozyme family of protein domains, for example, while the RNAase family's domain size tends to decrease. Thus, we can observe lineage- and domain-specific trends that are not universal across domains or families. However, that does not mean that universal trends in protein evolution are impossible to observe.

Perhaps the best way to look for a long-term trend then is to look at the history of a protein from the moment it is born. As we have seen in section 3.2, it is possible to infer the sequence of ancient proteins via maximum likelihood methods applied to a reconstruction of phylogenetic trees. However, this method attempts to infer what a protein was like millions of years ago when it just arose, while to test for long-term trends, we would like to know how a particular characteristic of the protein changes over time as it ages. To do this, we need to be able to measure how *old* an existing protein is. There is a method to do this, called *phylostratigraphy* (Domazet-Loso et al. 2007; Neme and Tautz 2013). This method relies on taking a gene in an existing organism (say, mouse), then following the phylogeny backward to figure out whether organisms that branched off the lineage in the past also have a version of that protein (called a "homolog"). So, for example, if both mouse and rats have the same (or very similar) version of a protein, this means that that protein must have existed before the split that gave rise to mouse and rat. If the gene is present in both rodents and primates, then it is older than the split between rodents and primates, and so forth. At one point, you hit a group of organisms that does not carry this gene: that would be the "birth instant" of that gene (see Fig. 5.25): the branching point in the phylogeny where one branch has the gene, while the other one does not.

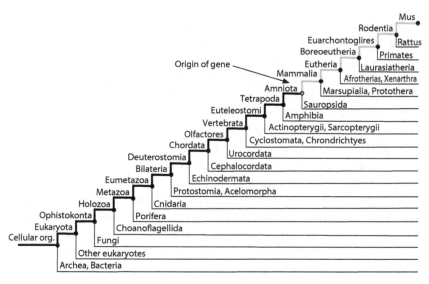

FIGURE 5.25. A schematic phylogenetic line of descent for the mouse, going all the way back to the origin of cells. Only the branching points leading to the mouse are shown. A particular gene's history is tracked in this phylogeny, and is colored gray when significant homology is detected in the gene family at that taxonomic level, and black if the gene cannot be found in that family. The point at which the gray line turns black is the time of origin of this particular gene. In the analysis of Wilson et al. (2017), all families earlier than the vertebrates are lumped together as "pre-vertebrates."

Box 5.2. Evolution of New Genes

In chapter 4 we witnessed the evolution of new functions in the digital and the biochemical realm. The emergence of the novel citrate exploitation in *E. coli* showed how the evolutionary process takes advantage of existing variation to cobble together a new function. How general is this process? Over the years, molecular biologists have collected many examples of the evolution of new functionality (reviewed in Long et al. 2003) and came up with a list of processes that contribute to the evolution of new function. They are exon shuffling, retroposition, gene duplication (with divergence in the duplicated copy), mobile elements, lateral gene transfer, gene fusion/fission, as well as *de novo* origination from previously noncoding genome regions. In exon shuffling, pieces of genes that code for proteins (the exons) that are interspersed with noncoding regions (the introns) are shuffled together to create a new gene. In retroposition, a messenger RNA in which the introns of a gene have been removed is reverse-transcribed to DNA. In the remarkable case of the evolution of the *jingwei* gene, three of these mechanisms can be seen at work. *Jingwei* is a young gene that emerged in an African subspecies of *Drosophila* about 2.5 million years ago (Wang et al. 2000). It consists of pieces of two other genes: an alcohol dehydrogenase gene *Adh* that was created by the reverse transcription of the messenger RNA of that gene in which all the introns had been removed (top part of Fig. 5.26), and a duplicated copy of the *Yellow emperor* gene, called *Yande*. *Yellow emperor* (bottom part of Fig. 5.26) remained functional in the fly so that *Yande* could diverge and be recombined with the *Adh* gene that was shuffled in-between the pieces of the *Yande* gene. The resultant *Jingwei* appears to have been immediately functional (Zhang et al. 2004), but needed to be optimized thereafter. Indeed, subsequent to the origination event, nine substitutions occurred in the *Jingwei* gene that "fixed" what needed fixing after the rough cobbling-together of gene pieces.

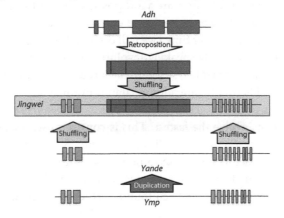

FIGURE 5.26. The *Drosophila* gene *Jingwei* is formed from pieces of the *yande* (*ynd*) gene, which itself is a duplicated copy of the *Yellow emperor* (*ymp*) gene. The two pieces of the *ynd* gene were shuffled together with a retroposed copy of the alcohol dehydrogenase *Adh*.

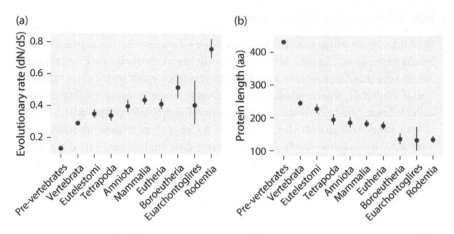

FIGURE 5.27. (a) Long-term trend in evolutionary rate determined by phylostratigraphy (Wilson et al. 2017) shows that young proteins evolve faster than old proteins. (b) Long-term trend in protein sequence length shows that older proteins are longer than younger proteins. Adapted from Wilson et al. (2017).

This analysis can in principle be performed for all the annotated 22,778 mouse genes (Wilson et al. 2017), so that each of those genes can now be assigned an age, starting with age "0" for the mouse genes that are not also in rats (meaning they were "just born"), all the way to the emergence of vertebrates.[7] Note that while ages can in principle be assigned all the way to the emergence of cells (see Fig. 5.25), the age assignment becomes fairly uncertain the older the genes are. For that reason, Wilson et al. (2017) only went back to the emergence of vertebrates, about 360 million years ago.

So what did they find? There are a few possibilities for trends for proteins that are easily imagined. For example, we know that evolution of a particular trait tends to slow down over time (we will discuss this feature in more detail in section 5.4.4). Generally speaking, evolution of a trait follows the *law of diminishing returns*: the first few steps are big strides, but as the trait improves it gets harder and harder to find additional improvements.

We can see two very basic trends in Figure 5.27. In panel (a) we see that the youngest proteins evolve the fastest. This is consistent with a law of diminishing returns: when a protein is "born" (see Box 5.2 for a short review of the mechanisms that can give rise to new genes), it is just barely suited for the job, meaning that there are many ways to improve it. Once the "quick-and-easy" improvements have been found, further changes require multiple

---

7. Because some genes in the mouse are unique to them, and some others are difficult to assign to only a single age category, the study used only 15,347 of the 22,778 genes.

substitutions, which will be comparatively more difficult to find. As a consequence, the rate at which improvements occur (the rate of evolution) slows down, to such an extent that the most ancient proteins barely change at all.

Another trend that is readily observed affects the length of a protein. In Figure 5.27(b), Wilson et al. plot the average length of the 15,347 genes they analyzed as a function of protein age, and find that on average proteins are "born short" and increase in length as they evolve to be better at what they do. This is completely consistent with our intuition, but we should keep in mind that these are trends for averages. It does not at all imply that proteins cannot be born long, or that for some old proteins the evolutionary rate cannot increase over time.

Let us look at another protein characteristic; one that gives us more insights about how new proteins form. What would happen if you would create a random RNA sequence and then translate it to a polypeptide? Proteins created from random RNA can be toxic to cells because they can fold into an inappropriate shape (Monsellier and Chiti 2007). These shapes include amyloid fibrils, the structures that create plaques like those thought to play a role in many neurodegenerative diseases such as Parkinson's and Alzheimer's disease. However, other random proteins have no well-defined shape at all. These "disordered" proteins do not seem to be toxic (Tretyachenko et al. 2017), but instead are highly flexible and are a decent starting point for evolution. It is possible to estimate how disordered a protein structure is from analyzing the amino acid sequence, using a measure called internal structural disorder (ISD) (Romero et al. 1998). Using this measure, we can test whether new proteins are highly disordered, and whether there is a tendency for proteins to become more ordered as they evolve to sharpen their function. Such an analysis has to be done carefully, as we already know that young proteins tend to be shorter, so that if it would turn out that shorter proteins are more disordered, we could be fooled by that correlation. Fortunately, it turns out that longer proteins are actually more likely to have a higher ISD (more disordered) than shorter proteins, so that the shortness of a gene cannot artificially make it look more disordered. In Figure 5.28, we can see the how ISD depends on protein age, showing clearly that the youngest proteins tend to also be the most disordered, and that as a protein ages it becomes more and more ordered, meaning that it acquires a compact, reliably folded structure.

Wilson et al. (2017) also asked how disordered a protein would be that is made from the sequences in-between genes, that is, introns. These regions are presumed not to be under selection[8] and therefore random (if you remove

---

8. While it is possible that introns contain sequences that are involved in regulation (and are therefore functional), this is only rarely observed.

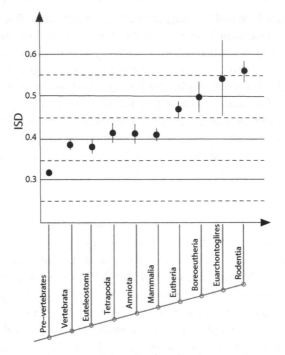

FIGURE 5.28. Internal structural disorder (ISD) values as a function of the clade in which the oldest detectable homologue of a gene can be found. ISD values are length-corrected to a standardized length of 179 amino acids. To minimize homology detection bias, the oldest phylostrata have been condensed into a single pre-vertebrate phylostratum. Adapted from Wilson et al. (2017).

repetitive sequences, as the authors have done). Proteins made from such sequences in fact turn out to be far more structured (have a significantly lower ISD value) than the young proteins, even lower than the oldest proteins studied (the prevertebrate class of proteins in Fig. 5.28). This means that intergenic regions are unlikely candidates for de novo genes, and indeed Box 5.2 discusses exonshuffling, not intron-shuffling, as one of the main mechanisms of de novo gene formation (even though sometimes introns can become exons via *exonization*; see Van Oss and Carvunis 2019).

Using phylostratigraphy as a tool to estimate the age of a protein, we can now make some general statements about how proteins evolve. According to this view, proteins are born in two main ways: via the duplication and subsequent divergence of an existing protein (the most common means of de novo protein origination during the last several hundred million years), and the generation of proteins from existing sequence fragments. In the latter case, the "just-born" protein is likely short and has a disordered flexible structure that makes it very malleable to take on new functions (and less likely to be toxic). As it adapts and is refined further, it grows in length, becomes better adapted to its task, and its fold becomes more compact and more predictable.

The trends in protein evolution that we just discussed—trends in evolutionary rate, sequence length, and disorder—are all long-term trends that can be observed on average, across many different lineages. But can we be more specific about what happens in any particular lineage? In the next section we will look at trends in fitness as a single lineage of organisms adapts to its niche.

5.4 Short-Term Trends in the Evolution of a Single Lineage

The central tenets of Darwinian evolution, as embodied by the "three principles" that we encountered in chapter 1, ensure that as long as the environment stays the same, fitness must either stay constant or be increasing. This "law of evolution"—vaguely reminiscent of the second law of thermodynamics that states that in thermodynamic equilibrium a system's entropy must be constant or increasing—can be rendered mathematically in terms of *Fisher's fundamental theorem*, which is due to the British statistician and geneticist Ronald Fisher (1890–1962). Let us take the time to study this theorem in more detail, as it offers a surefire trend that should apply to any lineage (not just the average over many lineages) as long as the niche to which the lineage is adapting is unchanged. However, we should keep in mind that the theorem only describes microevolutionary changes, that is, it says nothing about long-term trends.

5.4.1 Fisher's fundamental theorem

The fundamental theorem (Fisher 1930) describes the average change of fitness in a population in the absence of any mutations. Thus, it really is a statement about what happens to a population as it comes into equilibrium, subject to natural selection. Because of the absence of mutations, the theorem can be derived using only very few assumptions. The most central one is that the number of organisms of type i changes over time in proportion to the fitness of that type, which we call here w_i. So if we denote the number of organisms of type i at time point t by n_i and the number at time $t + 1$ (the next generation) by n_i', then we should find that

$$n_i' = w_i n_i. \tag{5.26}$$

When the numbers of each type i change over time, so does the mean fitness of the population. At time t it was (the sum is over all the different types in the population)

$$\bar{w} = \frac{1}{N} \sum_{i=1} n_i w_i \qquad (5.27)$$

where N is the total number of organisms at time t, while at time $t+1$ we would find

$$\bar{w}' = \frac{1}{N'} \sum_{i=1} n_i' w_i, \qquad (5.28)$$

where of course N' is the total number of organisms at generation $t+1$. Note that the w_i do *not* change, because as I mentioned earlier there are no mutations that could change a type's fitness. The sum runs over all the existing types with $n_i > 0$, a number that can decrease over time due to extinctions.

The total population size changes accordingly: all we have to do is sum Equation (5.26) over the existing types to find

$$N' = \sum_i n_i' = \sum_i w_i n_i = N\bar{w}, \qquad (5.29)$$

using Equations (5.26) and (5.27). These equations are sufficient to derive the change in average fitness (see Exercise **5.2**)

$$\Delta w = \bar{w}' - \bar{w} = \frac{1}{\bar{w}} \text{var}(w), \qquad (5.30)$$

where $\text{var}(w)$ is the variance of the fitness at time point t. Because the variance $\text{var}(w)$ is always larger or equal to zero, the fundamental theorem implies that the mean fitness of a population cannot decrease as the population reaches equilibrium. Of course, at equilibrium, there is only one genotype left: the one with the highest fitness.

While sometimes it is argued that the fundamental theorem is implied by the second law of thermodynamics,[9] this is not true because under natural selection in the absence of mutation, the Shannon entropy of the population must actually decrease (see Exercise **5.4**). This is easily understood because as the population equilibrates, its variance (and therefore the uncertainty about the genotype distribution) must decrease. In the limit

9. The second law posits that the entropy of a closed system almost always increases as the system approaches equilibrium (see, for example, the description in Landau and Lifshitz 1980).

$t \to \infty$, the entropy approaches zero, which incidentally mirrors the *third* law of thermodynamics.[10]

5.4.2   Fitness landscapes and the rise of fitness in the LTEE

In a sense, Fisher's fundamental theorem observes something trivial: if no new mutants are generated, competition between types will favor the best type, and its offspring will ultimately dominate the population. Almost everything that is interesting in evolution, however, involves the generation of new mutants. Can the theorem be extended to deal with new mutants?

The answer to this question is yes: this is the content of Price's equation (Price 1970; Price 1972) (see Box 5.3). But Price's equation, which connects the change in a trait value to changes in fitness, does not offer a simple insight. In particular, the equation does not imply that fitness must always be increasing in an unchanging environment. Instead, it allows us to separate the effects of selection (the covariance term) from the effect of mutations (the "expectation" term; see Box 5.3). This mode of selection is called "periodic selection" (sequential fixation of rare beneficial mutations), and is often interpreted in terms of single consecutive beneficial mutations leading to a path "upward," climbing the fitness peak. An analysis of the first 2,000 generations of Lenski's long-term experiment (see section 4.2) suggests just such a dynamics, as seen in Figure 5.30 from Lenski and Travisano (1994).

Whether or not fitness keeps increasing in a real evolutionary scenario is a difficult question, and in this section we will investigate it using both empirical data and theoretical arguments. In the simplest of settings (a fixed fitness landscape that does not change in time, with a single population inhabiting this niche, and where organisms do not interact) we expect fitness to increase by "jumps," where each jump represents the emergence of a new type that has taken over from the previous resident type.

Does the rate of increase slow down over time? And does this rise continue indefinitely? These questions were addressed within the long-term evolution experiment by Wiser et al. (2013), coming to a perhaps surprising conclusion.

10. The third law states that as the temperature approaches zero, the entropy of a closed system should approach the logarithm of the group state degeneracy, which is the number of inequivalent states that have zero energy. Usually, the ground state (the state of lowest energy) is not degenerate so the entropy should approach zero. In evolution at zero mutation rate, the entropy should approach the logarithm of the number of precisely neutral wild-type (highest fitness) genotypes. In a finite population, all but one of possibly many neutral genotypes would be lost to drift, so that again the entropy should approach zero.

Box 5.3. The Price Equation

The Price equation (due to the American population geneticist George Price) quantifies how a trait z (averaged over the population) changes from one generation to the next (here, from time t to $t+1$). It is generally written as

$$\bar{w}\Delta\bar{z} = \operatorname{cov}(w,z) + E(w,\Delta z), \tag{5.31}$$

where $\bar{w}$ is the fitness at generation t as defined in the text, z_i is a phenotypic (and measurable) trait attached to genotype i, and $\Delta\bar{z}$ is the average change of the trait: $\Delta\bar{z} = \bar{z}' - \bar{z}$. The covariance between the trait variable and fitness is defined as

$$\operatorname{cov}(w,z) = \frac{1}{N}\sum_i n_i(w_i z_i - \bar{w}\bar{z}), \tag{5.32}$$

while the expectation value of the trait change and fitness is

$$E(w,\Delta z) = \frac{1}{N}\sum_i n_i w_i(z_i' - z_i). \tag{5.33}$$

Here, mutations can change trait values $z_i \to z_i'$, as well as fitness values. In both Equations (5.32) and (5.33), the sum is over all genotypes at time t, but note that the number of genotypes at generation $t+1$ could be different. In mutation-selection balance, the trait value $\bar{z}$ is optimal and thus $\Delta\bar{z}=0$, which implies that the first and second term in Equation (5.31) must cancel (sum to zero). But when a population adapts to increase z, we expect $\operatorname{cov}(w,z) > E(w,\Delta z)$. In Figure 5.29 we can see the Price equation's left-hand side plotted against the two terms on the right-hand side separately, showing that in this particular case where z is the number of times an avidian digital organism (see section 4.3.3) has performed the "AND" operation, the terms Equations (5.32) and (5.33) are almost mirror images of each other (the covariance positive, the expectation value negative) because most of the time the population is in mutation-selection balance. The absolute size of each term is determined by the overall fitness level, which explains why there are "stripes" at different covariance levels in Figure 5.29.

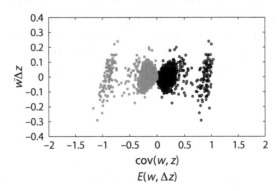

FIGURE 5.29. The left-hand side (LHS) of the Price equation (5.35) as a function of the first term of the right-hand side (RHS) $\operatorname{cov}(w,z)$ (dark gray dots), and the second term of the RHS (light gray dots). Plotting the LHS vs. the RHS gives a straight line, but plotting each of the terms individually reveals the strength of selection on the trait. From Mirmomeni (2015).

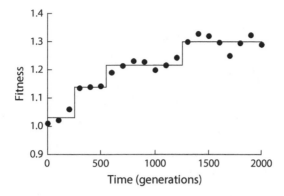

FIGURE 5.30. Analysis of the fitness trajectory in one population of *E. coli* during its first 2,000 generations of adaptation. Each point is the mean of ten fitness assays, and the solid line indicates the fit to a step-wise model. Adapted from Lenski and Travisano (1994).

To get a better insight into what factors constrain an organism's fitness and how fitness may be optimized under these constraints, researchers often use the concept of the "fitness landscape" to visualize those constraints. After all, it is not possible to optimize all characters at the same time: optimizing one may affect (and even impair) another. Dawkins made a visualization of one such landscape to illustrate the evolution of the eye in his book *Climbing Mount Improbable* (Dawkins 1986), which I reproduce in Figure 5.31. The idea here is that (as discussed in a little more detail in Box 4.4), the eye has evolved many times independently, and every time such a modification occurred, it would be optimized in a gradual process.

The idea of the fitness landscape of course predates *Climbing Mount Improbable*: it goes back to the American geneticist Sewall Wright (1889–1988) who first drew such landscapes in the context of population genetics, in particular to understand how phenotypic traits might change under natural selection. The general idea here is exactly what we discussed earlier: that optimizing one trait might come at the detriment of another. In his 1932 paper "The roles of mutation, inbreeding, cross-breeding and selection in evolution," Wright imagined how evolution may have to choose between different peaks that are separated by valleys (see Fig. 5.32 from that paper). In that figure, which depicts the fitness landscape of two trait combinations, peaks are denoted by "+," while dashed lines of "isofitness" denote combinations of characters that have the same fitness. Wright was fully aware that such a two-dimensional representation of the fitness landscape is misleading because it only represents two dimensions out of possibly thousands (Wright 1988), but

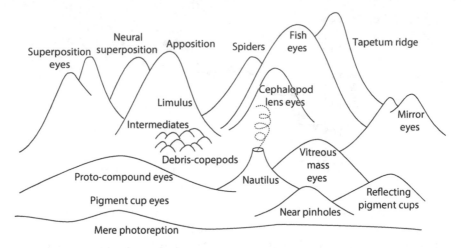

FIGURE 5.31. A schematic depiction of the evolutionary fitness landscape of eye evolution, originally drawn by Michael Land (Dawkins 1986). The compound eyes are depicted on the left side of the landscape, while the "camera-type" eyes are on the right.

also remarked that this simplification was necessary in order to depict processes that occurred on these landscapes. In a way, Wright's landscape can be seen as a depiction of a landscape such as Land's in Figure 5.31, but viewed "from the top."

A typical evolutionary process on the landscape is shown in Figure 5.33. There we can see a population (black solid dot) occupying a fitness valley that is located between multiple peaks. Evolution will take the population up one of the peaks by changing gene frequencies (in the absence of mutations), or else by changing trait values via mutations, so as to move the population toward higher values of fitness. Indeed, the first 2,000 generations of evolution in the Lenski experiment shown in Figure 5.30 can be seen as just such a climb up the fitness peak. And as dictated by the shape of the peak, progress is fast for the first few mutations, but as the population is close to the peak, progress must slow down (the slope of the peak is more shallow near the peak). If this depiction were at all realistic, it would predict that adaptation must ultimately stop as the population has reached the top of the peak. But how realistic is that?

Lenski's long-term evolution experiment is a phenomenal resource to study evolution and adaptation experimentally, and in particular we can use its long-term record of adaptation to answer some of these questions. To do this, Wiser et al. (2013) used populations from the first 50,000 generations of the experiment, and asked whether the increase in fitness recorded during that time

FIGURE 5.32. The first depiction of a fitness landscape (Wright 1932). Wright called this the "field of gene combinations," where the x-axis represents the quantitative value (or gene frequency) of one trait, and the y-axis the value or frequency of another. The elevation represents the "selective value" of this combination to the organism (Wright 1967), something we now call a "component of fitness." Wright called the dashed lines "adaptedness contours."

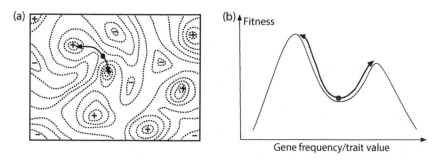

FIGURE 5.33. Two views of adaptive trajectories from an intermediate fitness value (solid black dot) to a peak. (a) "Top" view from Wright's landscape, and (b) the "side view" that shows that from a fitness valley there are often multiple different ways to go "up."

suggests that a peak will ultimately be reached, or if the record is consistent with a "peak-less" peak, that is, a landscape in which there is always a "way up." While this seems impossible for the simple landscapes that we can depict graphically, it is not inconceivable if thousands of genes are interacting to produce a complex phenotype. Wiser et al. first asked whether a functional form that is fit to the first 20,000 generations of the fitness (averaged over all twelve populations) would be able to also quantitatively describe the fitness up to 50,000 generations (the last datum considered in the paper).

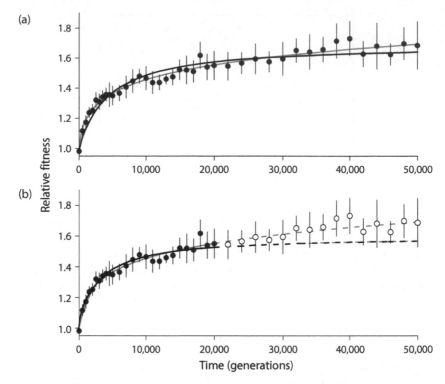

FIGURE 5.34. (a) Mean fitness (normalized to the ancestors' fitness, hence "relative fitness") of twelve replicate LTEE lines (solid dots), with 95% confidence intervals. The black line shows the hyperbolic fit to the data, while the gray line indicates the power-law fit (both are two-parameter fits). (b) Same data, but with fit to the first 20,000 generations only (solid lines), with model predictions up to 50,000 generations as dashed lines. From Wiser et al. (2013).

The answer to this question is "it depends": the standard functional form that is traditionally used to fit diminishing-returns fitness trajectories (a "hyperbolic" function) is not able to do this, but a different form (the "power-law" fit) actually can. Let us delve into this a bit more, because this observation teaches us important lessons about the evolutionary process and the nature of fitness landscapes.

In Figure 5.34(a), we can see (solid points) the mean fitness of the twelve populations,[11] and the fit of these data points to two different models in Figure 5.34(b). Let us discuss these two models in more detail. The

---

11. Only nine out of the twelve populations have complete fitness trajectories all the way to 50,000 generations. Three of the lines have shortened trajectories because those cells no

hyperbolic fit uses two parameters a and b to determine the relative fitness $\bar{w}$ of the two strains (the strain evolved to a particular generation, compared to the ancestor's growth rate):

$$\bar{w}(t) = 1 + \frac{at}{t+b}.$$ (5.34)

This form is normalized so that at $t = 0$ it gives $\bar{w}(0) = 1$, and in the long-term limit it tends to a constant $\bar{w} \to 1 + a$. Hyperbolic models have a strong tradition in modeling of biological processes, in particular in growth processes. For example, the Monod equation that relates the growth rate of an organism to the availability of substrate σ is a hyperbolic model:

$$w(\sigma) = w_{\max} \frac{\sigma}{\sigma_{1/2} + \sigma}.$$ (5.35)

In the limit of infinite resource availability, the growth rate $w(\sigma)$ is equal to its maximum, while the term $\sigma_{1/2}$ in the denominator indicates the concentration at which the population has attained half of its maximal growth rate. Lenski and Travisano (1994) used such a model to fit the average trajectory in the LTEE over the first 10,000 generations, but it turns out that the assumption that fitness will eventually "top out" is not warranted. Indeed, the best fit for the hyperbolic model has the parameters $a = 0.7007$ and $b = 4,431$ (Lenski et al. 2015), which implies a maximum average relative fitness of about 1.7, which is remarkably low given that some populations have achieved relative fitness exceeding 2.

Of course, there are infinitely many functions that could possibly fit the data with only two parameters, but an alternative model (the "power-law" model) stands out because it has a solid theoretical underpinning. The form of the law is

$$\bar{w}(t) = (1 + bt)^a$$ (5.36)

and the feature that distinguishes it the most from the hyperbolic model is that $\bar{w} \to \infty$ (albeit very slowly) as $t \to \infty$, as opposed to a constant. The power-law model (whose mathematical underpinnings we will examine shortly) basically says that fitness increases in proportion to a fractional power of time; in fact (according to the best fit), approximately the twelfth root of time. In Figure 5.34(a) and (b), the power-law fit is the gray line, and Figure 5.34(b) in particular shows that, treated as a model, Equation (5.36) predicts the future

---

longer produce reliable colonies on the agar plates used to measure relative fitness (see the Supplementary Information of Wiser et al. 2013 for details) even though the lines themselves are ongoing.

fitness evolution (based only on data for the first 20,000 generations) much better than the geometric model (black dashed line in Figure 5.34[b]).

What feature of evolution favors a power-law model? As Wiser et al. show, there are two main ingredients needed to derive such a model: the law of diminishing returns, and the theory of clonal interference. I briefly discussed the law of diminishing returns earlier: when approaching a performance level that requires the optimization of many different variables, the first few optimizing steps are easy, but fine-tuning all the variables in order to achieve optimal performance is difficult and time-consuming. The evolution of world-record performances is a typical example of such dynamics (we will discuss diminishing returns in more detail in section 5.4.4).

Wiser et al. modeled the difficulty of finding more improvements in fitness by studying a model for the likelihood of beneficial mutations. Generally speaking, beneficial mutations are rare, but it is also known that we can assume that the mutations are drawn from a probability distribution of a common form. For example, it is clear that mutations with a large benefit s (percent increase over the current best type) are exponentially more rare than mutations with a smaller effect. For that reason, the probability distribution

$$f(s) = \alpha e^{-\alpha s} \tag{5.37}$$

works well to describe this phenomenon, where α characterizes the mean beneficial effect:

$$\langle s \rangle = \frac{1}{\alpha}. \tag{5.38}$$

The general idea behind diminishing returns is that after every beneficial mutation is integrated into the genome, the mean average available effect *decreases* in proportion to the effect before the new mutation was found. In other words, the mean beneficial effect of a mutation depends on the fitness level attained, and therefore becomes time-dependent.

In a simple model (one that ignores the second element of the model, namely clonal interference) we could choose, for example,

$$\alpha_{t+1} = \alpha_t + g, \tag{5.39}$$

so that the distribution of beneficial effect sizes shifts to the left (with decreasing mean), as seen in Figure 5.35. Because of the relation $\langle s_t \rangle = 1/\alpha_t$, an increase in α translates into a decrease in mean beneficial effect:

$$\langle s_{t+1} \rangle = \langle s_t \rangle \frac{1}{1 + g\langle s_t \rangle}, \tag{5.40}$$

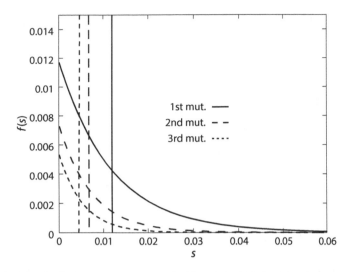

FIGURE 5.35. Distribution of beneficial effect sizes $f(s)$ according to Equation (5.41), with the first distribution (solid line) and its mean $\langle s_0 \rangle$ (solid vertical line). The second mutation will be drawn from the distribution with reduced mean shown as a long-dashed line (mean given by long-dashed vertical line), while the third mutation will sample the distribution shown as a short-dashed line.

so the larger g, the more severe the effect of diminishing returns becomes. A typical sequence of mean effect sizes is shown in Figure 5.36. To fit the average fitness trajectory, we must first determine what the effect of the first beneficial mutation is, which we call s_0. The data suggests that $s_0 = 0.1$, that is, the first mutation will produce on average a 10 percent benefit. According to Equation (5.40), the next mutation should have an average beneficial effect of

$$\langle s_1 \rangle = \langle s_0 \rangle \frac{1}{1 + g\langle s_0 \rangle}, \tag{5.41}$$

but in this analysis, we have assumed that if a beneficial mutation of size s occurs, it will ultimately establish itself in the population with a probability that only depends on s (and possibly the population size). In the parlance of population genetics, "establishing oneself" is called "fixation" and implies that a mutation that has previously occurred on a *single* individual in the population has risen in frequency to such an extent that it is shared by *every* member of the population. Another way to say this is that upon fixation, the organism in which the mutation first occurred ultimately became the ancestor of every single living individual. The mathematics of fixation is one of the oldest contributions to population genetics, and a short primer is given later.

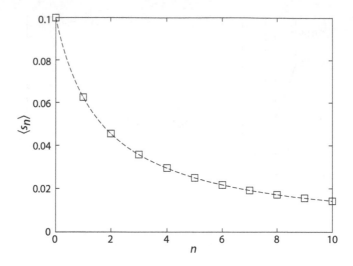

FIGURE 5.36. Decrease of mean beneficial mutation size $\langle s_t \rangle$ due to diminishing returns, as a function of the number of mutations that have already gone to fixation. The dashed line is the function $\langle s_n \rangle = s_0/(1 + gns_0)$ with $g = 6$ and $s_0 = 0.1$, which are the optimal values to fit the mean fitness trajectory of the six populations with complete trajectories and low ancestral mutation rate in Wiser et al. (2013).

Calculating the probability of fixation for a mutant with benefit s (see Box 6.1 as well as Exercises 6.1 and 6.2) assumes that it is the only mutant in the population that attempts to establish itself.[12] However, if the mutation rate is not infinitesimal, there may be other variants that the mutation must compete against. Of those, only one can ultimately win, which reduces each individual mutant's probability of fixation. This is the phenomenon of "clonal interference": multiple clones (mutants) interfering with each other while vying to become dominant.

To account for clonal interference, a sophisticated application of the theory of Gerrish and Lenski (1998) reveals that it is sufficient to correct Equation (5.39) to read

$$\alpha_{t+1} = \alpha_t + gQ, \qquad (5.42)$$

12. In Box 6.1 the fixation probability in the limit of small s is shown to be $P(s) = 2s$, the celebrated Haldane result. That formula is derived by using a Poisson distribution for the number of offspring (see Exercise 6.1). In binary cell division, which is the case for the LTEE we are discussing here, there can only be zero, one, or two offspring surviving into the next generation, leading to a different fixation probability $P(s) \approx 4s$ (see Exercise 6.2).

where $Q < 1$ is a complicated function of the mutation rate and population size that takes into account the competition between competing lineages fighting for fixation. The correction factor reduces the effect of diminishing returns, simply because clonal interference slows down adaptation. Consequently, the relationship between mean effect size $\langle s \rangle$ and α also includes the correction factor Q such that

$$\langle s_{n+1} \rangle = \frac{Q}{\alpha_n}. \tag{5.43}$$

With diminishing returns and clonal interference taken care of, the mean fitness as a function of generation time t can be estimated as (Wiser et al. 2013)

$$\bar{w}(t) \approx \left(2g\langle s_1 \rangle e^{g\langle s_1 \rangle} \frac{t}{\langle t_1 \rangle} + 1 \right)^{1/2g}. \tag{5.44}$$

In this Equation, the parameter g determines the parameter a in (5.36) via $a = 1/2g$, and b is related to the mean effect of the first fixed beneficial mutation $\langle s_1 \rangle$ determined in (5.41), and the time it takes for that mutation to go to fixation, $\langle t_1 \rangle$. Thus, the analysis provides a theoretical basis for the power-law fit to the average increase in fitness across replicates in the LTEE (Baake et al. 2018 later provided a more detailed derivation of the law that also investigates the effect of epistasis on the long-term fitness evolution).

The analysis of fitness trajectories in terms of a power law leads to the conclusion that—at least in circumstances where thousands of genes may contribute to fitness increases—an upper bound in fitness is nowhere in sight. Of course, we know that from a purely theoretical point of view this conclusion cannot be correct: after all, there are limits to the speed of replication that are set by physiology, never mind the laws of physics. How close are the cells in the LTEE to any physiological limit?

Wiser et al. did indeed engage in this speculation, and came to the conclusion that the cells still have a long way to go. The 50,000 generations of cellular evolution analyzed in that study unfolded in a period of about twenty-one years, which is on the order of the generation time of a typical human. What does the theory predict if we imagine instead that the experiment goes on for 50,000 generations of human lives instead, each overseeing their own 50,000 generations of cellular evolution, for a total of 2.5 million bacterial generations? Using Equation (5.44) fitted to the first 50,000 generations predicts that after 2.5 million generations, the average fitness of the twelve populations would have reached a level of about 4.7 with respect to the ancestor. Such a replication rate does not nearly violate any physiological limits, as it predicts a doubling time of about twenty-three minutes. Given that the ancestor's doubling time in minimal medium is approximately fifty-five minutes (not counting the lag phase or the stationary phase, see Figure 4.5), achieving a doubling time of twenty-three minutes (and also eliminating the lag phase) appears to be eminently achievable. In fact, there are microorganisms that replicate significantly faster than this. The marine bacterium *Pseudomonas natriegens* for example achieves a doubling time of less than ten minutes in a rich medium (Eagon 1962), and it appears that the bacterium has achieved such a small doubling time precisely by eliminating the lag phase, at least when the population size is not too large. Furthermore, there does not seem to be any reason to think that the doubling time of 9.8 min^{-1} cannot be improved upon further (Eagon 1962).

Granted, we will not be able to test the prediction of a twenty-three-minute doubling time for *E. coli* after about a million years of laboratory evolution, not the least because funding agencies usually favor shorter time horizons in the projects they support. But it is certainly awe-inspiring that an experiment conducted within the lifetime of a single scientist, on a dozen adapting lineages whose individual progress is governed by unpredictable events, when supplemented by theoretical insight can make predictions on the scale of millions of years.

5.4.3   Epistasis and valley-crossing in the NK-model of evolution

The story told by the twelve lineages of Lenski's LTEE over now more than 75,000 generations is not a simple narrative, and this book cannot do justice to the myriad of insights generated by this research. Yet, by design the experiment can only examine a narrow range of parameters and their effect on the evolutionary process. For example, the population size within the beakers is fairly large on average—even though it fluctuates significantly as each day the size is reduced to 1 percent and then allowed to grow back again. The effective population size of this population (a parameter that is used to characterize the overall features of the evolutionary process) is estimated to be about 330 million, so that the forces of evolutionary drift are severely mitigated, or even absent. Furthermore, the (beneficial) mutation rate is rather low, estimated to be about 1.7×10^{-6} per genome per generation for those lines that have maintained their ancestral rate. These numbers imply that at any point in time, about fifty-six beneficial mutations (the product of population size and beneficial mutation rate) coexist in the population, vying for supremacy. The vast majority will not succeed, of course, because fixation is a stochastic process (see Box 6.1).

But the evolutionary dynamics that we observe in these lines, where fitness appears to be an inexorable march upward without crossing through a valley of fitness like the one depicted in Figure 5.33(b), may not be representative of populations of smaller size, or instances when the mutation rate is significantly higher. After all, among biological organisms on this planet, the bacterium *Escherichia coli* boasts one of the most accurate information-copying systems ever recorded, with a rate of approximately one error per ten billion copied nucleotides (Barrick et al. 2009). Contemplate that number for a while, and then imagine the selective pressure for information preservation that drove this incredibly accurate copy-machinery.

In the absence of a long-term experiment with significantly higher mutation rates (higher than the 100-fold increase that is seen in the mutator lines of the LTEE) and smaller population sizes, computational experiments may have to fill the void. Some insights have indeed been gained by Avida experiments (see section 4.3.3), in which populations are significantly smaller and mutation rates significantly higher (Wiser 2015). In those experiments, it is possible to measure explicitly how mutations interact (a phenomenon known as *epistasis*) and how the degree of interaction (that is, how much the presence of one particular mutation can affect another mutation on the same genome), as well as the sign of that interaction, can affect the evolution of complexity (Lenski et al. 1999). Box 5.4 describes how the interaction between mutations can be measured quantitatively.

However, how mutations interact in Avida cannot easily be controlled because these interactions are simply a consequence of adaptation, that is, of the particular path that an adaptive trajectory has taken. If we want to study evolution when the degree of epistasis is fixed, we have to use a different model system. The go-to model to isolate the impact of epistasis on the mode and manner in which populations adapt (whether fitness trajectories will always be inexorable marches upwards) is the "NK"-model, invented by Stuart Kauffman and his co-workers (Kauffman and Levin 1987; Kauffman and Weinberger 1989). The NK model was originally designed to study adaptive landscapes, and in particular the impact of the "ruggedness" of the landscape on the adaptive "walks" that an evolutionary process would generate.

The model is defined by N "genes" G_i that can either be functional (with value $G_i = 1$, that is, "on") or nonfunctional with value $G_i = 0$ ("off"). The model really does not put any weight on the interpretation of the genes as on or off because in the standard description of the model, the mutation rate between the two states is equal, but for the purpose of imagining a network of interacting genes, the picture is useful.

The fitness of the "organism" is a function of the genome $G = G_1 \cdots G_n$ only, and depends on how the genes interact with each other. Each genome G

Box 5.4. Quantifying Epistasis

Two mutations (A and B) occurring somewhere on a genome with wild-type fitness w_0 are said to be *independent* if the fitness effect of both mutations together equals the product of the fitness effect of each of the mutations alone. Note the emphasis on fitness "effect": when it comes to mutations, we are interested how *changes* in fitness are modulated by changes in fitness due to other mutations. Thus, epistasis is always measured with respect to a particular fitness level. If the fitness effect of the double mutant is w_{AB}/w_0, while the fitness effect of each of the single mutations is w_A/w_0 and w_B/w_0, respectively, then mutational independence implies (see Fig. 5.37)

$$\frac{w_{AB}}{w_0} = \frac{w_A}{w_0}\frac{w_B}{w_0}.$$ (5.45)

Here we quantify epistasis as the deviation from this equality so that the measure of epistasis

$$\epsilon = \log\left(\frac{w_{AB}w_0}{w_A w_B}\right) = \log\frac{w_{AB}}{w_0} - \log\frac{w_A}{w_0} - \log\frac{w_B}{w_0}$$ (5.46)

vanishes when the combined effect of mutations A and B is the same as the product of the individual effects on organismal fitness. Epistasis can be measured in other ways. For example, Phillips (2008) as well as Costanzo et al. (2016) define

$$\epsilon = \frac{w_{AB}}{w_0} - \frac{w_A}{w_0}\frac{w_B}{w_0}.$$ (5.47)

The measures (5.46) and (5.47) are very similar as long as fitness effects are small.

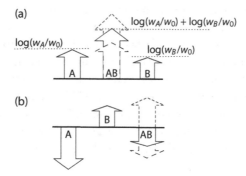

FIGURE 5.37. (a) With the baseline indicating zero, the height of an arrow indicates the log of fitness after one or two mutations. Here, both mutations A and B are beneficial but the joint effect of both mutations $\log(w_{AB}/w_0)$ can be larger than the sum $\log(w_A/w_0) + \log(w_B/w_0)$ (positive epistasis, short-dashed arrow), or smaller than the sum (negative epistasis, long-dashed arrow). If the joint effect is equal to the sum (solid arrow), the mutations are independent (vanishing epistasis). (b) If one of the mutations is detrimental, the joint effect of the two mutations AB can be additive (solid arrow), epistasis can be negative (long-dashed arrow), or positive (short-dashed arrow), in the latter case reversing the effect of the A mutation (sign epistasis).

consists of a binary string of length N, where the fitness $w(G)$ is given by

$$w(G) = \left(\prod_{i=1}^{N} w_i \right)^{1/N} , \qquad (5.48)$$

where w_i is the fitness contribution of locus i to the total fitness. Note the use of the geometric mean in Equation (5.48), as opposed to the arithmetic mean that is generally used in the NK model. This modification is not essential, but ensures that the fitness contribution of any locus multiplies the fitness of the rest of the organism, so that lethal mutations are in principle possible.

What is the fitness contribution w_i of locus i? This contribution depends on the allele at this position (a 1 or a 0), but in general also on the value of other alleles on the genome G. The number of other alleles that contribute to w_i depends on the parameter K of the NK model, which specifies how many other loci contribute to the fitness of locus i. The simplest case is $K = 0$. In that case, locus i does not interact with any other loci, and the fitness contribution w_i will simply be a random number, uniformly drawn from the interval $(0, 1)$. If the total fitness is the arithmetic sum, the fitness contributions will be normally distributed, with a mean $\bar{w}(G) = 0.5$ and a variance determined by the number of loci N, owing to the central limit theorem. A typical such distribution is shown in Figure 5.38(a), which is the expected fitness distribution for $N = 20$ and $K = 0$. Using the geometric average instead of the arithmetic one leads to the probability distribution shown in Figure 5.38(b) (for the same parameters $N = 20$ and $K = 0$). This distribution is not Gaussian, and is somewhat broader than Figure 5.38(a) with a mean below 0.5.

When loci can interact, the fitness contribution w_i depends on the value at locus i as well as K other loci. Figure 5.39 shows an NK string with $N = 16$ and $K = 2$, where each locus interacts with the two loci clockwise to its right. Because NK strings are circular, the "last" locus $N = 16$ interacts with loci 1 and 2. Which loci any particular locus interacts with can be set by the experimenter, but it is in fact unimportant unless strings are allowed to recombine during the evolutionary process.

Once we have decided which locus interacts with what other loci, we can create a table of random numbers (again between zero and one, drawn with uniform probability) for each of the possible multiplets. For example, with $K = 2$, the contribution of each locus to the organismal fitness is determined by a triplet (in general a $K+1$-plet) of alleles. It is easiest to illustrate this using a specific case. In Table 5.1 we can see the fitness contribution of loci as various random numbers for the eight possible triplets centered at each locus. Let us pick site 3, for example. The contribution to the overall fitness

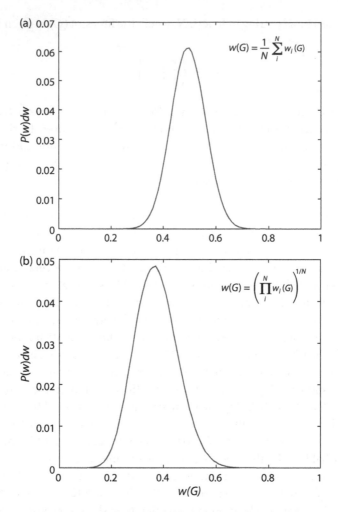

FIGURE 5.38. Probability distribution of fitness $w(G)$ for an NK string of length N with $K=0$ (no interaction between loci). (a) Normalized probability distribution $P(w)dw$ using an arithmetic mean $w(G) = \frac{1}{N} \sum_{i=1}^{N} w_i(G)$. (b) Normalized probability distribution of fitness when using the geometric mean Equation (5.55). The mean is somewhat smaller than the expected mean for the arithmetic average, and the distribution is somewhat broader. From Østman et al. (2012).

w_3 depends on the value at that site and (if we use the scheme depicted in Figure 5.39) the two sites to its right. Say locus 3 has a 1, while loci 4 and 5 both have a zero. Then the fitness contribution of locus 3 is obtained by looking into the $i = 3$ row of Table 5.1 and finding the column labeled "100," so $w_3(100) = 0.1973$.

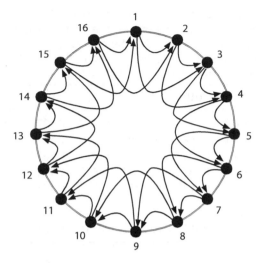

FIGURE 5.39. An NK "genome" binary (string) with $N = 16$ and $K = 2$, where we have chosen any particular locus to interact with the two loci clockwise to the right (indicated by arrows). From Østman et al. (2012).

The evolutionary process seeks to maximize organismal fitness, and of course this starts by maximizing the fitness contribution of each locus. A quick look at Table 5.1 shows that there are numerous ways to optimize the contribution w_3, for example by mutating the allele at $i = 3$ to zero because then the fitness contribution will be the number in the 000 column, namely $w_3(000) = 0.8703$. You may think that it is easy to find the best fitness by simply choosing the highest number in the table, but this does not work if loci interact. For example, choosing $w_3(000)$ affects the fitness of loci 1 and 2, which depend on the value at locus 1. Say site $i = 1$ has a "1," and $i = 2$ has a "1." When the $i = 3$ allele had the value "1" the fitness contribution of locus $i = 1$ was $w_1(111) = 0.9135$ (see the value in the first row, last column in Tab. 5.1). Mutating the $i = 3$ site from a "1" to a "0" in order to increase the fitness contribution from 0.1973 to 0.8703 will, at the same time, *decrease* the fitness contribution of locus $i = 1$ from 0.9135 to $w_1(110) = 0.0178$. It is clear from this example that the interaction between genes creates constraints that make finding the optimal fitness difficult. In particular, we witness precisely the phenomenon that we have discussed numerous times already, namely that improvements in a particular trait usually affect other traits negatively, which requires these traits to be "repaired" subsequently. We can therefore see that fitness can be optimized without constraints *only* if $K = 0$: this is known as a "smooth" landscape (with a single peak) that can be reached simply by following the slope of the hill upward. The most problematic case (from the point of

Table 5.1. The values of the fitness contribution $w_i(G_i G_{i+k} G_{i+j})$ in an NK model with $N = 16$ and $K = 2$ depend on the allele at site i as well as the alleles at site $i + j$ and at $i + k$ ($i \neq k$), where i and k range from 1 to $N - 1$. The genome is circular as in Figure 5.39, so that $i = 0$ is the site $i = N$, while $i = N + 1$ corresponds to site 1, and so on.

i	000	001	010	011	100	101	110	111
1	0.2314	0.4332	0.7125	0.5091	0.2467	0.6276	0.0178	0.9135
2	0.0945	0.6021	0.449	0.3166	0.714	0.5728	0.1737	0.8043
3	0.8703	0.5301	0.3591	0.6615	0.1973	0.1896	0.2252	0.5419
4	0.2965	0.098	0.1313	0.4268	0.4212	0.9844	0.3873	0.674
5	0.1103	0.3132	0.1643	0.7451	0.2073	0.7837	0.5241	0.6127
$\vdots$	$\vdots$	$\vdots$	$\vdots$	$\vdots$	$\vdots$	$\vdots$	$\vdots$	$\vdots$
16	0.8154	0.0012	0.4032	0.9261	0.8245	0.5645	0.2973	0.4455

view of evolution) is the most "rugged" landscape where $K = N - 1$. In that case, each locus interacts with *all* of the other loci on the genome, and optimization cannot follow any slope at all. Instead, each binary string of length N has its own random value as fitness, that is, the evolutionary process must find the highest value among all the 2^N random numbers in that particular table. This landscape is therefore often called a "random" landscape, and was studied (alongside the smooth landscape with $K = 0$) in the first publication that introduced the NK model (Kauffman and Levin 1987) (the landscape with arbitrary K was introduced later in Kauffman and Weinberger 1989).

For arbitrary K, the NK model interpolates between the perfectly smooth $K = 0$ landscape and the perfectly random $K = N - 1$ model. We can get a feel for the "character" of the landscape by plotting the *autocorrelation* function of that world. The autocorrelation function simply measures how the fitness of a sequence d mutations away from any particular reference sequence relates to the fitness of that reference sequence. If the landscape is smooth, for example, then the fitness a few mutations away from any particular reference point should bear some resemblance to that reference's fitness: they should be correlated. In a random landscape, on the contrary, a single mutation should destroy all reference (that is, all correlation) since fitness is random after all.

It is possible to calculate the correlation function $\rho(d)$ for the NK landscape as a function of N and K (Campos et al. 2002) to be

$$\rho(d) = \frac{(N - K - 1)!(N - d)!}{N!(N - K - 1 - d)!} = \binom{N - K - 1}{d} / \binom{N}{d}. \quad (5.49)$$

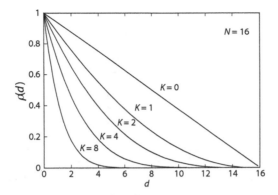

FIGURE 5.40. Autocorrelation function of an NK landscape with $N = 16$ and various values of K. For $K = 0$ (the smooth landscape) $\rho(d)$ is the linear function $\rho(d) = 1 - d/N$. From Østman et al. (2012).

The autocorrelation function measures to what extent the fitness at one site is correlated to the fitness at another site d mutations away. If the correlation is high, then fitness at one site is fairly predictive of the fitness at a nearby site. If the correlation function is zero (at a particular distance d), then this means that you cannot predict the fitness d mutations away from where you are in a particular landscape.

We can see in Figure 5.40 that for the smooth landscape $(K = 0)$ the autocorrelation function decays slowly (in fact linearly) with the mutational distance d. The larger K, the faster $\rho(d)$ approaches zero. For the random landscape $(K = N - 1)$ the autocorrelation function is zero already at the first step $(d = 1)$. The case where this happens $(K = 15)$ is not shown in Figure 5.40.

Let us now look at how evolution unfolds on such landscapes. In Figure 5.41, we can see evolutionary trajectories for four different cases: low and high mutation rate, as well as no epistasis $(K = 0)$ and a medium amount of ruggedness $(K = 4)$. For each experiment, first a random landscape with the requisite parameters N and K is generated, and then independent runs ("replicates") are performed on that same landscape (Østman et al. 2012). To generate the fitness trajectories, we create a line of descent backward from the best sequence in the population (we used populations with 5,000 individuals for these experiments). We first discussed lines of descent (and how to generate them) when we analyzed the experiment to study the rise of EQU consumers in section 4.4.1. In these asexual populations, each individual has only one ancestor, so there will always be one unbroken line of descent from the contemporary type all the way back to the ancestor. We can then plot fitness only for those sequences that find themselves on that particular line.

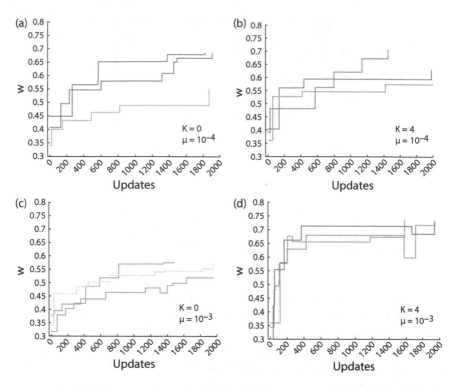

FIGURE 5.41. Fitness trajectories for four different experiments, on two different landscapes (both with $N = 20$). The two cases in the left column (a) and (c) have no epistasis (a smooth landscape with $K = 0$), while those in the right column (b) and (d) have a moderate amount of epistasis (a rugged landscape with $K = 4$). In the experiments depicted in (a) and (b), the mutation rate is small ($\mu = 10^{-4}$ per site), while in (c) and (d) the mutation rate is ten times as high. Trajectories are truncated when no more beneficial mutations arose before 2,000 updates (corresponding to 200 generations, see Østman et al. 2012).

We can see three independent fitness trajectories in Figure 5.41 (each obtained with the same experimental setting) to give us an idea of how diverse these trajectories are even when the fitness landscape (and all other parameters) is fixed. To initialize these experiments, we first generated a number of random NK binary sequences (recall that here, $N = 20$), and then chose as ancestor for the experiment a sequence at random from among those that were in the lower half in fitness. (That way, we could observe many beneficial mutations.)

We first notice how different independent evolutionary trajectories on the same landscape can be. Of course, all three runs in each of the panels in

Figure 5.41 were started with different ancestors, but after two thousand generations of adaptation, the peak of the landscape is not yet reached, even for the smooth landscape with $K = 0$ (Figs. 5.41[a] and [c]). Note that the highest fitness in the landscape is not fixed: every time a new landscape is created at random, there could be a new and different highest fitness. The theoretical highest fitness of course is known ($w = 1$) because that is the highest random number one can draw from the interval. In most landscapes, the highest fitness is significantly lower than that.

When the mutation rate is small (Figs. 5.41[a] and [b]), we see that there are exclusively beneficial mutations on the line of descent. The reason for this is that at these mutation rates, valley-crossings (where an inferior sequence goes to fixation, followed by a step to a higher peak) are essentially nonexistent. It is clear why this is the case: inferior types can only make it on to the line of descent (that is, go to fixation) via drift, and with a population size of 5,000 individuals, this would be an exceedingly rare event. Incidentally, the *mutation supply rate*, namely the rate at which new mutations are generated (determined by the genomic mutation rate μN times the population size), is of the order 10 to 1,000 in the experiments reported in Østman et al. 2012, which is smaller than the mutation supply rate in the LTEE (estimated to be about 24,000; see Barrick et al. 2009).

When the mutation rate is ten times higher (Figs. 5.41[c] and [d]), fitness trajectories are not a monotonic trek upwards anymore. Even for the smooth landscape with $K = 0$ (with only a single peak) the line of descent includes steps downward. However, these are not valley-crossings (as there are no valleys in that landscape) but are due to genetic drift, that is, the chance fixation of a neutral (or even deleterious) mutation. Such fixations are possible at a mutation rate this high because, due to the elevated rate, the effective population size is much reduced.[13]

For the much more rugged landscape with $K = 4$, where up to five "genes" interact to determine fitness (Fig. 5.41[d]), we can see several instances of valley-crossings, which the population uses to reach higher levels of fitness. When valley-crossings are impossible (at low mutation rates) the ruggedness of the same landscape seems to prevent the discovery of the higher peaks. Do landscapes with multiple peaks separated by valleys make it easier to reach high peaks? At first glance, this seems to be a contradiction. After all, valley-crossing is a rare event, which implies that fitness trajectories will get stuck on intermediate peaks most of the time. Instead, on a smooth landscape, the path uphill is unobstructed. However, the story is more complicated than that. What we will see is that when genes can interact, the peaks become higher and

13. We will study genetic drift in more detail in section 6.3.

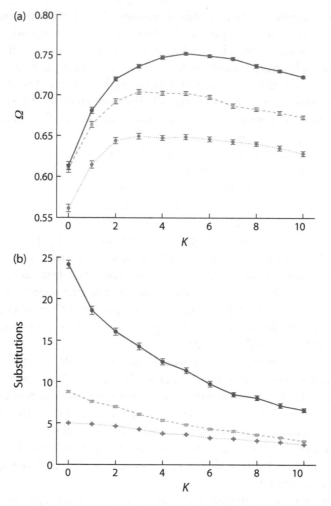

FIGURE 5.42. (a) Attained fitness Ω (largest fitness attained on the line of descent) as a function of K for 200 replicate populations with population size 5,000 and $N = 20$, for three different mutation rates. $\mu = 10^{-2}$ (solid line), $\mu = 10^{-3}$ (dashes), and $\mu = 10^{-4}$ (dots). (b) Average number of substitutions on the LOD as a function of K for three different mutation rates. Lines as in (a). From Østman et al. (2012).

the valleys get deeper. As a consequence, it is possible to reach greater heights as long as the mutation rate is high enough so that valleys can be crossed. Let us look at the evidence.

Figure 5.42(a) shows the highest fitness level attained, for an average of 200 lines of descent (LODs), as a function of the number of interacting genes K

(the ruggedness of the landscape), for three different mutation rates. Regardless of the mutation rate, the attained fitness first grows with K, reaches a maximum, and then drops. The drop at large K is easily understood: once the landscape becomes too rugged, the highest peaks in the landscape are unlikely to be found by the evolutionary process. What is less obvious is why the attained fitness rises with K initially. Should we not expect that because the smooth $(K = 0)$ landscape is easiest to climb, that it is those populations that reach the highest fitness, because they cannot get stuck? But this kind of thinking discounts that when many genes interact, higher levels of fitness can be *created* because cooperating genes can achieve more complex functions (leading to higher fitness) together, something that single genes would be hard-pressed to achieve.

This idea—namely that gene complexes with more genes have a higher fitness (on average) than complexes with fewer genes—is a common one in systems biology. For example, Costanzo et al. (2016) looked at the effect of mutations on yeast genes, and found that there is a strong correlation between how deleterious a mutation is and how many other genes that particular gene interacts with in an epistatic manner. Incidentally, Costanzo et al. showed that both negative and positive interactions are correlated with deleterious effects, something that was proposed theoretically much earlier (Wilke and Adami 2001). While negative interactions between genes imply that two gene mutations that are beneficial on their own have less of a benefit together than expected, positive interactions mean that two gene mutations together carry more of a "punch" than expected (see Box 5.4 for a quantitative definition of positive and negative epistasis). This is precisely what we also observe in the NK model. The reason for this is entirely mathematical (which does not mean that it could not also be physical, of course). Namely, if a gene interacts with K other genes in the NK model, that means that the fitness contribution w_i of that gene is one of 2^{K+1} random numbers (because there are 2^{K+1} possible binary combinations of $K + 1$ loci). The more numbers to choose from, the higher the chance that one of those numbers will be fairly high (this is, in essence, the law of large numbers: the probability that a random number will be larger than any particular cutoff increases with how many numbers are tried). In comparison, a single (binary) gene that does not interact with any other gene will have a fitness contribution that is one of two random numbers only. This is a much more constrained choice, leading to a much smaller maximum fitness, on average.

Figure 5.42(b) shows the average number of mutations that went to fixation on each of the lines of descent that achieved the fitness Ω in panel (a) of that figure. We note that this number decreases monotonically, which means that the higher K, the fewer mutations are fixed given the same number of generations. That also looks like a contradiction at first. As K increases, the attained fitness increases, but is this achieved with fewer mutations? It is not a contradiction because as we discussed earlier, the higher K, the bigger the average fitness increase of a mutation. We can see this in Figure 5.43(a), which shows the mean benefit s of a mutation (measured as a fractional increase in fitness $s = w_{new}/w_0 - 1$, where w_{new} is the fitness of the beneficial mutant). The increase of the beneficial effect size with K (open symbols in Fig. 5.43) documents that in the more rugged fitness landscapes the peaks are generally higher (due to the "cooperative genes" effect we discussed above), while the depth of valleys also increases with K (deleterious mutations, solid symbols).

We can also measure the average amount of epistasis on any particular line of descent, by averaging the pairwise epistasis ϵ_i of any mutation on the line of descent with the mutation that follows it. So, if there are n mutations on a particular line ($n - 1$ pairs of mutations), the mean pairwise epistasis is

$$\langle \epsilon \rangle = \frac{1}{n-1} \sum_{i=1}^{n-1} \epsilon_i , \qquad (5.50)$$

where ϵ_i is defined as in Equation (5.46) in Box 5.4. As we could have expected, the higher K, the larger the positive epistasis between subsequent mutations. And even though the degree to which mutations can cooperate in order to achieve higher benefits s increases with K, at some point the population size is too small for those mutations to be found in time (because these mutations also become more and more rare with increasing K). As a consequence, the achieved fitness in Figure 5.42(a) stops to increase with K, and then decreases.

The fitness landscape of the NK model is certainly a very abstract one, and nothing in its mathematical structure should give us confidence that it can describe broad features of the evolutionary process. However, in hindsight the landscape's inventor, Stuart Kauffman, had the right intuition that fitness is a function of multiple genes interacting in a positive as well as negative manner. In fact, the model reproduces some statistics from large-scale cellular mutation experiments to a surprising degree. For example, in their landmark paper "The genetic landscape of a cell," Costanzo et al. performed single and pairwise knockouts of the majority of yeast genes in order to better understand how genes interact (Costanzo et al. 2010). They then tested how the fraction of pairs of genes that interact epistatically depends on the mean deleterious effects of a mutation. The same procedure can be applied to an NK landscape. In particular, we tested the sequences ("haplotypes") of each of 679 peaks in a landscape with $N = 20$ and $K = 4$ and captured the deleterious effect as well as the pairwise epistasis with each of the 678 other peaks. The distribution

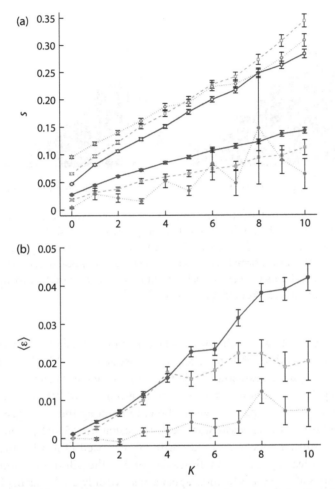

FIGURE 5.43. (a) Mean effect of a mutation s, as a function of the K of the evolutionary landscape, for the 200 replicate runs also shown in Figure 5.42. Beneficial mutations are shown as open symbols (with line styles determined by the mutation rate as in Fig. 5.42). Deleterious effect sizes are shown as solid symbols. Both the beneficial and the deleterious mutation effect size increase with K, indicating an increasing height of peaks and depth of valleys with landscape ruggedness K. (b) Mean epistasis $\langle \epsilon \rangle$ of all the mutation pairs on a particular line of descent as function of K, for the same set of 200 runs (line styles of different mutation rates as in Fig. 5.42). From Østman et al. (2012).

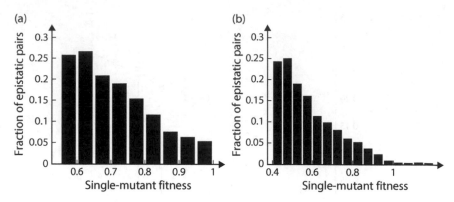

FIGURE 5.44. Fraction of epistatic pairs vs. deleterious effect. (a) Fraction of pairs of mutations of high-fitness NK haplotypes with epistasis $|\epsilon| > 0.08$, for all 679 peaks in a landscape with $N = 20$ and $K = 4$, as a function of the single-mutant fitness normalized to wild-type fitness. (b) Fraction of pairs of yeast gene knockouts with epistasis $|\epsilon| > 0.08$ as a function of the (normalized) single-mutant fitness. From Østman et al. (2012).

of the fraction of pairs with epistasis larger than a threshold (chosen to be $|\epsilon| > 0.08$ for both the NK pairs and the yeast gene pairs) looks fairly similar, even though the yeast study had many more genes and pairs. [There were 258,020 pairs with fitness larger than the threshold in the NK data set, while 5,481,706 yeast gene pairs survived the cutoff among the yeast gene mutations. In both cases, epistasis was measured using Eq. (5.46).] Note that in the yeast study, a knockout could create an increased fitness for the cell— occasionally leading to a mutant fitness larger than the wild-type—something that is impossible in the NK landscape because we only used peak haplotypes.

Historically speaking, the NK model is the first computational model with tunable ruggedness and epistasis (the fact that epistasis and ruggedness are coupled in this landscape is evidenced in Fig. 5.43), and evolutionary paths on this landscape depend strongly on this ruggedness (besides the mutation rate and population size) much like real evolutionary paths. Trajectories are almost never straight "uphill" paths, but instead—again depending on mutation rate and population size—there are steps up and down, there are trade-offs, and there are dead ends where adaptation seemingly stops for thousands of generations. Interactions between genes make it possible to achieve degrees of fitness that would be next to impossible to achieve using single genes only, but the price for these high peaks are deep valleys that make attaining those peaks all the more difficult. It may seem that, unless the mutation rate is unusually high, such high fitness peaks could never be reached. However, another feature of the NK landscape is that high peaks are usually surrounded by other high peaks (Østman and Adami 2013), which increases the likelihood that random trajectory starting-points will end up near the highest peaks in the landscape. Furthermore, in more realistic landscapes that allow for neutral mutations (such mutations are strictly impossible in the NK model, even though mutations of very small effect certainly do exist) there can be neutral paths that connect even distant peaks. Of course, this means that such fitness peaks are not really peaks in the conventional manner, but then again the evidence from the LTEE in section 5.4.2 suggests that such "isolated" peaks are really a mathematical aberration anyway.

In the next section, we'll compare the evolutionary paths obtained via simulation on the abstract NK landscape directly with a particular line of descent that was realized in an actual biological experiment: the long-term evolution experiment of Lenski.

5.4.4 Diminishing returns in adaptation

The analysis of how fitness increases in the long-term evolution experiment in section 5.4.2 showed a trend of diminishing returns: over time, improvements are harder to come by, so that with the same "investment" of evolutionary time, the "returns" on that investment (in terms of fitness gain) diminish. The concept of diminishing returns is a common one in economics of course (see, e.g., Nordhaus and Samuelson 2009), but we are used to the idea in everyday life too, whenever boundaries are being pushed. A typical example is the evolution of world records over time. Clearly, in most of these records physiological boundaries are being pushed, and with all else being equal (here, human genetics), we expect that improvements become more and more rare. A case in point is the evolution of the world record in marathon running (shown here for male athletes in Fig. 5.45). The "magical" achievement of covering the 42 kilometers and 195 meters[14] in two hours may at some time be possible (the Kenyan Eliud Kipchoge got tantalizingly close in 2018 and beat the mark in 2019 in an unofficial event), but progress beyond that mark will turn out to be excruciatingly slow. In biological evolution, diminishing returns are expected whenever a population climbs a fitness peak. However, we have also learned in section 5.4.2 that the idea of a fitness peak with a clear "summit"—beyond which no progress is possible—may not be a realistic picture after all. Indeed, the analysis of twelve replicate fitness trajectories in the LTEE suggested that fitness can increase in that experiment for another million years at least, with

14. 26 miles and 385 yards.

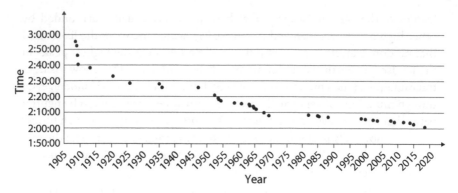

FIGURE 5.45. Evolution of the marathon world record (male athletes) from 1905 to 2018.

no summit in sight. If there is no summit, what causes the slow-down? A partial answer to this question was already given in section 5.4.2: the distribution of available beneficial mutations changes over time, as shown in Figure 5.36. The root cause of the decrease is often called "diminishing returns epistasis," and we'll discuss this concept in this section.

Diminishing returns epistasis is the observation that a set of mutations has less of an impact together than what each of the mutations on their own would have predicted. It is therefore, quite clearly, the opposite of the effect that we observed when studying lines of descent in the NK model in the previous section, when mutations worked together to achieve *more* than what each individual mutation could do on its own. Let us first define the concept of diminishing returns epistasis mathematically, then try to understand under what circumstances we expect to see the phenomenon, and how it can be avoided altogether.

Say you have a set of mutations M, each of which have fitness effect w_i/w_0 ($i \in M$). In this context, w_0 is the fitness of the organism without mutation i. In most cases, we will set w_0 to one, and simply write that the fitness effect of a mutation is w_i. We can then define the *epistatic return* E_M of a set of mutations M as

$$E_M = \log\left(\frac{w_M}{\prod_{i=1}^{M} w_i}\right),\qquad(5.51)$$

where w_M is the fitness effect of the entire set of mutations, applied to the background with fitness $w_0 = 1$. The denominator in (5.51) is the expected fitness of the set of mutations if they would not interact, given by the product of the individual fitness effects. The epistatic return can be positive (meaning that

Table 5.2. The first five mutations on the line of descent
of the Ara-1 population of the LTEE, and their individual
fitness effect (±95% confidence interval) measured
with respect to the wild-type (ancestral) strain REL606
(Khan et al. 2011). The gene name that carries the
mutation is listed in the first column, while the second
column shows the abbreviation.

Mutation	Abbreviation	Fitness Effect
Δrbs	r	1.012 (0.013)
topA	t	1.142 (0.023)
spoT	s	1.105 (0.017)
glmUS	g	1.027 (0.016)
Δpyk	p	1.000 (0.013)

the set M works cooperatively to increase fitness), it can be negative (implying diminished returns), or it can vanish, meaning that the group of mutations
as a whole shows no interaction.[15] In the limit of two mutations, this definition
is the same as the pairwise epistasis measure ϵ introduced in Box 5.4 (setting
$w_0 = 1$). Here and below, we will use the base-10 logarithm to calculate the
epistatic return.

To study epistatic return experimentally, Khan et al. (2011) isolated the
first five mutations on the line of descent of one of Lenski's LTEE populations
(specifically, the population termed Ara-1), and created all thirty-two possible combinations of them using genetic techniques. They then went about to
measure the fitness of each of these engineered genotypes, so as to measure
the epistatic return of each of these sets. By constructing pairs, triplets (and
so on) of these mutations (that is, inserting them on the REL606 ancestral
genome) and measuring the resulting genotype's fitness, the epistatic return
can be calculated. Khan et al. found as many positive E_Ms as negative ones
among the twenty-six multiplets,[16] but the difference was mainly due to one
mutation: the fifth one on the line of descent, Δpyk. This is a mutation in the
gene pykF that codes for a pyruvate kinase. The pyruvate kinase is necessary to
complete the final step in glycolysis, which it achieves by transferring a phosphate group to ADP, and in so doing making ATP (Siddiquee et al. 2004).
If that mutation is part of the multiplet then the epistatic return is mostly

15. Just because a group as a whole shows no interaction does not imply that there are no
interactions among any pairs, unless the group size is two.

16. Of the thirty-two possible combinations, five are single mutations and one is the
ancestral wild-type.

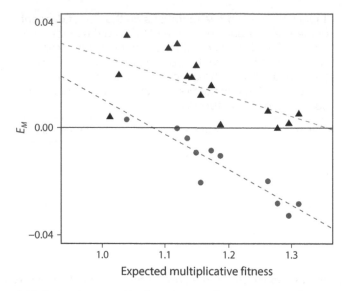

FIGURE 5.46. Epistatic return E_M for two sets of mutations as a function of the expected multiplicative fitness $\Pi_{i=1}^{M} w_i$ [see Eq. (5.58)]. The set of eleven multiplets without the Δpyk mutation p is shown as circles, and the fifteen multiplets that include mutation p as triangles. Adapted from Khan et al. (2011).

positive (see Fig. 5.46). Without that mutation, the return is mostly negative. For both sets, however, E_M decreases with increasing expected fitness (the multiplication of the single mutation fitness effects w_i): this is the signature of diminishing returns. Often the concept of diminishing returns is associated with negative epistasis (see, e.g., Kryazhimskiy et al. 2009; Kryazhimskiy et al. 2011), even though from Figure 5.46 it is clear that diminishing returns can be associated with both positive and negative epistasis. What determines whether epistasis is predominantly positive or negative in adaptation?

To investigate this question, let us look at a line of descent from an evolutionary experiment using the NK model fitness landscape in comparison. We will take the first five mutations of 100 replicate lines, and calculate the epistatic return for all twenty-six combinations of two to five mutations, and plot it against the expected multiplicative fitness, just as in Figure 5.46 (where the twenty-six mutations were separated into two groups of eleven and fifteen multiplets each). The result, for an NK model with $N = 20$ and $K = 4$, is shown in Figure 5.47. The signature of diminishing returns is obvious, but while some of the multiplets have a positive E_M, the majority have negative E_M, indicating diminishing returns via buffering. This finding is not altogether unsurprising: among a mix of mutations, we may find both positive and negative epistasis, but when climbing from the base of a peak, the first steps are

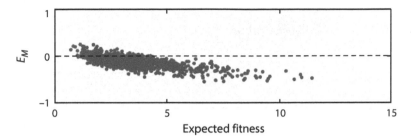

FIGURE 5.47. Epistatic return E_M for the twenty-six multiplets of two to five mutations, taken from the first five mutations on 100 replicate lines of descent each, in an NK model with $N = 20$ and $K = 6$. Unpublished results (A. Hintze, 2011, personal communication). Note that beneficial effects can be significant for the first few mutations in the NK model, so that the expected fitness is much larger than in Khan et al. (2011).

likely substantially larger than subsequent ones. Note, however, that testing for epistasis among multiplets is different from testing for epistasis on mutation pairs (mutations that follow each other on the line of descent), which we looked at in Figure 5.43(b). There we found that the sum of pairwise epistasis is positive, and correlated with K. Also, climbing a single peak may lead to more pronounced negative epistasis, while crossing valleys to adjacent higher peaks should produce more positive epistasis because in such a scenario, a step downward must be followed by a step upward.

Let us test what the epistatic return looks like if instead of focusing on the first five mutations, we look at five consecutive mutations a little further into the line of descent. In Figure 5.48 we can see the "diminishing returns plot" for mutations 1–5, but underneath it is the same plot but shifted forwards, that is, for mutations 2–6, 3–7, 4–8, etc.

It is clear from Figure 5.48 that as we move "into" the line of descent, the pattern of epistasis still follows the diminishing returns signature, but more and more of the multiplets shift toward positive epistasis. It thus appears that the idea that diminishing returns is associated with mostly negative epistasis is an artifact of only testing the very beginning of a line of descent, an observation that was made independently by Draghi and Plotkin (2013), also using the NK model.

In a similar vein, we could ask whether it even makes sense to test the epistatic return of *all* combinations of mutations from a given line of descent, when the most important mutation is always going to be the one that follows an existing mutation. In other words, perhaps it is more instructive to only study epistasis between pairs of mutations that immediately follow each other on the line of descent, and test whether diminishing returns can be observed

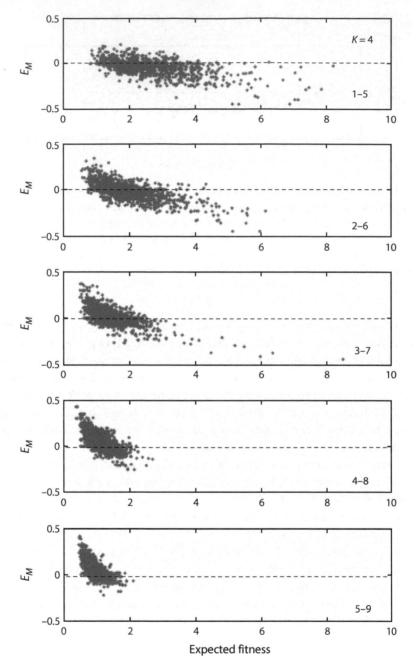

FIGURE 5.48. Epistatic return for sets of five mutation multiplets (twenty-six multiplets for each line of descent). The top plot shows the return for the first five mutations, while the plots below are sets of multiplets that are shifted by one mutation "forwards" on the line of descent. Average of 100 lines of descent each, in a landscape with $K = 4$. Unpublished results (A. Hintze, 2011, personal communication).

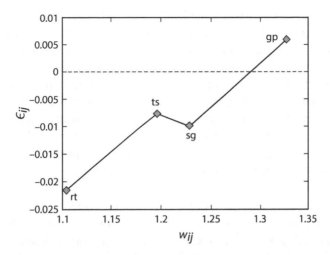

FIGURE 5.49. Pairwise epistasis ϵ_{ij} as a function of the achieved fitness w_{ij} for the first five mutations along the line of descent of Ara-1 (see Table 5.2 for abbreviations). The line of descent r-t-s-g-p is indicated by the solid line.

for those pairs. We have already made this test in the NK model and found that there is an excess of positive epistasis among those pairs (Østman et al. 2012). We can do the same test on the first five mutation of the Ara-1 line of the LTEE, and the result is shown in Figure 5.49.

We note that on this line of descent, pairwise epistasis starts out negative but actually tends to increase, turning positive for the last pair. Diminishing returns, it thus turns out, is a complex concept and care must be taken to define it. It certainly does not mean negative epistasis, but defining it in terms of the epistatic return plotted against the expected fitness also does not seem to be diagnostic, as this slope tends to always be negative. Perhaps the most instructive definition then would be to test for the distribution of pairwise epistasis for mutations that are directly adjacent on the line of descent. This distribution indeed skews negatively for the first five mutations in the NK model, but the distribution turns positive as a peak is climbed, in particular when new peaks are accessed via valley-crossing.

5.5 Summary

Several billions of years of evolution have shaped our biosphere to become the complicated, interdependent, hierarchical complex system we witness today. Among the parts and members of this system, there are certainly differences in complexity—some obvious to any observer, some less so. Structural complexity, the most intuitive of all measures, is notoriously difficult to define because there is no universal system to rank all possible physical structures. We have seen that automata theory and information theory allow us to quantify the complexity of a sequence in terms of its information content about the environment within which that sequence has evolved, and that this information is a good proxy for the functional complexity of the organism precisely because this information is used by the organism to function in a complex environment. Yet, an information-theoretic treatment of complexity will not allow us to determine whether any particular trait contributes to complexity or not, because it is not practically feasible to test whether or not a trait is informative about any of the potential environments the organism is adapted to (see also Box 10.1).

Measuring complexity in terms of information content is limited in other ways too, because so far it is only practical for short stretches of DNA or single proteins. A quantitative measure for the complexity of a whole genome using information theory will only be possible when a large number of complete genomes of closely related species is available. A limited number of such genomes is currently available, and Eugene Koonin of the National Center for Biotechnology Information (NCBI) has attempted just such an analysis (shown in Figure 5.50) which shows that the information content Equation (2.34) increases with sequence length for viruses and prokaryotes. For complex genomes, we have to rely on other measures such as network complexity measures, or an analysis of the information content of modules and motifs.

In this chapter we first discussed the information content of individual genes, and then showed using an artificial cell model that the complexity of a network of genes is also reflected in the genome complexity. We further examined long-term trends in evolution to address the question whether it is possible to determine if complexity tends to increase in evolution or not. We identified several long-term trends, but because it is not currently possible to accurately measure any of the standard complexity measures on a reconstructed line of descent, we must treat this question as remaining open.

We also discussed short-term trends that can be seen in the evolution of a single lineage (as opposed to the long-term trends that can be observed across lineages). One of the short-term trends is an increase in fitness over time; this particular one has been observed in the first 75,000 generations of evolution in the laboratory. And even though it is possible to make broad predictions for this trend to continue for millions of years in the future, it is important to keep in mind that this still constitutes a short-term trend, compared to the geological time scales involved in the trends studied in section 5.3. When comparing to computational models of evolution in artificial fitness landscapes whose ruggedness can be tuned (the NK model), we found that many of the characteristics of adaptation can be reproduced in these models,

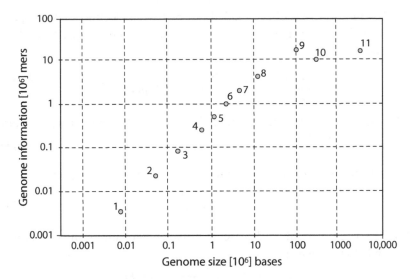

FIGURE 5.50. Estimate of information content of genomes as a function of genome size for viruses, prokaryotes and eukaryotes, calculated (except for *Drosophila* and *Homo*) using Equation (2.34), from an alignment of genomic data obtained from Genbank (Koonin 2016). 1: *Encephalomyocarditis virus* (RNA virus), 2: Lambda phage, 3: T4 phage, 4: *Mycoplasma genitalium*, 5: *Acanthamoeba polyphaga* (a giant mimivirus), 6: *Archaeoglobus fulgidus* (a free-living archaeon), 7: *Escherichia coli*, 8: *Saccharomyces cerevisiae*, 9: *Arabidopsis thaliana*, 10: *Drosophila melanogaster*, 11: *Homo sapiens*. For *Drosophila* and *Homo*, fraction of sites under selection was used instead.

but their predictions must fall short because the number of genes is so small in these models that features such as the "endless peak" of the LTEE cannot be observed.

Exercises

5.1 The Kraft inequality gives a necessary and sufficient condition for the existence of a *prefix code* (a code where no codeword can be the prefix of another codeword), so that the code is uniquely decipherable given a set of codeword lengths ℓ_i, from a set of n codewords. If D is the dimension of the alphabet that the codewords are composed from, then the condition for the existence of such a code is the inequality

$$\sum_{i=1}^{n} D^{-\ell_i} \leq 1. \tag{5.52}$$

Use the fact that the likelihood of a string s_i whose Kolmogorov complexity is $K(s_i)$ is given by $p(s_i) = D^{-K(s_i)}$, and the Kraft inequality, to show that

$$\sum_{s_i} p(s_i) K(s_i) \geq \sum_{s_i} p(s_i) \log \frac{1}{p(s_i)} = H(S). \qquad (5.53)$$

5.2 Derive Fisher's fundamental theorem $\Delta \bar{w} = \frac{1}{\bar{w}} \text{var}(w)$, where $\text{var}(w) = \frac{1}{N} \left(\overline{w^2} - \bar{w}^2 \right)$ is the variance of fitness, using the relationship between the probability distributions $p_i' = \frac{n_i'}{N}$ and $p_i = \frac{n_i}{N}$ given by

$$p_i' = p_i \frac{w_i}{\bar{w}}. \qquad (5.54)$$

5.3 The Kullback-Leibler divergence between probability distributions P and Q is defined as

$$D_{\text{KL}}(P||Q) = \sum_i P_i \log\left(\frac{P_i}{Q_i}\right). \qquad (5.55)$$

Show that the Kullback-Leibler divergence $D_{\text{KL}}(p_i'||p_i)$ between the distributions $p_i' = p_i \frac{w_i}{\bar{w}}$ and p_i is bounded by

$$D_{\text{KL}}(p_i'||p_i) \leq \log(1 + \frac{\Delta \bar{w}}{\bar{w}}), \qquad (5.56)$$

where $\Delta \bar{w} = \bar{w}' - \bar{w}$ and $\bar{w} = \sum_i p_i w_i$.

5.4 Show that the change in the Shannon entropy of a population during adaptation is negative ($\Delta S = S' - S < 0$), where $S = - \sum_i p_i \log p_i$ and $S' = - \sum_i p_i' \log p_i'$, using the relation between p_i' and p_i as given above. Hint: Write p_i' as a function of the time parameter t so that

$$p_i^{(t+1)} = \frac{w_i}{\bar{w}(t)} p_i^{(t)}, \qquad (5.57)$$

with the formal solution $p_i^{(t)} = \left(\frac{w_i}{\prod_{n=1}^t \bar{w}(n)} \right)^t p_i^{(0)}$. Show that in the limit $t \to \infty$ (where $p_i^{(t+1)} \approx p_i^{(t)}$) the entropy change ΔS takes on the form

$$\Delta S = \log \sum_i p_i^{(t)} w_i - \sum_i p_i^{(t)} \log w_i \qquad (5.58)$$

and then use Jensen's inequality.

6

Evolution of Robustness

However many ways there may be of being alive, it is certain that there are vastly more ways of being dead.

—R. DAWKINS, *THE BLIND WATCHMAKER*, 1986

When populations evolve and accumulate information about the environment in their genes, not all that information is about how to better exploit the environment. Some of that information is used to make sure that the organism can continue to survive even in extraordinarily challenging circumstances. In fact (while exact numbers are not known) it may very well be that a majority of genes are there to ensure *robustness* in the face of unpredictable conditions. In the budding yeast *Saccharomyces cerevisiae* for example, only about 1,050 genes are absolutely essential for survival, out of about 5,800 total genes (Yu et al. 2006).

This excess fraction of genes is reminiscent of robust engineering design (for example in airplanes), where many of the systems exist in a redundant manner to ensure survival of the aircraft even in dire circumstances. This means that in biological evolution there are many aspects of fitness that are counterintuitive simply because it is not immediately obvious how some of those genes further an organism's survival. One of the most blatant adaptations of this sort that we will encounter below is the counterintuitive evolutionary strategy of sacrificing replication speed for robustness.

Another odd adaptation that we will discuss is the evolution of a replication system that is so fragile that almost any mutation will lead to the organism's death. Such adaptive strategies seem to run counter to the idea that evolution always favors the fittest, but we'll see that it really just means that sometimes, fitness is a complicated concept. For example, imagine a heritable trait that does not increase the fitness in a particular environment, but it better prepares an organism to survive if the environment changes. If environmental changes

321

are rare but frequent enough, such a trait may make the difference in a competition even though there is no fitness effect in a constant environment (or in the extreme case, the trait could even be detrimental). The "bet hedging" strategy we discussed in Box 2.4 is a typical illustration. Traits of this sort are thought to enhance a lineage's *evolvability* (Wagner 2005; Lenski et al. 2006; Masel and Trotter 2010).

The phrase "survival of the fittest" is often used synonymously with natural selection, and often attributed to Darwin himself. However, it was coined instead by the British philosopher Herbert Spencer (1820–1903), who wrote this phrase in his book *Principles of Biology*, which appeared in 1864. Spencer was interested in how complexity emerges in biology even before he read Darwin, and only reluctantly converted to the Darwinian view after previously espousing Lamarckian mechanisms. The phrase "survival of the fittest" has become so common that we sometimes forget what "fitness" even means in the context of evolution, because the word has changed in meaning over time. Today the word fitness often evokes strength of body and endurance (i.e., what is known as "physical fitness") but within a Darwinian context it literally means "fitting its environment well." Thus, a "fit" organism is a well-adapted one, because it fits the niche it occupies. At the same time, it is sometimes erroneously claimed that "survival of the fittest" is a tautological statement, because the fittest "automatically survives because fitness means survival." But fitness is not, of course, synonymous with survival. The easiest way to see this is by noting that sometimes very unfit organisms can survive by chance alone, but they are unlikely to continue to survive in such a manner because luck is not a heritable trait. Natural selection, on the contrary, acts precisely on those traits that are heritable and have the capacity to increase the *likelihood* of survival in many generations to come.

When we look at evolutionary history using any unbroken line of descent, we usually can see adaptations occurring on top of previous adaptations, inexorably increasing the fit of the organism to its particular niche. This increase in fitness can even bridge, it seems, periods of time where the niche changes catastrophically, leading to an intermittently lower fitness. But the organisms whose ghostly imprints we see in the fossil records (or the not-so-ghostly ones stored in the freezers of the LTEE) are just a very biased sample of all the variation that previously occurred: they are the lucky ones to be visible, selected from all the lucky ones that survived. We can therefore ask: How often do increases in fitness (beneficial mutations) even occur? How often are mutations deleterious, or even lethal? And how likely is it that mutations do not affect the fitness of the organism at all, meaning that the mutation is neutral? Of course, this was not a question that either Darwin or Spencer could have answered (or even asked) as the concept of mutations, or even what mutations acted upon, was unknown at the time. But soon after the genetic basis of mutations was elucidated and the concepts surrounding *molecular* evolution were being explored, it became clear to an increasing number of researchers that beneficial mutations were exceedingly rare, and that most genetic changes either left the organism impaired in fitness, or had no effect. Once the concept of mutations was fully understood, including the idea that the effect of mutations spans the gamut from lethal to beneficial, it became imperative to understand the likelihood that they "go to fixation," meaning that at some future point in time, *all* organisms in the population will carry that mutation. The classical theory of population genetics (Haldane 1927; Fisher 1930; Wright 1931) provided a means to calculate that probability for mutations that increased fitness, but it took almost forty more years before people asked whether neutral mutations could play a role in evolution also.

6.1 The Neutral Theory of Evolution

How important are neutral mutations in the evolution of complexity? This question, central to the mathematical theory of population genetics, was first answered by the Japanese geneticist Motoo Kimura (1924–1994), who is credited with crafting the "neutral theory" of evolution. Kimura was able to show that the rate at which neutral mutations become fixed in the population is given entirely by the rate at which these mutations *appear* in the population. And if the majority of mutations are neutral (as is borne out experimentally for most genes), then most mutations that "make it" are in fact neutral, not beneficial. The reasoning behind this assertion is not difficult to comprehend. Say the neutral mutation rate is ν (ν for "neutral"). Then, in a population of N individuals, there will be νN such mutations at any given point in time, and each of these νN mutations will have a chance to "make it" (i.e., become fully established in the population at some future time). How many of those will actually "make it," on average? If all mutations are neutral (that is, we disregard for this exercise deleterious, lethal, as well as beneficial mutations) then each of these mutations has an equal chance, because after all those mutations are neutral. The likelihood of fixation of a neutral mutation is thus just $1/N$, so that the number of mutations that successfully are fixed even though they are neutral is $\nu N \times \frac{1}{N} = \nu$, that is, it is given entirely by the rate at which they occur, as claimed.[1]

---

1. In a sexual randomly mating population, N should be replaced by $2N$ because in such populations each gene occurs in two copies per individual. For the sake of brevity, we will be ignoring this complication here and discuss Kimura's fixation probability more quantitatively in Box 6.1.

Box 6.1. The Mathematics of Fixation

Calculating the likelihood that a mutation that freshly enters an adapting population goes on to become ubiquitous (that is, every member of the population has it) has a long history, and is one of the earliest applications of the mathematics of population genetics. Most textbooks on population genetics will have a derivation of this "probability of fixation," so we will only visit a few important milestones in this box. The earliest calculation of a fixation probability is due to the British mathematician J. B. S. Haldane (1892–1964), and concerned the simplest of cases: the fate of a single beneficial mutation in an infinite population. Let us first define the advantage of a mutation with fitness $w_\star$ in terms of the fractional increase with respect to the previously prevailing fitness level w_0 as $s = w_\star/w_0 - 1$. The probability of fixation $P(s)$ (for the simplest case of a haploid asexual population in which every organism produces offspring according to a Poisson distribution with mean equal to the fitness) becomes (in the limit that $P(s)$ is small, see Exercise 6.1.)

$$P(s) = \frac{2s}{(1+s)^2}. \tag{6.1}$$

For small s, Equation (6.1) turns into the celebrated result $P(s) \approx 2s$ (Haldane's rule), which is taught in every introductory population genetics class. If mutations are neutral ($s = 0$) this formula predicts a vanishing fixation probability, but we know that this must be wrong if the population size is finite. It was Kimura (1962) who first calculated the probability of fixation in case the population is finite, using a novel approach based on a diffusion process. His result, obtained for the case that the new mutation has frequency $1/N$ (where N is the population size) and where selection in a haploid population proceeds according to the Wright-Fisher process, reads

$$P(s) = \frac{1 - e^{-2s}}{1 - e^{-2Ns}}. \tag{6.2}$$

We can see right away that in the limit of an infinite population size (and s small) we recover Haldane's rule. But we can also see that in the limit of vanishing s Kimura's formula reproduces $P(0) = 1/N$, as we would have expected on general grounds. Kimura's formula was improved forty-three years later by Sella and Hirsh (2005) to read (for a mutation of fitness $w_\star$ fixing in a background of sequences with fitness w_0, again for a haploid asexual population)

$$P(s) = \frac{1 - \left(\frac{w_0}{w_\star}\right)^2}{1 - \left(\frac{w_0}{w_\star}\right)^{2N}} = \frac{1 - \frac{1}{(1+s)^2}}{1 - \frac{1}{(1+s)^{2N}}}. \tag{6.3}$$

By writing $1/(1+s)^2 = e^{-\ln(1+s)^2} \approx e^{-2s}$ for small s [and similarly for the denominator in Eq. (6.3)], we see that the Sella-Hirsh formula implies the Kimura formula. Indeed, computational simulations show that the Sella-Hirsh formula is more accurate in predicting the rate of fixation than Kimura's (Sella and Hirsh 2005).

Kimura's theory was first articulated in his 1968 article entitled "Evolutionary rate at the molecular level" (Kimura 1968)—his entire theory of neutral evolution appeared later in book form in 1983 (Kimura 1983)—and served to explain the observed rate of amino acid substitution in the human hemoglobin protein (about one change in ten million years, in a protein of 141 amino acids).

Kimura thus argued that as a consequence of the theory of fixation of neutral alleles, essentially *all* observed changes to the hemoglobin molecule must be effectively neutral.[2] Many people have argued after Kimura's insight that somehow he was arguing that *all* Darwinian evolution must be neutral. However, this is not an accurate representation of the neutral theory. Indeed, even if the majority of mutations that go to fixation are neutral when they occur, adaptation is still driven by the beneficial mutations that increase the organism's fit, even if rare. Furthermore, due to interactions with mutations that occur on the same sequence later (see Box 5.4), a mutation that was neutral (or nearly neutral) when it occurred and fixed can become highly beneficial later and therefore "entrenched" (Shah et al. 2015; Goldstein and Pollock 2017). Neutral mutations can have other uses still: some argue that "neutral walks" in fitness landscapes are essential for moving populations to different (unexplored) areas in genetic space where adaptive mutations might be more prevalent (see, e.g., Gavrilets 2004).

Besides leading populations to the "foothills" where selection can drive them "uphill," neutral mutations have more subtle uses. In the following, we will explore how populations might adapt to very stressful *evolutionary* conditions. The stress we discuss here is not the usual stress that an organism might be subject to (such as the stress of bacterial infection for a eukaryotic cell, or the stress of viral infection for a bacterial cell). Instead, they are stresses that jeopardize the *evolutionary* survival of the population. Two such stresses are usually recognized: the threat of high mutation rates and the scepter of small population sizes. We will discuss each one by one, and study how organisms have adapted to deal with them. The lesson we will learn is an important one: evolution is a process that takes place in a complex and changing environment. To survive, a population should adapt to the environment, but it must also adapt to the process *itself*, making evolution work better for the population under dire circumstances. This is the evolution of robustness.

6.2 Evolution of Mutational Robustness

Mutations are essential to the evolutionary process. As we saw in the very first chapter, without mutations, the evolution of novel structures and functions

---

2. In a curious twist of history, King and Jukes made essentially the same argument independently a year later (King and Jukes 1969).

cannot proceed. But, as with most things, too much of a good thing quickly turns into a nightmare. In the previous section we briefly discussed the process of adaptation in terms of the origination of mutations (given by the rate of appearance of mutations), and the probability of fixation of those mutations. Taken together, these two rates dictate the *rate of evolution* (see McCandlish and Stoltzfus 2014 for a review). When the rate of mutation is small, we can consider the process of mutation generation independently from the rate of fixation. In this limit, we wait until the newly generated mutation either is lost or goes to fixation before introducing a new mutation. This limit is known as the "weak mutation–strong selection" limit of population genetics: in this limit, mutations are rare, and selection acts supreme.

But imagine that you increase the rate of mutation while selection strength stays constant. Clearly, at some point we will be introducing mutations before the previous mutation's "fate" has been decided. At that point, multiple mutations exist in the populations at the same time, and (most importantly) new mutations might occur on a sequence whose future is still uncertain. This means that the fate of the original mutation is now unknown: if the second mutation is deleterious (as most non-neutral mutations are), then it likely blocked the path to fixation for the first mutation. If you increase the rate of mutations even further, it is easy to see that genomes will be overwhelmed with mutations that reduce fitness, completely swamping the rare beneficials that might exist. Thus, at high mutation rates, populations are in danger of *lethal mutagenesis* (Bull et al. 2007).

In the simplest model of this process of mutation production and fixation, we can imagine that a typical organism in the population has d offspring per generation. Suppose for a moment that all mutations (if they occur) are lethal. In that case, it is easy to calculate that the fraction of unmutated ("pristine") organisms is $e^{-\mu}$, where μ is the genomic mutation rate. (This is the celebrated result first mentioned by Kimura and Maruyama 1966; see Exercise **6.3**). Now, to ensure that a population survives, the mean number of replicating offspring must be at least one, so that a criterion for survival can easily be written as

$$de^{-\mu} \leq 1. \tag{6.4}$$

This relation puts an upper limit on the mutation rate $\mu \leq \ln d$, and short of increasing the number of offspring per generation, there would be little evolution could do to help organisms survive such extreme conditions. But not so fast. We made a crucial assumption when deriving this relationship—namely, that all mutations are lethal—which of course they are not. To remedy this, we can introduce another parameter: the likelihood that a mutation is neutral, which we will call ν. The overall mutation rate is still μ. The *lethal* mutation rate now becomes $\mu - \nu$, and the requirement for survival now is

$$\mu \leq \nu + \ln d. \tag{6.5}$$

Is it possible for evolution to increase ν in response to high mutation rates? The answer (we will see) is yes: all it takes is a rearrangement of the genome in such a way that the fraction of neutral mutations is increased. Note that under normal conditions (i.e., mutation rates that are not excessively high) there is no reason for the fraction of neutral mutations to be high, and as a consequence there is usually room to increase ν if the selective pressure requires it. But how can the fraction of neutral mutations be increased?

6.2.1   Selection for robustness in neutral evolution

To study how a population might increase the fraction of neutral mutations, we will study a greatly simplified situation that nevertheless captures the essence of the phenomenon. We are going to consider a model in which there are only two fitness values: fitness equal to one (the very best replication rate available) and zero (death). In such a model we can make neutrality variable by modeling evolution on a *neutral network*. The network view of a fitness landscape is somewhat complementary to the fitness landscape view that we discussed earlier in section 5.4.2.

In a neutral network, viable genotypes (with fitness 1.0) are represented by nodes, connected to other such (viable) nodes if they are separated by a single point mutation. Figure 6.1 shows an example neutral network made from RNA sequences that guide gene editing in the parasite *Trypanosoma brucei* that we will encounter again in section 6.4. Different genotypes can each have a different number of neutral neighbors. How would mutational robustness manifest itself in such a network? This question was first asked by Erik van Nimwegen and his collaborators in a 1999 publication that is now a classic (van Nimwegen et al. 1999). The mathematics of mutational robustness in that paper is (at least in the simplest incarnation) surprisingly simple.

Let us introduce a little bit of notation from graph theory first. Each node on the graph has the same fitness σ, and each node i (out of N, the total number of nodes in the network) has a particular number of neighboring nodes that can be reached by a single mutation: the node's *degree* d_i. We assume that there are differences in degree from node to node in the graph (graphs where each node has the same degree are called *regular*). A typical such graph is shown in Figure 6.1. We can also define the *degree distribution* of the graph $P(d)$, which is the likelihood that any particular node has degree d. Empirically, we can obtain an approximation of this distribution by setting $P(d) = n(d)/N$, where $n(d)$ is the number of nodes with degree d in the

FIGURE 6.1. A neutral network of 229 guide RNA (gRNA) sequences editing a region in the *T. brucei* mitochondrial protein RPS12 (Kirby 2019). Mutational neighbors are connected by a link. Sequences from the TREU 667 cell line (Sequence Read Archive accession number SRR6122539). The hub sequence has by far the most copies in the cell.

network. Naturally, the *mean network degree* $\bar{d}$ is given by

$$\bar{d} = \sum_{d=1}^{d_{max}} P(d)d, \tag{6.6}$$

where d_{max} is the largest degree in the network. Of course, because of $P(d) = n(d)/N$, (6.6) is equivalent to just summing up the degree of every node in the network and dividing by the size N of the network.

Let us introduce another important average quantity: the *mean population degree* of the network. As the name implies, this number reflects a population of organisms that is adapting to "live" on such a network. After all, each node stands for a possible genotype i, and we can ask how many organisms in a finite population of M individuals have this genotype i. We call this number m_i, with $\sum_{i=1}^{N} m_i = M$. The mean population degree then is

$$\langle d \rangle = \sum_{i=1}^{N} d_i \frac{m_i}{M}, \tag{6.7}$$

in other words, it is defined by weighting the degree of each network with how many organisms in an adapting population occupy this node. Typically we expect to find $\langle d \rangle \geq \bar{d}$ (van Nimwegen et al. 1999), with equality if (and only if) all nodes are occupied by the same exact number of organisms.

The difference between $\bar{d}$ and $\langle d \rangle$ is important to understand. That $\langle d \rangle$ takes into account population dynamics is clear, but let us imagine for a moment an evolutionary process where the mutation rate is exceedingly small. Let us also assume that enough time has elapsed that all inferior mutants (those with fitness less than σ) have been eliminated from the population. A new mutation then will be the only one of its kind. Whether or not it will ultimately strive (which in the parlance of population genetics means that it will ultimately be integrated in the genome of every future organism) depends on many factors, but most importantly on the fitness advantage it has compared to all the other types in the population (see Box 6.1).

Now, in the scenario we are contemplating, there are no fitness advantages (it is a neutral network), only fitness disadvantages (if you mutate to a genotype that is not *on* the network). According to Kimura's theory, the likelihood of a neutral mutation to go to fixation is $1/M$, where M is the population size.[3] Suppose a mutation is that lucky. In that case, we have reached a new equilibrium: every organism in the population is the same, only now incorporating that new mutation. In case the mutation had failed to go to fixation, we would just be starting over again, waiting for another mutation to occur. We can now see that when mutation rates are that low, the evolutionary process takes random steps in the space of genotypes, in the sense that the population *as a whole* explores one neutral mutation after the other. In this low mutation regime, the mean population degree $\langle d \rangle$ must be equal to the mean network degree $\bar{d}$ at all times. But what if the mutation rate is larger? In that case, the probability of fixation of the neutral mutant is still $1/M$, but the population is not uniform anymore because many different genotypes may be "attempting fixation" at the same time. As a consequence, we must find $\langle d \rangle > \bar{d}$ in this case.

As we increase the mutation rate, how high can $\langle d \rangle$ become? This question was answered by van Nimwegen et al. (1999) in an elegant fashion. They showed that in mutation-selection equilibrium, the population distribution function $\pi(m)$ (the likelihood to find in the population a genotype with frequency m) obeys an eigenvalue equation involving the *adjacency matrix* of the graph, and moreover, the largest eigenvalue (let us call it ρ) of that matrix is equal to the mean population degree. If the population has not reached equilibrium, then the mean population degree is $\langle d \rangle \leq \rho$.

3. In this description we use M for the population size (as opposed to N in Box 6.1) so as not to create confusion with the size of the neutral network N.

(a) 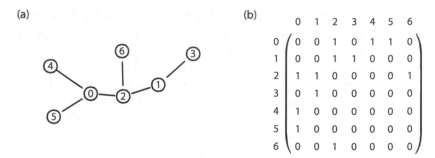 (b)

$$\begin{array}{c} \\ 0 \\ 1 \\ 2 \\ 3 \\ 4 \\ 5 \\ 6 \end{array} \begin{array}{ccccccc} 0 & 1 & 2 & 3 & 4 & 5 & 6 \\ \left(\begin{array}{ccccccc} 0 & 0 & 1 & 0 & 1 & 1 & 0 \\ 0 & 0 & 1 & 1 & 0 & 0 & 0 \\ 1 & 1 & 0 & 0 & 0 & 0 & 1 \\ 0 & 1 & 0 & 0 & 0 & 0 & 0 \\ 1 & 0 & 0 & 0 & 0 & 0 & 0 \\ 1 & 0 & 0 & 0 & 0 & 0 & 0 \\ 0 & 0 & 1 & 0 & 0 & 0 & 0 \end{array} \right) \end{array}$$

FIGURE 6.2. (a) An example neutral network with seven nodes. A link between nodes indicates that these genotypes are one mutation away from each other. (b) The adjacency matrix of the network shown in (a).

Let us briefly discuss the adjacency matrix because it is essential to understanding how mutational robustness is manifested in populations that evolve on neutral networks. The adjacency matrix **G** with elements G_{ij} of a graph is an $N \times N$ matrix constructed in such a manner that there is a 1 at every matrix element that connects two genotypes that are one mutation away:

$$G_{ij} = \begin{cases} 1, & \text{if } i, j \text{ adjacent,} \\ 0, & \text{otherwise.} \end{cases} \tag{6.8}$$

A simple example network of seven genotypes may help illustrate the concept. The graph in Figure 6.2 has a mean degree $\bar{d} = 12/7 \approx 1.71$, and the adjacency matrix of the network shown in Figure 6.2 has seven eigenvalues: three opposite sign pairs and a single zero eigenvalue. The largest of the eigenvalues $\rho \approx 2.053$ is known as the *spectral radius*[4] of the adjacency matrix and we'll colloquially talk about the "spectral radius of the network." To summarize, the mean network degree $\bar{d}$ does not depend on the evolutionary process taking place on the network and is smaller than the mean population degree $\langle d \rangle$, which itself is smaller or equal to the network's spectral radius, which again does not depend on the population.

Let us try to relate this to our earlier question, namely, how a population could adapt to increase the neutrality ν to satisfy Equation (6.5). What we called ν (the neutral mutation rate) can be related to network properties if we imagine that the genomes are sequences made from monomers drawn from an alphabet of D possible values, and that the sequences are at most L monomers long. There are then $(D - 1)L$ possible mutants for each genotype, but only d_i

---

4. Technically, the spectral radius of a matrix is the maximum of the set of absolute values of the eigenvalues of a matrix.

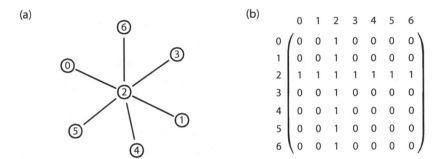

FIGURE 6.3. (a) A neutral network with seven nodes with maximal spectral radius. (b) The adjacency matrix of the network depicted in (a). Its largest eigenvalue is 3.

of them are neutral, so that the fraction of neutral mutations for this genotype is $d_i/(D-1)L$. The rate of neutral mutations is then this fraction multiplied by the mutation rate, or

$$\nu_i = \mu \frac{d_i}{(D-1)L}. \tag{6.9}$$

The mean neutral mutation rate is then

$$\nu = \sum_{i=1}^{N} \nu_i \frac{m_i}{M} = \mu \frac{\langle d \rangle}{(D-1)L}. \tag{6.10}$$

It is then clear that as a population equilibrates, it will increase ν up until $\langle d \rangle$ reaches ρ, the spectral radius of the network. What can evolution do now? The answer is, evolution can try to rearrange the topology of the network to increase ρ. For example, the "hub"-topology shown in Figure 6.3(a) (with corresponding adjacency matrix in Fig. 6.3[b]) has $\rho = 3$, which could significantly increase the neutral mutation rate. We can then understand that if the mutation rate is so large that a population is threatened with extinction for failure to comply with the bound

$$\mu(1 - \frac{\rho}{(D-1)L}) \leq \ln d, \tag{6.11}$$

then if increasing the number of offspring is not possible, changing the genomic architecture to increase the spectral radius might be an option.

Can such an adaptation to high mutation rates be observed in nature? Before we try to answer that question, let us first look at how this effect was discovered, using the digital organisms I described earlier.

6.2.2   Survival of the "flattest" in digital organisms

At the California Institute of Technology, where the discovery we will be exploring together took place, summer is a busy time when undergraduates look for research projects. The summer undergraduate research program at Caltech is called "SURF." It is a spiffy acronym, but obviously it would work better if Caltech actually had a beach (it does not). Claus Wilke (a postdoc in my lab at the time) and I were supervising a first-year undergraduate student participating in that program: Jialan Wang. When it came to assign her a research question, Wilke thought we might try to see if robustness could evolve in avidian genomes, even though the fitness landscape is not a perfect neutral network as in van Nimwegen's model, and as a consequence, we would not be able to measure how the spectral radius of a neutral network might change. However, we had preliminary evidence (also from Avida) that the neutral mutation rate will increase significantly when avidians are exposed to a high mutation rate (Ofria et al. 2003). Let us take a quick look at those results, before discussing the "survival-of-the-flattest" experiments.

The neutral mutation rate ν can easily be estimated for avidians by noting that the mean population fitness $\bar{w}$ is related to the wild-type fitness w_0 (that is, the prevailing best fitness in the population), by

$$\bar{w} = F_\nu w_0. \tag{6.12}$$

Here, F_ν is the *neutral fidelity* of replication (Ofria et al. 2003).[5] The neutral fidelity is just what the name implies: it is the likelihood that a genotype is unchanged at birth, or if it is mutated, then the fitness is unaltered. The likelihood that a genotype is not mutated at all is the fidelity of replication $F = (1 - R)^L$, where R is the mutation rate per-site and L is the sequence length. If R is small, then we can approximate F as $F = e^{-\mu}$, where μ is the genomic mutation rate $\mu = RL$. If ν is the likelihood that a mutation is neutral, then the neutral fidelity is just

$$F_\nu = e^{-\mu(1-\nu)}, \tag{6.13}$$

which allows us to measure the prevailing neutrality in the population simply as

$$\nu = 1 + \frac{1}{\mu} \ln\left(\frac{\bar{w}}{w_0}\right). \tag{6.14}$$

---

5. Even though this study came out four years after the 1999 paper by van Nimwegen et al., we actually had the experimental results before their paper came out, but we did not have the theory behind the effect.

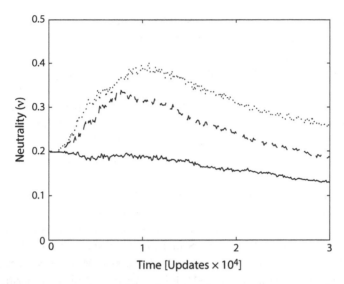

FIGURE 6.4. Increase in neutrality ν [measured in a population according to Eq. (6.17)], over time (measured in population updates). The solid line shows the mean of 100 replicate experiments at a fixed genome mutation rate of $\mu = 0.5$ (all genomes were constrained to $L = 100$). The dashed line shows the neutrality for $\mu = 1.0$, while the dotted line has $\mu = 1.5$. Data from Ofria et al. (2003).

In a set of experiments where the sequence length was being held constant (so that RL could be controlled by changing R only), we could observe that ν decreases if the mutation rate is low ($\mu = RL = 0.5$ in the experiments shown in Fig. 6.4), while it increases significantly when $\mu = 1.0$, and even more so for $\mu = 1.5$. We also see that after the initial increase for $\mu = 1.0$ and $\mu = 1.5$, the neutrality then tends to decrease. This is a result of keeping the sequence length capped at $L = 100$, while evolution "fills" the genome with information. No matter what the mutation rate or how this information is encoded, the neutrality is bound to decrease eventually. So we had clear evidence for "neutrality selection": the phenomenon where an increased mutation rate would favor an increased neutral fidelity. But was this neutrality "protective," in the sense that it provided a real fitness advantage? This was the question Wilke had asked Jialan Wang to test directly using digital evolution.

To test whether high mutation rates could drive organisms toward mutational robustness, Wang used a set of forty digital organisms that we had obtained in a previous study (Lenski et al. 1999). The set of forty was quite diverse in sequence length (ranging from 54 to 314 instructions) and fitness (performing between 20 and 30 different 2-input and 3-input logic operations), and had evolved for 50,000 updates. Wang then made a copy of the set, and let each of the two sets evolve some more, but now at a different (fixed) genomic mutation rate (Wilke et al. 2001). She did this by forcing sequence length to be constant, and adjusting the per-site mutation rate accordingly. The first set (set "A") was readapted for 1,000 generations to a low genomic rate of 0.5 mutations per genome per generation (note that for some genomes, for example the one with length 54, this rate is actually higher than the rate it experienced previously). The second set (set "B") was readapted at $\mu = RL = 2.0$ mutations per generation, a factor four times higher than the control set A.

Not much happened to the avidians in set A during re-adaptation, but the changes in set B were dramatic. After all, as avidians have only one offspring per generation on average, $d = 2$ in Equation (6.5).[6] If all mutations were lethal, a rate $\mu = 2.0$ would doom the population. Even though of course $\nu \neq 0$, there still is significant pressure to increase it. What was surprising to us was that many avidians achieved this increase in neutrality by sacrificing fitness! From the survival-of-the-fittest point of view, this hardly looked like an adaptation, that is, an increase in "fit." Were we just witnessing populations in decline, on the way to extinction? To test this, we decided to take twelve pairs of avidians (one from the A group and one from the B group that once had shared an ancestor) that had experienced a fitness change so dramatic that at the end of the re-adaptation phase, the A fitness was at least twice as high as the B-fitness. Then we pitted them against each other. If the fitness decline in the B group was just a sign of a population in decline, then the B types could never outcompete the A types, at any mutation rate. But if there was a mutation rate where the B type would win the battle (even though they replicate twice as slowly), then the B types must have adapted to the high mutation rate.

To perform this battle, we filled a population of 3,600 organisms half with A-types and half with the corresponding B-types. The battle would be over if only one of the types survives. And to make sure that the outcome of the contest did not depend on evolutionary vagaries, we repeated each battle ten times. The results were astounding. When competed at low mutation rates ($\mu = 0.5$, the "home" rate of the A-types), the A-types won handily, every time. You can see this in the upper left panel of Figure 6.5, where in all ten replicate experiments, the A-types dispatched the B-types swiftly.

This outcome was, of course, not surprising. The A-types are used to that mutation rate after all, and they are able to replicate at least twice as fast as the B-types. We then doubled the mutation rate, making it, in a way, "hotter" for

6. Because the progenitor genome does not die when an offspring is generated, making one offspring is more akin to cell division, hence $d = 2$.

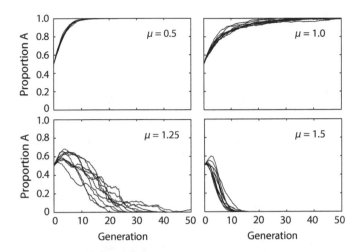

FIGURE 6.5. Result of competition experiments. Fraction of the population of A-types as a function of generation, for four different mutation rates. Adapted from Wilke et al. (2001).

the A-types. At $\mu = 1.0$, the A-types were still winning, but not quite as fast as at their "home temperature" (see upper right panel in Fig. 6.5). Raising the mutation rate just a bit more, to $\mu = 1.25$, revealed the surprise. At first, the A-types appear to be on their way to success, but soon the B-types, adapted for 1,000 generations to a mutation rate of $\mu = 2.0$ and shedding significant replication potential while doing so, were gaining the upper hand (lower left panel of Fig. 6.5) and ultimately defeated the A-types in all ten replicate competitions. At the "home rate" of B-types ($\mu = 2.0$), the end of the competition was swift and merciless for the A-types, as the lower right panel of Figure 6.5 documents. Even with a twofold (at least) disadvantage in replication rate, the B-types won easily. How is this possible?

An analysis of the competition at $\mu = 2.0$ shows convincingly what is going on: the A-types are producing offspring furiously, but at that mutation rate each suffers two mutations on average, and this is sufficient to kill many of the offspring. Dead sequences do not leave offspring, so for the A-types the furious rate of replication does not translate into growth of the clade. The B-types, on the contrary, reproduce more slowly, but a large fraction of their offspring (due to the increased ν) are viable. And it is this viability advantage that ultimately wins the battle.

It is a useful exercise to imagine the changes that affected the B-types during the period of adaptation to high mutation rate in terms of the type of peak they inhabit on the fitness landscapes. We visited the concept of fitness landscapes before in sections 5.4.2 but we did not really discuss the shape of fitness

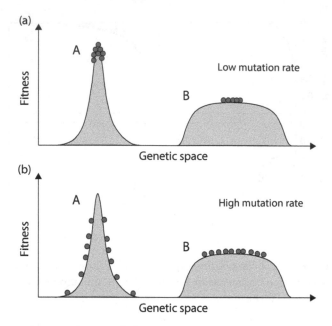

FIGURE 6.6. (a) Fitness peaks (fit) on the left, flat on the right, and the populations inhabiting them (A-types on the left, B-types on the right) at low mutation rate. (b) Fitness peaks and populations at high mutation rate. Adapted from Wilke (2005).

peaks specifically. As discussed earlier, the shape of the fitness peak is really a crude way to depict how evolution unfolds, but nevertheless the picture provides some insights, which we will try to take advantage of here. Pushing all these caveats aside, Figure 6.6 is an attempt to gain intuition about the process at work here. In Figure 6.6(a) we can see a "fit" peak (on the left) and a population (the A-types) inhabiting this peak, as the mutation rate is low. The mean fitness of the population is almost as high as it can be (this is achieved if all the organisms had the same exact peak fitness, which could only happen if the mutation rate vanishes.) On the right side of Figure 6.6(a) we see a flat peak, along with the population of B-types occupying it. The mean fitness of the B-types is lower simply because the highest fitness on the flat peak is lower.

Now we turn up the mutation rate significantly. Figure 6.6(b) shows the A-type population on the same fit peak, but the mutational pressure now forces many of the A-types to occupy states with very low fitness. The B-types on the flat peak in Figure 6.6(b) do not suffer nearly as much because this "flat" peak has a broad plateau so that the mean fitness of the population is barely reduced.

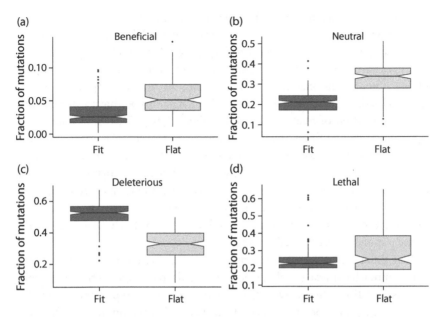

FIGURE 6.7. Relative proportion of mutations of different types on "fit" (dark gray) and "flat" fitness peaks (light gray) for 166 pairs of independent genotypes. (a) Beneficial mutations, (b) neutral mutations, (c) deleterious mutations, and (d) lethal mutations. Adapted from Franklin et al. (2019).

Clearly, if the mutation rate is high enough, the mean fitness of B-types could be larger than the mean fitness of A-types and outcompete that group.

Let us now quantify the shape of the fitness peak by counting the ways in which fitness can decrease from the peak. We will consider four main categories of mutations: beneficials (which of course are quite rare), deleterious mutations (not including lethals), neutrals, and lethals.

The most common mutations among digital organisms are usually neutral and deleterious mutations, followed by lethals and beneficials. However, the relative importance of these depends on whether a population inhabits a "flat" (i.e., a mutationally robust) or a "fit" peak. In Figure 6.7, we can see the relative proportions of the four different categories of mutations for the A-types (the "fit"-peak denizens) and the B-types (the "flat"-peak inhabitants).[7]

7. Note the data in Figure 6.7 comes from a reimplementation of the original 2001 experiment described in Wilke et al. 2001, some seventeen years later (Franklin et al. 2019). The latter study tried to measure the shape of the fitness peaks involved in more detail. It is gratifying to see that even though the underlying software implementing Avida has undergone drastic changes

First, we note that the rate of beneficial mutations is larger among the "flat" population compared to the "fit," which is understandable as the flat types are *descendants* of types with significantly higher peak fitness. This implies that the flat types have previoiusly *descended* from peaks with significantly higher peak fitness. Those additional beneficial mutations are, in a way, opportunities to gain back that fitness. Next, we can see that flat genotypes have a significantly larger proportion of neutral mutations (Fig. 6.7[b]). This is, of course, the signature of the flat fitness peaks: it is the essence of their "flatness." The difference in the deleterious mutation rate between the two types of genotypes that we see in Figure 6.7(c) is also fully expected: the fit genotypes suffer from extensive deleterious mutations. It is largely these mutations that are responsible for the demise of the fit competitors in the duel shown in Figure 6.5 at high mutation rates: those mutated types replicate so slowly (if at all) that the entire population is doomed to extinction.

The difference in lethal mutations between the two types is, on the other hand, somewhat unexpected. A priori we could have imagined that if the deleterious mutation rate is elevated among the nonrobust types, then the lethal mutation rate must be even worse. It turns out that the opposite is the case: the lethal mutation rate among the flat genotypes is higher. The difference is not large, but it is statistically significant (Franklin et al. 2019). But in hindsight, the reason why genotypes that are mutationally robust should increase the rate of lethal mutations is clear: to be robust to mutations, it is imperative to limit the effect of *heritable* deleterious variation. There are two ways to achieve it: one is to decrease the deleterious effect, all the way to making a mutation neutral. But if neither are possible, then that mutation should at least not be inherited, because if it were, then the next generation will be burdened by it. Thus, it is better for the population if a deleterious mutation is instead turned into a lethal one, as in this way it cannot be inherited anymore. This is precisely what we see in Figure 6.7(d), and this is particularly important when population sizes are small, as we will discuss in section 6.3.

It turns out that the survival-of-the-flattest effect is also a consequence of quasispecies theory (see Box 6.2), which itself turns out to be an extension of population genetics to high mutation rates (see Wilke 2005 for this correspondence). The quasispecies theory has been used extensively to describe viral dynamics, for the simple reason that viruses usually replicate in a fairly imprecise manner, leading to comparatively large mutation rates (Domingo et al. 2001; Domingo 2002). Standard population genetics cannot describe such dynamics as it generally considers mutation rates to be infinitesimal.

---

in the intervening years, to the extent that it is unlikely that a single line of code is unchanged, the results are qualitatively (and to a large extent quantitatively) the same.

Box 6.2. Mutational Robustness and the Quasispecies

The concept of the "quasispecies" precedes the idea of "neutrality selection," but encompasses many of the latter's components and predictions. The theory is due to chemists Manfred Eigen (1927–2019), who received the chemistry Nobel Prize in 1967 for discovering how to measure the rate of very fast chemical reactions, and Peter Schuster of the University of Vienna. The term "quasispecies" refers to a group of molecules that behaves as if that group was its own *chemical* species and is unrelated to the species concept of biology. By deriving a set of differential equations that a group of replicating molecules under mutation must obey, Eigen and Schuster were able to show that at high mutation rate the molecules form a stationary (that is, time-independent) "cloud" (Eigen 1971; Eigen and Schuster 1979). The idea is not that the same set of molecules persists over time, but rather that as some sequences are mutated (and thus disappear when they are present in low numbers), other sequences are mutated to replenish those that were lost. In other words, the cloud itself is stable, even when individual genotypes come and go. Rather than assuming that there are only two fitness levels (as in the neutral network model of van Nimwegen et al. 1999), the quasispecies model assumes that beyond the unmutated type (the "wild-type"), there are classes of mutant sequences (defined by the number of mutations they carry) that can be lower in fitness. Usually the fitness landscape is constructed so that the more mutations a sequence carries, the lower its fitness, but even sequences with a large number of mutations can still replicate (there are no

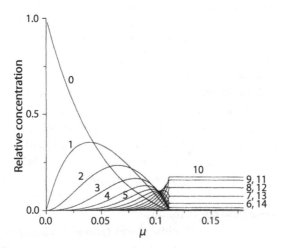

FIGURE 6.8. Relative concentration of sequences labeled by the mutational (Hamming) distance to the wild-type as a function of error probability μ, for binary sequences of length 20 where all mutant sequences have 10 percent of the fitness of the wild-type (after Tarazona 1992).

Continued on next page

Box 6.2. (*continued*)

lethal mutations in this version of the model). The most well-known result from the quasispecies model is the error catastrophe, which predicts the existence of a critical mutation rate (the "error threshold") above which selection fails to keep the mutant cloud stable, and the population instead becomes a random mixture of all possible sequences. We can see this transition in Figure 6.8, which is the depiction of molecular species abundances in a simple model with an error threshold at $\mu = 0.11$ (Tarazona 1992). A less heralded result of quasispecies theory implies that the quasispecies does not have to include the wild-type sequence (Schuster and Swetina 1988) (the sequence with the highest replication rate), which in hindsight is precisely the survival-of-the-flattest effect (Wilke 2005).

Even though the quasispecies theory of Eigen and Schuster (1979) (see also Eigen et al. 1988; Eigen et al. 1989) predates the theory of mutational robustness (van Nimwegen et al. 1999) and survival-of-the-flattest (Wilke et al. 2001), it correctly predicts either effect (Schuster and Swetina 1988) (as it must, given that it is the correct theory of population genetics at finite mutation rates), and it even works at finite population sizes (Forster et al. 2006).

While the origin of the survival-of-the-flattest effect in digital organisms is in hindsight intuitive, it is not immediately clear that the effect must be present in biological organisms too. The effect shows up readily in computational simulations where the organisms are simulated RNA ribozymes that fold into particular structures instead (Wilke 2001).

In that study, Wilke chose two RNA sequences of the same length (sixty-two nucleotides) and simulated their evolution in a flow reactor of 1,000 sequences at different mutation rates (the RNA sequences were folded using the Vienna package; Hofacker et al. 1994). The two different types of sequences he chose (shown in Fig. 6.9) had very different structures, however, which implied significantly different fractions of neutral mutations for each: fold-1 (on the left in Fig. 6.9) has neutrality $\nu_1 = 0.442$, while the more rod-like structure of fold-2 has fewer neutral mutations ($\nu_2 = 0.366$). In this study, Wilke chose to replicate both sequences with different rates: the "peak fitness" (the replication rate in the absence of mutations) of fold-1 (the robust type) was chosen as $w_1 = 1.0$, while the fit (and less robust) fold-2 had a 10 percent larger fitness: $w_2 = 1.1$. He then matched up fold-1 vs. fold-2 types in competition experiments similar to those we saw in Figure 6.5. The result for a population of 1,000 simulated RNA is shown in Figure 6.10. Just as in the digital life experiments, the flat types (here fold-1) overwhelm the fitter fold-2s at a critical mutation rate, which moreover can be calculated cleanly from theory (Wilke 2001).

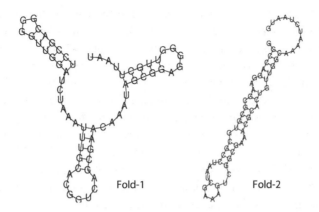

FIGURE 6.9. Secondary structures of simulated RNA sequences of sixty-two nucleotides each, with different fractions of neutral mutations. Fold-1 is more with robust with $\nu_1 = 0.442$, compared to $\nu_2 = 0.366$ of fold-2. From Wilke (2001).

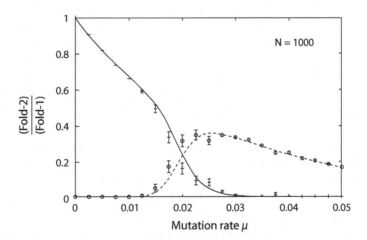

FIGURE 6.10. Relative abundance of fold-2 types over fold-1 types after 200 simulated generations of evolution as a function of the mutation rate μ. All competitions were initiated with an equal mixture of both types. All points represent an average over twenty-five independent simulation runs each, with error bars indicating the standard error. The solid and dashed curves are theoretical predictions. After Wilke (2001).

While this work proved that the survival-of-the-flattest effect was not limited to the digital avidians, and moreover that standard population genetics arguments could predict the effect, it was not clear whether it was observable in real life. It took not too long for such evidence to emerge, however. Likely inspired by Wilke's competition experiment with RNAs, Codoñer et al. realized that two RNA-based viroid species that are capable of infecting the same plant (the chrysanthemum *Dendranthema grandiflora*) have precisely the properties that Wilke used in his simulations: one type that replicates fast but has low neutrality (on account of a more rod-like structure, fold-1 in Fig. 6.11), and one that replicates slowly but with higher neutrality, which we call fold-2 in Figure 6.11. Viroids are RNA-based viruses that infect mostly plants, and can be thought of as the smallest replicating organisms on Earth (some are smaller even than avidians). Their RNAs are circular and fold into stable structures that allow them to infect and ultimately replicate in their host plants. The fit viroid CSVd usually achieves a twenty-fold larger concentration when coinfected with CChMVd (Codoñer et al. 2006; Fig. 6.12), which cements its status as a fast replicator.

To test whether there is a mutation rate that can allow CChMVd to eventually outcompete CSVd, Codoñer et al. conducted one competition experiment on the same plant under normal conditions, and one where the mutation rate was artificially increased via UV radiation. First, they noted that under UV radiation, CChMVd accumulated new mutations at a significantly higher rate than CSVd. Furthermore, they noted that the ratio of CChMVd to CSVd viral haplotypes that they culled from the plant after two, four, and six weeks of exposure was *increasing* under UV treatment, while it was decreasing under the control treatment. This is precisely what the survival-of-the-flattest effect would have predicted, and a computational simulation using the parameters measured from the plants cemented that finding (Codoñer et al. 2006). As far as I can tell, this constitutes the first time a biological experiment has validated an effect that first was discovered using an artificial life form.

Of course, if the effect is general, it should be found in all those forms of life where high mutation rates are a threat to an organism's survival. And while the example of RNA viroids was compelling, these viroids are fairly abstract molecules that seem closer to the simulated RNA loops of Wilke than they are to biological forms. However, a year after the viroid study came out, another team was successful with a study involving the *Vesicular Stomatitis Virus* (VSV), a common insect-transmitted virus that infects mammals. VSV belongs to the family *Rhabdoviridae* (which includes the rabies virus). It is a negative-strand RNA virus with an 11,161 nucleotide genome that encodes five major proteins. It turns out that the mutation rate of VSV is of the order of 10^{-4} per site per generation, which translates to a genomic mutation rate larger than 1.

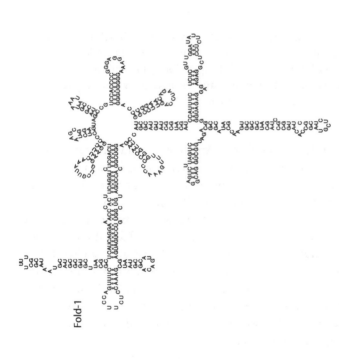

Fold-1

Fold-2

FIGURE 6.11. Secondary structure of two RNA viruses (viroids) that infect chrysanthemums. Fold-1 is the mutationally robust *Chrysanthemum chlorotic mottle viroid* (CChMVd), while fold-2 represents the fitter *Chrysanthemum stunt viroid* (CSVd). Adapted from Cho et al. (2013).

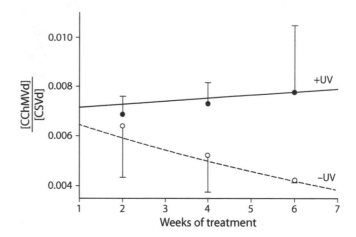

FIGURE 6.12. Competition experiments between robust CChMVd (fold-1) and fit CSVd (fold-2) under normal and mutagenic conditions. Solid dots represent treatment under UV radiation, while open dots represent the control treatment. Every dot is the median of the ratios estimated for three replicate plants. Adapted from Codoñer et al. (2006).

In order to do a survival-of-the-flattest experiment with VSV, ordinarily you should do an experiment where one of the strains is adapted to a high mutation rate for a significant amount of time so that mutational robustness can evolve. In the experiment of Sanjuán et al. (2007), the team instead used a pair of virus strains where one was a strain that was an artificial assemblage of a number of different strains that had never seen any laboratory evolution before—this will be strain A: the "fit one." Strain B, instead, is an isolate that had been replicating in human HeLa cells in the laboratory for hundreds of generations (Sanjuán et al. 2007). As both population size and mutation rate are different in the lab setting, the team hoped that strain B would show signs of mutational robustness. The two strains differed by fifty-four nucleotide substitutions throughout the genome.

To test the hypothesis that B is mutationally robust, Sanjuán et al. first tested the distribution of fitness for each strain by measuring plaque sizes for twelve randomly selected strains of each type. The resulting distribution shown in Figure 6.13 is rather telling: while the means in fitness between the two types is not significantly different, their distribution differs markedly: compare the variance of the log-fitness between types $\mathrm{var}(w_A)/w_A = 2.054$ vs. $\mathrm{var}(w_B)/w_B = 0.225$. While strain A displayed a large variance in its fitness distribution, the amount of sequence diversity within the strain was less than half (Sanjuán et al. 2007). This is surprising because usually a large variance in sequence diversity is mirrored by a large variance in fitness, but this

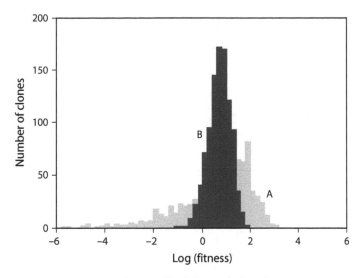

FIGURE 6.13. Distribution of 1,000 (log) fitness values for strains A and B based on plaque sizes (strain A is shown in light gray and strain B in dark gray). Adapted from Sanjúan et al. (2007).

was not the case here. Clearly some mechanism is limiting the expression of the increased diversity in strain B. But is it mutational robustness? To test this, Sanjúan et al. performed what is now the standard test: Can the (slight) fitness advantage of strain A be turned into a decisive disadvantage upon increasing the mutation rate?

Rather than subject the viruses replicating within animal cells (specifically, hamster kidney cells, a standard cell line from the American Type Culture Collection) to increased UV radiation, the team increased mutation rate via a mutagenic chemical agent, called 5-FU. This agent is actually an artificial nucleotide (fluorouracil) which, when incorporated into the nucleotide sequence during DNA synthesis, creates mismatches and thus mutations. To make sure that any differential reaction to the drug was not specific to the drug, the team also used another chemical agent (AzC, another nucleoside analog). Figure 6.14 shows that the relative fitness of the B strain divided by the A strain keeps increasing with increasing drug dose (for both mutation-inducing agents). This finding alone suggests that the B-strain, which after all replicates at $w_B/w_A \approx 0.75$ in the absence of mutagens,[8] takes over at larger mutagen concentrations where the B-strain grows about twice as fast the A-strain.

8. Note that the population growth rate is not the same as the mean fitness of samples (which was not statistically different in Fig. 6.13) because there is no competition between the samples.

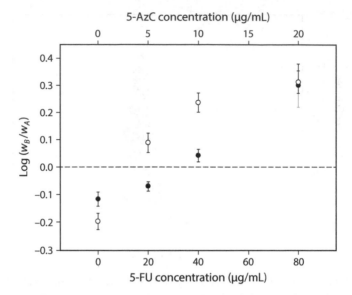

FIGURE 6.14. Competition experiments between B-strain and A-strain VSV clones. Data points are the log-fitness of the B-strain population relative to the A-strain population $\log(w_B/w_A)$, with error bars showing 95% confidence intervals. Open circles were obtained with 5-FU (lower scale), while filled circles represent data when changing 5-AzC concentration (upper scale). Adapted from Sanjuán et al. (2007).

However, another effect may be at work as well. Sanjúan et al. surmised that a population that was robust to mutations might also be more robust to genetic drift, that is, the loss of fitness due to the randomness of selection at small population sizes. Indeed, when passaging the different strains at the smallest population size that was experimentally feasible for twenty-five generations, they noticed that the B-strain hardly lost any fitness in twenty-four replicate experiments, while the A-strain (the "fit-but-not-flat" type) lost a significant fraction of fitness to drift (see Fig. 6.15). They thus found that the B-strain was not only robust to mutations, but also could withstand fitness loss via drift! How is this possible? Does drift robustness always go hand in hand with mutational robustness, or is it a different effect? We will explore this question in the next section.

6.3 Evolution of Drift Robustness

We have just seen that mutational robustness is achieved by genomes rearranging their genetic code in such a manner that the relative rates of mutations (neutral, deleterious, lethal) are changed toward more neutrals, fewer

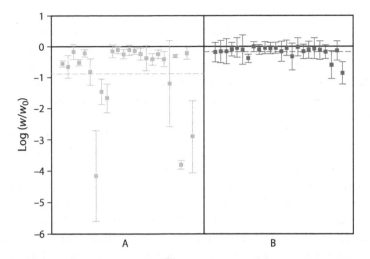

FIGURE 6.15. Change in log-fitness in mutation accumulation lines (small populations started with a single clone) for twenty-four replicate lines derived from a randomly sampled A-strain type (light-gray squares), and for twenty-four replicates of a randomly sampled B-strain type (dark-gray squares). Error bars indicate 95% confidence intervals from an extrapolation of fitness from plaque sizes. Dashed horizontal lines show the grand mean change in log-fitness for each of the strains, respectively. The solid line indicates no loss of fitness. Adapted from Sanjuán et al. (2007).

deleterious, and possibly increased rates of lethal mutation. Often, the increase in neutrality is achieved by losing code that achieves a fitness benefit (that is, losing information), thus trading fitness for neutrality. Can a change in the relative rates of different mutation classes also achieve robustness to drift?

To investigate this question, let us first define genetic drift. Basically, genetic drift is a process that leads to random changes in the genetic composition of a sequence as a consequence of relaxed (or absent) selection (Masel 2011). In drift, alleles are fixed irrespective of their fitness effect (within reason, depending on the size of a population). We began this chapter by discussing neutral evolution and saw that the likelihood that a neutral mutation goes to fixation is about $1/N$, where N is the population size. So, under neutral selection, alleles are fixed *by chance*. Therefore, genetic drift happens when alleles become common not due to their effect on fitness, but rather due to luck. What we will see is that when the population size becomes small, even mutations that lead to serious fitness defects can fix as if they were neutral.

While there are several conditions that can give rise to genetic drift in molecular evolution, the most prominent among them is a small population size. The reason for this is clear: natural selection is a population property. The smaller the population size, the weaker the strength of selection. Theoretically, even a population size of two individuals (in the haploid case) can give rise to selection, as there is competition between the two types at any point in time. However, in that case, selection is extremely weak because the fitter (in this case, faster replicating) type can be removed by chance effects half of the time. If the population size dwindles to one, however, natural selection ceases completely.

A good rule of thumb is that in a population of size N any fitness advantage (or disadvantage) of size $s \approx 1/N$ (or smaller) is all but "invisible" to selection. This means that beneficial mutations of this size are treated just as neutral mutations (and thus go to fixation with a rate $1/N$ instead of a rate $2s$), and at the same time deleterious mutations of size $s \approx 1/N$ will fix with a rate $1/N$, while under strong selection (very large population size), the probability of fixation of a deleterious mutation tends to zero. Thus under drift, beneficial mutations rarely go to fixation and achieved fitness can be lost (via the fixation of a deleterious mutation). Exposed to significant drift over time, populations can go extinct because they will lose one allele after the other (with barely a chance of recovering via a beneficial mutation in between). This is the principle of Muller's ratchet (Muller 1964; Haigh 1978; Lynch et al. 1993).

Is it possible that two different genetic architectures with essentially the same fitness can respond differently to drift, as Figure 6.15 seems to suggest? We can test this in digital evolution, using avidians adapted to survive in the harsh conditions of small population size. For this experiment (LaBar and Adami 2017), we evolved one hundred replicate populations at small population sizes (100 individuals) and another 100 populations at a large (10,000 individuals) population size. We aimed both of these populations to achieve about the same levels of fitness so that there is no bias due to the starting level of fitness. To achieve that, it was necessary to allow the smaller population to adapt longer, because the rate of beneficial mutations scales with the population size. Thus, we allowed the small populations to evolve for a million generations, while the large populations adapted for only 10,000 generations. (Rest assured, it turns out that giving both populations the same amount of time to adapt instead does not affect the results; see LaBar and Adami 2017).

For these experiments, adaptations were limited to optimizations in the gene for self-replication (the "copy-loop"), and no logical tasks were being rewarded. To test whether there was a difference in how each of these types (we will call them the "smalls" and the "bigs") respond to different population sizes, we took the most abundant genotype from each population (100 "smalls" and 100 "bigs") and measured how these types lost fitness when transplanted into an environment with strong genetic drift due to decreased population size. Figure 6.16(a) shows the fitness loss after 1,000 generations

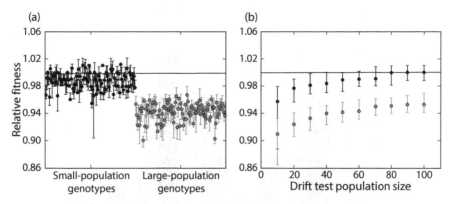

FIGURE 6.16. Change in fitness during the drift robustness test for small- and large-population genotypes. Black (gray) markers are for small-population (large-population) genotypes. (a) Relative fitness before and after 10^3 generations of evolution with test population size of fifty individuals. Circles represent the median value of ten replicates for one genotype; error bars are first and third quartile. (b) Same experiment as in (a), but across a range of test population sizes. Each circle is the median of 1,000 replicates (100 genotypes × 10 replicates), error bars as in (a). From LaBar and Adami (2017).

of evolution in a population of only fifty individuals. The "smalls," it turned out, barely lost any fitness at all even though the population size was half of what they were adapted to (black dots on the left in Fig. 6.16). The "bigs," on the other hand, clearly suffered (light gray dots in the panel on the right). Even though the "bigs'" fitness was essentially the same as the fitness of the "smalls" upon transplantation, drift during the 1,000 generations of adaptation removed over 5 percent of their fitness. When we repeated this experiment for different population sizes, from the native size of the "smalls" of 100 individuals down to only 10 (Fig. 6.16[b]), the trend was clear: the "smalls" had acquired a robustness to drift that the "bigs" did not have.

What makes the "smalls" so robust? The lesson that mutational robustness teaches us is to take a look at the mutation classes. In Figure 6.17 we can see the frequency of five different mutation classes for the "smalls" vs. the "bigs." The "smalls" have a somewhat increased beneficial mutation rate and a significantly increased rate of neutral mutations, compared to the "bigs." For the purpose of analyzing drift robustness, we split the deleterious mutations into two classes: deleterious mutations with small effect (less than a 5 percent change in fitness), and the rest, which we call large-effect mutations. The reason for considering these mutations separately is that, as we discussed earlier, small-effect mutations are the most dangerous for small populations:

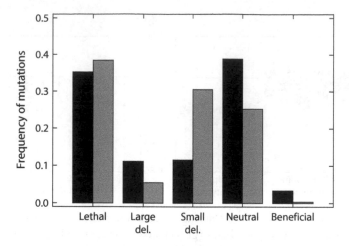

FIGURE 6.17. Differences in mutational effects between small-population genotypes (black) and large-population genotypes (gray). Here as before, "small" refers to a population size of $N = 100$, while large-population genotypes evolved in $N = 10,000$. From LaBar and Adami (2017).

once small advantages are lost to drift, they are unlikely to be gained back because such a mutation would have to have at least a size $s > 1/N$. Indeed, as Figure 6.17 shows, the fraction of small-effect deleterious mutations is significantly reduced among the genotypes that adapted to a small population size. The rate of lethal mutations is not significantly different between "smalls" and "bigs" in this experiment (but we could see an increase in lethals for small-population genotypes in different evolutionary settings).

That the decreased rate of small-effect mutations explains the drift robustness of the "smalls" can easily be shown by noting that the genotypes that lost the largest amount of fitness in the drift-test in Figure 6.16 are precisely those with the largest fraction of small-effect mutations (see LaBar and Adami 2017). But the increased fraction of beneficial mutations suggests that, perhaps, those drift robust genomes have also traded fitness for robustness. This would be an explanation for drift robustness that is similar to the selection for mutational robustness, in which populations shift from mutationally fragile (fit) peaks to less-fit but more robust peaks.

But in the case of drift robustness, there is an alternative explanation. Because mutations with a small benefit cannot permanently go to fixation in small populations (as they are effectively neutral), perhaps drift robustness is achieved by these small-population genotypes not by "ending up" on drift robust peaks after shedding fitness, but instead by preferentially *adapting to* existing peaks that are drift robust. To test this hypothesis, we can construct a simple model to tease out the differences between these two explanations. In this model (see Box 6.3) we assume that a population of genomes (all with the same fitness $w_0 = 1$) is faced with new opportunities in the form of two peaks: one with a steep slope that can be reached in one step (the "drift-robust" peak in black in Fig. 6.18[a]) and another peak with a shallow slope that, however, requires two mutations to climb. We also assume that the robust peak is somewhat less fit (peak height $w_4 = 1 + s - \epsilon$).

Obviously, if the population size is infinite, the population will end up at the taller peak with certainty. Is there a smaller population size that will favor the less fit but more robust peak? The mathematical model in Box 6.3 predicts that yes, this will happen. In particular, the model predicts that the bigger the fitness deficit (the larger ϵ), the smaller the critical population size at which the switch happens (the critical point is $R = 1$, where the equilibrium population size at peak 3 and at peak 4 are equal). Figure 6.19(a) shows the critical population size (where $R = 1$) according to Equation (6.19) as a function of the fitness deficit ϵ. As we would have guessed intuitively, the larger the fitness deficit of the robust peak, the smaller the population size where a switch to a lower-fitness peak becomes advantageous.

The other important parameter that determines the critical population size is the ratio between the beneficial mutation rate and the "slope" of the robust fitness peak s. In a sense, the "competition" between the robust and fragile peaks plays out as a trade-off between two competing pressures: a more shallow peak (with slope $s/2$) will lead to a higher probability of fixation and a higher likelihood for the spontaneous generation of such a mutation (because the distribution of beneficial mutation sizes are exponentially distributed; see Exercise **6.4**). However, if it takes n mutations to achieve the same ultimate fitness benefit (in a model where the slope of the fragile peak is s/n, see LaBar and Adami 2017) then these larger probabilities must be multiplied, and each being smaller than one, could potentially lead to a small overall likelihood of fixation. Indeed, Figure 6.19(b) shows that when the slope is more shallow (but at the same height, thus requiring more steps to reach), the critical population size is commensurately larger. In other words, the shallower the high peak, the more likely it is that a population will prefer the lower peak with steeper flanks.

We have now seen, both using digital experimentation and a simple population genetic theory, that populations can be robust to drift by adapting to the kind of fitness peak (with steep slopes) that prevents fitness loss in small populations. We have also seen, however, that this effect can occur as a by-product of adaptation to high mutation rates (see the drift-test experiment with VSV in Fig. 6.15). What if a population has to deal with both small populations *and* high mutation rates?

Box 6.3. Population Genetics of Drift Robustness

To calculate which peak is most likely chosen by a population adapting to a fitness landscape in which there are two alternative peaks to climb, we use a Markov process in which the transition probability T_{ij} between four different nodes in a network is determined by the likelihood u_{ij} that a mutation from type $i \rightarrow j$ takes place, multiplied by the probability π_{ij} that a successful such mutation goes to fixation (see Fig. 6.18[b]). In this landscape, the four nodes have fitness $w_1 = 1$, $w_2 = 1 + \frac{s}{2}$, $w_3 = 1 + s$, and $w_4 = 1 + s - \epsilon$. The peak with w_3 can be reached by two consecutive mutations of effect size $s/2$, while the lower peak with w_4 needs a mutation of about twice that size, and we therefore expect that peak to be more robust to drift. The question we ask is whether there is a critical population size N_{crit} below which evolutionary dynamics prefers the robust but somewhat less fit peak over the fit peak. In this Markov model, the change in population from time $t \rightarrow t + 1$ is determined by a transition matrix T. More precisely, let $\vec{x} = (x_1, x_2, x_3, x_4)$ represent the vector of concentrations at time t, with $\vec{x}(t=0) = (1, 0, 0, 0)$. The iteration $\vec{x}(t+1)$ is then obtained as $\vec{x}(t+1) = T\vec{x}(t)$, with

$$T = \begin{bmatrix} 1 - u_{12}\pi_{12} - u_{14}\pi_{14} & u_{12}\pi_{12} & 0 & u_{14}\pi_{14} \\ u_{21}\pi_{21} & 1 - u_{21}\pi_{21} - u_{23}\pi_{23} & u_{23}\pi_{23} & 0 \\ 0 & u_{32}\pi_{32} & 1 - u_{32}\pi_{32} & 0 \\ u_{41}\pi_{41} & 0 & 0 & 1 - u_{41}\pi_{41} \end{bmatrix}.$$

The equilibrium distribution $\vec{x}^* = (x_1^*, x_2^*, x_3^*, x_4^*)$ for which $\vec{x}^* = T\vec{x}^*$ is given by the (left) eigenvector of T with eigenvalue 1. Solving for the ratio between the population density at node 3 and node 4: $R = x_3^*/x_4^*$ gives

$$R = \frac{u_{41}\pi_{41}u_{12}\pi_{12}u_{23}\pi_{23}}{u_{14}\pi_{14}u_{21}\pi_{21}u_{32}\pi_{32}}. \tag{6.15}$$

Introducing the ratio of fixation probabilities (calculated using Kimura's fixation formula in Box 6.1)

$$P_{14} = \frac{\pi_{14}}{\pi_{41}} = e^{2(s+\epsilon)(N-1)}, \tag{6.16}$$

$$P_{12} = \frac{\pi_{12}}{\pi_{21}} = P_{23} = e^{s(N-1)}, \tag{6.17}$$

we obtain

$$R = \frac{u_{41}}{u_{14}}\frac{u_{12}}{u_{21}}\frac{u_{23}}{u_{32}}P_{12}P_{23}/P_{14} \equiv Me^{2\epsilon(N-1)}, \tag{6.18}$$

with $M = \frac{u_{41}}{u_{14}}\frac{u_{12}}{u_{21}}\frac{u_{23}}{u_{32}}$. Writing M in terms of the beneficial mutation rate $M = \frac{u_b}{s} \equiv \kappa < 1$ as determined in Exercise **6.4**, the critical population size (shown in Fig. 6.19) becomes

$$N_{\text{crit}} = 1 + \frac{\log \kappa^{-1}}{2\epsilon}. \tag{6.19}$$

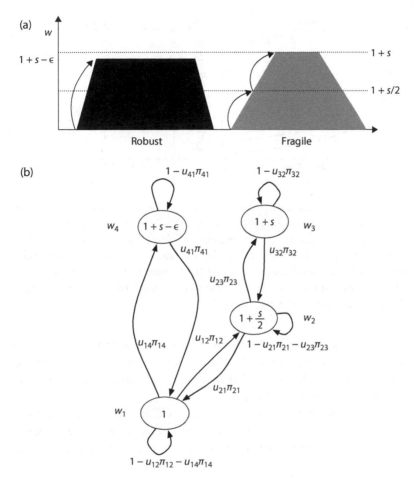

FIGURE 6.18. (a) Two fitness peaks in the drift robustness model discussed in Box 6.3. The robust peak has a steep slope that requires a mutation of benefit $s - \epsilon$ to reach the peak, while the fit-but-fragile peak has a shallower slope $s/2$ that requires two mutations to reach an equivalent fitness level. (b) In terms of a Markov model, the transition probabilities between the four different nodes of the network are indicated next to the arrows. Adapted from LaBar and Adami (2017).

Because the effects of mutational and drift robustness appear to be so intertwined and depend on how mutations interact, to disentangle the two we need a model with a complexity somewhere in between the simplicity of the Markov model in Box 6.3 and the comparative complexity of a digital experiment with self-replicating computer programs. Typically, this would be a

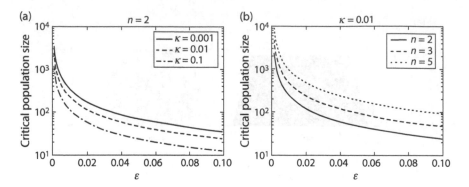

FIGURE 6.19. (a) Critical population size at which a population will switch from the fit to the drift-robust peak, as a function of the fitness deficit of the robust peak, ϵ, for three different ratios of the beneficial mutation probability and the effect size, $\kappa = p_b/s$. In this model, it takes two steps to reach the fit peak and only one to reach the robust peak. (b) Critical population size as a function of ϵ for landscapes that take two, three, or five steps to reach the peak, for $\kappa = 0.01$. Adapted from LaBar and Adami (2017).

model where we can quantify the fitness landscape in terms of basic mutational parameters such as the effect-size of a mutation, as well as how those mutations interact (the degree of epistasis). It is not difficult to construct such a model (Sydykova et al. 2020).

Consider a genome that consists entirely of L binary loci. We will assume that one particular sequence has fitness $w(0) = 1$ (the wild-type), whereas a k-mutant of that sequence (k mutations anywhere on the sequence of L) has fitness

$$w(k) = e^{-sk^q}. \tag{6.20}$$

In this model, all mutations are deleterious, and we are therefore discussing epistasis between deleterious mutations only.

In Equation (6.20), s is related to the one-mutant loss of fitness ($s = -\log w(1)$), and q is a parameter that characterizes the degree of *directional epistasis* between mutations. We discussed pairwise epistasis in chapter 5 (in particular, Box 5.3). Directional epistasis quantifies whether *on average* the pairwise epistasis of any two mutations is positive or negative. With a fitness function such as (6.20), the parameter q allows us to *specify* the average epistasis between pairs of mutations. Clearly, if $q = 1$, all mutations are independent (no epistasis between mutations) because $w(k) = (e^{-s})^k$, that is, the product of k single-point mutant fitnesses. If $q > 1$, the fitness effect of a pair of mutations is worse than what is expected from each of the single mutants, that is, epistasis is *negative* (fitness is less than expected from the product of single mutations). For example, if we calculate the epistasis parameter ϵ as introduced in Box 5.3 for a pair of mutations, we obtain

$$\epsilon = \log\left(\frac{w(2)}{w(1)w(1)}\right) = 2s(1 - 2^{q-1}), \qquad (6.21)$$

which is negative if $q > 1$. For interacting deleterious mutations, such negative epistasis is also sometimes called *synergistic* epistasis, as the mutations are enhancing their combined fitness effect. If $q < 1$, on the contrary, epistasis is *positive*: the joint effect of mutations gives a fitness that is higher than expected from the product of single mutations alone, and $\epsilon > 0$. When mutations buffer each others' deleterious effects, the resulting epistasis is called *antagonistic*.

We will now try to determine the equilibrium fitness of a population of sequences with fitness function (6.20), as a function of population size and epistasis parameter q. The difference to the calculation in the previous section is that there, we considered the competition between two peaks with different slopes, while here we are fixing the slope (all peaks have the same s) and are only changing epistasis and population size.

To calculate the mean fitness, we need to know the proportion of k-mutant sequences in the population $p(k)$, which in turn is related to the probability that the wild-type is replaced by a k-mutant, that is, the probability of fixation of a k-mutant. In the "weak mutation–strong selection" limit, that is, the limit where one mutation goes to fixation at a time, we can use the fixation probability of Sella and Hirsh (2005) to calculate that probability. According to that theory, the probability that a mutant with fitness $w(k)$ replaces the wild-type with fitness 1 is (for a haploid Wright-Fisher process)

$$P(0 \rightarrow k) = \frac{1 - 1/w(k)^2}{1 - 1/w(k)^{2N}}. \qquad (6.22)$$

Note that this formula for the fixation probability is just a special case of Equation (6.3) that we encountered earlier. Here we are looking for the probability that a deleterious mutation replaces a resident *fitter* type, that is, the likelihood of fitness *loss* via drift, rather than the likelihood that a beneficial mutation with advantage s takes over the population.

To calculate the stationary distribution of k-mutants $p(k)$, we have to use the transition probability $P(0 \rightarrow k)$ in a Markov process. This is equivalent to solving the detailed balance equations for genotypes in this system (see Exercise **6.5**), with the result

$$p_k = \binom{L}{k} w(k)^{2N-2}/Z, \qquad (6.23)$$

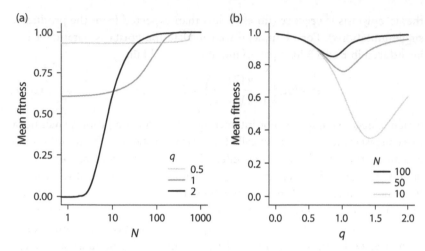

FIGURE 6.20. Mean expected fitness of a population at mutation-selection equilibrium, for a fitness landscape given by (6.20), as a function of population size and directional epistasis, for $s = 0.01$. (a) Equilibrium fitness [Eq. (6.25)] as a function of population size N for three different epistasis parameters: strong negative epistasis ($q = 2.0$, black), no epistasis ($q = 1.0$, gray), and strong negative epistasis ($q = 0.5$, light gray). (b) Equilibrium fitness as a function of epistasis q, for three different population sizes (light gray: $N = 10$, gray: $N = 50$, black: $N = 100$). Adapted from Sydykova et al. (2020).

where

$$Z = \sum_{k=0}^{L} \binom{L}{k} w(k)^{2N-2}. \tag{6.24}$$

The equilibrium mean fitness of the population is then

$$w_{eq} = \sum_{k=0}^{L} p_k w(k) = \frac{\sum_{k=0}^{L} \binom{L}{k} w(k)^{2N-1}}{\sum_{k=0}^{L} \binom{L}{k} w(k)^{2N-2}}, \tag{6.25}$$

with $w(k)$ given by Equation (6.20).

Let us study how this expected mean fitness of a population depends on population size and on the epistasis parameter q. Figure 6.20(a) tells part of the story right away. We'll first focus on the light gray line in that figure, which is the equilibrium fitness when directional epistasis is strongly antagonistic ($q = 0.5$). As population size drops, some fitness is lost to drift, but not much. Moreover, the fitness drop remains small no matter how small the population size gets. Clearly, this amount of positive epistasis has created a fitness peak with a very gentle decay: after a few mutations, fitness is almost constant. Peaks of that shape are reminiscent of the "flat" peaks we saw earlier in the context of mutational robustness.

If mutations are not epistatic (this is $q = 1$, the gray line in Fig. 6.20[a]), the effect of population size is much more pronounced. Without the buffering effect of positive epistasis, every mutation drops an equal amount of fitness s, and when the population size becomes really small (of the order of ten individuals), with $s = 0.01$ almost half of the initial fitness is lost. But that is not nearly the worst-case scenario: that is reserved for fitness peaks with strongly synergistic epistasis, because epistasis of that sort (in Fig. 6.20 you can see the case of $q = 2$) creates peaks that lose significantly more than a fraction s of fitness per mutation. It is very difficult to maintain fitness on such peaks, and we can see that as population size dwindles, fitness approaches zero: catastrophic extinction.

Let us look at this data from another angle. Given a particular population size, what is the level of average directional epistasis that is best for a population to thrive? Figure 6.20(b) reveals somewhat of a surprise: there is a *worst* amount of epistasis, a level that populations of a given size need to avoid. This "worst epistasis" is indicated by a minimum in the equilibrium fitness of the population. The location of that minimum depends on the population size, and this calculation implies that for large populations, the worst epistasis is on the antagonistic size ($q < 1$) while for small populations the minimum occurs for synergistic epistasis ($q > 1$).

This observation is quite peculiar. If there is a minimum in expected fitness for a population at equilibrium, we anticipate that evolutionary forces will attempt to increase the fitness by changing those parameters that led to the depressed fitness. In this case, the responsible parameter is not under direct control (there is, after all, no "epistasis trait"). But clearly, it should be possible for evolution to find alternative fitness peaks for which directional epistasis is different, simply because genetic space is vast and there are peaks of all kinds in the "vicinity" (in terms of mutational distance) of most peaks. But which way should a population go: increase q so that the slopes of the fitness peak become steeper thus increasing robustness to drift, or else decrease q so that the landscape becomes more neutral? It turns out that this decision is tantamount to asking whether you should be more robust to mutations, or more robust to drift. Let us explore this by letting evolution make that decision!

We already know how a population can change its robustness in this model, namely by changing the s in Equation (6.20), where higher s indicates lower robustness. The trick is to compress information as much as possible so that the remaining (unused) sequence is neutral to mutations or, if sequence length is not limited, to simply add neutrality by adding non-functional code to the genome. But how does a genome change q? In general, this is a simple

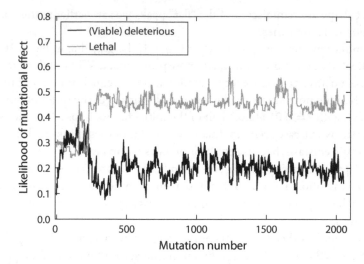

FIGURE 6.21. Fraction of deleterious (black) and lethal (gray) mutations as a function of the mutation number on the line of descent. A single mutation creates a change in architecture (around mutation number 200) that changes how other mutations interact with each other, increasing the fraction of lethal mutations significantly. Because this population was small (100 individuals), this increase in lethal fraction was adaptive and is maintained. Adapted from LaBar and Adami (2017).

question with a complicated answer. There are many different ways in which the same task can be genetically encoded, each with a different degree of epistasis. Sometimes single mutations can change how many sites in the sequence interact with each other, for example by significantly increasing the likelihood that a mutation is lethal. Figure 6.21 shows one such event from the set of evolutionary runs investigating drift robustness (LaBar and Adami 2017). In this particular evolutionary history (fitness as a function of the mutation number on the line of descent), a single mutation (about 200 mutations into this particular history) dramatically changes the effect of mutations for the whole genome, turning a significant fraction of mutations that were strongly deleterious into outright lethal mutations. Clearly, that mutation had a strong effect on a large fraction of the genome, which is why such mutations are said to have a "global" effect (Rajon and Masel 2011). There really is only one way to achieve such a global change in the deleterious effect of mutations, and this is by changing how mutations interact (because, as we have discussed several times before, robustness and epistasis are not independent; see Wilke and Adami 2001). In other words, that mutation significantly changed the q of the local peak.

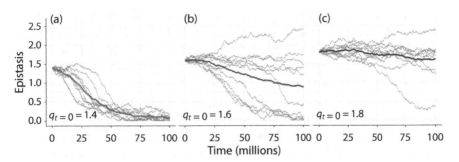

FIGURE 6.22. Results from experiments testing the dynamic evolution of epistasis, for a population size $N = 100$, a mutation rate $\mu = 0.01$, and a deleterious effect size $s = 0.01$ (for these parameters, the critical q at which the equilibrium fitness (6.33) is smallest is about $q_\star \approx 1.1$). The thick line is the average of all ten replicates. (a) Ten replicate runs starting with a population initialized with $q_{t=0} = 1.4$. (b) Ten replicates with $q_{t=0} = 1.6$. (c) Ten replicates with $q_{t=0} = 1.8$. Adapted from Sydykova et al. (2020).

In our simple model described by Equation (6.20), such subtle changes in average directional epistasis are described by a single parameter q, so for this model we will simply assume that q can change as if it was encoded by a genetic locus, that is, as if q was a genetic trait. If we treat epistasis like that, we can assume that the value of q can be changed by a mutational process, determined by a rate μ_q that changes the trait by a certain amount Δq, every time that q is changed. To test whether evolution prefers a high or low value of q depending on where you start (the hallmark of a bifurcation), we ran experiments where the population was started with a fixed chosen q for each individual, and then let the population adapt for a while (200,000 updates). After that, the value of q was allowed to change via mutations on the "q"-locus that can change q by small amounts, at a low rate. We can see ten replicate trajectories starting at different values of $q_{t=0}$, that is, different initial values in Figure 6.22. There does indeed seem to be a tendency for bifurcation (populations starting with a low q move toward lower q, while populations with a high initial q move toward higher q) but the critical point that, according to theory (for this set of parameters) should be about $q_\star \approx 1.1$, is different, and moreover not every trajectory that starts above the critical point moves toward higher q. The reason for this discrepancy is somewhat complicated.

Looking at Equation (6.20) we can see that it is immediately beneficial for each individual to *decrease* q, even if for the population as a whole such a decrease leads to a *decrease* in the mean fitness for the sequence's *offspring*. In a sense, this situation is akin to the dilemma of cooperation in evolutionary games, such as the "Prisoner's Dilemma" that we will discuss in detail in section 8.1 (see, for example, Axelrod 1984; Hofbauer and Sigmund 1998; Nowak 2006; Adami et al. 2016 and in particular Box 8.1), but it is not quite the same because if a sufficient number of genotypes have achieved a higher q, that population cannot be invaded by lower-q individuals, as the cooperators in the Prisoner's Dilemma can. Still, it appears that higher q is beneficial for the group as a whole, but not for each individual.

If we take the bifurcation at face value (even though the experimental evidence in Fig. 6.22 from Sydykova et al. 2020 is not fully conclusive) this suggests that when mutation rates are high and population sizes are small at the same time, what threat the population is responding to predominantly depends on where it finds itself, from the point of view of the prevailing level of epistasis. If epistasis is predominantly synergistic ($q > 1$), the population tends to move to protect itself from drift. But if q is not high enough, it appears to be more advantageous to decrease q even further, and in so doing increase neutrality to such an extent that the population becomes mutationally robust. Thus, from that point of view and under those circumstances, drift robustness and mutationally robustness are alternative fixed points for evolutionary dynamics.

6.4   Mutational and Drift Robustness in Trypanosomes

In this section I engage in the speculation that both drift robustness and mutational robustness are at work in protecting the genome of certain unicellular eukaryotic parasites from extinction, given that they face what appear to be dire evolutionary circumstances. The section is speculative because to make the case for this hypothesis conclusively involves difficult experiments that have either not yet been completed or have yet to be begun. But the evidence that we do have is quite suggestive, and the story is so fascinating that it is worth telling it here—even though the story is far from complete.

As we discussed in the previous section, a small population size can be very problematic for genetic information, in particular if the population size is small for a prolonged period, or when the small population size is recurring—as would happen in periodic bottlenecks. The risk for extinction via Muller's ratchet is particularly strong for genes that do not undergo recombination (because recombination can halt the ratchet by reconstituting lost genes, see for example Felsenstein 1974; Maynard Smith 1978; Takahata 1982). Indeed, this argument can be made mathematically (Lynch et al. 1993): according to those calculations, mutations with a small effect will lead to extinction within just a few hundred generations, as Figure 6.23 implies. That figure suggests that deleterious effects of the order of a tenth to 1 percent are particularly

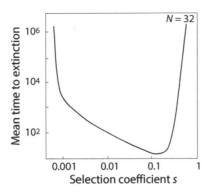

FIGURE 6.23. Mean time to extinction as a function of the deleterious selection coefficient s in an asexual population of $N = 32$ individuals, in a simple model where fitness after n mutations is given by $w(n) = (1 - s)^n$, after Lynch et al. (1993); Gabriel et al. (1993).

dangerous for small populations, leading to extinction within hundreds of generations (in this case for a population of $N = 32$ individuals). This prediction seemingly rules out the existence of a species such as *Trypanosoma brucei*, which undergoes bottlenecks significantly smaller than this, as we shall see. Yet, the species is one of the oldest lines of single-celled eukaryotes that we know of, with a lineage going back at least 300 million years (Haag et al. 1998), even though its current mode of transmission is more recent than that. How can these two findings be reconciled?

Let us spend some time getting to know *T. brucei*. It is a single-celled eukaryotic parasite that is the causative agent of sleeping sickness in sub-Saharan Africa. It has a genome of about 35 Mb (megabases) arranged in eleven major pairs of nuclear chromosomes (Jackson et al. 2010). While *T. brucei* is diploid and retains the genes necessary for a complete sexual (meiotic) cycle, it actually reproduces asexually (Weir et al. 2016) making it susceptible to the danger of extinction via Muller's ratchet if the population size is small. *T. brucei* is transmitted to mammals via the tsetse fly (see Fig. 6.24), and this lifestyle of essentially living "two different lives" (in the two different hosts) represents a major challenge.

First of all, the environment for the parasite in the insect, compared to the mammalian environment, could not be more different. Within the mammalian bloodstream, glucose is available in abundance and so the parasite can use glycolysis to generate energy. The energy landscape is very different when inside the midgut of the fly, however. There, glucose is unavailable and the cells instead have to gain energy from metabolizing the amino acid proline (van Weelden et al. 2003), via a process called "oxidative

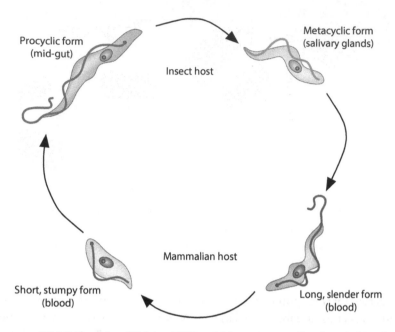

Procyclic form
(mid-gut)

Metacyclic form
(salivary glands)

Insect host

Short, stumpy form
(blood)

Mammalian host

Long, slender form
(blood)

FIGURE 6.24. Life cycle of *T. brucei*. When biting a mammalian host infected with *T. brucei*, the tsetse fly will carry the bloodstream form of the parasite (bottom left). In the fly's midgut, the parasites transform into the so-called "procyclic" form (top left) and multiply by binary fission. They then leave the midgut of the fly and transform into the "metacyclic" form (top right) to reach the fly's salivary glands. When the fly takes a blood meal on the mammalian host, the infected tsetse fly injects the metacyclic form into the host, passing it into the bloodstream. Inside the host the parasites transform into the long slender form (bottom right) where they continue replication by binary fission until they are taken up by a fly again in the form of the nondividing "stumpy" form (bottom left). The cycle in the fly takes approximately three weeks, but the parasite can stay in the mammalian bloodstream far longer, up to decades (Sudarshi et al. 2014). Modified from Lee et al. (2007).

phosphorylation" that requires the activation of the electron transport chain (ETC).

The genes enabling the generation of ATP via the ETC are encoded on the parasite's mitochondrion (as they are in all eukaryotes). We can now slowly appreciate the extraordinary obstacles faced by this parasite. First of all, the nuclear genes are in peril simply because the organism is asexual: according to standard evolutionary genetics, asexual organisms are doomed to fast extinction unless they figure out a way to maintain genes using other mechanisms (besides recombination), see for example (Maynard Smith 1986). It turns out that *T. brucei* may have found such a mechanism for the nuclear genes, namely *gene conversion*. In gene conversion, deleterious mutations are removed by replacing a mutated copy of a gene by the intact homologous version on the other chromosome. Incidentally, as gene conversion requires the existence of the meiotic apparatus, this may explain why the machinery is conserved even though the parasite does not engage in mating.

However, the genes encoding the elements of the electron transport chain reside on the mitochondrion, of which there is only one, so gene conversion cannot work. A closer look at the *T. brucei* mitochondrion reveals that this mitochondrion is not at all like the kind we are used to in eukaryotic cells. Compared to those, it looks like an alien structure. First of all, rather than being a single circular genome, the *T. brucei* mitochondrial genes are encoded on dozens of circles. The genes coding for the metabolic genes (along with two genes encoding ribosomal RNA subunits) are encoded on what is known as "maxicircles": circular segments of DNA of between twenty and forty thousand bases each. The maxicircle of *T. brucei* has 22 kB, but there are many copies of it: between twenty-five and fifty copies! These circles link up together like a chain mail and form a large structure called a kinetoplast (see Fig. 6.25). The circumference of this chain mail is lined by thousands of smaller circles of DNA: the so-called "minicircles." While each maxicircle contains the instructions to make all the mitochondrial proteins, the thousands of minicircles encode different "guide RNA" sequences, whose importance will become clear shortly.

Replication of the kinetoplast is coordinated with the replication of the nuclear DNA, but is a much more complicated affair. First of all, maxicircles and minicircles are replicated independently, most likely in a very different manner. This difference is crucial to understand genetic robustness, so let us spend some time discussing them.

Minicircles are replicated after they are released from the kinetoplast structure, using "theta structure" intermediates. In this form of replication, each circle is opened up creating two replication forks (which look like the greek letter θ, hence the name). Polymerization takes place independently at each of the forks, until the entire circle is copied. This mode of replication is very different from the "rolling circle" replication of circular DNA, which is more common in viruses. In that mode, a polymerase copies the circular DNA by repeatedly running along the circle, thus forming enough DNA for many circles. These two forms of replication imply very different population genetics, and it is important to understand these differences.

There are thousands of minicircles in each kinetoplast, encoding hundreds of different guide RNA sequences, compared to only several dozens of maxicircles. Because each minicircle is to some extent unique, there is no need

FIGURE 6.25. The kinetoplast is a DNA structure found in the mitochondrion of the *Kinetoplastidae*, a group of flagellated bacteria. The kinetoplast is formed by maxicircles (black) that are concatenated in the form of chain mail, and support the much smaller minicircles (gray) that are attached at the periphery.

to make more than one copy of each, so they can be replicated using the theta mechanism (Jensen and Englund 2012). And indeed, sequencing minicircle mRNA reveals enormous sequence variation, as expected if each circle is replicated independently (Kirby et al. 2016) where errors can accumulate. Maxicircle DNA is, on the contrary, extremely homogeneous (Shlomai 2004; Kirby et al. 2016), in the sense that only a handful of polymorphisms can be detected among the fifty or so copies. Such a lack of diversity is incompatible with individual replication via the theta structure, but perfectly in line with what we would expect if replication of maxicircles occurred via the rolling circle mechanism. After all, in that mode of replication, a single circle is picked randomly (among the fifty or so) to be the template for *all* the maxicircle copies in the next generation. And while early evidence did suggest replication of maxicircles in *T. brucei* via the rolling circle mechanism (Hajduk et al. 1984), later the literature concluded (using micrographic evidence) that maxicircles are also replicated via the theta structure mechanism (Carpenter and Englund 1995; Shlomai 2004). However, this conclusion is suspect from the point of view of population genetics, as we'll now see.

If maxicircles were replicated via the theta structure mechanism, each sequence must be replicated individually and—because fitness is determined by the fitness of the host organism—must therefore have the same exact fitness. As a consequence, mutations would accumulate neutrally in this pseudopopulation, putting the organism at risk of extinction by Muller's ratchet. Instead, all observations to date suggest that the population of maxi-circles is essentially uniform: any polymorphisms can be explained either by sequencing errors or by mutations incurred in the first-pass replication of a single maxicircle. But several other lines of evidence point to a different repli-cation mechanism for the maxicircles. The binding site for the protein that binds to the origin of replication is different between minicircles and maxi-circles (Shapiro and Englund 1995). But perhaps most importantly, while minicircles are first released from the kinetoplast before replication, there is no evidence for this for maxicircles. If maxicircles are all created from a single copy, it is then possible to produce all the necessary circles to create a second kinetoplast while the kinetoplast structure remains intact.

Replicating maxicircles via the rolling-circle mechanism (should this turn out to be the case) solves one big problem for *T. brucei*: mutations cannot accumulate in the many copies of the mitochondrial genes, while the copies can still serve as the structural element (the kinetoplast) that enables the replication of the minicircles.

But that creates another problem at the same time: since replication is done from a single copy and there is only a single mitochondrion in the cell, the mitochondrial genes are at risk of extinction if transmission occurs via a bottle-neck. You might think that because there are many copies of the mitochondrial genes in the maxicircle that there is plenty of redundancy, and indeed this is true: if a mutation occurs in one of the copies that are not used as the blueprint for the next generation, there is no harm to the parasite. However, if a copy *with* a mutation is picked instead, there is no back-up since all copies in the next generation will have the same mutation, leading to extinction.

And indeed, extreme bottlenecks are the norm for *T. brucei*: While in the midgut of the fly, the parasites can attain populations of half a million (Oberle et al. 2010), but during migration to the salivary gland, the population thins out dramatically. Using DNA tags to monitor population diversity, Oberle et al. (2010) found that just a few types actually make it all the way into the mammalian bloodstream. It is possible that this "thinning" out is akin to a selection process in which defective types are weeded out, but there is cur-rently no evidence for this. However, as infection of the mammal host can occur with just a single parasite, the bottleneck can be extreme: if even a sin-gle deleterious mutation makes it through the bottleneck, the genetic lineage would be cursed because extinction would occur within a few hundreds of generations (depending on the size of the deleterious mutation, according to Fig. 6.23). As a consequence, to survive, the parasite must eliminate (or, if possible, *reverse*) all deleterious mutations in the mitochondrial genes. How can it possibly achieve this? This is where the minicircles come in.

Although the fifty or so copies of the mitochondrial genes in maxicircles seem like a luxury, they have nothing on the minicircles. While the minicircles are smaller (as their name indicates: carrying on the order of a few hundred base pairs per circle) there are anywhere between 5,000 and 10,000 of those per mitochondrion. Each of the minicircles encodes between three and five guide RNAs (gRNAs), which direct how to *change* the maxicircle transcripts by inserting or removing U (uridine) bases in the transcripts. At first sight, such a statement seems preposterous. After all, one of the first lessons we learn in evolutionary genetics is the central dogma of molecular biology, which states that the flow of information in the cell is unidirectional: from DNA to RNA to proteins. How could information in one set of circles *change* the information stored in another?

It turns out, this editing of RNA transcripts is not uncommon in eukaryotes (Brennicke et al. 1999), in particular in mitochondrial genes (see Box 6.4), and since the original discovery of RNA editing in trypanosomes, RNA editing has been found in many other contexts as well, from slime molds, over animals, to plants (Koslowsky 2004). However, the complexity of the editing process—the incredible burden it imposes on the transcription of mitochondrial genes and in particular the apparent fragility of the system— has been viewed as mysterious and paradoxical. After all, an editing cascade such as that shown in Figure 6.26 appears to doom the organism carrying it. A mutation pretty much anywhere in the gene will, of course, lead to a corresponding change in the transcript, the mRNA. Such a mutation (unless it is corrected) is bound to interfere with the editing cascade, either by changing the anchor sequence where a guide RNA is supposed to bind or by changing the signal that the editing machinery (inserting and/or deleting uridine residues) responds to. In almost all cases, such a mutation would ultimately lead to a badly edited sequence, and therefore a failed protein. Unless, as alluded to above, mutations can be corrected. This is a big if: since mutations are random, such an error correction scheme would require there to be several alternative "programs" for every site that can be corrected by U insertion or deletion only. A correction would require a two-step editing process (one insertion and one deletion), but since RNA editing in trypanosomes is limited to U insertion and deletion, only a subset of errors could be corrected in this manner. The remaining errors therefore are likely to disrupt the editing cascade, and thus lead to failed translation of the corresponding protein.

But it turns out that for an organism that must deal with extreme bottlenecks and periods of weakened selection, this is the *only* way to survive: if deleterious mutations cannot be repaired, then they must be made lethal. And indeed, theory backs up this thinking. A look at Figure 6.23 reveals that when population sizes are small, there are *two* ways for the lineage to survive: either

Box 6.4. RNA Editing in Trypanosomes

RNA editing was first discovered in the lab of Rob Benne (Benne et al. 1986), who found that the DNA sequence coding for the COX2 protein (that we previously encountered in section 3.2) in *T. brucei* has a highly conserved frameshift mutation that should have resulted in a nonfunctional protein. Instead, the cell produces fully functional versions of this essential protein, leading those authors to conclude that the mRNA transcript must somehow have been edited before translation so as to "fix" the frameshift mutation. Soon after, several labs reported extensive editing of mRNA transcripts of several mitochondrial genes in *T. brucei* (Feagin et al. 1988; Shaw et al. 1988), sometimes creating a start codon where one has been missing, but sometimes changing more than half of the sequence. While at first it was a complete mystery how this editing is achieved, further work showed that the process is guided by a set of RNA sequences dubbed "guide RNAs" (Blum et al. 1990), which are most often found in the *minicircles* of the mitochondrial DNA (Sturm and Simpson 1990). Up to that discovery, the purpose of the minicircles was equally mysterious. The minicircles encode hundreds of guide RNA classes that differ in the "program" as well as in the recognition site with which the gRNA binds to the mRNA anchor (see Fig. 6.26). The circular part of the hairpin

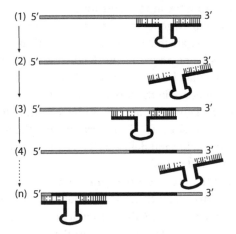

FIGURE 6.26. (1) A guide RNA molecule (the "hairpin" structure, black) binds to an eight- to twelve-nucleotide anchor sequence at the 3' end of the raw transcript (gray) via Watson-Crick pairs (solid lines) or sometimes G:U pairs (dotted lines). (2) Mismatches recruit enzymes that either add or remove a uridine (black edited region), after which the gRNA detaches. (3) The edits created a new anchor site that is complementary to another gRNA, which then binds there and (4) guides the addition or removal of more Us so as to match the gRNA template to the mRNA sequence. (n) This process continues until the 5' site is reached.

Continued on next page

Box 6.4. *(continued)*

contains the program: a sequence that is complementary to the "target" region, that is, to the sequence that is to be programmed into the mRNA. Specialized enzymes that together form the "editosome" remove and/or add uridine (U) residues in the mRNA sequence until that region is perfectly complementary to the program sequence, after which this part of the "reprogramming" is complete and the gRNA departs. The editing cascade begins with a gRNA that recognizes the anchor site at the $3'$ (the "downstream" end) of the mRNA sequence which, using its program, creates a new anchor that is complementary to a different gRNA program that prepares the sequence to bind to yet another gRNA and so forth, until the $5'$ end of the sequence is reached. For some transcripts, significantly more than half of the nucleotides change between the raw and the final edited form, which is now ready for translation. If the cascade does *not* finish, the all-important translation signal is not attached to the transcript and the protein will not be made.

---

reduce the effects of mutations (the "left branch" of the curve) or else make the effect of mutations large (the right branch). But what if the protein is not essential? Would this mechanism not imply that all nonessential proteins will eventually disappear from the mitochondrion?

While many of the edited genes in *T. brucei* are in fact essential in the insect stage (we recall that these are metabolic genes that activate the electron transport chain as well as being active in the Krebs cycle), those genes are *not* essential in the mammalian host, where energy is exclusively gained via feeding on the abundant glucose in the bloodstream of the animal. How can the editing cascade protect genes that are not under selection during this period, which after all can last decades? Is this observation not the nail in the coffin of the theory that the unwieldy and delicate RNA editing machinery only exists to protect genes from mutations?

We must concede that nonessential genes could not survive genetic drift under conditions of extreme bottlenecks without some protection. But where could this protection come from? Recent evidence suggests that RNA editing comes to the rescue here too. It turns out that besides fixing frameshift mutations, RNA editing can also *induce* them. The mRNA for the COX3 (cytochrome oxidase subunit 3) gene, for example, is normally "pan-edited," meaning that the entire sequence is edited from $3'$ to $5'$ by adding a whopping 547 U residues while deleting 41, with a final transcript size of 947 residues (Kirby et al. 2016). Because such heavy editing relies on a cascade of perhaps ten or more editing steps, it is not surprising to find partially edited sequences within a pool that is being sequenced: unfinished works, so to speak.

One of those sequences caught the eyes of Ochsenreiter and Hajduk (2006), because while only the first half of the sequence was edited (the second half was the unedited original transcript), a guide RNA had inserted an extra two U at the end of the first half, which changes the reading frame of the sequence. When analyzing the putative amino acid sequence, they realized that this edit was giving rise to a new protein: half of which (155 residues) was the sequence of the original COX3 (a transmembrane motif, as COX3 is normally embedded there), while the other half was a completely new sequence of fifty-nine residues that is able to bind DNA (Ochsenreiter and Hajduk 2006). Further study showed that the protein was indeed functional (knocking it out produced a fitness defect) and moreover they discovered that it was used as a link between the kinetoplast structure and the RNA editing mechanism. This is remarkable because COX3 is indeed just such a nonessential gene during the bloodstream phase of the life cycle of *T. brucei*, and as a consequence would accumulate mutations during that phase. By overlapping that gene with another (the "new" protein AEP-1, where AEP stands for "alternatively edited protein"), the RNA editing machinery rendered the gene essential after all, so that mutations on it (and therefore on the COX3 gene at the same time) are avoided. A later study found that as many as six other edited proteins that are encoded on the maxicircle have other proteins overlapping them (Kirby and Koslowsky 2017), suggesting that protecting genes that are not under selection is crucial for the long-term survival of the species.

But what about nonessential genes that are not edited? Indeed, Kirby et al. (2016) found that five of the twenty genes encoded on the *T. brucei* maxicircle are not edited. Of these, one codes for COX1 (another of the three subunits of the cytochrome oxidase complex encoded in the mitochondrion), while the other four make the subunits of the NADH dehydrogenase complex, which is an important part of the electron transport chain. All of these are genes that are important in the insect stage, but dispensable when in the bloodstream of the mammalian host. How these genes are able to withstand mutations is not fully clear, but there are hints. In particular, it appears that the NADH dehydrogenase complex in *T. brucei* is *multifunctional* (Panigrahi et al. 2008), that is, its components provide other functions beyond those associated with the Krebs cycle within the electron transport chain. Intriguingly, the evidence points toward mediating interactions between mitochondrial RNAs, that is, these proteins might be moonlighting to help in the editing process itself, making them indispensable after all.

Another such multifunctional enzyme is α-KDE1, a nuclear-encoded enzyme that is usually active in the Krebs cycle, but turns out to be essential in the bloodstream phase even though the Krebs cycle is not used there (energy is generated via glycolysis instead). In fact, in the bloodstream phase α-KDE1 appears to be associated with the glycosome (where glycolysis is carried out) instead (Sykes et al. 2015). We can thus tentatively conclude that mitochondrial proteins are protected from mutational decay by the editing cascade either by error correction (if possible), by aborting the cascade (if the gene is essential), by overlapping with essential proteins (when nonessential), or by multifunctionality (when the affected gene is nonessential and unedited).

But an even larger question looms. If the guide RNA system protects the genes encoded in the maxicircles, what protects the minicircles? After all, it is logically impossible to protect RNA editing genes via RNA editing. If this section's speculations about robustness mechanisms in *T. brucei* are to stand up, the robustness of minicircle-encoded gRNA sequences must be explained. First, we should ask: are the minicircles also as free of mutations as the maxicircles are? The answer: not at all; they are riddled with them! Indeed, Figure 6.1 shows the mutational network centered around one of the guide RNA sequences involved in editing the mRNA of the protein RPS12. Even though all the mutant gRNAs in that network are capable of performing the editing function (according to a bioinformatic analysis; Kirby 2019; Kirby et al. 2022), some versions are dramatically more abundant in the population than others (as determined by an analysis of RNA abundance in cell extracts).

How can we understand this distribution of variant gRNA sequences? Could the network of gRNA sequences actually form a quasispecies of sequences, and therefore be robust under mutations as a group as the theory in Box 6.2 implies? To answer this question, we need to measure the *frequency* of sequences at each node in a network like Figure 6.1. According to quasispecies theory (see Box 6.2), the mutant distribution of a quasispecies is *stationary* (time-independent). While we do not have access to two different time points for the same population of cells, we do have two independent cell lines that have been cultured via serial transfer. One of the cell lines (called TREU 667) was originally isolated from a bovine host in Uganda in 1966 (Hudson et al. 1980), while the other (EATRO 164) was isolated in 1960 from a type of antelope (Lichtenstein's hartebeest, Agabian et al. 1980). Both lines had been maintained for decades in the lab of Dr. Ken Stuart, Director of the Center for Global Infectious Disease Research in Seattle, Washington,[9] and therefore have been propagated independently for many decades. If the distribution looks similar in cell lines that have been separated for so long, then it is safe to assume that it is time-independent. To generate the abundance distribution,

---

9. The abbreviation TREU stands for Trypanosomiasis Research Edinburgh University, while EATRO abbreviates East African Trypanosomiasis Research Organization.

the procyclic form of the cells (the form *T. brucei* takes in the midgut of the tsetse fly, see Fig. 6.24) is ground up and RNA is extracted and sequenced (a procedure called RNAseq). Because the typical RNAseq run does not involve an amplification step, the number of times a particular sequence is detected represents its abundance in the extract.

Ideally, we would test the abundance distribution of gRNAs that edit each of the proteins individually. When Laura Kirby sequenced the gRNAs of both the TREU and EATRO cell lines in the laboratory of Dr. Donna Koslowsky at Michigan State University, she found that there were extraordinary differences in the abundance of gRNA sequences: between the two cell lines some had hundreds of thousands of copies, while others only appeared as singletons. Moreover, which sequence was the most abundant within groups of sequences that edited the same region of the same protein (groups like the one shown in Fig. 6.1) differed between cell lines (Kirby 2019; Kirby, Adami, and Koslowsky 2022).

When plotting the abundance distribution for each protein in each of the cell lines, there was a surprise: while the average abundance of gRNA sequences depended on the cell line, the distribution of abundances was very similar across proteins and even across cell lines.

Figure 6.27 shows the distribution of gRNAs that edit the mitochondrial protein RPS12 in cell lines EATRO 164 (Fig. 6.27[a]) and TREU 667 (Fig. 6.27[b]). Both are cell lines consisting of the procyclic form of *T. brucei*. While the overall abundance of the gRNAs is very different in the two cell lines, the distribution is remarkably similar, and is accurately fit[10] to a power law with an exponent of around 1.5. What gives rise to the remarkable similarity in distributions?

To answer this question, we must understand the mechanism of mutation and selection on the network of sequences. To simplify, we'll treat the number of sequences n of a particular gRNA type as a continuous variable (because after all there can be many thousands of each particular type in a cell), and assume a process in which mutations will increase or decrease this number by chance.

If we assume n gRNAs of a particular type in a mother cell, which are subsequently doubled to $2n$ just before division, unequal division produces $n + i$ and $n - i$ copies in each daughter cell. For a continuous variable, this process gives rise to random drift, so that the distribution function $\rho(n)$ (the probability to find n copies of the type in a cell) obeys the diffusion equation (keep

---

10. The data in Figure 6.27 are binned using the threshold binning method described in the Appendix of (Adami and Chu 2002), to allow for variable bin sizes. Error bars are counting error.

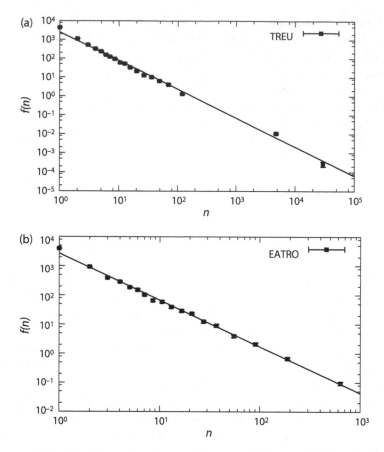

FIGURE 6.27. Abundance distribution of connected clusters of gRNAs of the protein RPS12. (a) The distribution $f(n)$ is the number of times a variant with n copies was found in the sample of gRNAs editing the mRNA for mitochondrial protein RPS12, extracted from the cell line TREU 667. The distribution is fit to the power law $f(n) = \frac{2514}{n^{1.52}}$. (b) The distribution $f(n)$ for gRNAs editing the same protein, but in the cell line EATRO 164 (Sequence Read Archive accession SAMN04302078). The fit is to the power law $f(n) = \frac{2697}{n^{1.6}}$.

in mind that n is a continuous variable in this approximation)

$$\frac{\partial \rho(n)}{\partial t} = D\frac{\partial^2 \rho(n)}{\partial n^2},\qquad(6.26)$$

where D is a diffusion coefficient that is related to the average change in numbers i (the "step size" of the random walk) at each division, and the rate of cell division. However, the solution of a pure diffusion equation gives a very poor fit to the observed distribution in Figure 6.27. This implies that the gRNA numbers cannot just change randomly up and down; they also must be under selection, which will add a term to Equation (6.26). In reality, when the number of gRNA sequences becomes low, cells suffer because they cannot edit the mRNA sequences fast enough (Wang et al. 2002).

As a consequence, there is a selective pressure for types with low abundance to *increase* in abundance, while types with high abundance feel almost no such pressure. In particular, should the type vanish completely $(n = 0)$ we assume that the cell will die because the essential protein (here, RPS12) cannot be made. We can model such a selective pressure using a force term $F(n)$ that acts on the distribution, giving rise to the equation

$$\frac{\partial \rho(n)}{\partial t} = -\frac{\partial}{\partial n}(F(n)\rho(n)) + D\frac{\partial^2 \rho(n)}{\partial n^2}. \tag{6.27}$$

Equation (6.27) is a Fokker-Planck equation, and if the force $F(n)$ can be written as the gradient of a potential $F(n) = -\frac{\partial}{\partial n}\phi(n)$, the equation has a unique stationary (that is, time-independent) solution, given by (see, for example, Risken 1989)

$$\rho(n) = \frac{1}{Z}e^{-\phi(n)/D}. \tag{6.28}$$

The normalization constant Z is related to the smallest n that can be tolerated by the cell. Because our continuous approximation breaks down for such small n, we will simply treat Z as a parameter to be fit.

What is the appropriate "repulsive" potential that gives rise to distributions of the type seen in Figure 6.27? It turns out that a simple logarithmic potential does the trick:

$$\phi(n) = \alpha \log n \tag{6.29}$$

works remarkably well, as the solution $\rho(n)$ then becomes

$$\rho(n) = \frac{1}{Z}\frac{1}{n^{\alpha/D}}. \tag{6.30}$$

We can see potentials with different values for α in Figure 6.28. It is clear that for large n, there is barely a force on the abundance (as the gradient of the logarithm quickly becomes tiny), while small n will give rise to a sizable restoring force. Comparison to the distributions shown in Figure 6.27 then suggests that $\frac{\alpha}{D} \approx 1.5$.

If any of these speculations are confirmed, we can see a remarkable assembly of robustness-conferring features at play in the genetics of *T. brucei*. The mitochondrial genes in its maxicircles are under assault because the organism periodically undergoes dramatic bottlenecks where sometimes only a

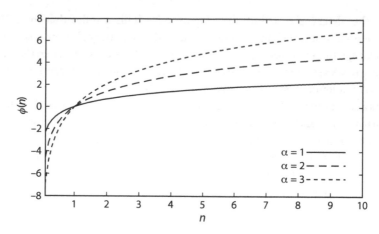

FIGURE 6.28. Potentials $\phi(n) = \alpha \log(n)$ for candidate solutions to the Fokker-Planck equation with different parameters α.

single copy of the organism becomes the ancestor of the lineage. Even though there are many copies of those genes within the maxicircles, they really represent just a single copy that serves as the blueprint for a copying process that creates exact copies to be fashioned into the kinetoplast structure. Many of the mitochondrial genes are pan-edited by a set of guide RNAs stored in minicircles. Because mutations in the mitochondrial genes almost certainly arrest the editing cascade (and the unedited transcript cannot be translated) those mutations (if they cannot be corrected) are lethal and therefore rejected (they cannot accumulate). Proteins that are not essential are either protected by having their sequence overlap with that of an essential protein (a process also directed by RNA editing), or else by being multifunctional, with an alternative function that is essential. The guide RNAs themselves appear to be exposed, however, because they are not translated. Instead, it appears that the set of gRNAs is mutationally robust, in the sense that the distribution of variants is time-independent, much like a molecular quasispecies.

The group of gRNAs is different from a molecular quasispecies, however, as the set is already "group-selected" by default: after all, all gRNAs in a cell are replicated together, regardless of their function. The quasispecies effect appears to occur via an asymmetric division process that creates changes in the number distribution, along with a selective force that discourages very small copy numbers. As a consequence, the distribution of gRNA sequences is stationary (time-independent) with a long tail (a power law).

6.5 Summary

While in chapter 5 we discussed the processes that contribute to an increase in complexity and information via adaptation, in this chapter we focused on the many (likely the majority) of genetic changes that are in fact neutral. But neutrality—a genetic change that does not translate into a change in fitness—does not mean nonadaptive, because the environment is more than just the functional niches that offer ways to make a living. Since evolution takes place in a physical environment, the genome also needs protection from hazards other than parasites, viruses, or droughts. A high mutation rate, for example, could spell doom for a population unless changes in genetic architecture are available that make elevated mutation rates tolerable. And while those changes do not translate into higher fitness, by mitigating the average effect of a mutation, those changes are certainly adaptive. In the case of the survival-of-the-flattest effect, a population carrying this protection can outcompete a "fitter" population that cannot protect its genome in this manner.

Small population sizes can also present a hazard as genes can be lost to drift, in an irreversible ratchet-like manner. It turns out that there are also mutations that can protect from drift, by interacting with other mutations so as to amplify otherwise small-effect deleterious mutations, effectively preventing such small-effect mutations from becoming fixed. In this manner, populations can become robust to drift, even though none of the mutations that confer drift robustness necessarily have an effect on wild-type fitness.

It is perhaps worthwhile to point out that this view of the evolution of robustness is different from the perspective of *constructive neutral evolution* (CNE) (Stoltzfus 1999; Stoltzfus 2012). According to this theory, neutral mutations can build up complexity when there are many more ways to create disorder than there are to create order, as long as there is a global mechanism that prevents those "disordering" mutations to affect organismal fitness. In fact, CNE was invoked early on to explain the peculiar kinetoplastid RNA-editing machinery (Stoltzfus 1999), implying that the Rube Goldberg–like complexity is a useless feature that the cell got stuck with, as it does not confer any fitness to the organism. Quite to the contrary, the theory of drift robustness suggests that by rendering the vast majority of mutations in mitochondrial genes lethal (if they cannot be reverted by editing), the RNA-editing machinery protects the organism from extinction via gene loss. Thus, while CNE focuses on the ratchet-like accumulation of mutations with no effect, drift robustness instead focuses on the effect of those mutations on *other* loci, preventing the ratchet-like loss of important genes. The same is true for mutational robustness: a mutation that confers mutational robustness can be neutral (or even decrease wild-type fitness), but by having an effect on a large number of *other* sites via epistatic interactions with them, they can save the entire population from extinction.

Exercises

6.1 Show that in asexual haploid replication, the probability of fixation $P(s)$ of a mutation with beneficial effect s is

$$P(s) = \frac{2s}{(1+s)^2}. \tag{6.31}$$

Hint: An organism with a beneficial mutation of size s on a background with fitness $w_0 = 1$ has fitness $w = 1 + s$. If the distribution of offspring is Poisson, the likelihood $\pi(n)$ that the organism has n offspring is given by

$$\pi(n) = e^{-(1+s)} \frac{(1+s)^n}{n!}. \tag{6.32}$$

Write the probability of extinction $1 - P(s)$ of a mutant as the sum over the likelihood that the mutant has n offspring that will all go extinct, to obtain the self-consistent formula

$$1 - P(s) = e^{-(1+s)P}. \tag{6.33}$$

Estimate $P(s)$ by expanding the exponential to second order.

6.2 Show that for a binary division process, the fixation probability for a mutant with benefit s is instead

$$P(s) = \frac{4s}{(1+s)^2}. \tag{6.34}$$

Hint: Instead of (6.32), use an offspring probability distribution where $1 - p$ is the probability that no offspring make it into the next generation, $2p(1-p)$ is the likelihood to end up with one offspring, and p^2 is the probability that the next generation sees both offspring. Show that for a population with mean fitness w, we must have $p = w/2$ (Gerrish and Lenski 1998), and calculate the relevant self-consistent formula for the extinction probability $1 - P$ as before.

6.3 Show that the mutational load of an adapting asexual population in the small mutation rate limit is (Kimura and Maruyama 1966)

$$L = 1 - e^{-\mu}. \tag{6.35}$$

Take into account that the mutational load is given by $1 - F$ (where F is the fidelity of replication) and use the expression for fidelity from Equation (6.13).

6.4 In Box 6.3 we derived the critical ratio R in terms of the ratio of mutation probabilities

$$M = \frac{u_{41} \, u_{12} \, u_{23}}{u_{14} \, u_{21} \, u_{32}}, \qquad (6.36)$$

where u_{ij} is the likelihood of mutation from node i to j. Because this is a beneficial mutation, we can write this probability as $p_b(s) = u_b \mu \rho(s)$, where $\rho(s)$ is the distribution function of mutations with benefit s, u_b is the likelihood that a mutation is beneficial, and μ is the overall mutation rate (we can also assume that the "back mutation" rate u_{ji} is simply given by the background rate μ). Show that if the distribution function of mutations of beneficial effect size s is given by the distribution function (Gerrish and Lenski 1998; Fogle et al. 2008; Good et al. 2012)

$$\rho(s) = \frac{1}{\bar{s}} e^{-s/\bar{s}} \qquad (6.37)$$

(here $\bar{s}$ is the average beneficial effect) then for ϵ small $\frac{u_{14}}{u_{41}} = p_b(s - \epsilon) \approx p_b(s)$, while $\frac{u_{12}}{u_{21}} = \frac{u_{23}}{u_{32}} = p_b(s/2)$, so that

$$M = [p_b(s/2)]^2 / p_b(s) = \frac{u_b}{\bar{s}} \equiv \kappa < 1. \qquad (6.38)$$

6.5 Show that the detailed balance condition combined with the Sella-Hirsh fixation probability implies the steady-state distribution of k-mutants

$$p_k = \binom{L}{k} w(k)^{2N-2} / Z. \qquad (6.39)$$

Hint: Detailed balance implies that in a process where transformation $i \to j$ and $j \to i$ are possible, the number of changes $N_{i \to j}$ equals the number $N_{j \to i}$. If we are interested in the steady-state distribution $p(k)$, we need to calculate how many k-mutants are produced (go to fixation) from the wild-type sequence with fitness w_0, and compare this to the rate at which k-mutants are replaced by the wild type. The rate of fixation of a k-mutant with fitness $w(k)$ is, according to Equation (6.22)

$$P(0 \to k) = \frac{1 - 1/w(k)^2}{1 - 1/w(k)^{2N}}. \qquad (6.40)$$

FIGURE 6.29. In detailed balance, the
flux from the wild-type node (solid-
white circle) into its k-mutants must
equal the flux from all the k-mutants
combined into the wild-type node. The
outward rate is indicated on the arrow
(the flux is rate times density). The density of k-mutants $p(k)$ (the quantity that
we want to solve for) is the sum over the
densities of all the individual k-mutant
densities (each of the $\binom{L}{k}$ gray nodes).

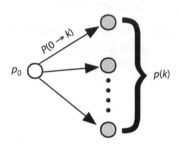

The "reverse" rate then is

$$P(k \to 0) = \frac{1 - w(k)^2}{1 - w(k)^{2N}}.$$

(6.41)

The detailed balance condition is then just (see Fig. 6.29)

$$p(0)\binom{L}{k}P(0 \to k) = p(k)P(k \to 0).$$

(6.42)

In Equation (6.42), $p(0)$ is the equilibrium density of the wild-type, while
$p(k)$ is the (combined) equilibrium density of all individual k-mutants.
Use the unknown $p(0)$ to normalize the density $p(k)$.

7

The Informational Origins of Life

But if (and oh what a big if) we could conceive in some warm little pond with all sorts of ammonia and phosphoric salts,—light, heat, electricity &c. present, that a protein compound was chemically formed, ready to undergo still more complex changes, at the present day such matter would be instantly devoured, or absorbed, which would not have been the case before living creatures were formed.

—C. DARWIN, LETTER TO JOSEPH HOOKER, FEBRUARY 1, 1871

That Darwin was speculating about the origin of life before evolution (in this letter to his close friend Hooker, botanist and director of the Kew Botanical Gardens in London) shows us how profoundly he understood the ramifications of his ideas. He understood that evolution could not proceed in a vacuum: very special circumstances were required to launch the process. But exactly how daunting the odds are against such a process certainly was not clear to him.

Evolution, as the previous chapters have repeatedly emphasized, is an extraordinary process that can take simplicity and, over time, turn it into complexity. We have also seen that the evolutionary process fundamentally operates on information. The "Darwin Demon" (Maxwell's demon applied to biomolecular sequences) ensures that the information content of the sequences coding for a given type tends to increase, on average. And because information content is a proxy for complexity, the latter by and large also increases.

What makes evolution extraordinary from a mathematical point of view is the *copying of information*. When copying is not possible or has not yet been invented, information is (as we will see in more detail below) intrinsically rare. Depending on the amount of information, we can say it is usually *astronomically* rare. Of course, information is ubiquitous on Earth as we all can see from our day-to-day experience. This is because biology sometime somewhere figured out how to copy *genetic* information, which ultimately led to people, who in turn figured out how to write, print, and copy *any* information (Mayfield 2013). In the absence of that first event (the first copying of genetic information), information would still be astronomically rare today.

How likely was that first copying event? Can information theory give us some guidance about the likelihood of the spontaneous emergence of information? It turns out that it can, precisely because information is a mathematical measure that does not depend on the chemistry (that is, the alphabet) within which the information is written. For this reason, an information-based approach to study the origin of life has many advantages over traditional approaches.

The origin of life on Earth took place sometime between the formation of our planet some 4.5 billion years ago[1] and the first documented traces of life. What constitutes "traces of life" is, as you can easily imagine, quite controversial. However, the consensus is that life emerged on Earth sometime between 4.1 billion and 3.5 billion years ago (Bell et al. 2015). So what is the likelihood of that momentous event, the emergence of copying?

Everyone agrees that it is difficult to estimate the likelihood of spontaneous emergence of life because so much is unknown about the chemistry of the early Earth. But even if we knew exactly what chemicals were available on early Earth, along with the external conditions (such as temperature, pressure, acidity, etc.) for each and every environment, there would still be enormous uncertainties. For example, if we assume (as I do in this book) that life is equivalent with "information that encodes the means for its own replication," then all forms of life must be based on a set of molecules that can be used as an alphabet. We know, of course, what this alphabet is on Earth: it is the four nucleic acids G, C, A, and T. However, we also know that DNA cannot have been the first alphabet, as it relies on protein machinery for replication, and DNA and proteins are two very different molecules. As a consequence, scientists have hit upon the idea that the "first" living molecule must have been both information and "replication machine" at the same time: an RNA molecule. Let us delve into this hypothesis a bit deeper, because even if the speculation turns out to be wrong, there are important lessons about our origins to be found.

7.1   The RNA World

What we usually call RNA is a polymer made from several or many ribonucleotides. To appreciate what they are (and understand a little bit how they are different from DNA), we might as well become a bit more accustomed to the

---

1. To be exact, $4.54 \pm 0.05 \times 10^9$ years (Dalrymple 2001).

biochemistry of these molecules. These nucleotides, after all, are the alphabet of life.

We are all familiar with the four bases that make up DNA (we encountered them earlier when we were dealing with the information content of these molecules in section 2.1), called adenine (A), cytosine (C), guanine (G), and thymine (T). RNA molecules use the same first three letters, but use uracil (U) instead of T. You can think of the letters as if they were Scrabble pieces: to understand what follows, their biochemistry is less relevant. But just like in Scrabble, you need something to hold the letters up. That is where the "ribose" part of "ribonucleotides" comes in.

Ribose is a sugar. It is known as a "simple" sugar, consisting of a ring of five carbon atoms, with five water molecules hanging off of it: $C_5H_{10}O_5$. Well, not water molecules technically, because the hydrogen H and the hydroxyl group HO actually hang on two different sides of the carbon atom, as in Figure 7.1(a). Figure 7.1(b) shows the sugar in its ring form, with the carbon atoms of the ring as well as some Hs omitted. You can see from the picture in Figure 7.1(b) that the ribose ring is not symmetric: there is a mirror image of the molecule called "L-ribose" (as opposed to the "D-ribose" shown here) where the CH_2OH group and the OH group are flipped. L-ribose does not occur at all in nature, even though obviously it could in principle. This observation is called "homochirality," and is ubiquitous in biochemistry: not just sugars but also amino acids are homochiral (but they only exist in the L-form).

To form the backbone of RNA or DNA, ribose molecules are sewn together using a phosphate group PO_4 that attaches to the fifth carbon (the 5'-carbon, counting clockwise starting with the carbon at the three o'clock position, the so-called 1'-carbon), as seen in Figure 7.2. The phosphate group is attached to the 3' carbon of the next ribose ring, whose 5' phosphate attaches to the 3' of the next, and so on. This forms the backbone chain, and the "letters" (the nucleobases) are all attached to the 1' carbon, as seen in Figure 7.2.

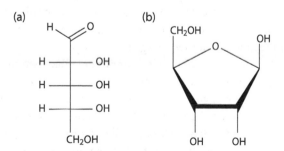

FIGURE 7.1. (a) Ribose molecule in the open chain form. (b) The "D"-form of the ribose molecule as a ring.

(a)

(b)

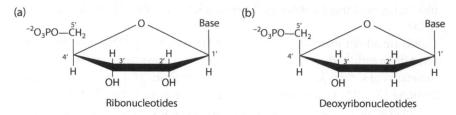

Ribonucleotides

Deoxyribonucleotides

FIGURE 7.2. (a) Ribonucleotide (with "OH" at the second carbon of the ribose ring). (b) Deoxyribonucleotide, with a missing O at 2′, hence "de-oxy."

This trick works for both DNA and RNA: the DNA backbone's only difference is a missing oxygen at the 2′ position (see 7.2[b]). This change looks innocuous, but the consequences for all of life are profound. Because of the OH group, RNA molecules are quite flexible: there are two other OH groups on the ring, and there are therefore several alternatives of how different nucleotides (attached to the ring) can bind to each other. In principle, the RNA molecule (which readily forms a double-helix just like its DNA cousin, unlike what some textbooks claim) could branch into two, and it can even form triple- and quadruple helices. A DNA molecule, because of the missing OH group in the ribose ring, does not nearly have that much freedom. Instead, the nucleobases have to stack in perfect sequential order, exactly what is needed for the storage of information. Indeed, the DNA molecule is also much more stable than its RNA counterpart: it takes a lot more energy to "mess it up."

It turns out that ribose, this "little sugar molecule that could," is central to all of biochemistry. Not only does it form the backbone of RNA (and in the deoxydized form, DNA), it is also used in the manufacture of several amino acids and plays a major role in metabolism (in the pentose-phosphate pathway). Moreover, when the nucleobase adenine is attached to ribose just as in the helix, and a phosphate group is attached to the ribose (again, just as in the helix), you have one molecule of AMP (adenosine monophosphate). String two more phosphates on, and you have ATP: the "energy currency molecule" of life. ATP quite literally shuttles energy from one place in the cell to another: when it is used up, it is recycled into its component. Take two riboses stitched together with phosphate, hang an adenine on one and nicotinamide (which is a structural analogue of guanine) on the other, and voilà, you have NADH, another one of life's crucial energy molecules. We thus see that via the single molecule ribose, information storage and energy processing are intimately coupled: it is as if the Scrabble rack that is keeping the letters in place was literally made out of food.

We can imagine how information—first stored in RNA either in single strands or double-helices—is, through a set of evolutionary adaptations that came later, transferred to a more stable substrate, the DNA molecule. In such a picture of life's early stages, the four letters of RNA constitute the first alphabet. And because the RNA molecules can form molecular machines (that is, enzymes)—in part due to the flexibility that the RNA chemistry affords—it is an enticing idea that RNA chemistry gave rise to all that we are now.

The idea that RNA constitutes the chemical cradle of all of life on Earth is called the "RNA world" hypothesis (Woese 1967; Orgel 1968; Crick 1968; Orgel and Sulston 1971; White 3rd. 1976). The idea is enticing because it solves a number of problems: information storage and replication machinery could be one and the same. However, the road to self-replication is—as far as our current evidence informs us—a difficult one, even if the monomers A, C, G, and U, along with the backbone molecules ribose and phosphate, can form easily. In order to obtain life—that is, polynucleotides that encode the information necessary to replicate that information—it is necessary for specific sequences of ribonucleotides (or at a minimum the information encoded therein) to serve as a template for the synthesis of another such sequence. In present life, this duplication is achieved via enzymes, but before there is life such enzymes (even those made out of RNA, so called ribozymes) simply are not there, as the smallest highly active self-cleaving ribozymes are of the order ~ fifty nucleotides (Ferré-D'Amaré and Scott 2010). Sequences of that length cannot arise by chance, as we will see further below. Thus, they cannot be there to help in the emergence of life.

However, it is possible for templates to be copied *passively*, via a process where an RNA duplex opens up, and monomers aggregate on a strand: a C on a G, a U on a T, and so on, thus re-forming the duplex. This process is slow and error prone, because mismatched nucleotides (a T settling on a G rather than a U, for example) do occur. If the rate of production of correct copies was equal to the rate at which incorrect copies are made, then template-based copying is of no help at all: information could simply not emerge.

Fortunately, mismatches cause a stalling in the assembly process (Rajamani et al. 2010), so that sequences with many errors take longer to finish, and thus those that lucked out with fewer errors finish first, effectively replicating faster. It has been known for a while (Kun et al. 2005) that stalling due to mismatched bases is a feature of enzymatic polymerization (there, the mismatched nucleotide creates a suboptimal conformation for the polymerase), but Rajamani et al. (2010) showed that stalling occurs in nonenzymatic template-base polymerization as well, with "stalling factors" of the order of ten to one hundred. This discovery opens up the prospects that pieces or parts of the produced polymer might help (or hinder) the correct incorporation of

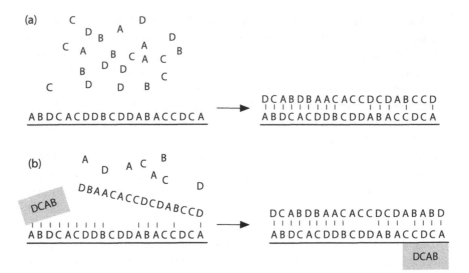

FIGURE 7.3. (a) Schematic view of a template-directed polymerization process where a sequence (composed of the arbitrary alphabet A, B, C, D) held in place by a scaffold gives rise to an approximate copy, assuming the preferential pairings A-D and C-B. (b) The same polymerization process, where a piece of the polymer DCAB binds to the sequence and the scaffold, to increase the accuracy of polymerization.

nucleotides (see Fig. 7.3). If an oligomer (a piece of the polymer that is being copied) binds in some way to the assembly with the result that copy-fidelity is increased (Fig. 7.3[b]), then because such sequences carrying that motif are copied more quickly, we can say that the sequence itself now carries information about how to copy the sequence. It might not be a lot of information at first, but in a way we could say that information about how to copy a sequence has seeped from the hardware (the clay-like scaffold that keeps the strands in place and makes template-based polymerization possible) into the software: the sequence being copied. Note that such a process only leads to the preferential amplification of the sequence carrying the information if the information is not equally shared between all other competing polymerization reactions, in other words, if the information can be kept somewhat private.

There are many open problems in the quest to understand how an RNA world may have emerged on an early Earth (reviewed lucidly by Szostak 2012), but none of these problems appear like they would be insurmountable roadblocks.

In the next section we will take a very different look at the problem of how life may have emerged. Rather than focusing on a specific chemical realization, we will discuss the likelihood of the emergence of life from a probabilistic point of view.

7.2 The Likelihood of Information

In the previous section we had a glimpse of how life *could* have emerged on Earth. This scenario requires just the right environmental circumstances: a high temperature (70–$200°C$), an alkaline PH, freely available phosphates, as well as formaldehyde so that ribose can form (Kopetzki and Antonietti 2011) (but note that ribose and other sugars can also form in cold environments, such as in the icy grains that end up forming asteroids and comets; Meinert et al. 2016). Because any incipient replication system needs to be powered, it would be highly advantageous if the reactions took place in a nonequilibrium environment that easily delivers energy to any ongoing reaction. Incidentally, there are candidate environments of this sort still on Earth, such as the hydrothermal vents of the Lost City in the mid-Atlantic ocean. These vents produce significant amounts of hydrocarbons spontaneously (Proskurowski et al. 2008) and provide an abundant flux of hydrogen that could be used to power a rudimentary metabolism. Incidentally, phylometabolic reconstructions of the ancient bacterial metabolism suggest that those ancient types used a reductive version of the citric acid (or TCA) cycle, which is essentially the standard cycle but run in reverse (Braakman and Smith 2012, 2013, 2014). This type of metabolism must use H_2 (but could use H_2S), and is found almost exclusively in bacteria that live near hydrothermal vents.

We can imagine that the right environmental conditions can significantly increase the chance of the spontaneous emergence of life. But how do you calculate this likelihood? In the following, we will first ask what that likelihood is in the absence of any other specific enabling factors, and then study how favorable conditions can increase that chance exponentially.

It is easy to calculate the likelihood of finding any particular polymer (a sequence of monomers) that is assembled in a random process. Let us investigate this problem using the alphabet of the English language at first. I mentioned earlier that the power of information theory is that it can handle any alphabet, so let us study one that we are all intimately familiar with. The procedure I will develop using English is easily adapted to arbitrary alphabets. We will first restrict the English alphabet to the twenty-six lowercase letters a - z (for reasons that will become apparent in a little while). If you fish in a well-mixed bowl of these letters, the chance that you will come up with the letter "o," for example, is $1/26$. The chance that you obtain the word "or" when fishing for two-letter words is the square of the probability of obtaining any single letter, so $P(\text{or}) = \left(\frac{1}{26}\right)^2 \approx 1.5 \times 10^{-3}$. What is the chance that you generate a single seven-letter word (such as "`origins`") by chance? Simple: the answer is $P(\texttt{origins}) = \left(\frac{1}{26}\right)^7 \approx 1.3 \times 10^{-10}$, or about one in eight billion fishing expeditions. You can see right away that the chance of finding any particular long sequence of letters by chance quickly becomes astronomically small.

The mathematical problem of generating random sequences by chance is sometimes cast in the image of monkeys blindly hitting the keys of typewriters. This metaphor [often traced back to the French mathematician Émile Borel (1913)] is used to formulate the "infinite monkey theorem," which stipulates that the eponymous stochastic dactylic monkey would be able to type just about any text (including all of Shakespeare's works), if given an infinite amount of time.

However, the theorem is misleading, because "infinite" is a concept quite different from "really extraordinarily long." Even correctly typing out the first thirty characters of Hamlet's soliloquy in the Shakespearean play ("To be or not to be . . .") cannot occur during the time our universe has been around (about 4.36×10^{17} seconds), as Hamlet's 30-mer occurs by chance about once in 3×10^{42} attempts. Compare this to the ridiculousness of typing out *all* of *Hamlet*. There are a little more than 130,000 letters in that play (we are not counting spaces and capitalization here). To get all of *Hamlet* by chance on the first fishing expedition is now one in $26^{130,000} \approx 3.4 \times 10^{183,946}$. This means that to get the whole text flawlessly about once, $3.4 \times 10^{183,946}$ attempts have to be made.

To put these numbers in perspective, keep in mind that there are "only" about 10^{80} atoms in the entire universe. As discussed earlier, just about 4.36×10^{17} seconds have elapsed since the Big Bang (assuming an age of the universe of 13.8 billion years; Planck Collaboration 2016). If we would assume that you can do a single "try" of hammering out *Hamlet* within a single second, it would take a single monkey about 183,946-17 = 183,929 ages of the universe to get *Hamlet* right once. If there were as many monkeys as there are atoms in the universe furiously typing out *Hamlet* attempts once every second (a preposterous assumption, as monkeys must be made out of atoms), this still does not make a dent in the odds (just subtract 80 from 183,929). For any finite universe—finite in space and time—the likelihood to obtain *Hamlet* by chance is zero. Utterly zero.

Granted, *Hamlet* is a big piece of information, but this exercise gives us a glimpse of the scale of the problem. Let us investigate the chance of the spontaneous generation of a more reasonably sized piece of information, say, 50 mers of information (which, if it is encoded in a quaternary alphabet, is equal to 100 bits; recall section 2.3.3). How likely is that? Before we answer this question, let us look at our assumption of how information must be encoded: in a linear chain of letters.

It might appear at first sight that focusing on the self-replication of linear molecules with fixed length is a rather strong assumption, and to some extent it is. For example, is it not possible to store information in other ways than linear sequences (that is, polymers)? In fact, there is no doubt that information *can* be stored in other ways. For example, it is possible to store information in a biased distribution of monomers in a well-mixed ensemble of monomers, a bag—or a soup—as it were.

Imagine, for a moment, an urn that holds different letters (as opposed to the more common colored balls). If the urn holds N letters total, and there are N_c different types (the number of colors), then there are $\binom{N+N_c}{N_c}$ different ways to apportion those (see Exercise **7.1**). If each of the different compositions of the bag is used to encode information, the entropy $\log \binom{N+N_c}{N_c}$ is large enough to encode the information for a ribozyme (if N_c and N are large enough). This is, in essence, the idea behind the GARD (Graded Autocatalysis Replication Domain) model of the origin of life (Segré et al. 2000; Segré and Lancet 2000; Shenhav et al. 2003). The advantage of such "compositional genomes" over the more traditional sequence-based genomes is that their replication is much simpler: it only requires the (catalyzed) duplication of the monomer concentrations in the assembly of molecules, and a subsequent splitting of the assembly.

However, the decoding of the information contained in the assembly is a much harder problem: while sequence-based information can be decoded "step by step" (Cover and Thomas 1991), this is not possible for compositional genomes. As there are no specific letter locations, information is carried by the entire assembly. So to decode an assembly, knowledge of the concentration of every molecule in the "bag" is required. The decoding of linear genetic information, on the contrary, can proceed codon-by-codon, so that even a huge molecule such as titin (a protein composed of 244 individually folded protein domains, encoded by between 27,000 and 33,000 amino acids) can be constructed from the information encoded in the DNA sequence.

Assuming therefore that the self-replicator is a linear molecule is indeed a constraint, albeit one that has strong support from coding-theoretical considerations. Assuming that this linear molecule has a fixed length also seems like a drastic assumption, but is not in reality. Rather, it is simply a shorthand for the total amount of information contained in a group of sequences. It is clear that modern life is encoded in a myriad of molecules that depend on each other for replication. The information that produces these molecules, however, is contained in just a few strands of information (the set of chromosomes), or one (if there is only one chromosome). That the information is not actually encoded in a single sequence of given length is immaterial for the information-theoretic analysis: for the purpose of mathematics we can just replace the total information (even if fragmented across multiple sequences) by a single sequence that contains all of the information that is inherent in any of the molecules participating in the process.

Let us thus focus on molecules of fixed length ℓ, with monomers taken from an alphabet of size D. We can readily calculate the likelihood L that any particular sequence of that length is assembled by chance: it is (just as in the simian production of Shakespearean poetry) given by

$$L = D^{-\ell}. \tag{7.1}$$

This equation (7.1) is *not* the likelihood to produce ℓ mers of *information* by chance, because as we saw in section 2.3.3, information is *embedded* in a sequence $X = X_1 X_2 \cdots X_\ell$, and the amount of information stored within a sequence is guaranteed to be smaller than (or equal to) the length ℓ of the sequence (if both of them are measured in the unit "mer"). In fact, smaller by the amount of entropy in the sequence [confer Eq. (2.50)]:

$$I = \ell - H(X|e). \tag{7.2}$$

Here $H(X|e)$ is the entropy of the ensemble of sequences X under mutation-selection balance, given the particular environment e that the sequence finds itself in. We are reminded again that information content is contextual: If the sequence of letters means nothing in that environment (if no active molecule is formed from it) then mutations of the sequence will be inconsequential, and the entropy $H(X|e) = H(X)$ will be approximately ℓ, that is, the information vanishes on account of (7.2). If, on the contrary, every monomer is so meaningful that any alteration ruins the sequence's ability to produce offspring, then the entropy vanishes (no alternatives are tolerated), and the information equals the sequence length. This is true, for example, for the string `origins`: barring our brain's ability to infer meaning even in altered strings, any mutation of the string does not spell `origins`, and thus eliminates the string's "meaning." But this is not true for biological sequences, of course: there are many sequences that differ in multiple letters but still have the exact same function.

Calculating information content via Equation (7.2) looks innocuous enough, but is (in most practical cases) intractable. The entropy $H(X|e)$, after all, is a sum over all sequences (of the particular length ℓ chosen) that encode the information—given the particular environment—and calculating that sum requires us to know how likely it is to find any such sequence in an infinite population.

Let us call that particular environment variable $E = e$, and N_e is the number of sequences that encode that e-dependent information. We can further simplify the problem by assuming that all the sequences encoding a particular function are *equally likely*. In that case the likelihood to find a sequence with information I by chance is

$$L = \frac{N_e}{N} = D^{-I}, \tag{7.3}$$

where $I = \ell - H(X|e)$ and $H(X|e)$ is the entropy of the potentially informative sequences. Since all the informative sequences are all equally likely by our assumption, i.e., $p_i = 1/N_e$, $\forall i$, we obtain

$$H(X|e) = \log_D(N_e). \tag{7.4}$$

Of course, sequences encoding a particular information are not all equally likely, so (7.4) is only an approximation of the true conditional entropy.

We thus see from Equation (7.3) that informative sequences are still exponentially unlikely, but not as unlikely as the sequence length might suggest as $I \leq \ell$. But Equation (7.3) still makes the spontaneous generation of a significant amount of information (say 100 bits), extremely unlikely: the likelihood of 100 bits of information (a ribozyme of fifty informative mers) $L_{50} = 2^{-100} \approx 10^{-30}$.

How many sequences can be "tried," that is, how many sequences can randomly polymerize per second, on an early Earth? Of course we do not know the answer to this question, because we do not know how many could be tried on the early Earth in parallel. The fact that the Avogadro number (the number of molecules in a mole of a particular substance) is large (about 6×10^{23}) does not help here, because we have to rely on very specific molecules to polymerize: those that constitute our alphabet, and they are spontaneously synthesized in much smaller numbers. Even if we (preposterously) assume that there are a billion places where these "experiments" take place in parallel, this is not nearly sufficient because from the formation of the Earth to the first appearance of life (say 500 million years), "only" 1.5×10^{16} seconds have passed. Even if we assume one polymerization per second (which is at least an order of magnitude too fast because the polymerization is not assisted by catalysis, of course), we are missing five orders of magnitude.

Is the spontaneous generation of information thus impossible? The answer to this question is no because the equation (7.3) relies on an assumption that is unrealistic and also leads to a dramatic underestimate of the actual likelihood. It assumes, namely, that during the random polymerization, each monomer is available with equal probability. If you recall, we applied the same formula to calculate the likelihood with which blind monkeys would type the word `origins` by chance, and they had an equal likelihood of typing any of the twenty-six letters. But it is almost certain that the rates of formation of the letters of the biochemical alphabet by spontaneous organic synthesis reactions are all very different. For example, purines and pyrimidines are formed under laboratory conditions at very different rates (Basile et al. 1984).

Let us investigate what happens if, by chance, the availability (meaning the rate of spontaneous formation) of monomers is close to the rate at which you would find them in a functional informative molecule. We can test this first with English, as the letters in English text also do not appear with equal likelihood, of course. Instead, we find "e" more often than "t," followed by "a," and so forth (see Fig. 7.4). If we use a "biased typewriter" (Adami and LaBar 2017) that produces letters with the probabilities shown in Figure 7.4, generating the word `origins` suddenly becomes much more likely. We can estimate this enhancement using the following arguments (Adami 2015).

Let us first imagine that rather than creating each monomer (randomly) with equal likelihood $1/D$, monomer i is produced with probability π_i, with $\sum_i^D \pi_i = 1$, but with no other assumption about that probability distribution. The uncertainty at each position is then

$$H_a = -\sum_{i=1}^{D} \pi_i \log_D \pi_i, \tag{7.5}$$

which is guaranteed to be smaller than 1 by the laws of information theory. Perhaps a reader is concerned that, given the arguments advanced here, an

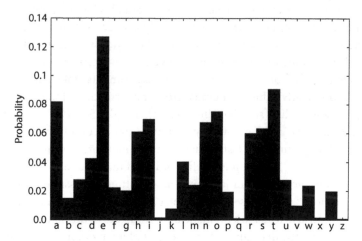

FIGURE 7.4. The probability distribution of letters in the English language. Data from Lewand (2000).

entropy that is smaller than 1 (here, smaller than the maximal entropy for a monomer) must be a conditional entropy, even though this entropy is clearly the entropy of a random ensemble. This is indeed true: (7.5) is a conditional entropy, but the condition here is not one that tests for functionality, but instead one that constrains the abundance distribution of monomers via differences in the rate of formation of monomers.

The entropy of sequences of length ℓ produced using such a process is ℓH_a, where the subscript "a" stands for abiotic because these are abiotically synthesized sequences. As only sequences with this entropy are produced, they represent the denominator of the likelihood equation (7.3), and thus

$$L_\star = D^{-(\ell H_a - H(X|e))},\tag{7.6}$$

compared with the unbiased likelihood

$$L_0 = D^{-(\ell - H(X|e))}.\tag{7.7}$$

In the limit where the bias disappears, naturally $H_a \to 1$ and $L_\star \to L_0$. Now let us try to estimate the entropy of replicator sequences H. Previously I just wrote $H(X|e) = \log N_e$, but a more precise formula is[2]

$$H = -\sum_{j=1}^{N_e} p_j \log p_j,\tag{7.8}$$

where p_j is the probability to find the jth-type replicator in a (very large) ensemble of sequences. The estimate $H = \log N_e$ is obtained from (7.8) by assuming that each replicator has the same probability $p_j = 1/N_e$.

Equation (7.8) is useless for estimating the entropy, as the probabilities p_j are unknown. But it is possible to rewrite H in terms of the entropy of monomers, pairs of monomers, triplets, etc., using a formula that is credited to the computer scientist and information-theorist Robert Fano (Fano 1961)

$$H = \sum_{i=1}^{\ell} H(i) - \sum_{i>j}^{\ell} H(i:j) + \sum_{i>j>k}^{\ell} H(i:j:k) - \cdots\tag{7.9}$$

Incidentally, we have already seen the first three terms of this equation when we introduced the information content of a protein, as Equation (2.45).

---

2. I drop the reference to the environment e from now on, since all entropies that are not maximal should always be understood to be conditional on the environment within which the sequences are functional.

Using this equation appears not to help matters greatly, as it has an enormous number of terms: there are ℓ terms in the first sum, $\frac{1}{2}\ell(\ell-1)$ in the second, and so on. The last term in the sum is the mutual entropy between all sites $(-1)^{(\ell-1)}H(1{:}2{:}3{:}\cdots{:}\ell)$ (the terms in Equation (7.9) are alternating in sign).

If all these terms significantly contribute to the entropy H, then we would be no better off than using (7.8). Fortunately, it appears that the higher-order corrections (triplets or higher) in biomolecular sequences can mostly be ignored,[3] while the pairwise terms $H(i{:}j)$ can be important as we have seen in section 2.3.3.

For our purposes here, let us ignore even the pairwise terms to get a first estimate. We would then approximate H as

$$H = \sum_{i=1}^{\ell} H(i), \qquad (7.10)$$

and this equation now gives us the approximate average entropy-per-site of biological, that is functional, sequences that we call H_b,

$$H_b = \frac{1}{\ell} \sum_{i=1}^{\ell} H(i). \qquad (7.11)$$

We can then write the information in terms of the difference of abiotic and biotic entropies as

$$I = \ell(H_a - H_b), \qquad (7.12)$$

which is guaranteed to be smaller than (7.2) because $H_a < 1$. How much smaller can I become, and how does this affect the likelihood of that information's emergence?

Let us take a closer look at H_b. According to (7.11), this is the average entropy-per-site, given the particular environment. The entropy per site is obtained by performing sequence alignments of functional molecules as we recall from chapter 2, and at each site the entropy is determined by the probability distribution of monomers at this particular site. We can think of the *average* entropy per-site as determining the *average* likelihood to find any of the D monomers at any site, much like the distribution shown in Figure 7.4

---

3. We should keep in mind that this statement is for the most part hopeful, as statistics to cement such a finding are very difficult to come by.

is the *average* likelihood to find any of the twenty-six letters at an arbitrary position in a word.[4]

Let us call this "biased" monomer distribution of functional replicators q_j with $j = 1, \ldots, D$, and the average biotic entropy thus

$$H_b \approx - \sum_j^D q_j \log_D q_j, \qquad (7.13)$$

where the $\approx$ instead of $=$ reminds us of the caution in the footnote. A quick sanity check shows us that if $q_j = 1/D$ (that is, you can find each monomer with equal probability), then indeed $H_b = 1$ and $I = 0$ according to (7.2) and writing $H = \ell H_b$. Using the abiotic and biotic entropy estimates, we can now write the information (7.12) as

$$I = \ell(- \sum_j^D \pi_j \log_D \pi_j + \sum_j^D q_j \log_D q_j). \qquad (7.14)$$

Remember that to obtain (7.14) we only assumed that random sequences are produced with a biased distribution π_j, and functional sequences occur on average with monomer probabilities q_j.

Now let us imagine that, by sheer chance, the probability distribution with which monomers are produced; in other words, π_j, happens to be close to the probability distribution of monomers in *actual* replicators q_j. This is, to wit, the assumption that we made when creating text with the same biased probability distribution of English text above. We'll imagine that the two probability distributions are close as described by a parameter ϵ, so $q_j = \pi_j(1 + \epsilon_j)$, with small $\epsilon_j \ll 1$ parameters that are symmetrically distributed around zero, and that satisfy $\sum_j \epsilon_j = 1$. It is then straightforward to show that

$$I = \ell(- \sum_j^D \pi_j \log_D \pi_j + \sum_j^D q_j \log_D q_j) \approx \ell(\sum_j^D \pi_j \epsilon_j^2) + \mathcal{O}(\epsilon^4). \quad (7.15)$$

Note that because we assumed here that the biotic distribution is *derived* from the abiotic one ($q_j = \pi_j(1 + \epsilon_j)$), the information (7.15) is guaranteed to be positive.

---

4. Note that the average entropy is not automatically equal to the entropy of the average probability distribution, but unless the distributions are very skewed, the two quantities are usually close.

How big of a difference in the likelihood of spontaneous generation of information could such an innocuous (and lucky) change in the probability distribution make? We will see that it can be quite significant, and possibly dramatic. First, let us test the predictions made by this approximation for the likelihood to find the word `origins` when randomly sampling the English alphabet. As we discussed earlier, the unbiased likelihood is

$$L_0 = 26^{-7(1-H_b)} = 26^{-7} \approx 1.3 \times 10^{-10}, \tag{7.16}$$

or about one in eight billion (H_b is zero in this case, because only a single word of $\ell = 7$ spells `origins`).

The estimate for the biased likelihood $L_\star$ using the distribution π_i of English letter usage (shown in Fig. 7.4) involves the entropy of that distribution, which turns out to be $H[\pi_i] = 0.89$. While this value is not dramatically smaller than the entropy of the unbiased distribution, the likelihood according to this theory should be

$$L_\star \approx 26^{-7*0.89} = 26^{-6.23} \approx 1.53 \times 10^{-9}, \tag{7.17}$$

an enhancement of $L_\star/L_0 \approx 12$. And indeed, when we actually searched one billion sequences of length 7 for the word `origins` using the biased distribution, we found it twice (Adami and LaBar 2017), in line with the estimate (7.17).

Compared to the astronomical orders of magnitude that we have been discussing, an enhancement of a single order of magnitude seems tame. But keep in mind that this enhancement pertains to the likelihood to generate a single word of $\ell = 7$ (a little less than 33 bits). Consider instead a protein of ninety-nine monomers, such as the protease of the HIV-1 virus. The likelihood to find such a protein by chance is not 20^{-99} of course, because there are many sequences of that length that are fully functional proteases. We do not know how many there are in total, but we can infer this number from the information content of the protein, which turns out to be around 76 mers (see Fig. 3.20 and Gupta and Adami 2016[5]).

If we were to use formula $L_0 = D^{-I}$ to calculate the likelihood of spontaneous emergence of such a protein, we would infer $L_0 \approx 1.32 \times 10^{-99}$, an absurdly low probability. But if the distribution of amino acids was just somewhat skewed in the direction of the probability distribution of amino acids in typical proteins (say from a uniform distribution to one with entropy 0.9

---

5. This number is uncertain because we ignored the first fifteen and last nine residues in (Gupta and Adami 2016) because sequence statistics is poor at the beginning and the end of the protein due to limitations of sequencing technology.

per monomer), the likelihood to discover such a protein by chance would still be astronomically low, but about fourteen orders of magnitude more likely ($L_* \approx 2 \times 10^{-85}$). In the extraordinarily lucky circumstance that the abiotic entropy per site is reduced to 0.5 from 1, the likelihood becomes a mere $L_* = 20^{-99(0.5-0.2323)} \approx 3.33 \times 10^{-35}$, an enhancement of sixty-four orders of magnitude! (Here, $23/99 \approx 0.2323$ is just the average entropy per mer of a 99-mer protein with 76 mers of information.) Of course we discussed the example of a physical protein here just for the purpose of illustration, as we know full well that the HIV-1 protease emerged via evolution, not spontaneous abiogenesis.

On paper, then, a biased probability distribution could work wonders. Yes, the environmental conditions to pull this off are probably very rare, but finding such conditions is still far more probable than the likelihood of finding a large piece of information by chance. But does this theory actually hold water when tested?

7.3   Experiments with Digital Abiogenesis

Testing the theory of the "biased typewriter" in biochemistry is, of course, absurd (but see further below for experiments to enhance the likelihood of randomly finding molecules with increased *functional capacity*, as opposed to enhancing the likelihood of self-replication). We can, however, test this using Avida, the digital life system that I described in previous sections. We begin evolution experiments using Avida with a hand-written ancestor for a very good reason: self-replicating sequences are also extremely rare in the avidian "logical" chemistry. They are rare for the same reason that biochemical self-replicators are rare: they need to encode information about how to self-replicate in that computer world, and information is by definition rare.

7.3.1   Unbiased search for replicators

A typical avidian ancestral replicator sequence in Avida is a 15 mer with an alphabet size of $D = 26$ (coincidentally, you might think, the same number as letters in the English alphabet that we have been using). Finding a 15 mer by chance in such an alphabet is a low probability event, as we have already established: $L_0 = 26^{-15} \approx 6 \times 10^{-22}$. Just to give an idea how rare any particular 15 mer in this language is, I calculated that if you were able to test one million sequences per second on a single CPU, and ran 1,000 such CPUs in parallel (this is easily done using today's computers) it would still take on average about 50,000 years to find that sequence (Adami and LaBar 2017). However, it is clear that the 15-mer handwritten self-replicator has significantly less than 15 mers of information, so we should be able to find one much faster than that. How fast depends on how much information is contained in it. We could estimate the information content of the replicator using the methods outlined in chapter 2, but we can get a much better estimate by simply trying to find it! According to Equation (7.3), the information content is given by

$$I = -\log_D \left(\frac{N_e}{N} \right),$$ (7.18)

where N_e is the number of functional (here, self-replicating) sequences among the N that were tried.

Using a uniform probability distribution to generate the twenty-six monomers, we searched for self-replicators by generating one billion random strings, and found fifty-eight that were able to replicate (Adami and LaBar 2017). Mind you, these were not all perfect replicators (sequences that make exact copies when the mutation rate is zero). Most of them created shorter sequences that created flawed copies that were able to create other flawed copies and so on. Our only criterion for replication here was that the sequence was "colony-forming" (at zero mutation rate), meaning that it could produce an exponentially increasing number of sequences. But this number 58 immediately allows us to estimate the information content of the 15-mer replicators as

$$I(15) = -\log_{26}(58 \times 10^{-9}) \approx 5.11 \pm 0.04 \,\text{mers}.$$ (7.19)

So just over 5 mers of the 15 instructions were information! After we obtained this result, we did what surely any reader would immediately have tried: are there replicators made of only 5 mers?

There are only 11,881,376 different sequences of length 5, and they are easily tested: none of them can form colonies. Next we tested if perhaps 6-mers could replicate. There are 308,915,776 such sequences, and we tested them all: none of them can form colonies. There are over eight billion different 7-mers, and none of them are replicators either!

When sampling a billion 8-mers, however, we found six unique self-replicators among that set, giving us an estimate of

$$I(8) = -\log_{26}(6 \times 10^{-9}) \approx 5.81 \pm 0.13 \,\text{mers}.$$ (7.20)

At first sight this result is surprising: this is quite a bit larger than the information content of the 15-mer that we estimated in Equation (7.19). Should not the information necessary to replicate be independent of the sequence that this information is embedded within? It turns out that this is only true as a first approximation, and we will discover the reasons for the deviation below.

Before we get to that, let us look at some more empirical results. We searched a billion candidate sequences of length 30, and found 106 replicators (Adami and LaBar 2017), suggesting

$$I(30) = -\log_{26}(106 \times 10^{-9}) \approx 4.93 \pm 0.03 \text{ mers.} \qquad (7.21)$$

This trend suggests that the longer the embedding sequence, the smaller the information in it, but a continuing trend would be absurd, of course. To test this, we checked sequences of length 100. This takes much longer (because it takes so much longer to run programs with 100 instructions compared to those with between 6 and 30 instructions), so we only checked 300 million sequences instead of the billion. We still found seventeen replicators amongst that set, giving us the estimate

$$I(100) = -\log_{26}\left(\frac{17}{3} \times 10^{-8}\right) \approx 5.10 \pm 0.09 \text{ mers,} \qquad (7.22)$$

which is significantly higher than $I(30)$. You can see the dependence of the information content on the length of the embedding sequence in Figure 7.5. That figure shows that the information (as a function of embedding sequence length) first decreases, then increases again. This seems odd at first as Equation (7.18) does not depend on the sequence length, but this "violation" can be explained quite naturally.

The argument that the embedding sequence length can in fact influence the information content runs as follows. Consider embedding the 7-mer sequence origins into an 8-mer. The only way we can keep the information intact is by adding a monomer to the beginning or the end of the sequence. Adding a random letter to the beginning gives us twenty-six sequences that spell origins, and adding a letter to the end gives us another twenty-six so that $N_e(8) = 52$, that is, there are fifty-two sequences that spell origins among 8-mers. But the number of *possible* 8-mers N has only increased by a factor 26, so that the ratio N_e/N has increased, which implies that the information has decreased. This decrease of information, however, cannot persist as sequence length increases. The reason for that is also instructive.

While the meaning of the sequence origins is not influenced by other letters that may appear on the string that it is embedded in (unless such a letter interrupts the sequence), this is not true for biochemical or avidian sequences. Imagine that among the instructions that make up the set of twenty-six, one instruction can stop self-replication. In truth, there is not any such instruction among the avidian instruction set, but some instructions are quite "poisonous" and are generally avoided (Dorn et al. 2011). For the sake of the argument, imagine that there is such a detrimental instruction, and that it

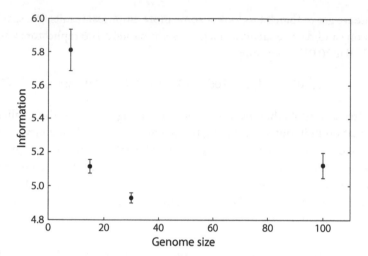

FIGURE 7.5. Information content of self-replicators (in units mer) found in a random search as determined by Equation (7.18), as a function of the embedding genome size, estimated from sampling 10^9 sequences (except for $\ell = 100$, where we sampled only 3×10^8 sequences). Error bars are 1σ standard deviations. Adapted from Adami and LaBar (2017).

is so detrimental that its mere presence in the code spells doom for the organism: a `kill` instruction. In that case, a successful replicator would be much harder to find among long sequences, where the `kill` instruction would appear about once every twenty-six instructions. In other words, to have a self-replicator, you not only need the right sequence of instructions, you also need the *absence* of the `kill` instruction. From the point of view of information theory, the absence of something can be just as much information as the presence of something, so we expect the information necessary to self-replicate to increase as the embedding sequence increases in length, which is precisely what we see in Figure 7.5.

7.3.2   Biased search

While a search for self-replicators in an unbiased manner (where each sequence tested has an equal likelihood of occurring by chance) seems intuitive, it is also a very inefficient procedure. We know for a fact that functional molecular sequences *cluster* in genetic space, which means that if you have found one, you have found many others. Very many others. Other areas of genetic space are a barren empty desert. If only there would be a way to avoid the sterile areas and search only the fertile ones!

It turns out that the search with a biased monomer distribution that we discussed above is precisely that trick. To test whether it works with the search for replicators, we have to bias the distribution with one that is biased toward self-replicators, just as we biased the search for an English word with a probability distribution appropriate for English words. We can test this easily because we do have plenty of self-replicators that we found via a random search. For example—as discussed above—we found fifty-eight self-replicators of length $\ell = 15$. We can use them to find out how often any instruction appears on average in this functional set. We can then use this frequency distribution to define a biased probability distribution of instructions in replicators, meaning we would design a random process that produces monomers at precisely the frequency with which it appears in this set of fifty-eight. Here, we assume that these fifty-eight replicators are somehow "typical" of any self-replicators. Of course, the more replicators we have at our disposal, the more accurate that probability distribution becomes (until you have discovered all replicators at that length).

So what happens if we use that probability distribution—that "biased typewriter"—to write the random sequences? The answer is that you find many many more self-replicators. When we tested a billion sequences of length 15 using the biased prior, we found 14,495 replicators, compared to the fifty-eight in the unbiased search. According to our information content formula (7.18), this corresponds to a measly 3.42 mers. If we agree that the minimum information in a self-replicator is about 4.8 mers from inspection of Figure 7.5, the biased distribution provides at least 1.4 mers of information *for free*. Indeed, the "enhancement factor" E, which is the ratio of replicators found using a biased distribution divided by the number found for an unbiased prior, is $E = 14,495/58 \approx 250$, which is worth 1.4 mers (1 mer gives an enhancement of 26, of course).

It is worth asking how the "theory of the biased typewriter" that I described in section 7.2 holds up to the empirical evidence. The theory predicts that the unbiased likelihood $L_0 = 26^{-(\ell-H)}$ (with $\ell = 15$) is enhanced to a likelihood $L_* = 26^{-\ell(H_a-H)}$, where H_a is the abiotic "prior" (and biased) entropy, which is smaller than 1. The probability distribution (Fig. 7.6) has an entropy $H_a \approx 0.91$ mers, giving rise to a predicted enhancement $E = L_*/L_0 \approx 81$, about three times smaller than the actual enhancement. There are many factors that can explain the theory's underestimate of the enhancement (given that the theory was crude at best) but we will not pursue those here. We should instead be encouraged that a biased distribution can achieve so much. In fact, we could ask, how much more can it achieve? For example, if we cobble up a probability distribution from the small set of replicators we found at $\ell = 8$ (all six of them), that distribution $p_*(8)$ is bound to have low entropy. Indeed,

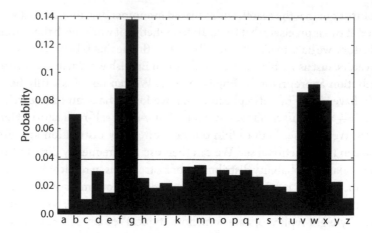

FIGURE 7.6. The biased probability distribution of Avida instructions obtained from the genomes of fifty-eight randomly generated $L = 15$ replicators (the meaning of each letter is described in Table 4.1). The solid black line represents the uniform distribution. From Adami and LaBar (2017).

that distribution (shown in Fig. 7.7[a]) is fairly sparse, but still with entropy $H_\star(8) = 0.71$ mers (Adami and LaBar 2017). Our crude theory predicts an enhancement of

$$E = L_\star/L_0 = \frac{26^{-\ell(H_\star - H_b)}}{26^{-\ell(1 - H_b)}} = 26^{\ell(1 - H_\star)}. \tag{7.23}$$

Plugging in $H_\star(8) = 0.71$ predicts an enhancement $E \approx 1,918$, but instead we only found an enhancement of about $E = 383$. For sequences of $\ell = 8$, the theory thus *overestimates* what we actually find. It is very likely that this overestimate is due to the fact that the low entropy $H_\star(8)$ is low only because the probability distribution $p_\star(8)$ was constructed using such a sparse data set. The low entropy does not reflect low uncertainty; it reflects insufficient evidence!

You can also test how searches for replicators with $\ell = 30$ might be enhanced by using the probability distribution of monomers of randomly generated replicators. Recall that we found 106 such sequences per billion. The probability distribution from this set is shown in Figure 7.7(b). The entropy of this distribution is fairly close to 1 ($H_\star(30) = 0.98$ mers) because the information to replicate only occupies about 5 mers of information. This means that the rest of the sequence is free to mutate, driving the entropy up. An entropy this high predicts a weak enhancement of $E = 26^{30(1 - 0.98)} \approx 7$. Empirically, however, we find an enhancement of almost 22 (see Fig. 7.8).

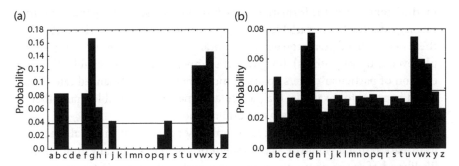

FIGURE 7.7. Probability distribution for avidian instructions in self-replicators. (a) $p_\star(8)$ obtained from the replicators of length $\ell = 8$, giving rise to an entropy $H_\star(8) = 0.71$ mers. (b) $p_\star(30)$ obtained from the replicators of length $\ell = 30$, giving rise to an entropy $H_\star(30) = 0.98$ mers. The solid horizontal line denotes the uniform probability distribution $1/26$ in both panels (from Adami and LaBar 2017).

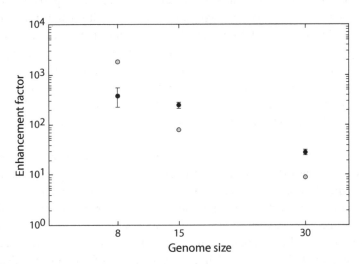

FIGURE 7.8. Empirical enhancement factor (black dots, with 1σ counting error), along with the predicted enhancement factor using the entropy of the distribution based on Equation (7.23) (gray dots) for $\ell = 8, 15, 30$. From Adami and LaBar (2017).

7.3.3 Consequences for spontaneous abiogenesis

What do these results imply for the prospects of a chance emergence of life on this planet (or any other, for that matter)? First, we have learned that the likelihood of information does not just depend on the sequence, it also depends on the likelihood of the formation of each monomer, something that is pretty much guaranteed to be variable in a complex biochemistry. More importantly, these rates depend greatly on environmental conditions (such as temperature, pH level, and many others). Imagine if there are processes that enable the production of particular dimers (pairs of monomers) at an enhanced rate. This could happen, for example, if a particular dimer is more stable than others, and thus does not degrade as fast as other dimers.

Such a prior, even though it is completely agnostic of the final piece of information that it would be a component of, could dramatically improve the likelihood of spontaneous emergence over and above a bias in the monomer production. For example, let us return to the likelihood of producing the word $origin$ by typewriting simians. Beyond the distribution of letters we saw in Figure 7.4, particular pairs of letters are also significantly more likely. Not surprisingly, the dimer "th" is much more common (at 3.56 percent in English) than the letter frequency of "t" and "h" multiplied. The next most common dimers in English are "he" (at 3.07 percent) and "in" (2.43 percent). These do not help much in making "origin" more likely, but it turns out that "or" is enhanced in English, at 1.28 percent (keep in mind that under the uniform assumption, any dimer should occur at the frequency of 0.15 percent). The combination "ig" is also somewhat enhanced, at 0.255 percent. If we thus would use a "dimer typewriter" that produces pairs of letters with the frequency that they occur in actual English, the word "origin" would likely be formed fairly fast. This is a good computational exercise that is left to the reader as Exercise **7.2**.

It turns out that these considerations carry over to the avidian sequences too. For example, in the replicators of length $L = 8$, even though we found only six of them while scanning a billion, the dimer gb appeared significantly more often than the product of the likelihood of g and b might suggest. Indeed, gb is a common sequence motif in self-replicating avidians (LaBar et al. 2016). Thus, if local conditions could influence dimer formation favorably, and these dimers happen to be the kind of dimers expected in informational molecules, then in those locales the rate of spontaneous emergence of information is exponentially amplified. Of course, none of these arguments guarantee the emergence of self-replicators in any particular local chemistry: the hard work of the planetary geochemist is still needed. However, the arguments provide a principled way to gauge the likelihood of spontaneous emergence, and the factors affecting them. Thus, mathematics does not rule out that life emerged in "some warm little pond with all sorts of ammonia and phosphoric salts," but rather points into the direction of some very specific warm little ponds.

7.4   The Fitness Landscape before Evolution

The central problem that we have to surmount when it comes to understanding the origin of life is that there appears to be an insurmountable gap between the minimum amount of information it takes to self-replicate, and the amount of information that can appear by chance, even if we account for a biased monomer distribution that can deliver a significant amount of information "for free," as it were, via the enhancement factor the bias provides. In order to bridge this gap, we need to uncover a dynamical process that will drive information to higher levels in a nonequilibrium fashion, perhaps in a ratchet manner (see the sketch Fig. 7.9, in which nothing is to scale). To this day, no one has identified such a process (but Fig. 7.3 sketched out one such possible path). But even if we could identify it, can such a process reach the flanks of the fitness peaks of the primordial fitness landscape: the landscape of the simplest self-replicators? What does this landscape even look like? Can we even ask this question? What is the fitness landscape before evolution?

Clearly, we cannot answer this question if we do not even know what alphabet was at the origin of the first self-replicators here on Earth (or elsewhere). But in this book we have used a surrogate digital world and associated fitness landscape several times in order to make predictions about biological evolutionary processes and landscapes. Can we do the same thing here? In the previous sections we discovered that within the avidian chemistry, the

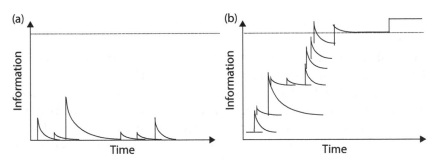

FIGURE 7.9. (a) Information dynamics in the absence of replication. Any spontaneous emergence of information is unstable and will decay, according to the second law of thermodynamics. The dashed line represents the minimum information necessary to achieve self-replication. (b) Spontaneous generation of information in a putative (as yet to be discovered) nonequilibrium process will build upon existing spontaneous information faster than the second law will destroy it, ultimately reaching the Darwinian threshold (dashed line), after which Darwinian evolution will lead to the accumulation of information via the Maxwell demon discussed in section 3.1.

simplest self-replicator is an 8-mer (we found six of them by random sampling and many more by biased sampling). Given that 8-mers are the simplest self-replicators in the avidian "primordial" landscape, we could now ask the question: what is the landscape of 8-mer self-replicators? In order to study this, we would have to generate all 26^8 possible 8-mers and find whether those self-replicators create a fitness landscape on which evolution is possible. Even though 26^8 represents over 200 billion sequences, it is possible to exhaustively test them for replication. Once we have done this, we will be able to, for the first time ever, look at a primordial fitness landscape that was not shaped by evolution: a landscape that is the consequence of statistics and computational chemistry only.

Among the more than 209 billion sequences of length 8 in the avidian world, we found exactly 914 sequences that could self-replicate in such a manner that colonies (groups of related replicators) of that sequence would form (C G et al. 2017). Note again that in this context, it is important to distinguish "replication" from colony-forming, because while some sequences can create other sequences via replication, if those sequences are not able to create "fertile" copies themselves, such sequences will not be able to generate a colony, and in so doing preserve information going forward. In turn, the concept of "colony-forming" also does not insist on perfect error-free copies, as long as the information necessary to replicate is preserved. Using Equation (7.18) we can immediately calculate the exact amount of information in $L = 8$ replicators, and find

$$I(8) = -\log_{26}(914/26^8) \approx 5.9 \text{ mers},\qquad(7.24)$$

that is, almost 6 of the 8 mers are information [compare this to the estimate from sampling a billion sequences only, in Eq. (7.20)].

It turns out that the 914 replicators are not at all equally distributed across genetic space. Rather, we found that replicators are likely to be found close to other replicators, in "fertile" areas of sequence space that lie within vast barren regions where no replicator can be found anywhere (C G et al. 2017). The 914 replicators we found fall into only two broad groups of replicators that form a total of 13 clusters. These two groups coincide with two basic ways in which self-replication among avidians can be achieved, and can be identified from the monomer and dimer distribution functions. Earlier in section 7.3.2 we saw that self-replicators in Avida have a very biased distribution of monomer abundances (see in particular Fig. 7.7[a]). This bias is partly due to the algorithm that the sequences use in order to create copies of themselves. It turns out that in Avida, there are two main strategies to make copies, which can be distinguished by certain di-instruction motifs in the genome: the fg and

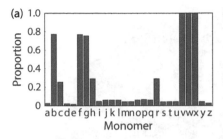
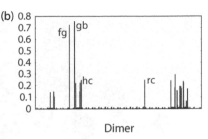

FIGURE 7.10. (a) Distribution of monomers within the set of 914 $L = 8$ self-replicators (proportion of self-replicators containing a given monomer). (b) Distribution of pairs of instructions (dimers). Dimers are ordered lexicographically on the x-axis. From C G et al. (2017).

the hc replicators (the letters refer to the instruction code, refer to Table 4.1, and see LaBar et al. 2016 for a detailed dissection of these different replication algorithms).

The monomer distribution function of the 914 replicators is shown in Figure 7.10(a). To construct this distribution, we plotted how often a particular instruction is present across the set of 914 replicators. As a consequence, instructions that are absolutely essential by themselves have probability 1.0, such as instructions v, w, and x (copy, allocate, and divide), while useless instructions will have low frequency. The di-instruction distribution reveals a number of highly amplified pairs of instructions, such as fg and gb, which are the hallmarks of one of the types of replicators (we call them "fg-replicators" for short), as well as the hc and rc motifs that occur in the group we call the "hc-replicators." The fg-replicators and hc-replicators form two distinct nonoverlapping groups, in the sense that no hc-replicator is a mutational neighbor of an fg-replicator. Figure 7.11 shows the complete set of replicators where the distance between two nodes is a reflection of their Hamming distance,[6] rendered as a multidimensional scaling (MDS) coordinate (see, for example, Hout et al. 2013). Each of the groups is composed of highly connected clusters of sequences that are mutational neighbors of each other, which suggests that perhaps an evolutionary process could take hold on such a prebiotic landscape. We can visualize this landscape by plotting fitness over the MDS clusters in Figure 7.11. The first surprise we find is that the fg-replicators are consistently higher in fitness than the hc-replicators.

6. Hamming space is the high-dimensional space of sequences. The distance between any two sequences (Hamming distance) is the smallest number of mutations that takes one sequence into the other.

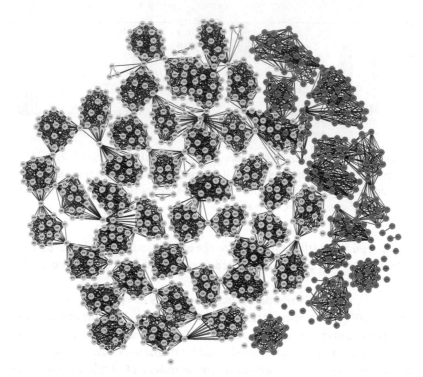

FIGURE 7.11. Complete set of all 914 length-8 replicators, shaded according to the class of motifs they contain (fg-replicators in light gray, hc-replicators in dark gray). Sequences that are one-mutant neighbors of each other are linked by an edge. The distance between any pair of nodes reflects their distance in Hamming space, rendered via multidimensional scaling. Dark- and light-gray clusters are never linked by an edge. One isolated fg-replicator (light gray) is close to an hc-replicator cluster (dark gray), but is not connected to it. From C G et al. (2017).

In Figure 7.12, the low fitness hc-replicators are on the periphery (just as in Fig. 7.11), and there are several high-fitness peaks that are close to each other, suggesting that evolution on this landscape could take off. Let us test this.

We now ask the following question: What if, in some process that we do not currently understand, all length 8 progenitors are generated with equal probability, and therefore evolution can take hold? Which of the 914 sequences would end up the ancestor of all life henceforth? We tested this by creating 200 replicate populations of these 914 ancestors, and recording how often any of the ancestral sequences (initially with only one copy in the population)

FIGURE 7.12. Ancestral fitness landscape of all primordial self-replicators of $L = 8$. The height of the peaks corresponds to the replication rate, while the $x - y$ coordinates are the MDS coordinates, as in the network shown in Figure 7.11. From C G et al. (2017).

became fixed, after a period of 50,000 updates (a few thousand generations on average). The results mirror closely the likelihood expected from the theory of fixation (outlined in Box 6.1). No sequence from the hc-replicator periphery ever became the ancestor of all life, but those sequences that represent the highest peaks in Figure 7.12 had a high likelihood of succeeding.

Standard fixation theory cannot perfectly predict what sequence ultimately becomes the ancestor, because fixation depends not only on the fitness advantage of the sequence, but also on the sequence's neighborhood. A peak that stands alone, for example, is unevolvable, and cannot lead to a successful evolutionary trajectory, while peaks that are surrounded by other peaks will have a much higher chance of succeeding, as predicted by the quasispecies fixation theory (Wilke 2003). Figure 7.13 shows the network of 914 ancestral self-replicators, with the sequences that became the ancestor of all life five or fewer times rendered in white, while the nodes in gray never became fixed even once across all 200 trials. The three nodes shaded black are the eventual winner in 130 or more competitions combined, that is, in 65 percent of all competitions, suggesting that evolution on a primordial fitness landscape is largely deterministic: for the most part, the same high-fitness sequences that are surrounded by other high-fitness sequences will ultimately carry the day.

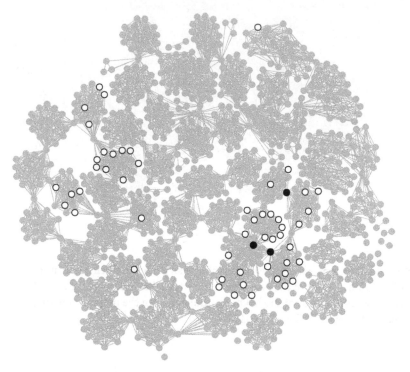

FIGURE 7.13. Network of 914 potential ancestors ($L = 8$ replicators) shaded according to their likelihood to become the ancestor of all life in 200 replicate evolution experiments in which all 914 were competing against each other for 50,000 updates. Replicators that were never the ancestor genotype of the final population are colored gray, while those that outcompete all other genotypes in fewer than 6 (out of 200) competitions are in white. The three genomes that eventually become the ancestor of all digital life in over 130 competitions are shown in black. From C G et al. (2017).

The results we found with $L = 8$ self-replicators will likely carry over to the landscape of more complex replicators, as the $L = 8$ self-replicators form the core of most more complex replicators. Indeed, an exhaustive analysis of all 5.4 trillion sequences of length 9 (see Box 7.1) appears to suggest just that.

7.5 Summary

How life originated, here on Earth—or for that matter anywhere else in the universe—is one of the most fascinating and at the same time thorny questions in all of science. Traditionally, this question is asked within the realm of chemistry (and in particular biochemistry), but once the problem is detached

Box 7.1. The $L = 9$ Landscape in Avida

With modern computers it is possible to push the boundaries of what was previously thought to be computationally feasible. After publication of the study that analyzed all possible replicators of length $L = 8$ in the avidian landscape (C G et al. 2017), some wondered whether it was possible to map the complete $L = 9$ landscape, and whether we can learn lessons about how the $L = 8$ landscape is "embedded" in the $L = 9$ landscape. For example, it is conceivable that a good fraction of the $L = 9$ replicators have $L = 8$ replicators at the core, and simply adding an instruction somewhere in many cases does not affect the replicator's function. Searching the $L = 9$ landscape for replicators requires testing over 5.43 trillion sequences, which, if you could check one sequence every millisecond (feasible on modern computers) would still cost about 172 years of CPU time. Fortunately, it is now possible to run such jobs in a massively parallel manner, which has allowed us to map the full $L = 9$ landscape in a few weeks. We found exactly 36,171 unique replicators, translating to an information content of (C. G. and Adami, 2021)

$$I(9) = -\log_{26}(36{,}171/26^9) \approx 5.77 \text{ mers}. \tag{7.25}$$

This number is consistent with the "law" of decreasing information in embedding seen in Figure 7.5. The monomer and dimer distribution across the set of $L = 9$ replicators (see Fig. 7.14) is similar to the distribution we have seen from the $L = 8$ replicators seen in Figure 7.10, but there are also differences. For example, the hc-class of replicators are much less pronounced among the $L = 9$ set, which implies that the b, f, and g instructions are in almost all replicators. At the same time, the rc dimer peak that was prominent in the $L = 8$ replicators has all but disappeared.

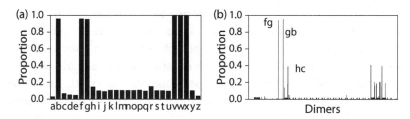

FIGURE 7.14. (a) Distribution of monomers across all 36,171 $L = 9$ primordial replicators. (b) Distribution of $L = 9$ dimers. Data courtesy of Nitash C G.

from any particular chemistry, we realize that the fundamental mystery is really about the origin of information. How can a chunk of information large enough so that it encodes its own replication emerge by chance? Naive arguments that assume that each monomer of the alphabet is formed with equal likelihood, and that strict thermodynamical equilibrium reigns everywhere, suggest a likelihood that is so small that it is effectively zero in a finite universe that has existed a finite amount of time. However, neither of these assumptions is warranted. A biased production rate—when biased in just the right way—can enhance the likelihood of spontaneous emergence of information by many orders of magnitude. At the same time, nonequilibrium processes (likely to be found on an early Earth) could give rise to a cascade of information accumulation that bests the destructive force of the second law. And perhaps most importantly, a slow process in which information about replication that is initially passive (meaning it is present in hardware in the form of a template-based error-prone replication system on a surface) gradually seeps into the sequence (that is, into software) because that information promotes faster copying, which could lead to information growth that ultimately will spark evolution. However, whether this is a viable path to the origin of life is not known, and it is very difficult to test experimentally.

Perhaps digital experiments could lead the way here too. Indeed, recent work with self-replicating computer programs [in a world called Amoeba (Pargellis 2001; 2003) that is similar, but also different from Avida] seems to suggest that a self-organizing process can lead from biased nonreplicators to the emergence of building blocks of self-replicators, to ultimately self-replicating sequences (Greenbaum and Pargellis 2017), in a process similar to the sketch in Figure 7.9(b).

What does a fitness landscape of informational replicators look like just before Darwinian evolution takes hold? If we knew this, the origin of information would not be a mystery anymore, but this is not a question we can answer in biochemistry. However, it is a question we can ask and answer in "digital chemistry." Even though digital self-replicators are vastly more likely than biochemical ones, they are still exceedingly rare, and an exhaustive search for the smallest self-replicators in Avida still takes a significant amount of time. The "primordial" landscape of the default chemistry in Avida reveals that just under 6 mers of information is necessary for replication, and that the landscape of all possible minimal replicators is already fairly complex, forming two main groups of replicators that fall into thirteen large clusters. Of the two types, one type has significantly higher fitness and is much more likely to be the root of the ensuing evolutionary tree. While even the digital realm does not give us hints at what the nonequilibrium dynamics are that bridge the "Darwinian gap," digital experiments strongly suggest that if the information can be found, Darwinian evolution can take hold immediately so as to shape the complexity of the life to come.

Exercises

7.1 (a) Show that the number of different configurations M for an urn containing N balls of n different colors is

$$M = \binom{N+n-1}{N}. \qquad (7.26)$$

Hint: For n colors, M is the number of solutions to the equation $x_1 + x_2 + \cdots + x_n = N$, where x_k is the number of balls of color k.

(b) Compare the potential information in the ensemble encoding from (a) to the potential information in an encoding using a linear polymer of length N using n colors.

7.2 Calculate the enhancement factor that applies to finding the word `origins` by chance when using the observed di-letter frequencies in English for `or`, `ig`, and `in` given in the text, above and beyond the enhancement coming from the single-letter frequencies.

8

Information for Cooperation

Tit for Tat. God knows it's an honorable attempt at stabilizing. In fact, it's the strategy of choice in the best textbooks.

—RICHARD POWERS (1988), *THE PRISONER'S DILEMMA*

Cooperation is found everywhere in life. We observe cooperation between cells in multicellular tissues (Michod and Roze 2001), we can see cooperation among bacteria of the same (Bassler 2002) and even different species (Keller and Surette 2006), and we can witness cooperation between animals in a multitude of complex and intricate forms. It is often said that cooperation is difficult to understand from the point of view of evolution because evolution rewards selfish behavior that maximizes the immediate benefit to the individual, while the benefits of cooperation are delayed, and may not directly benefit the cooperating individual. We will see in this chapter that, on the contrary, cooperation is very much selected for by evolution (in the right circumstances) and that in the overwhelming number of cases the benefit of cooperation *does* accrue to the individual, so that evolution can act to increase the propensity of that behavior.

The study of the emergence of cooperation in biology rose to prominence in the early 1960s. One of the first to point out that the ubiquity of cooperation may be at odds with evolutionary theory was the British evolutionary biologist William Hamilton (1936–2000). In his 1963 paper (Hamilton 1963), Hamilton observed that Darwin's theory certainly explains the evolution of any behavior that directly benefits the organism itself, but that it was less clear how to explain the evolution of behaviors that benefit others:

In particular this theory cannot account for any case where an animal behaves in such a way as to promote the advantages of other members of the species not its direct descendants at the expense of its own.

Indeed, this simple observation can cause consternation. Darwinian evolution works because an advantageous mutation promotes the propagation of precisely that mutation in the lineage of the individual where it occurred. If it were to promote somebody else's survival, how could that mutation ever reach fixation, as it would compete with the mutations of those that it helped to thrive? One explanation for the evolution of altruistic behaviors that has sometimes been advanced is called "group selection," a concept where the advantage of altruistic behavior somehow selects for groups in which this behavior occurs, and which would outcompete other groups in which cooperation was absent. However, it quickly became apparent that there is no mechanism by which groups could be selected for (as it is not groups that replicate, but individuals). Thus, even when an altruistic behavior benefits the group as a whole, cheaters will ultimately outcompete cooperators within any particular group leading to extinction of the cooperators.[1] As a consequence, explanations of altruism being selected for because they "benefit the species" are unworkable (Fisher 1958).

It turns out that another British population geneticist, John Haldane (who we have met earlier in this book in Box 6.1), had thought of the same problem in evolutionary theory before, and realized that it could be overcome with what is now known as "Hamilton's rule." Haldane argued that it would pay for him to sacrifice himself for "two brothers or eight cousins," implying that since he shared half of his genes with his siblings, saving two of those would make up for the loss of his genome (since after all in evolution, it is not the individual that counts, but rather information). While counting information in terms of genes shared is perhaps a naive way of measuring information, Hamilton made the idea quantitative and argued that as long as the "relatedness factor" between the individual engaging in the altruistic act and the recipient of that act exceeds the ratio of benefit to cost (to the altruist), then that behavior would be selected for. Because this factor is one-half for siblings, then if cost and benefits are the same, it would be necessary to save two siblings. For cousins the relatedness is one-eighth, hence the rule that you would readily "save eight cousins." Hamilton's rule later came to be known as the concept of "kin selection."

In 1971, the evolutionary biologist Robert Trivers suggested an alternative solution to the cooperation conundrum. From the observation of cooperative

---

1. Technically, this is only true for interactions where there is a "cooperative dilemma" (we will study these in detail below). There are other situations (as in symbiotic interactions) where a dilemma is absent and where frequency-dependent selection can give rise to an equilibrium between types. Because an absence of a dilemma means that "cheating" is impossible, those kinds of interactions are not considered to be difficult to explain by evolution.

behavior between animals (in particular "cleaning symbiosis" in fish, and "alarm calling" in birds), Trivers concluded that cooperation could occur via reciprocal altruism, where the altruistic behavior is being repaid by the recipient of the altruistic act. According to this theory, as long as the reciprocal act occurs as often as the act itself during an individual's lifetime, evolution can efficiently amplify the genes responsible for altruism. The dynamic between cleaning fish and their host especially intrigued Trivers, in particular because that behavior cannot be explained by kin selection (the cleaner and the host are different species). The cleaning fish are usually much smaller than the host and typically clean the gills of the host fish from ectoparasites, often entering the mouth of the host fish to do so. There is little doubt that this behavior is intrinsically risky for the cleaner fish, yet the host lets it happen and allows the cleaner to escape even after the host is fully cleaned. Both fish thus obtain an advantage in this symbiosis but it is a true cooperative interaction because, at least from the point of view of the host fish, withholding cooperation would lead to a short-term benefit.

Within two decades, the two competing (and seemingly incompatible) theories of altruism were on the road to unification. First, Queller (1992) showed that Hamilton's rule can be seen to follow from Price's equation (recall Box 5.2). Applied to an altruistic behavioral trait, that equation (Queller's rule) predicts that as long as the *covariance* between that trait and fitness is large enough, then that trait is selected for. We can write this rule in terms of genetic variables and the ratio of cost-to-benefit c/b as

$$\frac{\mathrm{cov}(G_x, P_y)}{\mathrm{cov}(G_x, P_x)} > \frac{c}{b},\tag{8.1}$$

where G_x and G_y refer to the genotype of two players x and y, respectively, while P_x and P_y stand for their phenotype. In case the genotype is sufficient to predict the phenotype of a player, then the denominator of (8.1) simply becomes the variance of player x's behavior.

Mathematically speaking, as long as the relatedness between x and y is large enough, a sufficiently large covariance $\mathrm{cov}(G_x, P_y)$ to overcome a large cost-to-benefit ratio could in principle be achieved. Reciprocal altruism can also increase that covariance, leading to the unification of both theories (Fletcher and Zwick 2006). But while it is clear how relatedness increases the correlation between the altruistic behavior of the recipient of the altruistic act and the genes responsible for this behavior in the one delivering it, it is not immediately apparent how reciprocal altruism works. What mechanism gives rise to the increased correlation? As we will see, the answer is *information*, more precisely, communication between the participants.

8.1 Evolutionary Game Theory

To understand the importance of communication for cooperation a bit better, we first need to delve into evolutionary game theory, a theoretical framework built to study the evolution of cooperation, inspired by game theory proper. The latter theory is due to the economist Oskar Morgenstern (1902–1977) and the physicist and mathematician John von Neumann, whom it seems we keep meeting in these chapters (von Neumann and Morgenstern 1944). Game theory was designed to describe rational decision-making (in particular in the economic field), but some of its foundations were used by John Maynard Smith and George Price (he of the Price equation, see Box 5.3) to understand how animal behavior evolves (Maynard Smith and Price 1973) by casting behavior in terms of *strategies*. In the simplest case, a strategy is simply one of several possible moves a player can make. There are numerous excellent textbooks on evolutionary game theory (EGT), beginning with John Maynard Smith's slim volume (Maynard Smith 1982), and the reader is invited to peruse any one of them for an in-depth introduction to EGT (Axelrod 1984; Weibull 1995; Dugatkin 1997; Hofbauer and Sigmund 1998; Nowak 2006). Here, I will focus only on some simple examples, to drive home the importance of communication in cooperation.

The quintessential game that is used to illustrate evolutionary game theory is the *Prisoner's Dilemma* (often abbreviated as PD). One form of this game was invented in 1950 by two mathematicians at the RAND Corporation, which at the time was heavily involved in applying von Neumann and Morgenstern's theory of games to conflict situations in politics and the battlefield. The game introduced in (Flood 1952) as "A Non-cooperative Pair," however, is not a symmetric game and was only later called the Prisoner's Dilemma.

The game was introduced in its current form by Axelrod and Hamilton (1981). The usual backstory of the game (there are numerous versions of this; the details matter less than the payoffs) is given in Box 8.1. Let us dig even deeper, by analyzing the game mathematically. Rather than using a currency of prison sentences, Axelrod and Hamilton described the game in terms of payoffs to the player, where a good outcome is equivalent to a higher payoff (as opposed to a reduced sentence). When the payoff is used for reproduction, we can think of the relative payoffs as relative fitnesses for the trait that is embodied by the strategy. The payoff when both players cooperate is called R for "reward," while when both players defect they each get P: the "punishment" for defection. There is a temptation to defect when the other player cooperates: this value is termed T for the obvious reason, while the player on the wrong side of the deal has to be satisfied with the "sucker" payoff: S.

Box 8.1. The Prisoner's Dilemma

The Prisoner's Dilemma is usually introduced in the literature with a somewhat comical backstory. Two individuals that are suspected of having committed a major crime are apprehended by police, and put into solitary confinement (with no means of communicating). Each already has a record, and the police have enough evidence to prosecute and convict either of them of a lesser crime. Prosecutors now offer the criminals a deal: betray your fellow hoodlum and provide evidence that they committed the major crime in exchange for freedom (if the fellow is convicted). There are then four outcomes:

(a) Prisoner A and B both stay silent. Each receives a one-year sentence for the minor conviction.
(b) Prisoner A betrays B, while B stays silent. A goes free and B gets three years.
(c) Prisoner B betrays A, while A stays silent. B goes free and A gets three years.
(d) Both Prisoner A and B squeal. Both end up serving two years.

These outcomes are best visualized in diagrammatic form (a "payoff matrix") as in Figure 8.1, where the payoff is the length of the sentence. We now understand why this is a "dilemma." Clearly, both players are better off if they stay silent (this is the "cooperative" move). But each can better their situation if they—assuming the other player will stay silent—choose to "tell all" instead. Any player that squeals while the other stays mum goes scot-free. This incentive is too much to ignore, and as a consequence the rational move for both players is to defect (leading to a two-year sentence) even though they could have walked after one year, if only they had trusted each other. Trust, however, is irrational in the absence of communication.

Prisoner A \ Prisoner B	Stays silent (cooperates)	Betrays (defects)
Stays silent (cooperates)	1 year / 1 year	Goes free / 3 years
Betrays (defects)	3 years / Goes free	2 years / 2 years

FIGURE 8.1. Payoff table ("matrix") for the Prisoner's Dilemma game, in terms of the length of sentences for the four possible outcomes of the game. The game is called "symmetric" because what player A obtains for an action given player B's move is also what player B gains given player A's choice.

Player B

C   D

$$\begin{array}{c} \\ \text{Player A} \end{array} \begin{array}{c} C \\ D \end{array} \left(\begin{array}{cc} R & S \\ T & P \end{array} \right)$$

FIGURE 8.2. Payoff matrix describing the payoff to player A (the row player) in the Prisoner's Dilemma game, using the abstract values R, S, T, P. Whether or not the game is in the Prisoner's Dilemma class depends on the relative values of the payoffs.

We can then write the four entries in Figure 8.1 in matrix form, where the payoff listed is the one given to the "row player," that is, the player listed on the left of the matrix.

Inspecting the payoffs in Figure 8.2 makes it clear that the dilemma does not exist for all values R, S, T, P. Crucially, if $R > T$, then there is really no temptation (assuming S is low): both players will immediately cooperate as the cooperative move *dominates* defection in that case. Axelrod and Hamilton determined that the Prisoner's Dilemma is defined by the conditions $T > R > P > S$, as well as $R > (T + S)/2$. The typical numbers they chose (and which are now sometimes called the "Axelrod payoffs"), are $R = 3, T = 5$, $P = 1$, while the sucker gets nothing: $S = 0$. For simplicity we'll be adopting these values throughout (except when explicitly stated).

I previously argued that the rational strategy in this game is to defect, but let us make that more precise. Let us imagine that we have a population of players that are each defined by a "gene" that encodes the player's behavior. Naturally, since there are only two behaviors (cooperate or defect), there are only two alleles. Suppose that this population of players is initially composed of a mix of C players (cooperators) and D players (defectors). Who will ultimately win the competition? This question can be answered in several different ways. For one, we can write down a differential equation that quantifies how the number of C players changes over time (this is called the "replicator equation" approach; see for example Hofbauer and Sigmund 1998).

The replicator equation describes evolutionary dynamics in a well-mixed, infinitely large population, so it is quite abstract. It also only describes evolution on short time scales, because the all-important "mutation" part is missing. A key ingredient in this equation is, of course, the fitness of a particular strategy. Let us first calculate the fitness of a cooperator in terms of the payoffs; say, a matrix such as Figure 8.2. If the population is well mixed, the density of cooperators will determine how likely it is that a cooperator plays another cooperator, or a defector instead. Say n_C is the number of cooperators in a finite population of size N. We can then define the "density" of cooperators simply as $x = n_C/N$, while the density of defectors then is just $1 - x$. The mean "take-home pay" of a cooperator (which we'll take as the fitness w_C) will then be

$$w_C = xR + (1 - x)S \qquad (8.2)$$

while the defector's fitness will be

$$w_D = xT + (1 - x)P. \qquad (8.3)$$

We can calculate the mean population fitness $\bar{w}$ directly from Equations (8.2) and (8.3) and the densities x and $1 - x$ as

$$\bar{w} = xw_C + (1 - x)w_D = x^2 R + (1 - x)^2 P + x(1 - x)(S + T). \quad (8.4)$$

The replicator equation prescribes that the change in density of x over time $\frac{dx}{dt} \equiv \dot{x}$ is given by how many cooperators there already are times the fitness w_C, minus how many cooperators are being removed per unit time, which is given by the density times the average fitness $\bar{w}$

$$\dot{x} = x(w_C - \bar{w}). \qquad (8.5)$$

We can easily calculate $w_c - \bar{w}$ from the equations above to find

$$\dot{x} = x(1 - x)\big(x(R - T) + (1 - x)(S - P)\big). \qquad (8.6)$$

Because the number of C players plus the number of D players equals the population size, this equation automatically also gives the change in the number of D players.

Now let us check whether $\dot{x}$ is positive or negative. If it is positive, then the fraction of cooperators will increase and overall cooperation will ensue. But if it is negative, defectors will win the day. Since $T > R$ by the definition of the game given above, the first term within the big bracket in (8.6) is negative, and since $P > S$, so is the second. As a consequence, $\dot{x}$ is negative (unless it vanishes), which implies that cooperators cannot survive in this game. This is one of the central observations in evolutionary game theory: for games in which the temptation to defect is high $(T > R)$, defection is the rational strategy, even though two players can achieve a higher payoff (and therefore fitness) if they cooperate with each other. That is the "dilemma" of EGT.

Of course, the assumptions behind the replicator equation are rather restrictive, but if we simulate finite populations using agent-based methods (Adami et al. 2016), the result of the simulation is the same (if the population size chosen for the agent-based simulation is large enough): you end up with a population composed *only* of defectors.

This "endpoint" of evolution has several different names in the literature. It is sometimes called the Nash equilibrium point, after the mathematician John Nash (1928–2015), who first proved that every two-strategy game has at least one equilibrium point (Nash 1951). This work ultimately garnered Nash the economics Nobel Prize in 1995. (Incidentally, the latter publication is almost word-for-word Nash's Ph.D. thesis from Princeton University!) John Maynard Smith (along with George Price) later showed that this equilibrium point does not have to be an *attractive* equilibrium point (a point that all evolutionary trajectories are attracted to) but can also be a repulsive fixed point. Maynard Smith went on to define the notion of an *evolutionarily stable strategy* point (or ESS) which, in the words of Maynard Smith and Price (1973), is

> a strategy such that, if most of the members of a population adopt it, there is no "mutant" strategy that would give higher reproductive fitness.

It is actually simple to tease out which strategy (or strategies, as there can be more than one, or none) is the ESS for payoff tables with an arbitrary number of strategies. All you have to do is to look at how a strategy you are testing for ESS status (the "test" strategy) plays against itself (these values are written on the diagonal of the table) and check whether that payoff is larger than any number in the test strategy's particular column (see Exercise **8.2**). If the payoff is larger than what any other strategy receives against the test strategy, than the latter is an ESS. If the payoff against itself of the test strategy is as large as what some other strategy gets against against it (that is, there is a number in the test strategy's column that is as large as the payoff against itself), then the test strategy is still an ESS as long as what the test strategy receives against that other strategy is larger than what the other strategy obtains when playing against itself. Of course, all ESS must also be Nash equilibria, but clearly the opposite is not true, as some Nash equilibria are not stable.

Now that we have seen that the rational strategy in the PD game is in fact also the stable evolutionary strategy, this question has become more acute: If defection is the endpoint of evolution, how come we see so much cooperation in biology? More precisely, we should ask: What is it about the Prisoner's Dilemma as a model of cooperation that is fundamentally different from what occurs in biology? The answer is disarmingly simple. Remember the description of the PD in terms of the backstory in Box 8.1? According to that story, the alleged perpetrators are prevented from talking to each other before they are being interrogated. Surely if they could communicate, they would form a pact that allowed them to cop to the lesser charge only? But how do you model communication between strategies?

It turns out that making strategies communicate with each other is not very difficult. In principle, an intelligent strategy could use any and all hints or cues that give away a future move, in the same way a poker player tries to intuit whether an opponent is bluffing. But for such a strategy to work, it is imperative that any pair of players face each other more than once, because otherwise the information gained from a play will be of no use. To better understand strategies that can use information, we have to study *iterative* games, where two strategies face each other for an undetermined number of times. Why undetermined? Well, imagine that I know that I will play exactly n rounds against a particular opponent (and my opponent knows this too). It is clear then that I will defect on my last move, because my opponent will not be able to retaliate. Of course, my rational opponent will think the same, so I then determine that since we will both defect on the last round, then I should defect on the next-to-last round also, as there is no possible retaliation for that move either since my opponent will already be defecting. Needless to say, my opponent will think the same way, and this kind of thinking leads to both players defecting on all the moves. As a consequence, neither of us can be told how many times we will play against each other.

Now that we know that we'll be facing the same player many times, we need to think about how we can guess the player's next move. In poker, players are looking for "tells," that is, information about what the player will do next. What is a good "tell" in the Prisoner's Dilemma? What kind of things can we even take into account? The easiest (and simplest) way to infer the opponent's intention is to take a look at the opponent's past play given the player's own play. Does the opponent have a tendency to take advantage of my cooperative moves by throwing in a defection, for example? Strategies that play by taking into account the past play are called *conditional strategies*, where a decision (to cooperate or to defect) is made conditional on the opponent's past play.

Historically, the existence of conditional strategies presented an opportunity for computer scientists to start developing *intelligent* strategies. After all, strategies that always play C or D are extremely simple. What if a strategy were to toss a coin at each move and play randomly instead? What if you allowed strategies to take advantage of *any* feature or information they could? To find an answer to this question, Robert Axelrod (a professor of political science at the University of Michigan) decided to host a tournament to find the best strategies for playing the PD game. The invitation to submit strategies to this tournament went out in 1979, and fourteen strategies were submitted. In the tournament, those fourteen strategies competed against each other, as well as against a copy of themselves and a program that played randomly, as a control.

In each encounter the program played 200 moves, after which the total score was tallied and the winner of the encounter determined (Axelrod 1980).

The winner of the tournament turned out to be one of the simplest strategies to play the game, but it also quickly became the most famous. This strategy was called "Tit-for-Tat" (TfT), and was submitted by the Ukrainian-born mathematician Anatol Rapoport (1911–2007). Rapoport's TfT strategy could be programmed in just a few lines. It would start each series of games by cooperating, and then would play exactly the move that its opponent did the last time: tit-for-tat. While TfT did not actually record a single win against any of the strategies that were entered, it did come out the overall winner on account of having racked up the most points.[2] But as Rapoport et al. (2015) convincingly demonstrate,[3] who the ultimate winner is in such a tournament depends very much on winning criteria as well as tournament design.

8.2   Strategies That Communicate

What is the secret of TfT's success? Obviously it is copying the opponent's past move. Why is this a successful strategy? Let us analyze the interaction between the TfT strategy and its opponent in terms of information theory. The TfT strategy is simple enough, but probably too simple to take advantage of a sophisticated opponent. For this, more information is needed. Another source of information is the strategy's *own* past play. Basing one's move on both the opponent's and one's own past play creates the opportunity to react in a more sophisticated manner to past plays, much like the lac operon in section 2.4 could react to a complex set of sugar concentrations.

For example, we can parameterize all possible "memory-one" strategies[4] by writing the strategy in terms of the *probability* to engage in cooperation, conditional on the previous move

$$S(P) = (P_{CC}, P_{CD}, P_{DC}, P_{DD}), \tag{8.7}$$

where P_{XY} is the probability to cooperate given that in the previous move, the strategy played X and the opponent played Y. For the deterministic strategies that we have been discussing up to now, the probabilities are of course

---

2. Because TfT reacts to the opponent, by definition it can never win. Even the program RANDOM managed to win one encounter, and thus ranked higher in wins than TfT.

3. The study reevaluating the design of the Axelrod tournament was led by Amnon Rapoport, who also submitted a strategy to the tournament and who had earlier published the TfT strategy (Rapoport 1967). While Amnon was Anatol's postdoctoral fellow, they are unrelated (Kopelman 2019).

4. These strategies are called "memory-one" because they act based on the moves one time-step back, but in principle a strategy could take into account longer game histories.

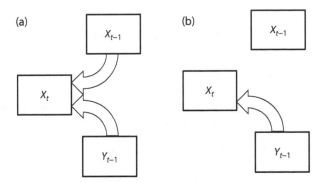

FIGURE 8.3. (a) The channel in which information comes from both the opponent's (Y) and the agent's own (X) previous move. (b) The channel where X's
decision is only influenced by the opponent's previous move.

either zero or one. The "All-C" strategy translates to $(1,1,1,1)$ for example,
while "All-D" is $(0,0,0,0)$. In this notation, TfT is $(1,0,1,0)$. Clearly, there are
sixteen different deterministic strategies, but while we can run all pairwise
"duels," this does not determine who would win when competing in a population, as we will see shortly. But first, let us calculate how much information
these strategies truly have about each other. To do this, we'll define random
variables X_t for the player at time t (which can take on the two values C and
D of course), as well as the variable X_{t-1} to record what the player did on
the previous move. We'll also define the opponent's variables Y_t and Y_{t-1}.
The communication channel can be sketched as in Figure 8.3(a), which indicates that the X-player uses both their own and their opponent's past play to
determine the next move. A simpler channel has the X player using only the
opponent's move to make up their mind, which is in fact precisely what the
TfT strategy does. But since we'll be discussing the more complex channel
later, it is best to introduce it right away.

The information that X has at time t about the previous move can be
written as the shared entropy between X_t and the joint variable $X_{t-1}Y_{t-1}$

$$I(X_t : X_{t-1}, Y_{t-1}) = H(X_t) - H(X_t|X_{t-1}Y_{t-1}). \qquad (8.8)$$

In Equation (8.8) I wrote this information as a difference between an uncon-
ditional entropy $H(X_t)$ and the conditional one $H(X_t|X_{t-1}Y_{t-1})$ just as we
have been doing over and over again; see for example Equation (2.29), where
you just insert X_t for X and $X_{t-1}Y_{t-1}$ for Y.

To calculate the information $I(X_t : X_{t-1}, Y_{t-1})$, we need to calculate the
two terms in (8.8). To get $H(X_t)$, all we need is the probability for X to cooperate, which we will write as p_C. We'll get to this probability in a moment, but

let us first calculate $H(X_t|X_{t-1}Y_{t-1})$, which is arguably easier. First, we introduce a convenient notation for the conditional cooperation probabilities P_{XY} in Equation (8.7):

$$P_{CC} = p_1, P_{CD} = p_2, P_{DC} = p_3, P_{DD} = p_4. \tag{8.9}$$

It turns out that these conditional probabilities are precisely those that go into the evaluation of $H(X_t|X_{t-1}Y_{t-1})$, since

$$P(X_t = C|X_{t-1} = C, Y_{t-1} = C) = P_{CC} = p_1 \tag{8.10}$$

and so on. So, for example the entropy of X_t given that the previous play was CC (both players played C in the previous move) is

$$H(X_t|X_{t-1} = C, Y_{t-1} = C) = -p_1 \log p_1 - (1 - p_1) \log(1 - p_1) \equiv H[p_1], \tag{8.11}$$

where I used the notation for the binary entropy function shown in Figure 2.20. Clearly we can calculate the other three conditional entropies in a similar manner, but to get the *average* conditional entropy [see Eq. (2.25)], we need to know how often any of the four plays (CC, CD, DC, and DD) occur, that is, we need to know the *play prevalence*.

Let us define the play prevalence π_{XY} as the probability to find a pair of plays XY at any particular time in an equilibrated population, and follow the notation introduced in (8.9) by writing

$$\pi_{CC} = \pi_1, \pi_{CD} = \pi_2, \pi_{DC} = \pi_3, \pi_{DD} = \pi_4. \tag{8.12}$$

Recalling the definition of the average conditional entropy (2.25), we can obtain $H(X_t|X_{t-1}Y_{t-1})$ by multiplying terms like (8.11) that are conditional on a particular play by the likelihood to observe such a play, so that

$$H(X_t|X_{t-1}Y_{t-1}) = \sum_{i=1}^{4} \pi_i H[p_i]. \tag{8.13}$$

Now let us return to calculate $H(X_t)$, the entropy of plays in the population. There are two ways to calculate this quantity: we can either calculate it from the *probability* that the X player will cooperate, or we can assume that the population is in equilibrium, so that all players (both X and Y) are the same. In the latter case, we can obtain the frequency of plays directly from the prevalence of pairwise plays, as the prevalence of cooperation π_C is related to the four prevalences π_i by

$$\pi_C = \pi_1 + \frac{1}{2}(\pi_2 + \pi_3). \tag{8.14}$$

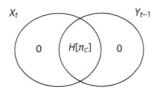

FIGURE 8.4. Entropy Venn diagram between TfT players. If the opponent does not play TfT, its conditional entropy $H(Y_t|X_{t-1})$ could be different from zero.

On the other hand, we know that a player with strategy $\vec{p} = (p_1, p_2, p_3, p_4)$ will, in a well-mixed population, cooperate with probability

$$p_C = \sum_{i=1}^{4} \pi_i p_i, \tag{8.15}$$

that is, the sum of the conditional probabilities p_i multiplied by the likelihood of the occurrence of that play. If we assume that a cooperative play will appear in the population with the probability that it is played, then

$$H(X_t) = H[p_C] = H[\sum_i \pi_i p_i]. \tag{8.16}$$

Now that we have calculated the unconditional as well as the conditional entropy, we can write down the information as the difference between the two:

$$I(X_t : X_{t-1}, Y_{t-1}) = H[\sum_i \pi_i p_i] - \sum_{i=1}^{4} \pi_i H[p_i]. \tag{8.17}$$

Let us check this formula for a few example strategies. First, let us see what we obtain for a noncommunicating strategy such as "All-D," which (as the name implies) always defects no matter what the opponent does. Its strategy vector is naturally $(0, 0, 0, 0)$, which implies $p_C = 0$ and every $H[p_i]$ vanishes as well. As a consequence, no information is transferred between strategies, as of course we expected. Let us instead test TfT, with $\vec{p}_{TfT} = (1, 0, 1, 0)$. Because all the $H[p_i]$ vanish, there can only be a contribution to Equation (8.17) from $H[p_C]$, and with $p_C = \pi_1 + \pi_3$, we find (see entropy Venn diagram between TfT players in Fig. 8.4)

$$I(X_t : X_{t-1}, Y_{t-1}) = H[\pi_C]. \tag{8.18}$$

If the population consists of half cooperators and half defectors, the information is exactly one bit, and incidentally that is also the capacity of the information transmission channel since, as we recall from the capacity formula of the binary symmetric channel (2.65), the capacity is obtained by maximizing the information processed by the channel over the source distribution. For TfT, the maximum occurs at $\pi_C = 0.5$, but this is not generally the case for communicating strategies, as we will see.

Now we know what is the power wielded by TfT. Even though the strategy can never win any battle outright, it accumulates scores better than any other because it makes *informed* decisions. But how best to wield the power of information in a population? This question was not addressed by Axelrod, as he only studied what strategy plays the PD game efficiently in pairwise encounters between strategies. Could the process of Darwinian evolution, instead, produce the overall winning strategy, in a scenario where strategies have to compete against a myriad of other strategies in a population?

It turns out that this is a simple question that does not have a simple answer. It seems unlikely that TfT would emerge victorious in an evolutionary setting because TfT is deterministic. Deterministic strategies have an obvious weakness. For example, once a TfT strategy is on a run of defection against another deterministic strategy, there is no way to break out of such a "rut." A superior strategy is the "generous Tit-for-Tat" (gTfT) strategy, defined by

$$S(\text{gTfT}) = \left(1, \tfrac{1}{3}, 1, \tfrac{1}{3}\right), \tag{8.19}$$

which implies that gTfT will (generously) respond to an opponent's defection with cooperation on average once every three times. Let us calculate the information that gTfT gathers. According to Equation (8.17), the information $I(X_t : X_{t-1}, Y_{t-1})$ turns out to be (see Exercise **8.3**)

$$I(X_t : X_{t-1}, Y_{t-1}) = H\left[\tfrac{1}{3} + \tfrac{2}{3}\pi_C\right] - (1 - \pi_C)H\left[\tfrac{1}{3}\right] \tag{8.20}$$

which is less than the maximal information that TfT can extract (and is maximal at a $\pi_C > 0.5$). Nevertheless, gTfT fares better than TfT in noisy environments, which tells you that it is not just the information that counts; it is also how you use it. Let us delve more deeply into the different ways that strategies might be using the information they gather.

8.2.1 Zero-determinant strategies

Strategies like gTfT that play certain moves with a probability are called "probabilistic," or "stochastic" strategies. In fact, such strategies have been investigated since the 1990s (Nowak 1990; Nowak and Sigmund 1990), but for a long time the space of strategies spanned by $S(P)$ in Equation (8.7) was baffling. A breakthrough occurred in 2012 when William Press and Freeman Dyson announced that they had discovered a set of memory-one stochastic strategies that could win any pairwise confrontation with another memory-one strategy by unfairly exploiting the opponent (Press and Dyson 2012). Such a strategy—one that guarantees to its wielder that they will dominate any opponent no matter what their action—would have seemed like the ultimate jackpot to the employees of the RAND Corporation, the global policy think tank where Nash, Flood, and Dresher worked out the mathematics of noncooperative games in the early 1950s. These coercive strategies were not discovered there, however, but instead by two novices in the field of evolutionary game theory. While novices in *this* particular field, the two certainly were powers to be reckoned with. William Press is an accomplished astrophysicist and computational scientist at Harvard University who is perhaps best known for what used to be the "bible" of computational physics (Press et al. 1986). Dyson (who passed away in 2020 at the age of ninety-six) was a mathematical physicist and one of the founders of quantum field theory who, working at Princeton University at the age of eighty-nine at the time these strategies were discovered, had apparently not lost a step.

William Press knew that it was possible to find the "winner" in a probabilistic game when decisions were made unconditionally: the winner was Maynard Smith's evolutionarily stable strategy, but where the stable population fraction (in case the ESS is a mixture of two strategies) is replaced by the probability to engage in that action. So, for example, if in the deterministic game the ESS is a population mixture of 70 percent cooperators and 30 percent defectors,[5] then the probabilistic ESS is a strategy that cooperates 70 percent of the time and defects otherwise. But what strategy is the ESS if you have strategies such as Equation (8.7) that have a one-move memory?

There are infinitely many different probabilistic zero-memory strategies, since the probability to cooperate is a continuous variable. In that sense, having memory-one strategies labeled by four probabilities does not seem to be more complex, but what Press realized by investigating the space of all memory-one strategies numerically is that there is a lot of *structure* in that space. Specifically, he found that there were subsets of strategies [in this four-dimensional space, where the four dimensions are given by the four conditional probabilities in (8.7)] where the opponent's payoff is forced to be a fixed number *independent* of that opponent's payoff. At first sight, such a state of affairs makes no sense. How can the opponent's actions not affect their payoff?

What Press had discovered was, in a nutshell, an evil power of communication. A communicating strategy can tune itself to the opponent's choices

---

5. Such mixtures can be ESS, for example in the "snowdrift" game, defined by a different set of parameters than those that define the Prisoner's Dilemma game shown in Figure 8.2.

in such a manner that all the opponent's actions are automatically negated. Now, there are many ways to fix the opponent's payoff, but the meanest ones are those that fix the opponent's earnings to be as low as possible. How is this achieved? And what is the role of Freeman Dyson in this saga?

When Press noticed this strange set of strategies in his results, he turned to Dyson for help to make sense of it. Dyson responded a day later[6] with the calculation that is now equations (1-7) in (Press and Dyson 2012). Dyson had realized that the condition that gave rise to the set of unfair strategies came about from setting a particular determinant to zero, which resulted in the name "zero-determinant" (ZD) strategies.

Among the ZD strategies studied by Press and Dyson there are two main classes: the "equalizers" that unilaterally set the payoff of the opponent, and the "extortionists" that fix the relative payoff between the extortionist and the opponent in such a manner that when the opponent attempts to increase its own payoff by changing strategy, the opponent's payoff increases by the same amount instead. We'll focus here on the "equalizers," which are defined by setting the probabilities p_2 and p_3 [given in Eq. (2.12)] as a function of p_1 and p_4. This constrains the equalizer strategies to lie in a subspace of all possible memory-one strategies, determined by

$$p_2 = \frac{p_1(T-P) - (1+p_4)(T-R)}{R-P},$$

$$p_3 = \frac{(1-p_1)(P-S) + p_4(R-S)}{R-P}, \qquad (8.21)$$

where $R, S, T,$ and P are the payoffs of the Prisoner's Dilemma game shown in Figure 8.2. We should also keep in mind that the choice of p_1s and p_4s is somewhat constrained so that the resulting p_2 and p_3 are between zero and one.

Let us look at a typical such strategy, using the Axelrod payoffs $R = 3,$ $T = 5, P = 1,$ and $S = 0$. We have the freedom to choose two probabilities, so we will pick $p_1 = 0.99$ and $p_4 = 0.01$: a strategy that will almost always cooperate if the previous move was CC, and almost always defect if the previous move was DD. According to (8.21), the ZD strategy is then given by

$$\vec{S}_{\text{ZD}} = (0.99, 0.97, 0.02, 0.01). \qquad (8.22)$$

What makes ZD strategies "mean" is that they take away the freedom of an opposing strategy to have any way to influence their own payoff. At first this appears absurd: intuitively you expect that by changing your own

---

6. This is recounted by Bill Press in a conversation with William Poundstone at *Edge Magazine* on June 18, 2012.

probabilities to engage in plays conditional on the previous outcomes you will affect the score you receive! Say an opponent O has a strategy vector $\vec{S}_O = (q_1, q_2, q_3, q_4)$ with the probabilities q_i determining their play depending on the previous play. We can use these probabilities, along with the probabilities (8.21) to calculate the payoff that each player receives if they play infinitely often against each other (remember that if they only play a finite number of times, then they should not know when the game is going to end).

Dyson calculated that the payoff that strategy O receives against ZD is given by

$$E(O, ZD) = \frac{(1-p_1)P + p_4R}{1 - p_1 + p_4} \equiv f(\vec{p}), \qquad (8.23)$$

that is, it is a function $f(\vec{p})$ that does *not* depend on any of the q_i. What ZD receives against O *does* depend both on $\vec{p}$ and $\vec{q}$

$$E(ZD, O) = g(\vec{p}, \vec{q}), \qquad (8.24)$$

but the general form for the function $g(\vec{p}, \vec{q})$ is too complicated to write down here,[7] and we do not need it to make sense of anything that follows. We can complete the effective payoff matrix of ZD playing any other strategy by defining the payoff that the other strategy receives against itself, which we call $h(\vec{q})$, so that

$$E = \begin{array}{c} \\ ZD \\ O \end{array} \begin{array}{c} ZD \qquad O \\ \begin{pmatrix} f(\vec{p}) & g(\vec{p}, \vec{q}) \\ f(\vec{p}) & h(\vec{q}) \end{pmatrix} \end{array}. \qquad (8.25)$$

So, how successful is this devious strategy? Well, it depends. If we pick the ZD strategy in Equation (8.22) as an example, it will make sure that the opponent always receives $f(\vec{p}) = 2$, which is Equation (8.23) evaluated with the probabilities given in (8.22). But ZD actually receives less than that against the notorious defector All-D given by $\vec{q} = (0, 0, 0, 0)$, as $g(\vec{p}, \vec{0}) = 0.75$. In fact, All-D is evolutionarily stable against this ZD, as we can see from inspecting the matrix

$$E = \begin{array}{c} \\ ZD \\ All-D \end{array} \begin{array}{c} ZD \quad All-D \\ \begin{pmatrix} 2 & 0.75 \\ 2 & 1 \end{pmatrix} \end{array}. \qquad (8.26)$$

But let us check how ZD fares against one of the champions of conditional strategies, called "Win-Stay-Lose-Shift" (WSLS), introduced by Nowak and

---

7. The much simpler form where the Axelrod payoffs have been inserted for the general R, S, T, P is shown as Equation (15) in Adami and Hintze (2013) and is an equation that spans nine lines.

Sigmund (1993). WSLS is defined by the vector $S_{\text{WSLS}} = (1, 0, 0, 1)$. The strategy gets its name from the fact that it keeps cooperating if the previous encounter was beneficial, but it shifts play if the previous encounter was detrimental. It is also sometimes called "Pavlov" (Kraines and Kraines 1989), after the eponymous dog that learns via positive and negative reinforcement.

In the Prisoner's Dilemma, CC plays and DC plays are beneficial (to the player mentioned first: the "row" player), while CD and DD are detrimental. For example, $p_2 = 0$ means that the player will shift from cooperating (in the previous move) to defection, while $p_3 = 0$ says that the player will continue to defect after a DC play, which is highly beneficial. $p_4 = 1$ implies that the player will attempt a cooperation after a DD play, something that TfT would never do, and which proves to be TfT's main drawback.

It turns out that the ZD strategy obtains a fairly high payoff against WSLS[8]: $E(\text{ZD}, \text{WSLS}) = 11/27 \approx 2.455$, while WSLS is forced to receive $f(\vec{p}) = 2$ (like any other strategy, of course). That is unfair: ZD will as a consequence win absolutely every pairwise encounter between those two strategies. But does this mean that ZD will prevail in the end? Not so fast.

As nefarious as these ZD strategies may seem, they have an Achilles' heel. If a ZD strategy can force any opponent strategy to accept a lower payoff, how would it fare against itself? This is not a question that RAND strategists would ask; in their world, after all, it was always us against them. But in real populations we have to interact with one of "our own" as often as we might play against "others." And the answer to the first question is clear: a ZD strategist playing against another ZD strategy has to accept a low payoff also, leading to a poor result for both. WSLS, on the other hand, is a cooperator, and obtains a payoff of 3 against itself. The full matrix of average payoffs then reads

$$
\begin{array}{cc}
 & \begin{array}{cc} \text{ZD} & \text{WSLS} \end{array} \\
\begin{array}{c} \text{ZD} \\ \text{WSLS} \end{array} & \begin{pmatrix} 2.0 & 2.455 \\ 2.0 & 3 \end{pmatrix}.
\end{array}
\tag{8.27}
$$

It is easy to determine the eventual winner from this matrix using the rules described above (see also Exercise **8.2**): it is WSLS. Indeed, because the payoff against ZD is the same for every strategy, ZD can never be more than "weakly stable," meaning that it has the same payoff against itself as another strategy receives against it. According to Maynard Smith's rule, a strategy can still be ESS if what it receives against the other strategy is larger than what the other strategy obtains against itself. But WSLS receives 3 against itself (it cooperates

---

8. See Equation (17) in Adami and Hintze (2013) for the payoff of arbitrary equalizer ZD strategies against WSLS.

against itself) while ZD only receives 2.455 on average against WSLS (Adami and Hintze 2013). As a consequence, ZD is not even weakly evolutionarily stable. So ZD wins every battle against the others, but it loses the overall war and will ultimately be driven to extinction: winning is not everything (Adami and Hintze 2013)!

But WSLS is a "designed" strategy, and it is not even stochastic. What stochastic strategy will ultimately emerge victorious in Darwinian evolution? It turns out that this depends on the mutation rate that is chosen for evolution (Iliopoulos et al. 2010). Previously, we had not concerned ourselves with mutations: the replicator equation (8.6) famously does not have such a term. We could in principle think about letting all possible stochastic strategies battle it out against each other, until we realize that because the probabilities are continuous, there are an infinite number of those strategies.

The better way to look for evolutionarily dominating strategies is to implement a true evolutionary process, where a "genotype" encodes the strategy in a few loci, and mutations change the value at these loci. To simulate a population of such genomes undergoing evolution, we would pick a strategy and let it play against a number of randomly chosen opponents, and collect the total payoff as an estimate of that strategy's fitness. Then, the next generation of genomes will be populated preferentially with the offspring of the strategies that achieved the highest payoff.

Let us then encode stochastic memory-one strategies by creating a genome of four loci that each encode one of the probabilities in (8.7), and supplement it with one locus for the probability to cooperate on the first move. For infinitely long games, that first-move probability is unimportant, but for finite games, this has to be specified. At first glance the mutation rate itself should not determine what strategy ultimately prevails. After all (we might say), should not the winner be determined by the payoffs only? However, that is only true when mutations are absent (or the rate of mutation is very low). The reason for this is that mutations introduce uncertainty into game play: every time there is a mutation on the genome of an opponent, the opponent's strategy changes, and as a consequence, what was the right move against the unmutated opponent might be the wrong move against the mutated version. That might not be a problem if the mutation rate is low, but as it increases, it makes the opponent less and less reliable. In a way, when the mutation rate becomes sizable, a player cannot depend on the accuracy of the information that it obtained anymore, and this holds for the accuracy of the opponent's play just as much as the strategy's *own* play, as the player's *own* genome might have changed. Clearly, this implies that the information $I(X_t : X_{t-1}, Y_{t-1})$ that a strategy has about previous plays must decrease with increasing mutation rate. If it is *this* information that allowed cooperation to be maintained in the first place, then at some mutation rate we should expect cooperation to cease, and defection to take hold again.

When we simulate the evolution of strategies at finite mutation rate in a well-mixed population of 1,024 individuals, we find that low mutation rates indeed select for cooperators (a strategy will emerge that we term "GC" for "general cooperator" in the following), while high mutation rates favor defectors. This is good news in general: now we do not have to worry anymore about why cooperation is observed pretty much everywhere in the biosphere even though the game defined by the payoffs of the Prisoner's Dilemma leads to mutually assured defection. The players in the PD story (told in Box 8.1) were not allowed to communicate; but if you let them, they will cooperate with abandon!

What is so special about strategies like GC? And how do they fare against the mean ZD strategies? Let us first look at what the ZD strategies have in common with GC. Both are strategies that can communicate (just like TfT discussed earlier); that is, they have opened up a communication channel that allows them to better predict their opponent's behavior. The selfish ZD strategies use the information to manipulate their opponent into submission (forcing them to accept a low payoff), while the GC strategy uses the information to avoid cooperating with defectors.

While at any mutation rate a cooperating strategy transmits more information than a defecting strategy, there is a critical point at which not enough information is gathered in order to sustain cooperation, at which point it pays to switch to defection—in a manner reminiscent of a phase transition, see Figure 8.5 and Iliopoulos et al. (2010). We should, however, keep in mind that merely communicating does not guarantee a win; just as in real life you want to send the right signals, and act appropriately based on the signals that you receive.

For example, let us test what happens if we seed a population with a ZD strategy. We find that ZD quickly "devolves," even though the ZD strategies communicate: they are ultimately outcompeted by nicer strategies, as Figure 8.6 demonstrates very effectively. (The particular ZD strategy that we used to seed that population is given in the caption of Fig. 8.6.) The strategy that ZD ultimately evolves into turns out to be the GC strategy discussed above, given approximately by the average strategy[9]

9. Because of the stochastic nature of the evolutionary process, the mean values for the GC strategy depend not only on the mutation rate, but also on how long we run evolution, and how many replicas (runs with the same set of parameters) are taken. The values cited here are from Adami et al. (2016).

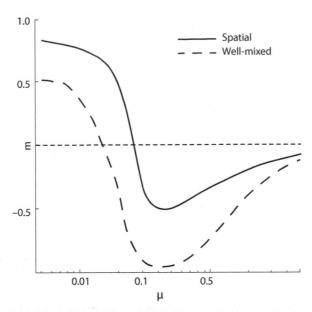

FIGURE 8.5. The "order parameter" m for a population of strategies evolved at fixed mutation rate μ for a spatially structured (solid line) and a well-mixed population (dashed line). The order parameter is simply the fraction of CC plays minus the fraction of DD plays, divided by the sum of those plays. When everybody cooperates (All-C strategy) this fraction is 1, while when everybody defects (All-D), the fraction is -1. A random strategy has $m = 0$. As the mutation rate increases the order parameter changes sign, indicating a transition from cooperation to defection. Figure adapted from Iliopoulos et al. (2010).

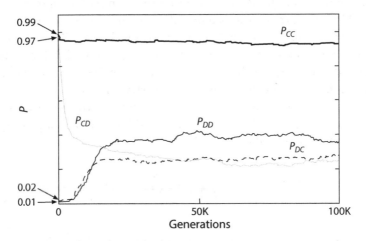

FIGURE 8.6. Evolution of the four probabilities (8.7) P_{CC} (black bold line), P_{CD} (gray line), P_{DC} (dashed line), and P_{DD} (black) along an averaged evolutionary LOD of a well-mixed population of 1,024 agents, seeded with the ZD strategy $\vec{S}_{ZD} = (0.99, 0.97, 0.02, 0.01)$. Lines of descent are averaged over 200 independent runs. Arrows indicate the initial four probabilities of the seeding ZD strategy. Mutation rate per gene is $\mu = 1\%$ (adapted from Adami et al. 2016).

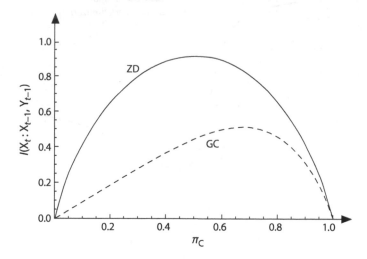

FIGURE 8.7. Strategy information (information used by a strategy to choose a move) as a function of the cooperation prevalence π_C (the likelihood to encounter a cooperator in the population). Solid line: strategy information for the ZD strategy (8.22); dashed line: information gathered by the GC strategy.

$\vec{S}_{GC} = (0.91, 0.25, 0.27, 0.38)$. Does this mean that GC gathers more information than ZD?

We can test how much information each strategy is using by plugging $\vec{S}_{ZD}$ and $\vec{S}_{GC}$ into the formula for the strategy information (8.17). Figure 8.7 shows that ZD in fact uses significantly more information than GC (albeit for nefarious purposes), so it is not *that* difference that makes the difference. Let us look at the payoff matrix between the two strategies instead. It is possible to calculate the average payoff between a GC player and ZD that play long games against each other using the genotypes of the strategies only.

These "effective" payoffs can be used to write down an "effective payoff matrix" just like we did when we checked how ZD would play against "win-stay–lose-shift" in the matrix (8.27)

$$
\begin{array}{cc}
 & \begin{array}{cc} \text{ZD} & \text{GC} \end{array} \\
\begin{array}{c} \text{ZD} \\ \text{GC} \end{array} & \left(\begin{array}{cc} 2.0 & 2.125 \\ 2.0 & 2.11 \end{array} \right).
\end{array}
\qquad (8.28)
$$

According to this matrix, the ZD strategy is weakly stable (because $2.125 > 2.11$), while GC is not stable. However, we have seen that when we seed a population with the ZD strategy and allow it to adapt at a mutation rate of 1 percent it evolves *into* GC (see Fig. 8.6)! How is this possible?

At first glance, it should not be. The payoffs in matrix (8.28) unambiguously predict the winner in an infinite-size well-mixed population in the weak mutation-strong selection limit (that is, the limit $\mu \to 0$). Then, which of these conditions (infinite, well-mixed population, zero-mutation-rate limit) is violated? The experiments for Figure 8.6 had 1,024 individuals in them, which is large enough so that finite population-size effects are expected to be small. The population was also well-mixed. This leaves the zero-mutation-rate limit. But how could a finite mutation rate lead to a violation of the ESS condition for stability?

The ESS condition (as defined by Maynard Smith in the quote in section 8.1, page 420 and in Exercise **8.2**) says that a strategy is evolutionarily stable if no mutant strategy can outperform it (in terms of replicative fitness). However, we have seen in chapter 6 that when mutation rates are high enough, groups of strategies that are mutationally related (the "quasispecies") can outperform single strategies even if those single strategies have higher replicative fitness (as measured at zero mutation rate). Could this phenomenon (the "survival-of-the-flattest" effect) be behind the success of GC? Could it be that GC is not a single strategy, but rather a group of strategies that is mutationally robust, that is, a *quasi-strategy* in the sense of the molecular quasispecies we discussed in section 6.2.2? If true, this would represent a radical departure from Maynard Smith's ESS concept.

8.3 Quasi-Strategies

We can test for the existence of a quasi-strategy by analyzing the strategy *distribution* in a population. For stochastic memory-one strategies, there are actually five distributions: one for every probability in the set (8.7), plus the distribution for the probability to be used on the first move (when there is no history yet that can be used to make a decision). Because in long games the no-memory probability is used so rarely, the distribution is not under selection and the trait drifts (Iliopoulos et al. 2010). For this reason, we will not consider it any further. The remaining four probability distributions (for P_{CC}, P_{CD}, P_{DC}, and P_{DD}) represent a single strategy if each of the distributions is sharply peaked at some value of P.

Ultimately we would like to test whether quasi-strategies can, if the mutation rate is sufficiently high, outcompete even the best "clonal" strategies, where "clonal" refers to populations of copies (clones) of a single strategy with given probabilities P_{XY}. A candidate for the "best possible" memory-one strategy happens to be a ZD strategy, but it is not among the group of devious strategies that we discussed earlier. Rather, it plays nice with others: it is called a "generous" ZD strategy, which we will discuss in more detail below. However, because that strategy uses a different payoff matrix (the payoffs of the "donation game") than the one we have been using up to now, we need to evolve a quasi-strategy with that set of payoffs in order to be able to run the competition.

The "donation game" is actually in the same class as the Prisoner's Dilemma, but the payoffs are slightly different from the standard Axelrod payoffs. In the donation game, the payoffs are determined by two positive numbers only: the benefit b and cost c, for a payoff matrix (payoffs go to the row player, as always)

$$
\begin{array}{c}
 \\
C \\
D
\end{array}
\begin{array}{cc}
C & D \\
\left(\begin{array}{cc}
b - c & -c \\
b & 0
\end{array} \right).
\end{array}
\tag{8.29}
$$

The payoffs imply that mutual cooperation carries a cost and delivers a benefit, but if one of the players defects, the cooperator still pays the cost while the benefit only accrues to the defector. We assume that $b > c$, which implies that defection is the ESS even though cooperation gives the highest payoff, creating the same dilemma as in the Prisoner's Dilemma.

Figure 8.8 shows the probability distribution for the four "genes" encoding the four probabilities (8.7) to engage in cooperation. Of the four, only the distribution for P_{CC} is strongly peaked (at $P_{CC} = 1.0$), while the other three distributions are fairly broad. To obtain the results in Figure 8.8, we used $b = 3$ and $c = 1$ in (8.29), which is equivalent to a PD game with Axelrod payoffs $(R, S, T, P = 2, -1, 3, 0)$. The distribution represents values taken from the line of descent of 200 replicate populations after equilibration. It reflects the distribution present in a single equilibrated population.

The strategy population distributions seen in Figure 8.8 persist in time: they are *stationary*. Indeed, these distributions have been sampled after 1.75 million generations of evolution, and averaged over 200 independent replicate populations. Furthermore, if you sample the strategies from the line of descent (after equilibration) rather than from the population, you obtain the *same* distributions (a property known as "ergodicity"). The ergodicity property along with the fact that the distributions are stable when averaged over many independent instantiations indicate that the distribution is indeed stable over both time and space. Such a distribution is consistent with the molecular quasispecies that we encountered earlier (see Box 6.2). These groups of sequences can outcompete fitter types when the mutation rate is high enough because the "fit" types are unstable under mutation (they "fall apart," so to speak; see section 6.2.2). Is this what is going on here too?

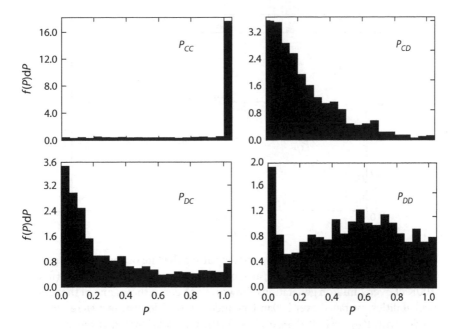

FIGURE 8.8. Distribution functions $f(P)dP$ of probabilistic conditional (memory-one) strategies in an adapted population (population size $N = 1,024$, mutation rate $\mu = 0.5\%$) after 1.75 million generations, using the donation game payoffs, averaged over 200 replicate populations (from Adami et al. 2016). The average strategy is the "general cooperator" strategy with payoffs $b = 3$ and $c = 1$: $\vec{S}_{\text{GC}_d} = (1.0, 0.27, 0.3, 0.6)$.

To test how well a strategy distribution does against a clonal strategy, we are going to pit GC not against ZD, but instead against an even more formidable foe: a *generous* ZD strategy that is said to be the ultimate winner among all possible memory-one strategies (Stewart and Plotkin 2013). Generous ZD strategies were not discussed by Press and Dyson, but they are a subset of the "good" strategies described by Akin (2012). Stewart and Plotkin (2013) were able to show quite generally (albeit in the weak mutation limit) that these "robust" ZD strategies they termed ZD_R (they are robust to invasion by any other stochastic memory-one strategy) should win all competitions. The generous ZD_R strategy should dominate any other memory-one strategy, yet it succumbs (and evolves into) the GC strategy (Adami et al. 2016), as we can see in Figure 8.9. In order to find out whether the eventual demise of ZD_R really is due to its failure to thrive at high mutation rates, we can check how strategies play against mutants of themselves. After all, for a quasi-strategy to

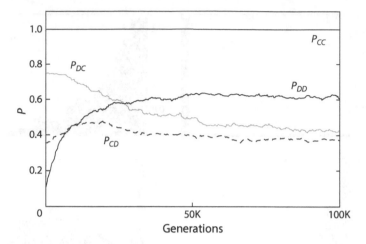

FIGURE 8.9. Evolution of the ZD_R strategy along the line of descent, in a population of $N = 1,024$ individuals, at a mutation rate of $\mu = 1\%$ per locus (first 100K of over 2 million generations shown). Each curve shows one of the four probabilities, averaged over 200 independent lines of descent that were each seeded with $\vec{P}_{ZD_R} = (1.0, 0.35, 0.75, 0.1)$ (the unconditional "first-move" probability is kept at 0.5 throughout). Bold black: P_{CC}, dashed: P_{CD}, gray: P_{DC}, and black: P_{DD}. The strategy evolves into the general cooperator quasi-strategy in the donation game, GC_d, shown in Figure 8.8. Adapted from Adami et al. (2016).

exist, the mutants of the wild-type strategy *must* be evolutionarily stable (at least weakly) and their mutants must be also, and so on.

Let us see how robust ZD_R really is to mutations. Ideally, we'll have ZD_R play against all of its one-mutants, but of course that is not possible because the probabilities are continuous. To approximate those continuous values, we discretize the probabilities into twenty possible discrete values (zero and one, as well as nineteen equally spaced probabilities in between). Then, we can create *all* $21^4 = 160,000$ mutant strategies, and let them play against ZD_R to get the effective payoffs $E(ZD_R, ZD_R(1))$ and $E(ZD_R(1), ZD_R)$, where $ZD_R(1)$ stands for the set of one-mutant strategies of ZD_R.

Since we know that ZD_R cooperates with itself (and thus receives payoff $b - c = 2$ given the values used here), we just have to measure how the mutants play against themselves to fill in the effective payoff matrix for ZD_R (8.30). By averaging the payoffs of all the mutants, we obtain the effective payoff matrix

$$
\begin{array}{cc}
 & \mathrm{ZD_R} \quad \mathrm{ZD_R(1)} \\
\begin{array}{c} \mathrm{ZD_R} \\ \mathrm{ZDR(1)} \end{array} &
\begin{pmatrix} 2.0 & 1.7 \\ 1.77 & 1.66 \end{pmatrix}.
\end{array}
\tag{8.30}
$$

We can see from (8.30) that $\mathrm{ZD_R}$'s peak is really quite steep (the payoff drops from 2.0 to 1.77 for the mutant, or 11.5 percent off the peak), but the mutant is unstable and will be eliminated from the population (because $1.66 < 1.7$). In other words, the "robust ZD" strategy is not very robust to mutations!

Let us compare $\mathrm{ZD_R}$'s payoffs to the effective payoff matrix for the $\mathrm{GC_d}$ strategy that replaces $\mathrm{ZD_R}$ at higher mutation rates (as shown in Fig. 8.8),

$$
\begin{array}{cc}
 & \mathrm{GC_d} \quad \mathrm{GC_d(1)} \\
\begin{array}{c} \mathrm{GC_d} \\ \mathrm{GC_d(1)} \end{array} &
\begin{pmatrix} 2.0 & 1.847 \\ 1.92 & 1.851 \end{pmatrix}.
\end{array}
\tag{8.31}
$$

We see that, first, the peak is much flatter (the fitness loss is only 4 percent) but more importantly, $\mathrm{GC_d(1)}$ is evolutionarily stable! We expect (but have not shown because it requires much more effort) that if we look at the effective payoff matrix between one-mutant strategies $\mathrm{GC_d(1)}$ and two-mutants $\mathrm{GC_d(2)}$, we would find that both mutants are ESS, and so on to higher and higher mutants, since otherwise the distribution shown in Figure 8.8 could not be stationary.[10]

We can therefore conclude that GC truly is a quasi-strategy that survives at high mutation rates due to the survival-of-the-flattest effect, an effect we discussed in much more detail in section 6.2.2.

8.4 Summary

We have seen that evolution favors cooperation as long as a communication channel exists that allows players to direct cooperation only toward those opponents that they predict (using the acquired information) to reliably return the favor. Indeed, this analysis suggests that there is a general theory of cooperation that is entirely information-theoretic in nature, and in which Queller's rule (8.1) (see also Queller 1992) is replaced by one that relates the ratio of an information and an entropy to the ratio of cost-to-benefit (c/b) of that behavior. Finally, we saw that the concept of evolutionarily stable strategies (ESS) needs to be extended to take into account groups of strategies that

---

10. As a technical point, the equilibrium between the mutants is not an attractive fixed point as in the snowdrift game, but rather is a dynamic one, with the mutant strategies forming a permanent hypercycle (Hofbauer and Sigmund 1998).

are stable distributions under mutational pressures: the quasi-strategy. Such quasi-strategies can replace otherwise unbeatable strategies when mutation rates are large enough, and are the game-theoretic equivalent of the molecular quasispecies.

Exercises

8.1 Show that for a general payoff matrix of a 2×2 game (two players, two moves)

$$
\begin{array}{c}
\quad\quad C \quad D \\
\begin{array}{c} C \\ D \end{array}
\begin{pmatrix}
a & b \\
c & d
\end{pmatrix}
\end{array}
\tag{8.32}
$$

the fixed points of the replicator equation (the points at which $\dot{x} = 0$) only depend on $a - c$ and $b - d$, so that the fixed points of the most general payoff matrix for 2×2 games effectively only depend on two parameters.

8.2 Maynard Smith's criterion for an evolutionarily stable strategy (ESS) states that a strategy S is an ESS if the payoff $E(S, S)$ when playing itself is larger than the payoff $E(O, S)$ between any other strategy O and S, that is, S is ESS if $E(S, S) > E(O, S)$. In case $E(S, S) = E(O, S)$, then S is an ESS if at the same time $E(S, O) > E(O, O)$, that is, that the payoff that the strategy receives against the opponent strategy is more than what the opponent gains when playing itself (in that case S is said to be "weakly stable"). Use this rule to determine what strategy (or strategies) is the ESS from simply inspecting the payoff matrix in the following games. Note that sometimes no strategy is an ESS, and sometimes multiple strategies are.

$$
\text{(a)} \quad
\begin{array}{c}
\quad\quad C \quad D \\
\begin{array}{c} C \\ D \end{array}
\begin{pmatrix}
3 & 0 \\
5 & 1
\end{pmatrix}
\end{array}
\quad
\text{(b)} \quad
\begin{array}{c}
\quad\quad C \quad D \\
\begin{array}{c} C \\ D \end{array}
\begin{pmatrix}
2 & 3 \\
2 & 1
\end{pmatrix}
\end{array}
\quad
\text{(c)} \quad
\begin{array}{c}
\quad R \quad P \quad S \\
\begin{array}{c} R \\ P \\ S \end{array}
\begin{pmatrix}
0 & 1 & -1 \\
-1 & 0 & 1 \\
1 & -1 & -0
\end{pmatrix}
\end{array}
$$

8.3 Calculate the information that generous Tit-for-Tat (gTfT) strategies (which play with probabilities $\vec{p}_{gTFT} = (1, 1/3, 1, 1/3)$) have about each other, using Equation (8.17). Assume that the population is equilibrated so that the play fractions π_i can be written in terms of the cooperative fraction π_C only, that is $\pi_1 = \pi_C^2, \pi_2 = \pi_C(1 - \pi_C)$, and so on. At what level of cooperation is this information maximal?

9

The Making of Intelligence

I visualize a time when we will be to robots what dogs are to humans. And I'm rooting for the machines.

—CLAUDE SHANNON (1987, IN AN INTERVIEW
WITH *OMNI* MAGAZINE)

The display of intelligence in animals (including us humans) never fails to impress. Intelligence has many facets that each contribute to what we perceive as intelligent behavior. Among those elements are categorization, learning, reasoning, prediction, and planning, to name just a few. Because the different facets contribute differently to overall intelligence, and because they interact with each other in nonlinear ways, it is unlikely that a single number will ever describe intelligence quantitatively.

But most of the time, when we see intelligence at work, we can recognize it. Among the many different elements of intelligence, the one that perhaps impresses us the most is the capacity to anticipate events; to plan for the future. From an evolutionary point of view, being able to predict the future better than a competitor has enormous fitness benefits. Just think about the advantage a predator gains if they can predict the trajectory of their prey so as to better intercept them. Then amplify that idea and think about how the capacity to anticipate a changing climate can affect the future of all of humanity.

We have repeatedly encountered the notion that information is that which allows the bearer to predict the state of a variable with accuracy better than chance. In a way we can see intelligence as a trait that will allow an agent to predict its *own* future, along with the future of the environment it finds itself in, together with all the other agents in it—and then planning actions according to this predicted future (Hawkins and Blakeslee 2004). Perhaps a good quantitative measure of intelligence then is the extent of the time horizon that accurate predictions can be made. Whether or not such a "temporal radius of prediction" is a good proxy for intelligence, we can be certain that information is crucial for making those predictions. However, intelligence (as we will see) goes beyond just information: exactly how that information is processed and used for prediction is crucial for survival.

9.1   Intelligence by Design or Evolution

Before we delve into the information-theoretic view of intelligence, perhaps a few higher-level remarks are in order. Historically, human intelligence has been studied both empirically and theoretically, of course. The empirical arm is enormously important, as it provides the behavioral data that any theoretical analysis is measured against. In the twentieth century the theoretical approach has focused mostly on understanding what the cognitive psychologist John Robert Anderson called the "architecture of cognition" (Anderson 1983), namely understanding the structural and mechanical basis of intelligence. Such a "systems" view of intelligence was thought to be a necessary precursor to the creation of artificial intelligence—after all, how could we construct something without knowing its blueprint? While the "structure and mechanisms" school of cognitive science has by no means died out, a different view has begun to appear at the end of the twentieth century, sometimes called the "adaptationist" view. Also championed by Anderson (1990) (but anticipated by the Cybernetics movement that we will discuss shortly), the adaptationist framework focuses instead on "optimality of function," rather than on structures and design. In that view the researcher asks what a system *should* be doing, rather than how it is doing it. In this "rational analysis" approach, we are to think about the brain as an adaptive system that has solved an information-processing task, in a near-optimal manner.

Now at the beginning of the twenty-first century, I believe we have come to a point where we can synthesize the architecture and the adaptive views of intelligence. This unification occurs by focusing on the computational implementation of elements of intelligent behavior, and an information-theoretic analysis of the processes that drive that behavior. But while a computational implementation of elements of intelligence is of course the standard approach of artificial intelligence (AI) research, the twenty-first-century-approach uses an adaptive algorithm to search for these computational implementations: Darwinian evolution of intelligence in the computer. Such an approach has two obvious advantages: First, evolution can discover mechanisms in the absence of any blueprint. Second, Darwinian evolution is the ultimate "optimizer": it is a process that can guarantee that the evolved solution is adaptive in the given setting. The drawback of using evolution as the designer is also immediately clear: to generate useful digital brains via evolution, we need to be able to construct the complex fitness landscapes within which intelligent behavior provides an advantage, but there is also no blueprint that tells us how to construct these landscapes. Furthermore, evolution is a slow process that requires millions of generations acting on very large populations. Thus, in any particular evolutionary experiment, we are not guaranteed to find these extraordinarily rare solutions in an effectively infinite search space. But the greatest appeal of the evolutionary approach is that we can be certain that it *can* work, because it worked at least once (and according to some, many times over in different species; see Roth 2013, for example).

In a nutshell then, in computational cognitive neuroscience (Kriegeskorte and Douglas 2018) we can test which mechanisms (embedded in what structure) give rise to the behavior we are interested in, and using an adaptive process (Darwinian evolution) allows us to find precisely those mechanisms that are adaptive, and that are therefore best suited to "do the job" given the particular environment that the organism finds itself in. While we are quite a ways away from achieving human-level intelligence using this approach, it is promising that for tasks that humans do quite well in the laboratory (in the case we'll discuss below, comparing the length of a tone to the tones in a rhythmic sequence), evolved digital brains not only perform the task at the same level as people, but are fooled by tonal manipulations in the same way as people are (Tehrani-Saleh and Adami 2021). But generating behavior is one thing; it is quite another to understand and analyze it. Information theory appears to present us with precisely the tools necessary to achieve this analysis, as we'll see in this chapter.

That the computations performed by the neocortex—the part of the mammalian brain responsible for higher-order brain functions—should be analyzed within the language of information theory is, of course, not a new idea. After all, the Cybernetics movement that was started by the mathematician Norbert Wiener (1894–1964) with his eponymous book (Wiener 1948) had information theory as the central unifying framework. However, with the rise of artificial nervous networks, the emphasis on information theory faded, and ultimately the excitement around Cybernetics waned (see Box 10.2).

A resurgence of information theory as the key to unlocking brain function occurred in the early to mid-1990s [even though the influential analysis of visual perception in terms of compression by Barlow (1961) occurred much earlier]. In that revival, information theory was mostly used to understand neural coding (Bialek et al. 1991; Rieke et al. 1997) (see also the review by Borst and Theunissen 1999), but also to quantify the synergistic firing patterns of groups of neurons (Tononi et al. 1994).

Box 9.1. The Rise and Fall of Cybernetics

Wiener coined the word cybernetics from the Greek $\kappa\upsilon\beta\epsilon\rho\nu\acute{\eta}\tau\eta\varsigma$, meaning steersman or governor, envisioning the brain as a control organ that steers the body through a complex and changing world. He had in mind a framework that would allow him to describe "Communication in Animals and Machines" (the subtitle of his book), and information was central to this approach. Indeed, prediction is a cornerstone of Wiener's cybernetics, and so his book puts significant emphasis on time-series prediction as well as filtering (to remove noise from signals), along with chapters on feedback and ruminations about the nature of computing in the nervous system. The book—as famous as it is—now reads as a set of somewhat disconnected chapters that introduce a hodgepodge of ideas. Some are discussed in mathematical language, but mostly the book is philosophical rather than technical. Wiener dominated the movement by the strength of his personality, eclipsing other members of the core group, among which were von Neumann, Turing, Shannon, W. Ross Ashby, and Warren McCulloch (to name only the most famous). During the annual meetings of the group under the auspices of the Macy Foundation, information theory became the dominant framework to discuss all aspects of control and communication in animals and machines. Expectations were high that within a few decades at the most, computers could rival humans in terms of intellectual achievement. Gradually, the neuropsychologist McCulloch's role in the movement became more prominent, leading to a greater emphasis on neural networks as the foundation of machine intelligence [his paper with Walter Pitts (1943) had shown that simple automata with thresholds can implement arbitrary complex logical calculations]. The focus on these artificial neural networks led to even more ambitious expectations under the name of "Connectionism," which ultimately yielded to disappointment. Along with the disillusionment that followed, cybernetics as a movement faltered. While some blame the focus on artificial neural networks at the expense of other approaches for the downfall of cybernetics, others fault Wiener's personality and in particular the gradual de-emphasis of information theory. In the introduction to his book, Wiener obliquely hinted that he had developed the framework of information theory independently of Shannon (and before even Kolmogorov) and perhaps became increasingly annoyed at Shannon's mounting fame. Privately, Wiener would refer to "Wiener-Shannon theory," arguing that the theory "belongs to both of us." Shannon's colleague John Pierce would write later that

> Wiener's head was full of his own work. Competent people have told me that Wiener, under the misapprehension that he already knew what Shannon had done, never actually found out. (J. Gleick, "The Information," 2011, p. 235).

Given that Wiener often wrote that information was just the negative of entropy (which it is emphatically not), this assessment by Pierce rings true.

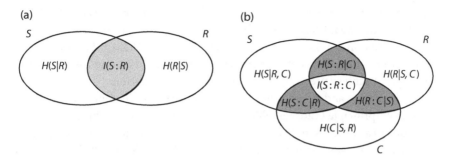

FIGURE 9.1. (a) Entropy Venn diagram describing a signal/reponse channel in the brain, where S represents sensory neurons and R is a set of response neurons. The white areas of the Venn diagram represent conditional entropies that an adaptive process would attempt to reduce, so as to increase the information processed by the channel (light-gray area, given by $I(S:R)$). (b) Entropy Venn diagram including a "contextual field" C: a set of neurons that has memory of past events, and stores representations of elements of the environment within which the organism operates. The entropy of the "decision maker" R is composed of four terms: a "noise" term $H(R|S, C)$ that an adaptive process should decrease, as well as the coherent information $I(S:R:C)$ (center) and two shared conditional informations (shaded dark gray).

9.1.1   Intelligence as information processing

Generally speaking, we can try to characterize cognitive processing in terms of a complex mapping from sensory inputs to actions. For simplicity, imagine a sensory signal S and a "response" R (such as a muscle movement). Information theory can describe the relationship between presence or absence of an object in the sensory organ (say, the eye) and a grasping reflex mediated by a muscle: the response R. If presence of object is associated with grasping, then the information between R and S [the center of the Venn diagram in Figure 9.1(a), given by the information $I(S:R)$] would be high. The white areas in the diagram Figure 9.1(a) are conditional entropies, and they play important roles in the quest to optimize the information. In fact, we recognize this construction as a simple example of the information transmission channel introduced in section 2.4 [see for example Equation (2.74)].

The quantity on the left in Figure 9.1(a) (given by $H(S|R)$) is the entropy of the signal given the particular response and is a measure of the signal *redundancy* (roughly speaking, the logarithm of the number of sensory inputs that give rise to the same exact response). Early on, Barlow (1961) suggested that in an adapted brain this redundancy (it is also called the *channel loss*) should be minimized, so that the brain can use the available capacity in an optimal manner. Another important quantity in a channel is the *channel noise* $H(R|S)$. Evidently, this quantifies the variability of the response given the same exact input, so that in a noiseless system we would have $H(R|S) = 0$. Indeed, the optimal encoding of a genetic signal that we discussed in section 2.4.3, page 90 attempts to achieve precisely this redundancy minimization.

Redundancy and noise reduction can be seen as elements of Linsker's "Infomax" principle (Linsker 1988), which is a well-known technique of optimization in artificial neural networks where the information between adjacent layers of the network is optimized (see also Bell and Sejnowski 1995 for an application in the absence of noise). But this approach rewards fairly simple "reactive" behavior in which the output is essentially predicted by the input, without taking into account input from memory, for example. According to the "two streams" hypothesis (Goodale and Milner 1992; Goodale and Milner 2004), our brain processes visual information via two separate pathways: an ancient one that is mostly reactive (the "where" pathway that runs along the dorsal side of the brain) and the more recent "what" pathway (on the ventral side of the brain; see Fig. 9.2).

The "where" pathway is involved in controlling motion ("action") based on visual input: for example, it allows you to quickly move away from a falling tree or swat away a bug, without being aware of what was falling, or what kind of bug was attacking. The "what" pathway is involved in perception, and relies on long-term stored representations to situate objects into complex contextual scenes. It is the "what" pathway that recognizes objects and scenes in relation to their environment, and helps plan actions based on those scenes.

The "what" pathway's computations are far more complex than those carried out by the "where" pathway, and in particular rely on long-term stored representations as well as memories of past events. These representations and memories provide context to the input, and play a crucial role in learning and the acquisition of knowledge as we'll discuss in more detail in section 9.2 below.

Phillips and Singer (1997) describe the relationship between sensors S, response neurons R, and contextual neurons C as in Figure 9.1(b), and propose a more complex optimization algorithm (compared to Linsker's) where any of the four components that make up the entropy of the response set of neurons (which Phillips and Singer call the "local processor") could be an element of a fitness function. For Linsker's Infomax, the term $I(S:R) = H(S:R|C) + I(S:R:C)$ would be optimized by jointly increasing $H(S:R|C)$ and $I(S:R:C)$, while Phillips and Singer propose that the information jointly shared by the sensors, the response, as well as the context (the center of the diagram in Fig. 9.1[b]) is optimized by itself: an adaptive process that the authors call "coherent infomax."

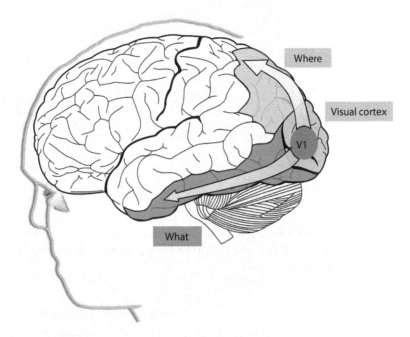

FIGURE 9.2. Schematic representation of the two streams of visual information processing in the neocortex. The dorsal processing stream (light gray,
"where" pathway) and the ventral stream (dark gray, "what" pathway) originate
from a common source in the visual cortex (area V1), but arrive at different
destinations. From Wikimedia (CC BY-SA 3.0).

The main idea behind this construction is that signals acquire meaning
only within the context of stored representations of the world (acquired in the
past), and that these stored representations provide *predictions* about expected
behavior. According to this view, sensor/response pairs that are correctly predicted by the context are reinforced, while incorrect predictions will modulate
the prediction instead (Phillips and Singer 1997). In that way, a simple learning rule can be implemented information-theoretically, as long as the coherent
information is maximized.

While this construction is obviously more powerful than the simpler infomax rule, we will see shortly that there are many important situations where
coherent infomax cannot possibly be the target for optimization, simply
because the coherent information in those cases is (and must be) negative. An
instructive example of this sort is provided not by the sensory-motor circuitry
of the brain, but instead by the sense/response circuit of the lowly bacterium,
which we will study now.

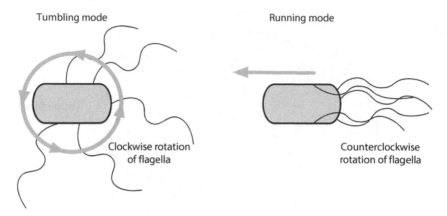

FIGURE 9.3. Tumbling motion (left) and swimming (right) is effected by either turning the motor connected to the flagella clockwise (tumbling) or counterclockwise (swimming).

We know that every organism, no matter how simple, uses the information stored in its genes to make predictions about its environment. A bacterium, for example, uses its genes to make "bets" about what environment it is likely to encounter (as we discussed in Box 2.4). But bacteria also use their genes to make predictions about the future. One of the most well-studied bacterial behaviors is *chemotaxis*: the tendency of bacteria to swim toward nutrient sources by preferentially swimming along the gradient of increasing nutrient concentration. In a way, we can see this behavior as the bacterium using the gradient to make a prediction about the best place to feed. It is worth studying chemotaxis in a bit more detail because while it is a simple behavior, it is interesting in many respects and allows us to discuss several important aspects of intelligence in a model system.

Bacteria use their flagella to move around—mostly within watery environments, but sometimes also in viscous fluids like mucus. While different bacteria have different numbers of flagella, most use them by alternating between two behaviors: swimming and tumbling; see Figure 9.3. This behavior is controlled by the flagellar motor which, when turning clockwise, gives rise to random tumbling where the direction of motion constantly changes. When the motor turns counterclockwise, however, the bacterium swims straight into the direction it was pointing to. Goal-directed behavior occurs by alternating between these modes guided by the *gradient* of nutrient concentration in the medium.

The algorithm controlling this behavior is simple enough, but its implementation is not. For example, it is not simple for the bacterium to measure

Table 9.1. Probabilities for a bacterium to swim, given sensed concentration and context. The sensed variable S is the concentration at time t ($S = [S]_t$), while the context is the (remembered) earlier concentration, $C = [S]_{t-\Delta t}$. The third column shows the probability that the response variable R takes on the swim state.

S	C	$p(R = \text{swim}|S, C)$
H	L	1
L	H	0
L	L	0.5
H	H	0.5

a nutrient gradient. For it to perceive changes, it is necessary to compare two concentrations sampled at different points in space or time. Because the bacterium is small it must use temporal changes, and to implement this it must have some sort of memory. Without going into the molecular details (see, e.g., Wadhams and Armitage 2004 for those), this memory is achieved by a system in which the detected concentration gradually reduces its influence on the "downstream" molecular machinery that reacts to the detection of the concentration by turning the motor counterclockwise and therefore swimming. If the concentration does not change, the weakened influence will result in tumbling instead. This kind of control is sometimes called "integral feedback control" (Dufour et al. 2014), and is implemented by a regulatory apparatus that relays the sensory output to the flagellar motors (the actuators) via a response regulator, to control the probability to change direction. The result is that the bacterium "remembers" prior concentrations for a few seconds, which is sufficient to drive a biased random walk that will move the bacterium in the direction of the nutrient gradient, toward the maximum.

A simple model of chemotaxis can illustrate how the bacterium uses information from the environment as well as memory to drive the swim/tumble decision. Suppose a bacterium can sense nutrient levels $[S]_t$ using a sensor that can have two states: high (H) and low (L), and that the bacterium can store those values in memory so that it can compare those values to the current ones. In the model, an increase in concentration drives the motor $[R]_t$ counterclockwise with probability one (in reality the probability is lower) leading to swimming, while when the concentration decreases over sensed time, the motor turns clockwise (leading to tumbling; see Fig. 9.3). Furthermore, to prevent the bacterium from getting stuck in constant concentrations, we will assume that the bacterium swims with probability 0.5 in unchanging conditions; see Table 9.1. If we assume that L and R are sensed on average with

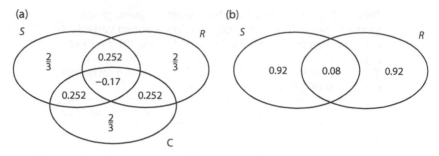

FIGURE 9.4. (a) Entropy Venn diagram for the chemotaxis model studied in Exercise **9.1**, in which the sensor S reflects the molecular concentration of a nutrient source at time t, $[S]_t$, influencing the action of the motor R at that time, controlled by the concentration $[R]_t$ of a molecule that drives the motor. The context C for the relationship between $[S]_t$ and $[R]_t$ is given by the *past* sensed concentration $[S]_{t-\Delta t}$. *Given* the context, sensor and motor share about a quarter bit. (b) In the absence of the context (that is, ignoring C altogether), very little information is relayed to the motor from measuring the concentration, as evidenced by the small $I(S:R)$.

equal likelihood, and that the concentration changes after a period Δt with probability 1/3, then the logic in Table 9.1 gives rise to the entropy Venn diagram shown in Figure 9.4 (see Exercise **9.1** for mathematical details of the model). This simple model illustrates that about a quarter bit of information drives the flagellar motor, but only when the context $[S]_{t-\Delta t}$ is given. In the absence of this context, $[S]_t$ and $[R]_t$ are statistically almost uncorrelated: $I([S]_t : [R]_t) = 0.08$ bits. This relationship between the sender and receiver variable is reminiscent of the Vernam cipher (discussed briefly in Exercise **2.6**), in the sense that the context provides the cryptographic key to unlock the information between the sensed value and the response. However, unlike in cryptography, in this model even in the absence of the contextual key there is some information in the sensed value (but not much). But clearly, coherent infomax cannot be the optimizing principle at work here, as the coherent information $I(S:R:C)$ is negative.

9.2   Elements of Intelligence

Models like the one we just studied can be made much more general, and in particular information theory can be used to show that evolution tends to select for strategies that best predict changing environments, using inherited information as well as information acquired from the environment (Rivoire and Leibler 2011). Being able to use acquired information to make decisions is crucial when dealing with environments that change in complex ways, but this acquired information must build on a foundation of robust and unchanging knowledge about how to live in that particular world. Indeed, many aspects of the biotic and abiotic environments an organism faces are fairly constant, and to take advantage of those more or less unchanging features, an organism can rely on the information coded in its genes.

If the environment changes quickly (on the order of seconds for example), predictions must rely on information that is sensed (using sensors whose readings must be continually updated) along with a "model" that uses those readings to predict the future. These models can be as simple as a linear or nonlinear function that just translates the input signal into an output/action, or they can be extraordinarily complex. But whether these models are simple or complex, they must be stored within the organism itself: they are "internal models" of some aspects of the world, and they are also time machines: they tell tales about possible futures. In the following we will look at six elements of intelligence that work together to achieve what we call intelligent behavior in a complex changing world. These elements are not exhaustive (nor even nonoverlapping), but they all have information-theoretic underpinnings. Following this exposition, we will see how those elements are used by digital brains that were evolved to solve specific tasks.

1. Prediction. In many ways, the ability to predict the future using observations and internal models of the world is the essence of intelligence. Naturally, species differ in the precision of this prediction, as well as in the time horizon up to which that prediction is more accurate than chance. Bacteria, for example, use the chemotaxis system just described to predict the future state of their environment a few seconds ahead. Other organisms can predict much further out in time. A typical example is the squirrel, which collects nuts in the fall so as to have food in the winter where nuts cannot be found. It may seem like an obvious prediction (and one that is almost certainly genetically encoded), but it is a prediction nonetheless. Other animals are even more sophisticated with their predictions. Manufacturing tools, like for example the hook tools made by the New Caledonian crow (Hunt and Gray 2004), is a prediction that such tools might be of use in the future. Even more spectacularly, certain birds (in this case, scrub jays) have been found to hide food from other birds, but only if they had been observed by those other birds caching (that is, storing) the food, and when they themselves had been stealing food from other bird's caches before (Emery and Clayton 2001). It is difficult to evade the conclusion that these birds are acting on internal models of the world that allow them to predict that if another bird has watched them cache the food—and that other bird is like them—that the food will be stolen later.

Prediction is not possible without information, because information is that which allows prediction with accuracy better than chance. A prediction that is true by chance is not a prediction at all: it is a random guess. Mathematically speaking, a prediction is a guess that a particular random variable will take on a particular state (or sets of states) among all the possible states, with elevated probability. But where do these states come from in the first place? How many different states are there and how do we recognize their differences? It is clear that it is our brain that creates these distinct states, by putting objects and concepts into *categories*.

2. Categorization. The kinds of predictions that the scrub jays are making—sometimes called *experience projection* (Emery and Clayton 2004)—suggest that animals can have a theory of mind, namely the ability to not only represent their own thoughts, but predict that other animals like them have similar thoughts. Can we quantify such mental states using information theory? Can we measure the power of thought?

It is clear from the assumptions behind Figure 9.1(b) that brains can store information within internal states (it is, of course, also clear from knowing that we can remember things). But we need to keep in mind that just the storage of information itself is not something that is automatically given. Information storage is so common in our everyday world that we scarcely ever think about it. But before we can even store (and later retrieve) information, we must first create a symbolic alphabet. This is a requirement for storage of information in computers of course (where the alphabet is zeros and ones, typically instantiated by different voltages in semiconductors). It is also necessary for the storage of letters on pages, of information in genes, or even for the early pictographic languages such as hieroglyphics and cuneiform scripts. In each case, something that occurs in many different forms in the outside world must be *categorized*, and assigned a particular abstracted symbol before it can be stored and remembered. How exactly this categorization proceeds is probably different for each symbol, but we may perhaps think of it as a process akin to the evolution of the glyph for "head" in the Sumerian cuneiform script, shown schematically in Figure 9.5. In this evolution (which took two thousand years) a more or less faithful representation of a head morphs from being referential to symbolic. During that process, the other symbols in the alphabet evolve as well, and it is likely that the number of used symbols changes at the same time, with those that are rarely used being abandoned.

This process of categorization is an important element in most machine learning applications, because it is an essential step in taking objects from the external world and internalizing them as stored caricatures of reality:

3000 BC ————————————————————————→ 1000 BC

FIGURE 9.5. Two thousand years of evolution of the pictogram for "head" into the corresponding glyph in the Sumerian cuneiform script, after Kramer (1981).

representations.[1] From this point of view, categorization is a prerequisite for prediction, and it allows us to quantify the amount of information necessary to make such a prediction. Categorization is also a prerequisite for information storage, because the number of states within a category defines the maximum entropy per symbol. But how *is* this information stored over time then?

3. Memory. Before the advent of computer memory or even paper, storing information over time was a problem that needed to be solved. As we discussed in much greater detail in chapter 7, storing information about how to copy that information is the essence of life itself, and a problem that (here on Earth) ultimately had a very elegant solution: our DNA.

Storing information in the brain is perhaps even more difficult, as the carriers of information (neurons) do not store information in their state as nucleotides do. Instead, memories appear to be formed by patterns of neuronal firings that can be triggered by other such patterns: temporal series of patterns, if you will. We will encounter one form of such neuronal memory when analyzing "digital brains" that were evolved in the computer (a bit further below), but it is safe to say that at this point in time we do not have a general understanding of the different ways in which information is stored (over the short and long term) in the brain.

Memory makes it possible to retain information, but how is information acquired in the first place? We discussed at length the acquisition of information via Darwinian evolution in chapter 3. This is a slow process, as this information must be built up mutation by mutation. The obvious advantage of genetic information is that it can be inherited, but in a fast-changing world

1. One might object to the importance of categorization in the forming of memory by noting that we can recall scenes that we have lived as if watching a replay of the scene in our "mind's eye" (a recall that is sometimes called "episodic memory"). However, it is possible to take a point of view (espoused below) that such vivid detailed memories are really illusions created by our brains, and that episodic memory does not have a different substrate than general memory.

it is necessary to retain information on very much smaller time scales. Because that changing information is valuable at that particular time only, there is no need for it to be inherited. In response, evolution has given us the mammalian brain (Allman 1999), with a practically infinite capacity of information storage. Of course, none of that stored information can be inherited, but over time animals—in particular people—have developed methods to retain some of that information over generations; first by language, then finally (in the case of people) via writing systems.

4. Learning and inference. The mammalian brain is uniquely suited for learning, and many books and thousands of papers have been devoted to understanding the algorithmic, physiological, and neuroscientific underpinnings of this process. We cannot do justice to all that literature here, but Sutton and Barto (1998) is a good introduction to the algorithmic side, and Thompson (1986) gives a concise overview of the neurophysiological aspects of learning. It is useful to examine the learning process from the point of view of *Bayesian inference*, which is a statistical method that allows you to attach likelihoods to scientific hypotheses (perhaps the best book on Bayesian inference is the incomparable Jaynes 2003). Without going into details, the main idea of Bayesian inference is that we should decide among a set of hypotheses by choosing the one with the highest *likelihood*, which is a probability formed from a prior probability (basically, a set of *a priori* beliefs) and current evidence. In learning, those a priori beliefs ("priors," for short) are constantly updated in the light of incoming evidence, into what are called "posterior" probabilities. This means in a nutshell that learning is the process of changing ones "before the facts" beliefs into "after the facts" beliefs (see Vilares and Kording 2011 for a concise review of Bayesian models of inference in the brain).

Where do these a priori beliefs come from? Generally speaking, priors are formed from the internal models that we carry with us. These models are shaped in part from eons of Darwinian evolution (and therefore etched into our DNA), and in part they are formed from day-to-day experience, that is, they are themselves learned, and stored in our brain. As Jaynes (2003) emphasizes in his book (and Phillips and Singer corroborate in their essay), information theory succinctly describes learning as the accumulation of information via inference, and within Phillips-Singer theory the priors are specifically given by the coherent information shown in Figure 9.1(b). According to this point of view, we learn by creating models of our environment, and these models are stored in the connections between neurons of our brain. But how exactly are these models stored, and how can we measure how much we know about the world in which we live?

5. Representations. As we have now seen emphasized multiple times, internal models that represent physical objects and processes in the world are essential for learning, memory, and prediction. The concept of representation is ubiquitous both in cognitive science as well as machine learning, but the word means somewhat different things in those two disciplines. Let us first focus on the concept of representation as it is usually used by cognitive scientists and philosophers of the mind, but keep in mind that it is not difficult to imagine that even among cognitive scientists there is no general agreement about what exactly qualifies as a representation.

The philosopher and cognitive scientist Andy Clark makes the case that a representation is a set of internal states that not only reflect possible states of the environment, but that these states actually *stand in* for features of the environment, so that these states can be used *instead* of the real external features in order to make plans and predictions (Clark 1997). We often use the word "model" to describe such abstractions: model airplanes, for example, allow us to test aspects of the plane that are difficult to test with the real thing, while models of climate (even when highly abstracted) give us some ideas of how, for example, human activity can influence global temperatures. In the first example, the model only shares certain structural features with the object it is supposed to represent, is manufactured from a different material, lacks internal structure, and is altogether the wrong size. In the second example, the model is (in general) purely mathematical, can only deal with aggregate data, and can only make predictions for averages.

Is it possible to measure how much of the world a brain represents within its memory? To do this, we would have to define representations mathematically first, and information theory appears to be the right candidate for that endeavor. Let us first set up the random variables we need. Like we did when we quantified how much a DNA sequence knows about the environment within which it needs to function, we define a "world" variable. While perhaps we should use W for this variable, we'll use the same letter E as earlier to make it clear that this world is the environment within which the brain must make a living. But while in section 2.3.1 we assumed that the environment was constant, here we must assume that the variable E can take on many states over time and space. After all, brains have evolved precisely because a fast-changing world cannot be tracked with genes whose expression levels can change on the order of tens of seconds at its fastest (see, for example, Jékely 2011).

We also need to define the state of the brain B, where B can be seen as the joint state of all neurons. We can then quantify how much the brain knows about the world as

$$I(B : E) = H(B) - H(B|E), \qquad (9.1)$$

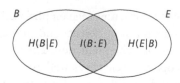

FIGURE 9.6. Entropy Venn diagram showing the relationship between the world random variable E and the brain B. The information that the brain has about the world is shaded gray.

that is, the entropy of the brain (independently of what environment it needs to function in) minus the average entropy of the brain given the world. The entropy Venn diagram that helps us keep track of information, entropy, and conditional entropy is shown in Figure 9.6. However, a little thought reveals that $I(B:E)$ cannot be a good measure of representation after all: representations should be models of the world that are independent of what appears in the brain's sensors. To wit, representations should not change in response to changes in the image that we perceive (even though our model could change later, based on what we just saw). In other words, changes in our brain states that are directly correlated with changes in what we perceive do not count as representations: they do not model anything; they are just reflections. Representations should continue to exist even when we close our eyes.

To tease apart correlations between brain states that just reflect the world and brain states that are modeling world states, we need to introduce the sensor variable S, which stands for all those neurons that receive information directly from the world. For the visual system, this would be the photoreceptors, for example. The entropy Venn diagram for the variables E, B, and S is shown in Figure 9.7, where we now have explicitly excluded the sensory neurons from B. This diagram may appear confusing at first, and superficially it might look like the Venn diagram between sensor neurons, actuator neurons, and contextual neurons we discussed in section 9.1.1, specifically Figure 9.1. However, this is a very different diagram because one of the variables (E) is *external* to the brain, while all variables in Figure 9.1 represent neurons inside the brain.

In Figure 9.7, the information that the brain has about the world, shaded gray in 9.6, is now shown to have two parts: the part of the information that is shared with the sensors $I(B:E:S)$ and the part that is not shared with the sensors: $H(B:E|S)$.[2] The shared entropy $I(B:E:S)$ measures how the brain B

2. Technically, $H(B:E|S)$ is a "mutual conditional entropy." We have encountered the mutual conditional entropy before in Figure 2.12, where we quantified how much information is shared between a particular nucleotide sequence and the polypeptide it encodes, given the particular selective environment. It also made an appearance in the Venn diagram 9.1(b), where

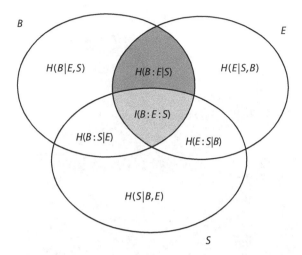

FIGURE 9.7. Entropy Venn diagram between the brain B, the world E, and the sensors S. The dark-gray area is the information the brain has about the world, given the sensor state: our definition of representation. The light-gray area is the information about the changes in the world that are relayed directly to the brain without consulting any models. The shading reflect the ventral and dorsal streams of information-processing shown in Figure 9.2.

changes in response to the world E's changes, as seen through the changes in the sensor's states (what Andy Clark calls "environmentally correlated inner states"; Clark 1997). These changes are important as they allow us to react quickly to changing events, but they are not internal models of the world, which have to be independent of what is currently perceived. Rather, this is information that is relayed via the dorsal pathway shown in Figure 9.2: the "where" pathway that is entirely reactive.

World models (representations) are stored in the ventral "what" pathway, which allows us to integrate the sensory information with what we already know about the world. The extent of these world models is measured by $H(B:E|S)$, which is our information-theoretic definition of representation (Marstaller et al. 2013). This definition squares well with what philosophers like Andy Clark call representations in cognitive science (Clark 1997): those internal models that allow us to solve problems "off-line," that is, with our eyes closed—simply by imagining the world. We discuss representations in more detail in section 9.2.

---

it is shaded dark gray. The quarter bit of information in Figure 9.4 is also a mutual conditional entropy.

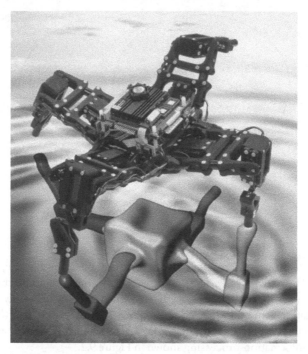

FIGURE 9.8. The "Starfish" four-legged robot studied in Bongard et al. (2006), imagined contemplating its own internal model (courtesy Viktor Zykov).

It is important to keep in mind that those models are not intended to stand in for *every* aspect of the object or process they are representing; rather, they are useful for predicting very specific elements only. This idea is illustrated neatly in a computational example, where robot brains can learn representations of their own morphology (Bongard et al. 2006; Fig. 9.8): by having a model of what they look like, those robots can change how they walk when they are injured. In this particular study, the robots attempt to infer their own morphology by narrowing down the set of all possible models (constrained to have at most four legs, and at most two joints and two joint-angle sensors per leg). The robot achieves this by carrying out an arbitrary motor command, and recording the resulting sensory data. A "model synthesis" module then constructs fifteen possible self-models to explain the observed sensory-action causal relationship. To choose among those fifteen models, the robot "imagines" what would happen (meaning, which sensors would record a touch, and what angles the joints would take) if a particular action were carried out, *given* the particular model. In this manner, the robot can choose precisely the action that distinguishes the fifteen models the most—in other words, it can choose the *most informative* move. After going through sixteen such cycles of choosing the most consistent set of self-models, then generating test moves to rule out the alternatives, the robot can then choose the action that it predicts would move it forward the most, given what it believes is its most likely shape.

Bongard et al. were able to show that robots that were able to choose moves based on self-models were performing significantly better (moved forward farther) than robots without such "self-knowledge." Even better, they showed that if the robot was damaged (by, say, removing a leg), it was able to recover very quickly by correctly synthesizing the changed robot morphology, and planning moves taking this incapacity into account. A similar approach was used to create robust soccer-playing robots by another team (Gloye et al. 2005). In a sense, we can imagine that when the robot is choosing among models, it really is "dreaming up" possible actions and examines their consequences in "the mind's eye" (Adami 2006c). This example shows that it pays to have an internal model of oneself because it allows you to test out moves on the model, without actually having to perform them in the real world.

Could a robot make better plans by modeling not just itself, but also the world in which it lives? It is clear that we are making models of the world "in our mind's eye" all the time, and so do animals. A particularly well-studied example of environment-modeling occurs in many mammals, which are able to represent the physical space in which they live within a set of specialized neurons called "grid cells." The discovery of these cells resulted in the Nobel Prize in 2014 (Burgess 2014) and has revealed the existence of an internal representation for location, direction of movement, and head-direction within the animal brain. What is remarkable about this representation is that while it is learned during the animal's lifetime using landmarks, the representation of this space persists even in darkness (Fyhn et al. 2007), that is, it is not bound to the direct sensory experience of these locations. Thus, it qualifies as a true representation in the sense of Clark (1997).

To understand how grid cells encode location, we'll study a simple algorithm that allows us to encode a location in a 2D space using only three grid-cell modules, shown in Figure 9.9. In principle, the exact location of an animal in the example 27×27 grid in Figure 9.9 could be encoded in ten binary neurons (using a standard binary code), but this representation would not be very robust. For example, if the pattern 0011111 encodes position 31 using a binary code, moving to position 32 would result in pattern 010000, a change in six neurons. The grid-cell encoding in Figure 9.9 instead guarantees that the firing pattern of the grid-cell neurons varies only slightly if the animal moves slightly (mostly in the Module 3 neurons only), while the patterns of Module 1 and 2 remain persistent. Neuron 6 of Module 3 in Figure 9.9, for example, will fire anytime the animal is in one of the locations indicated by the dark-gray color. By itself, therefore, this neuron cannot encode the location of the animal. Together with the neurons in Module 1 and 2, however, location is perfectly encoded.

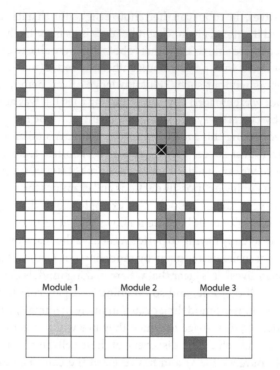

FIGURE 9.9. In this example the physical location of an animal in a 27×27 grid is encoded using three grid-cell modules that each indicate the location at different scales. This representation is effectively a base-9 encoding, where each module represents successively more significant location information. The first module (Module 1) encodes the least significant (that is, most coarse) digit. If we number the cells from left to right and top to bottom with the decimal numerals 0–8, this least significant digit of the location shown as a light-gray square would be 4. Module 2 records location information at a higher resolution, and in the case depicted it indicates 5. Module 3 encodes the highest resolution (most significant digit), here a 6 (dark-gray square). Together, the location of the black square marked 'X' corresponds to the base-9 sequence 654, encoded in twenty-seven binary neurons organized in three modules of nine neurons each.

It is instructive to look at the information theory of grid-cell encoding. In this case, the location variable (call it X) has $\log_2(27 \times 27) \approx 9.51$ bits of entropy. Specifying one module M_i (any module) reveals about 3.17 bits because the location uncertainty $H(X|M_i) = \log_2(81) \approx 6.34$ bits: $I(X:M_i) = 3.17$ bits for all three modules. Given one module, the second module reveals another 3.17 bits: $H(X:M_i|M_j) = 3.17$ for any pair of modules. Finally, giving the state of all three modules reveals all the information about X: $I(X : M_1M_2M_3) \approx 9.51$ bits. The beauty of this encoding is that there is no encryption: the three modules M_1, M_2, and M_3 provide the relevant information independently, that is, additively (see Exercise **9.2**).

In real animals, grid-cell neurons do not tile space in 3×3 regular lattices, but instead form hexagonal lattices with a variable number of neurons per module and enough modules to cover the foraging range of the animal. For example, the rat has a foraging range of between one hundred meters and one kilometer, but must be able to remember distances down to a scale of about five centimeters (Fiete et al. 2008). In fact, Fiete et al. (2008) estimate that just twelve modules, with a total of about 50,000 neurons in the medial entorhinal cortex (an area of the brain that is crucial for memory, navigation, and the perception of time), is enough to cover a foraging range of two kilometers at a resolution of five centimeters. To appreciate the economy of this representation, we can compare this to how a different code might encode this location information. Indeed, there is a different set of neurons in the hippocampus that encodes location information using a simpler code, where a group of about ten neurons fire only when the rat is in one specific location. It would take about ten billion neurons to encode all the specific locations in a 2 km $\times$ 2 km area to a resolution of 5 cm, even though there are only about a million neurons in the rat hippocampus (Amaral et al. 1990). In a sense, place cells can be viewed as holding a temporary map of the current local environment the animal finds itself in, and are remapped when the animal leaves that area. Place cells are thought to receive inputs from the grid cells in the entorhinal cortex, which is indeed directly adjacent to the hippocampus in the temporal medial lobe of the brain. However, exactly how grid cells influence and guide place cell firings is not yet fully understood (Neher et al. 2017).

Of course, to function in a complex world, the brain must create models not only of the physical space it finds itself in, but also of the events and processes it encounters. These events take place in time, and to recognize such events as important or not, the brain must be capable of perceiving time, and in particular patterns in time. That the perception of time is not at all simple can be seen by studying a simple example: recognizing a particular temporal pattern of binary symbols. We can think of recognizing such a pattern as a simple example of recognizing a melody: we are able to do this effortlessly even after listening to the tune only once or a few times, and we can recognize it by hearing just a short sequence of it, from anywhere within the sequence. How is this temporal memory created?

Let us imagine a simple circuit where a single sensory neuron S receives the acoustic stimulus, and a single motor neuron M that fires if a particular sequence of tones is recognized (we can imagine that this neuron controls a muscle that leads to us raising our hand if we recognize the tune). In the

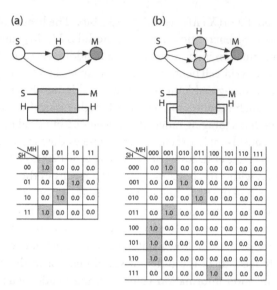

FIGURE 9.10. (a) Top: Simple model of a time-sequence recognition circuit that can recognize at most one out of four possible two-bit sequences. Middle: The 2 × 2 logic connecting the three neurons must involve the hidden neuron as a memory unit that keeps time. Bottom: A logic table that recognizes the sequence 10. In this table, the input sequence consists of the signal S and the value of the hidden unit H at the *previous* point in time (rows), while the outputs are the value of the motor unit M and the hidden neuron H, at the subsequent time point. The entries in the table are the probabilities with which the output states are generated given the input state (the values in each row must sum to one). This table implements deterministic logic, but in general the transition matrix is probabilistic. With the hidden state initially at 0, a 1 in the sensor will give rise to the output state 01 with probability one, meaning that the sequence is not yet recognized, but the hidden state is flipped. If the next input is 0, the input to the gate is now 01 and the output therefore 10, which signals recognition of the sequence and re-initialization of the hidden unit. Inputs of 00 and 11 simply re-initialize the detector, so that it is always expecting a 1. (b) Simple model that can recognize sequences of length 4, in this case the sequence 0001.

simplest model, there is only a single intermediary ("hidden") neuron H as in Figure 9.10(a) that keeps track of "where," in time, we are. Such a circuit can recognize at most one out of four different two-bit sequences.

To recognize longer temporal sequences, we need more memory. In Figure 9.10(b) we can see a logic gate with two hidden neurons connected to input and output via a 3-in-3-out logic gate. The logic table shows the values of a gate that recognizes the four-bit sequence 0001, or the binary equivalent of the introductory theme (the "faith motif") of Beethoven's fifth symphony (if we identify 0 with G and 1 with B♭ and ignore the rhythm).

Clearly, the structure of the logic table implements a model of the sequence to be recognized, and indeed models of this sort are very common in bioinformatics, and used to recognize biomolecular sequences with specific properties called "Hidden Markov Models," or HMMs), or else in speech recognition (Gales and Young 2007). In general, the parameters of such a model must be learned from data, with fairly time-consuming methods such as the Baum-Welch algorithm (see, e.g., Durbin et al. 1998). However, we will see later that it is also possible to "learn" such models either via Darwinian evolution (committing such models into DNA) or with reinforcement learning (committing them to memory).

6. Planning. Armed with models of the world and the temporal processes within them, a brain can make predictions of the future and make plans that maximize the organism's future success. For example, it is immediately clear to anyone who is familiar with Beethoven's opening theme that—upon hearing those first four tones—it is almost impossible to *not* immediately produce the next four, thus using the model to make a prediction of the future series of tones.

While the internal representation of spatial location is learned from sensory cues (usually visual or olfactory), those representations are maintained in the absence of these cues, for example at night. It is this feature that allows animals to solve the "vector-navigation problem," by using the internal representation of space to navigate directly to goal locations without following explicit sensory cues. In a sense, the animal moves "virtually" within its mind, and updates its location there with every step. Specifically, the representation of a goal location can be combined with the representation of the current location to deduce the vector between them. Following this vector can give rise to a novel trajectory the animal has never taken before, passing through areas it has never previously visited.

Another example of using internal models to plan actions are the saccades of the visual system. The human vision system is accurate (20/20) only within the central $1°-2°$ of the visual field, namely the fovea. Outside of this narrow cone (about the size of your thumb if held outstretched in front of you), vision is blurry. To get an impression of a scene, the eye *saccades* across the visual field, about two to six times per second. In the example in Figure 9.11, six saccades are shown starting from a random location. It is clear from just this example that the saccades are attempting to focus on the most important aspects of the scene: the location of the figure on the left and the faces of all those in the room. But how does our brain know where to look, which parts of the image to pay attention to?

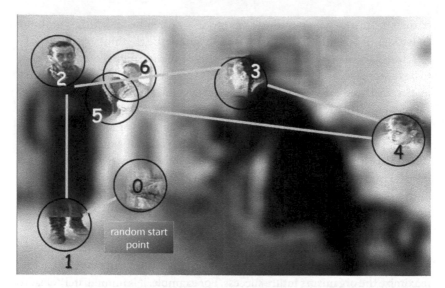

FIGURE 9.11. Eye movement within the first two seconds of inspecting a scene through six saccades, based on data from Yarbus (1967). In this case, the eye saccades from a random starting point to the feet of one of the people in the scene, after which they eye saccades from face to face. Modified from H.-W. Hunziker CC BY 3.0 (Wikimedia Commons).

It turns out that saccades are driven both by visual cues and by memory. The visual-guided saccades respond to cues in the image (for example, edges, or high-contrast elements). Memory-guided saccades do not respond to any particular cues in the image, but instead consult stored representations of similar scenes that suggest where the important parts of the scene might be found. From the point of view of information theory, the brain's internal models would therefore drive attention toward where the most *informative* parts of the scene are likely to be found, and saccades to them.

Such a view of attention in perception is actually different from more established views that maintain that attention is driven by the "salient" parts of the image, that is, those that stand out (are in contrast) with the background (Itti and Koch 2001). I will discuss this model of attention in more detail below, but let us first look at an example of information-driven attention that can help identify a scene from looking at just a small percentage of the information in it. In this example, we will try to identify which out of ten numerals (coded in 8×8 bit maps, shown in Fig. 9.12) we are looking at, using a 2×2 camera that saccades over the image. As there are ten different images, we need $\log_2 10 \approx 3.32$ bits to specify any particular one. The camera itself can potentially see sixteen different patterns (shown in Fig. 9.13), but the entropy of camera images is not four bits because the frequency distribution of patterns (given the ten

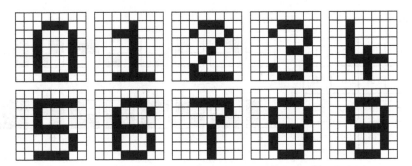

FIGURE 9.12. Simple numerals are represented by 5 × 7 pixels in an 8 × 8 frame.

images) is highly skewed (for example, the empty view occurs disproportion-ally often), giving an entropy of camera images of only about 2.66 bits. This means that a single snapshot could potentially reduce the entropy from 3.32 to 0.66 bits, and it is not hard to imagine that a judiciously chosen set of two saccades can unambiguously identify the image even when looking at only eight of the sixty-four pixels. This is indeed the case, as we will now see.

Let us construct a matrix that associates camera locations with sensed sym-bol, for every numeral in the set. In matrix (9.2), the ten columns stand for the ten different numerals, while the rows correspond to the sixteen camera loca-tions shown in Figure 9.13(a). Each entry in the matrix is the hexadecimal symbol associated with the camera's image defined in Figure 9.13(b).

$$
\begin{array}{c|cccccccccc}
 & 0 & 1 & 2 & 3 & 4 & 5 & 6 & 7 & 8 & 9 \\
\hline
0 & 0 & 0 & 0 & 0 & 0 & 0 & 0 & 0 & 0 & 0 \\
1 & 4 & 0 & 4 & 4 & 3 & 9 & 4 & 9 & 4 & 4 \\
2 & 9 & 3 & 9 & 9 & 0 & 9 & 9 & 9 & 9 & 9 \\
3 & 0 & 0 & 0 & 0 & 0 & 3 & 0 & 3 & 0 & 0 \\
4 & 0 & 0 & 0 & 0 & 0 & 0 & 0 & 0 & 0 & 0 \\
5 & A & 5 & 1 & 1 & 7 & 7 & 7 & 0 & 7 & 7 \\
6 & 0 & 7 & 4 & 0 & 0 & 0 & 0 & 4 & 0 & 0 \\
7 & 7 & 0 & 1 & 7 & 0 & 0 & 1 & 1 & 7 & 7 \\
8 & 0 & 0 & 0 & 0 & 0 & 0 & 0 & 0 & 0 & 0 \\
9 & 7 & 0 & 4 & 0 & C & 5 & E & 0 & A & 2 \\
10 & 0 & 7 & 1 & 5 & C & 5 & 5 & 7 & 5 & 5 \\
11 & 7 & 0 & 0 & 3 & 3 & 3 & 3 & 0 & 3 & 7 \\
12 & 0 & 0 & 0 & 0 & 0 & 0 & 0 & 0 & 0 & 0 \\
13 & 6 & 9 & C & 6 & 0 & 9 & 6 & 0 & 6 & 6 \\
14 & 9 & C & 9 & 9 & 7 & 9 & 9 & 7 & 9 & 9 \\
15 & 1 & 3 & 3 & 1 & 0 & 1 & 1 & 0 & 1 & 1 \\
\end{array}
\qquad (9.2)
$$

(a)

0	1	2	3
4	5	6	7
8	9	10	11
12	13	14	15

(b)

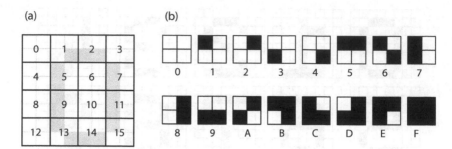

FIGURE 9.13. (a) The 2 × 2 camera can be positioned over sixteen different locations of the image. (b) The sixteen possible images that a 2 × 2 camera can potentially record, with their hexadecimal symbol.

If our goal is to reduce our uncertainty (the entropy) as much as possible, the ideal strategy (we will call this the "maximum entropy" or "maxent" strategy) should search out the camera location with the highest entropy as the starting point. A look at matrix (9.2) shows that this is location 9, where eight different symbols may occur.

Of the eight possible symbols that the camera might record at location 9, seven are actually immediately diagnostic of the numeral and the saccade ends (see Fig. 9.14). For one of the symbols (0), another maximum entropy saccade must be performed. The symbol 0 at position 9 is consistent with three numerals: 1, 3, and 7. Given that we have observed symbol 0, matrix (9.2) tells us that there are seven locations with maximum entropy that can now be reached: locations 1, 5, 6, 7, 13, 14, and 15 (see Fig. 9.15). Because each of the possible symbols at each of the new locations is uniquely associated with one of the three remaining possible numerals, the sequence ends after two saccades.

The maxent algorithm will choose the correct location to saccade to based on knowing matrix (9.2), meaning that given the current location and the image obtained there, a brain that uses such an algorithm will know where to look next, and saccade to that location. In fact, the grid-cell code discussed earlier is ideal for that task, as it allows us to extract the distance and direction of the saccade just from comparing the grid-cell code at the current location with the code at the desired location. Notably, there is now evidence that directed saccades from one position in the image to the *expected location* of the next are indeed controlled by grid cells (Bicanski and Burgess 2019).

Incidentally, there is circumstantial evidence that such a "maxent" algorithm is at work in people playing the computer game *Tetris*. Apparently, when deciding where to place a particular "tetromino" (one of the seven different shapes that are randomly generated and fall toward the filled rows below),

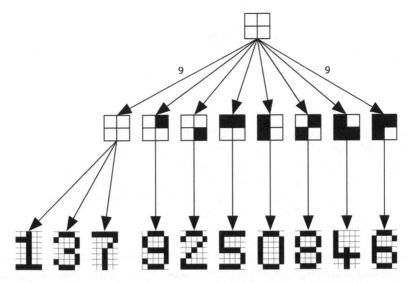

FIGURE 9.14. A saccade to position 9 on the numeral gives rise to eight possible symbols, seven of which are uniquely associated with a numeral so that the saccade ends there. If the symbol 0 is read, another maximum entropy saccade must be performed to disambiguate between the numerals 1, 3, and 7, described below.

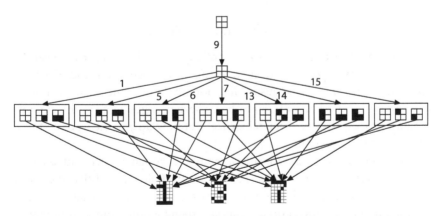

FIGURE 9.15. If the first saccade to the maxent position 9 yields symbol 0, a second maxent saccade to one of seven other locations must be performed. Each of those locations (indicated on the arrow) will reveal a symbol that can unambiguously determine the numeral.

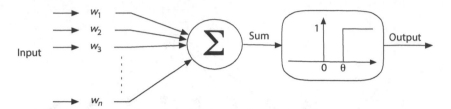

FIGURE 9.16. The input to the McCulloch-Pitts neuron is given by several neurons with binary activation levels (0 or 1) that are multiplied by weights w_i that are positive for excitatory connections, and negative for inhibitory connections. All excitatory inputs carry the same positive weight, and all inhibitory connections carry the same negative weight. By adjusting the threshold parameter θ, all possible Boolean logic functions can be implemented.

players make a number of decisions prior to releasing the tetromino that are not geared at scoring (called "pragmatic" actions), but rather at extracting as much information about the nature of the falling block (by rotating it) as possible, as quickly as possible (called "epistemic" actions) (Kirsh and Maglio 1994). A detailed analysis of what rotations are performed given any particular "camera view" of the tetromino reveals that players create an optimal decision tree just like the one shown in Figures 9.14 and 9.15.

9.3   Evolution of Intelligence

In the previous section I discussed a set of important elements of intelligence: prediction, categorization, memory, learning, representation, and planning. Of course this is not an exhaustive list, nor are these elements independent. However, we expect all of them to be important ingredients in the computational process that we call intelligence. But how do you incorporate these elements into a program?

In section 9.1 I described early attempts to design artificial intelligence using artificial neural networks (ANNs) where I also hinted that there is a complementary approach that uses Darwinian evolution instead of design. Before I describe those efforts, we should spend a little time with the more conventional ANN approach to AI, to appreciate the differences to the approach we will take here.

The standard artificial neural networks of today largely use variations of the original McCulloch-Pitts (MP) neuron model from 1943 (McCulloch and Pitts 1943), shown in Figure 9.16, where neuron activation levels are digital, and are processed via a threshold gate. The MP model has several well-known drawbacks, not the least that both the input and the output are restricted to be digital (binary). Extensions of this model allow for continuous-valued inputs that can be either positive or negative, as well as more general threshold functions (such as the hyperbolic tangent, or else piecewise-linear functions) that are designed to handle continuous-valued inputs. By moving from digital to continuous-valued inputs and outputs, we circumvent the problem of how to encode analog variables into digital ones; however, moving from digital to continuous values for information processing *within* the brain also carries drawbacks. If the activity of a neuron is a real number between -1 and 1 (say), then the state of the brain is a complicated but differentiable function. This function can be used for *training* the network via gradient descent: the standard "backpropagation" method.

Backpropagation is a method for optimizing the weights in the network so as to minimize the error between expected and actual activation pattern. Without going into the mathematical details (see, for example Goodfellow et al. 2016) backpropagation requires calculating derivatives of vectors of activation patterns (multiplied by the matrix of connection) for each layer, starting with the last layer, going backward all the way toward the sensory layer. These computations are expensive, and because the error in deep layers drops exponentially with the number of layers, increased precision requires even greater computational resources. Because graphics processing units (GPUs) can multiply large matrices very fast, such architectures now dominate the field of deep learning.

Backpropagation is intrinsically a *supervised learning* approach: to calculate the error function, it is necessary to compare expected and actual results. Such a method cannot be used if expected results are not known. Moreover, because the function is defined in a very high-dimensional space, it is difficult to implement behaviors that are noncontinuous, such as when a small detail in the environment should dramatically change the outcome of the computation. In other words, the "law of large numbers" that makes sure that the variance of the mean is small works against you when you need small changes to have big effects.

For digital neurons, instead of a differentiable function acting on firing rates, we can simply define a discrete-time Markov chain, by specifying transition rules between brain states (activation patterns). These rules are given directly by the logic that connects the neurons (much like the logic tables in Fig. 9.10), but because a differentiable gradient function does not exist (unless we use stochastic rather than Boolean logic), we cannot use backpropagation to optimize the brain's logic.

One alternative to backpropagation is *neuroevolution* (Stanley et al. 2019), that is, using the process of Darwinian evolution as an optimizer. In this method, populations of candidate solutions compete against each other, over many thousands of generations, where the fitness of each potential solution is evaluated by testing its performance in a complex world. Since backpropagation is, at its heart, a stochastic gradient-descent search method, in principle other search methods could equally well find optimal sets of weights. While no search algorithm can guarantee finding optimal solutions for all possible problems (Wolpert and Macready 1997), some algorithms perform better on some subset of problems than others. Darwinian evolution, for example, has evidently succeeded in creating structures of astounding complexity using a simple quaternary encoding of information, albeit over vast eons of time. In particular, we know that our mammalian brains, with their astonishing complexity, are precisely the product of this algorithm. Neuroevolution attempts to port this algorithm [generically termed "genetic algorithm" or GA (Michalewicz 1999), already discussed in section 1.1.4] within our standard computers, that is, in silico.

Neuroevolution is conceptually very different from backpropagation. First, many possible solutions are studied at the same time in a population (as opposed to the optimization of weights in a single "individual"). Second, solutions are modified randomly, as opposed to the directed (gradient-based) modification in backpropagation. Third, in backpropagation an error function (or other suitably defined figure of merit) is used to define the *fitness* of any particular individual (low error implies high fitness), while in neuroevolution selection acts on the population by amplifying those individuals with high fitness, and suppressing those with low fitness, in the next generation. Thus, neuroevolution proceeds through the cycle depicted in Figure 9.17, generation after generation, until a stopping criterion is reached.

It is important to note that neuroevolution can occur in an *unsupervised* manner: as long as the fitness of an individual can be determined by evaluating the performance of the solution in context, neuroevolution can move toward the fitness maxima in the landscape. While backpropagation and neuroevolution essentially try to solve the same problem—to find extrema in a high-dimensional complex landscape—the two processes have very different properties. The most important one is obvious from the outset: because error functions cannot be computed unless we know the "right" solution, backpropagation cannot be used when the correct response to an input is not known in advance. For this reason, backpropagation methods are usually not used when searching for brains that control the behavior of agents (or robots, for that matter), but they are enormously successful when labeled training data is available. Neuroevolution, on the contrary, excels at finding control structures (brains) that induce *embodied agents* (agents that control a simulated body in a simulated physical world) to behave appropriately in a complex world. But the genetic algorithm has an Achilles' heel. The changes applied to any particular candidate solution (the mutations) are random, that is, they are undirected. Any particular mutation can increase or decrease fitness, and

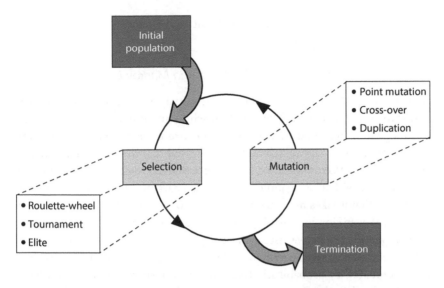

FIGURE 9.17. Schematic representation of the genetic evolutionary algorithm. An initial population of candidate solutions is subjected to *selection*, which changes the frequency of each of the candidates according to a figure of merit, usually called "fitness." There are different methods of selection, with the most common ones listed in the adjacent box. The new generation of solutions is subjected to *mutations*, which change the solutions according to mutation processes that are inspired from genetics. The most common mutation operators are point-wise mutation, recombination, and segmental duplication (but there are many others). After the mutation step, the modified set of solutions is reevaluated, and their frequencies again adjusted in the selection step: this completes one generation. This cycle continues until a cut-off criterion is met, usually either a fixed number of generations, or when a threshold in solution quality is met.

when a fairly high level of fitness is already reached, the most likely change is detrimental. A guided change only occurs at the population level, and as such is far less effective than backpropagation, in which the computed gradient explicitly predicts which way the parameters of any particular solution should be changed.

In the following, I will briefly describe an architecture that allows us to use neuroevolution to create digital brains, that is, computational control structures that operate on digital inputs, and where all information is stored within sparse digital representations. Because the logic of these brains essentially implements a Markov process, we call these brains "Markov Network Brains," or "Markov Brains" (Hintze et al. 2017) for short. Those networks were first introduced to study the evolution of short-term memory in traversing sequential T-mazes (Edlund et al. 2011), but we'll look at two other important applications further below.

9.3.1 Markov Brain states and Markov Logic

In many ways, Markov Brains resemble the initial construction of McCulloch and Pitts (McCulloch and Pitts 1943), who envisioned a brain assembled from digital neurons that are wired together so as to perform logic operations on sensory inputs. McCulloch and Pitts showed that the threshold function shown in Figure 9.16 is universal, in the sense that any logic operation could be implemented with it, using judiciously chosen parameters. A drawback of this method is that it takes many neurons—and connections between them—to implement even simple 2 → 1 logic such as the "exclusive OR" (XOR) operation. In Markov Brains, we instead implement logic *directly*, using explicit logic gates specified by a set of parameters. Specifically, the logic of Markov Brains[3] is implemented by probabilistic (or, when appropriate, deterministic) logic gates that update the Brain states from time t to time $t + 1$, which implies that time is discretized not only for Brain updates, but for the environment as well. Whether or not brains perceive time discretely or continuously is a hotly debated topic in neuroscience (VanRullen and Koch 2003), but for common visual tasks such as motion perception (VanRullen et al. 2005), discrete sampling of visual scenes can be assumed.

For Markov Brains, the discreteness of time is a computational necessity. Because no other states (besides the neurons at time t) influence a Brain's state at time $t + 1$, the Brain possesses the *Markov property* (hence the name of the networks). Note that even though the Markov property is usually referred to as the "memoryless" property of stochastic systems, this does not imply that Markov Brains cannot have memory. Rather, memory can be explicitly implemented by gates whose outputs are written into the inputs of other gates, or even the same gates (i.e., to itself), much like the gates that recognize temporal sequences that we saw in Figure 9.10.

Consider as an example a small Brain of only eight neurons, depicted in Figure 9.18(a). The states of the neurons in this example are updated via two logic gates only. Figure 9.18(b) shows one of those gates, which takes inputs from neurons 0, 2, and 4 and writes the output into neurons 4 and 6. Thus, the gate connects neurons 0, 2, 4, and 6 by a "wire," and neuron 4 acts as a bit of memory. As opposed to the all-to-all connectivity pattern between layers of an

---

3. From here on out, I will use the capitalized "Brain" to mean "Markov Brain," and use "brain" to refer to biological brains.

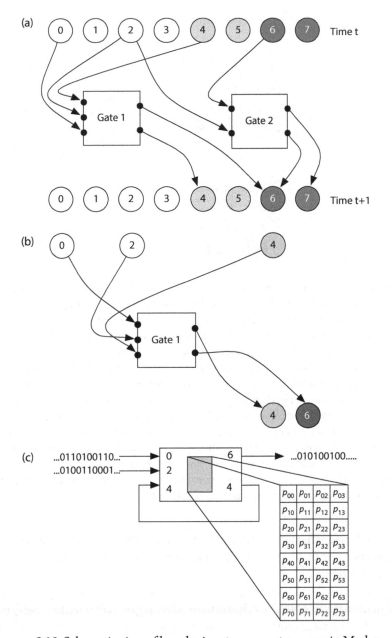

FIGURE 9.18. Schematic view of how logic gates connect neurons in Markov Brains. (a) A simple Brain with four sensory neurons (white), two hidden (internal) neurons (light gray), and two output (motor) neurons (dark gray). Two gates are depicted that show how the state of the neurons is updated over time. (b) Because Gate 1 reads from neuron 4 and writes into the same neuron, information can be stored over time. A neuron that does not become activated returns to the quiescent state. (c) The logic table for Gate 1 has 8×4 entries that fully specify any possible $3 \rightarrow 2$ stochastic logic gate. Because one of the outputs of the gates is fed back into it, this particular gate actually implements a $2 \rightarrow 1$ gate with memory.

(a)

Sensory (input) layer

Hidden layers

Motor (output) layer

(b)

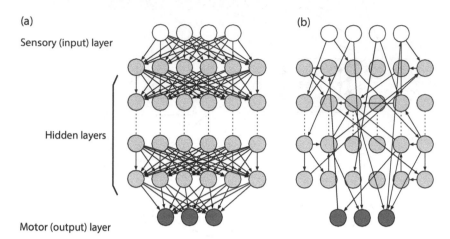

FIGURE 9.19. Architectures of neural networks. (a) In the standard fully connected feed-forward architecture of ANNs, the input layer (white, only four neurons shown) is connected to several hidden layers (light gray, only six neurons per layer shown) via all-to-all connections between layers. The last hidden layer is connected to the output layer (dark gray). In such networks, no information can travel backward through the layers, but the weights that determine the strength of the influence between neurons are trained via stochastic gradient descent in the form of backpropagation. (b) In a recurrent sparse neural network, only a small fraction of all possible connections between neurons exist, and neurons can connect across layers both forward and backward, so that the impression of a layered architecture disappears. Loops between neurons can be used to store memories, as well as represent concepts in the external world. Such networks are usually trained using supervised methods such as gradient descent or unsupervised methods such as neuroevolution.

ANN (shown schematically in Fig. 9.19[a]), the logic of Markov Brains only connects a few neurons at a time, creating very sparse networks, as depicted in Fig. 9.19(b).

There are other key differences between our Markov neuron variables and the neurons in an ANN. First, because an ANN is generally used to classify patterns, they undergo only a few time steps: a pattern is applied to the input layer, which is then propagated toward the output layer. In Markov Brains, there is no general predefined layered structure (except for designating some neurons as inputs and some others as outputs) and as a consequence, the brain can be updated as long as it takes to obtain a satisfactory response. Second, when a Markov neuron is in the active ("firing") state 1, it will return to the quiescent state 0 in the next time step unless another neuron activates it. In that way, a neuron may fire many times and thus contribute to the computation over time.

ANN neurons, by comparison, will remain at a particular activation level after the signal is propagated through the network, since "its work is done" after one use.

The logic gates of a Markov Brain are fully specified by listing the neurons they read from, as well as the set of neurons they write to, along with *probabilistic logic tables* that prescribe how inputs are transformed into outputs. For example, in the probability table shown in Figure 9.18(c), p_{52} specifies the probability of obtaining output state $(N_4, N_6) = (1, 0)$ (a state with decimal representation '2') given input state $(N_0, N_2, N_4) = (1, 0, 1)$ (decimal translation '5'), that is,

$$p_{52} = P(N_0, N_2, N_4 = 1, 0, 1 \rightarrow N_4, N_7 = 1, 0).$$

Since this gate takes three inputs, there are 2^3 possible patterns that are shown in eight rows. Similarly, this probabilistic table has four columns, one for each of the 2^2 possible outputs. The sum of the probabilities in each row must equal 1: $\sum_j p_{ij} = 1$. When using deterministic logic gates instead, all the conditional probabilities p_{ij} are zeros or ones. In general, Markov Brains can contain an arbitrary number of gates, with any possible connection patterns, and arbitrary probability values in logic tables (Hintze et al. 2017). However, in order to maximize sparsity while maintaining the capacity to create complex logic we usually limit the input/output capacity of logic gates to a maximum of four in, four out.

The trade-offs between gate complexity, wiring complexity, sparsity, and the number of parameters are an important issue to consider. In principle, the logic of a Brain with n neurons can be implemented with an $n \times n$ logic gate requiring no wiring, but the number of parameters of such a gate scales exponentially as $2^n \times 2^n$. On the other end of the spectrum, we could limit all logic to $2 \rightarrow 1$ gates, which might require complex wiring over multiple Brain updates.

As we will see when discussing how logic gates are genetically encoded, multiple gates can connect (and therefore write into) a single neuron. Having multiple neurons all affect the state of a single neuron "downstream" is very common in neurobiology (Hawkins and Ahmad 2016), and facilitates the modulation of behavior by complex context. There are several ways in which signals coming from multiple gates can be aggregated in a single neuron. One is to simply use the OR function, so that if at least one of the inputs to the neuron is 1, then the target neuron's state becomes 1 (Hintze et al. 2017). Such a choice is problematic, however, if many (more than three, say) gates write into a single neuron, as all those signals quickly become redundant and the Brain loses the ability to respond to specific contextual stimuli. One way to resolve multiple inputs is to treat a 0 as an inhibitory signal, while a 1 is treated

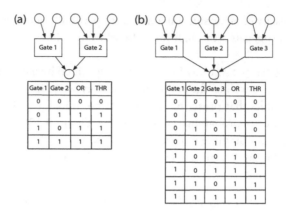

FIGURE 9.20. (a) If two gates write into the same neuron, the OR function is the same as the THR function, which treats a 0 like a −1 when summing up the inputs and using a threshold of zero. (b) For three gates, the OR function activates the target neuron if only a single input neuron is on, leading to an unbalanced output pattern. The THR function with threshold zero gives rise to a balanced output pattern.

as an excitatory signal. Then, the number of excitatory and inhibitory signals is summed (treating a zero as a −1), and the neuron is only activated if the sum total is nonnegative (the function THR; see Fig. 9.20). If only two gates write into a neuron the THR function is the same as OR (see Fig. 9.20[a]), but for three or more gates, the THR creates a more balanced output pattern. (For three gates, half of the outputs are 1, as opposed to 7/8 for the OR function, see Fig. 9.20[b]). By treating 0/1 inputs as inhibitory/excitatory, the logic at the confluence of signals becomes similar to the transfer function of a single McCulloch-Pitts neuron (McCulloch and Pitts 1943) (shown in Fig. 9.16, receiving multiple inputs from excitatory and inhibitory neurons) except that in the MP neurons, excitatory neurons may have a different weight than the inhibitory neurons.

9.3.2   Markov Brain evolution

Even though a Markov Brain is a network, we do not evolve the network itself. Instead, the rules to *build* the network are encoded in a set of bytes that we call "genes," and we evolve these rules instead (in the same way that our genes are rules for building an organism that can survive optimally in a given environment). Because a Markov Brain's structure and function are determined entirely by the transfer function, it is sufficient to encode all the parameters necessary to specify this function. As the transfer function is determined by the set of gates in Figure 9.18(a), it is therefore sufficient to encode all gates (one gene for each gate).

A gate is fully specified by giving the addresses of the neurons it reads from, as well as the address(es) of the neurons it writes to, along with the parameters specifying the logic (Fig. 9.18[c]). The set of genes that determines all the gates (and therefore the transfer function) is called the "chromosome" (Fig. 9.21[a]), and each gene on the chromosome specifies one gate (Fig. 9.21[b]). Executing these rules creates the network shown in Figure 9.21(c).

Because genes are delimited by specific symbols (the equivalent of the start codon of genetics, here the first two bytes in Fig. 9.21[b]), the chromosome can harbor noncoding regions, as well as overlapping reading frames.[4] In biology, only a small fraction of the genome (about 1.5%) is organized into recognizable genes (called open reading frames), but about 5 to 8 percent of the genome is actively under selection, meaning that it contributes directly to the fitness of the organism (ENCODE Project Consortium 2012). However, according to the ENCODE consortium up to 80 percent of the genome contributes in one way or another to the function of the organism. In the same manner, genes that are not actively participating in function in the Markov Brain chromosome can nevertheless contribute to robustness, and act as a reservoir of genetic material that may become useful in the future.

Boolean (possibly stochastic) logic gates are not the only ways in which neurons can be functionally connected. Markov Brains are implemented using a versatile platform called *Modular Agent-Based Evolver* (MABE) that allows the user to efficiently swap out worlds, genetic architectures, and genetic encodings, while setting up experiments and controls using a scripting language (Bohm et al. 2017). Within MABE, different types of connecting logic (implementing, for example standard artificial neural networks, convolutional neural networks, etc.) can be specified by using different start codons that change how the following byte string is interpreted; see (Edlund et al. 2011) or (Hintze et al. 2017) for details on how genes encode function in these digital brains. MABE also allows "gene-free" encodings of logic gates (sometimes called "direct encoding") in which logic gates are mutated by directly changing the parameters of an existing set of gates. Using a direct encoding avoids the complexity that comes with the indirect genetic encoding (for example gene duplication, overlapping reading frames, and gene fusion). While these mechanisms may be important for the evolution of complexity in some circumstances, they make the analysis of evolutionary trajectories, and in particular the control of mutational mechanisms, more difficult. As a

---

4. Markov Brain genetics in principle allows diploid chromosomes; see Hintze et al.(2017).

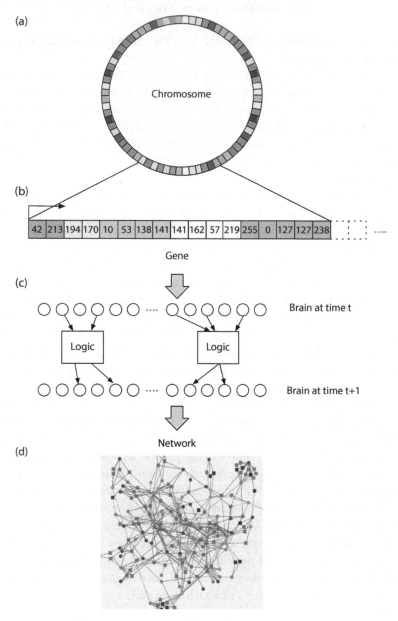

(a)

Chromosome

(b)

| 42 | 213 | 194 | 170 | 10 | 53 | 138 | 141 | 141 | 162 | 57 | 219 | 255 | 0 | 127 | 127 | 238 | | | ..... |

Gene

(c)

Logic   Logic   Brain at time t

Brain at time t+1

Network

(d)

FIGURE 9.21. (a) A chromosome encodes all the genetics of creating the Brain. (b) Individual genes encode the logic that updates the state of a subset of neurons over time. The genes specify the type of Brain substrate (first two bytes) the number of neurons involved (following two bytes) the connectivity (bytes following), as well as the logic (at the end of the gene). (c) Each gene controls the logic of how a subset of neurons changes its state over time. In this manner, genes encode both the mechanisms and the structure of the Brain. (d) The totality of all genes creates the neural network architecture. In this depiction, a Brain's neurons are circles while squares are logic gates that connect neurons.

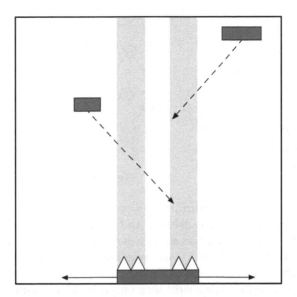

FIGURE 9.22. The setting for the active categorical perception task is a 20 × 20 arena where the agent (dark gray block at bottom) can move laterally while looking upwards using four sensors (white triangles) that are arranged in such a way that a blind spot prevents accurate categorization of the diagonally falling blocks using only a single measurement. From Marstaller et al. (2013).

consequence, most of the development work with Markov Brains is carried out using a direct encoding.

Once a Brain is constructed by interpreting the chromosome, the network is tested for performance in a specific environment. Brains that perform best according to a suitable fitness metric are rewarded with a differential fitness advantage. As these genomes are subject to mutation, heritability, and selection, they evolve in a purely Darwinian fashion (albeit asexually). Let us take a look at how this works with a particular example: the evolution of active perception in a simple world. We have encountered active perception before, namely in our simple example of the recognition of digitized numerals with a camera smaller than the numerical field (see Fig. 9.13). But rather than considering a static situation (saccading across the numeral), we ask whether an agent can classify a falling object with a simple camera with a blind spot (see Fig. 9.22). The arena is a 20 × 20 board in which large (3 × 1) and small (2 × 1) blocks (light gray in Fig. 9.22) fall diagonally (one unit per update). The agent (dark gray) must identify whether the block that is falling is large or small, and then make its decision known by catching the larger block (meaning it positions itself underneath the falling block so that the agent and block are

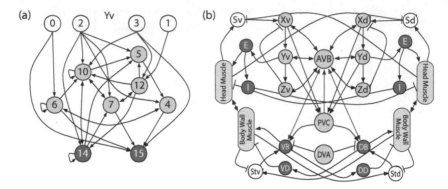

FIGURE 9.23. (a) Network structure of an evolved Markov Brain with perfect fitness, correctly catching all small blocks and avoiding all large ones. Nodes in white are sensors; motor variables are dark gray and internal neurons are light gray, from (Marstaller et al. 2013). Arrows indicate causal influence via a logic gate and double arrows represent reciprocal causal connections. Nodes with arrows that point to themselves write their output back into their input and may work as memory. Internal nodes can read from motors, giving rise to proprioception or, more precisely, kinesthesia: the ability to sense one's own motion. (b) Model of the forward locomotion module of the *C. elegans* neural network shows sensory, hidden, and motor neurons connected via repressing (⊣) or activating (→) interactions (from Karbowski et al. 2008). Neuron names are standard (White et al. 1986). Because the worm has bilateral symmetry, so does the neural network controlling it.

touching after twenty updates have passed) but avoiding the smaller one. This task is made difficult by the agent's imperfect upward-looking camera that has a blind spot in the middle, so that large or small blocks cannot easily be distinguished. Instead, the agent must move around and make the categorization based on integrating several views of the block (which requires that previous views are kept in memory), all under time pressure as the block continues to fall. When the agent has made its determination, it must plan to either catch or avoid the block contingent on the particular finding.

This task, even though it clearly is a simple one, does seem to use all the elements of intelligence (except learning) that we discussed earlier. Brains that perform perfectly (they catch or avoid both types of blocks started from any of the twenty locations, dropping either left or right) evolve readily using deterministic logic only, and can be analyzed (Marstaller et al. 2013). The Brain's network structure is simple and—in the case of the network shown in Figure 9.23(a)—appears to rely on a master controller neuron (neuron 10 in Fig. 9.23[a]) that integrates information from the sensors as well as from the agent's own motion. This structure is vaguely reminiscent of the forward locomotion network of the worm *C. elegans*, shown in Figure 9.23(b). Now, the worm brain is extremely simple compared to the mammalian brain (302 neurons in the worm compared to the 86 billion neurons in the human brain, with 16 billion in the neocortex alone) and the mini-brain to solve the active categorical perception task is even simpler. However, there are also striking similarities: the networks are sparse, and use input from various sensors (including from the muscles) to situate themselves in the world, and moreover use single neurons as master controllers. For both brains, history and context are important, which is reflected in their brain architecture.

However, even though the network structure of the Markov Brain in Figure 9.23(a) is fairly simple, it is nearly impossible to figure out the algorithm behind it, even though a description in terms of Boolean logic is available (Marstaller et al. 2013). Short of a full-fledged algorithmic analysis, there are several ways in which we can make sense of how the Brain works. For biological brains, the standard methods are single-cell recording, neuroimaging, and lesion techniques (where particular parts of the brain are impaired or inactivated in order to study the effect on behavior). We can use very similar methods to investigate artificial brains, creating a "Virtual Brain Analytics" toolbox (Hassabis et al. 2017).

In one method, we can use information theory to relate firing patterns in time to changes that happen in the world, as well as to the patterns that appeared in the sensors. In a way, this is very similar to how modern neuroscience analyzes brains, stemming from the realization that behavior is linked not to the firing pattern (in time) of single neurons, but rather to how groups of neurons fire, something advocated early on by Hebb (1949) who introduced the idea of "population codes." While recording from large groups of individual neurons is difficult (see, e.g., Buzsáki 2004) we can make full use of this technique when analyzing Markov Brains. To understand how the Brain shown in Figure 9.23(a) stores and processes information, we can partition the Brain into the sensory neurons $S = S_0 S_1 S_2 S_3$ (in white in Fig. 9.23[a]) and the "internal" (or hidden) neurons $B = B_4 B_5 B_6 B_7 B_{10} B_{12} B_{14}$. We include the actuators (dark gray neurons in Fig. 9.23[a]) in the set B because they actively participate in the computation.

To characterize how the environment changes as the agent behaves, we can introduce a variable $E = E_0 E_1 E_2 E_3$ that summarizes four aspects of the environment that we hypothesize could be important to the agent: E_0 is a binary variable that reflects whether or not any of the four sensors are activated by a falling block, while E_1 is "on" if the falling block is to the left of the agent. Variable E_2 marks whether the falling block is small or large, and E_3 tells us if the block is falling diagonally toward the left or to the right. Even though the variable E can take on only sixteen states (while there are exactly 1,600 possible world states; Marstaller et al. 2013) it is clear that many of these different world states are effectively the same for the agent. Of course, it is possible to choose different salient "bits" than those four in E, but in hindsight they cover many of the features that are important to the agent.

We can now ask: how do changes in E correlate (and predict) the state of the Brain B, and how is this correlation related to what the Brain actually perceives of the environment via its sensors S? To answer this, we construct the entropy Venn diagram for these three variables. Quantifying how much the brain knows about the world *given* the sensor's state is very different from quantifying what the Brain knows about the world irrespective of the sensor, as we discussed earlier in section 9.2. If we measured the latter (the information $I(E:B)$), we would pick up correlations between the Brain states and the sensors.

How much does the agent "know" about the world, and how is this knowledge correlated with how well the agent does, that is, its fitness? Let us measure how much of the world is represented in the Brain's internal states, according to the measure $R = H(B:E|S)$, the measure of representation that we introduced earlier (see Fig. 9.7). Using the rules of information theory, we can rewrite R in terms of unconditional entropies as (see Kirkpatrick and Hintze 2020 for a succinct proof)

$$R = H(E|S) - H(E|S, B) = H(E, S) + H(S, B) - H(S) - H(E, S, B). \quad (9.3)$$

The entropies in (9.3) can easily be calculated from recording the temporal history of the three variables E, B, and S, and constructing the joint probability distribution $p(e, b, s)$, where

$$p(e, b, s) = P(E = e, B = b, S = s). \quad (9.4)$$

The joint pairwise and single-variable probability distribution can be obtained from (9.4) by marginalization (see section 2.1, page 43).

In Figure 9.24 we can see fitness (black solid line) increasing over evolutionary time (this is an average over 200 independent evolutionary histories), approaching the maximal fitness $w = 1.0$. Fitness increases from its minimal value $w = 0.5$ at the beginning of the run, which is seeded with a random Brain. Random Brains do not move, so such an agent will get about half of all perception tasks right, as it simply will correctly avoid almost all small blocks. At the same time, R increases from zero (random brains have no model of the environment at all) toward ever-increasing values (dashed line in Fig. 9.24). Indeed, R is a very good predictor of w, which means that an internal model of the environment is an essential element driving the performance of the agent.

But information theory can reveal to us much more than how much of the world is represented within the Brain. Because the environment variable is composed of a number of concepts ("Is the object visible?," "Is it to my left

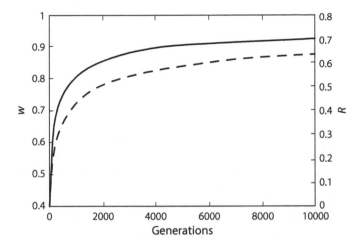

FIGURE 9.24. Evolution of fitness (black line, scale on the left) over evolutionary time (in generations), and evolution of representation $R = H(B : E|S)$ (dashed line, scale on the right). Fitness is measured as the fraction of eighty different perception trials (twenty starting positions, for two block sizes and two block-movement directions) correctly judged. Because random Brains will get about half of all trials correct, fitness starts at $w = 0.5$. Those Brains, however, have $R = 0$. Both curves are an average of 200 independent evolutionary runs. After Marstaller et al. (2013).

or right?," "Is it large or small?," "Is it moving left or right?"), we can measure how much the Brain knows about any of these particular concepts, for example by measuring $R_2 = H(B : E_2|S)$: what the Brain knows about the size of falling blocks. At the same time, B is a composite variable that allows us to dissect exactly *where* in the Brain information is stored. So, for example, $R_{2,10} = H(B_{10} : E_2|S)$ measures how much information about block size is stored in neuron 10. We find that information about particular concepts is not stored in individual neurons, but rather is smeared across groups of neurons (Hintze et al. 2018). At the same time, we find that individual concepts (like those encoded by the variables E_0 or E_1) are not stored in single neurons. Rather, the Brain represents amalgamated concepts (such as "the large block that moves leftwards") into groups of neurons. But which combination is encoded into which part of the Brain is different in different evolutionary histories.

It is clear that an information-theoretic approach to measuring brain activity during behavior has the power not only to find out which concepts are being used to make decisions, but even which areas of the brain are involved

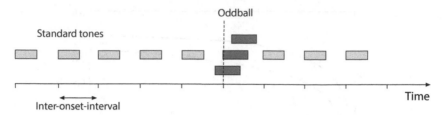

FIGURE 9.25. Schematic of the auditory oddball paradigm. The oddball tone is placed within a rhythmic sequence of tones ("standard tones"). Standard tones are shown in gray while the oddball tone is shown in black. The oddball tone duration may be longer or shorter than the standard tones. Both the time between tones (the "inter-onset-interval," IOI), and the length of the standard tones can vary. Two standard tones that are advanced and retarded with respect to the rhythm are also shown. Adapted from Tehrani-Saleh and Adami (2021).

(as long as we specify the concepts beforehand, as this method cannot discover concepts we do not test for). Developing these tools using artificial brains therefore is a harbinger of the power that information theory can bring to our understanding of biological brains, once methods for recording from significantly larger groups of neurons (over longer periods of time) are available. For example, in a recent study (Bartolo et al. 2020), the team recorded from up to 1,000 neurons in macaque monkeys that were rewarded for learning the identity of a preferred image ("What" information), or learning a preferred direction of gaze ("Where" information). The team found that the internal representation of "What" information was more complex than the representation of "Where" information, but a proper information-theoretic analysis of these representations could have revealed more details about how and where "What" and "Where" information was stored.

To illustrate another method of virtual brain analytics, we'll now study a different behavioral task: the judgment of the length of tone intervals. The perception of time, and particular judging time differences, is crucial to the survival of an organism (Richelle et al. 2013), and so our brains have developed sophisticated mechanisms to be able to predict arrival times from observations, as well as to judge the length of time intervals with respect to a given rhythm (Buzsáki 2006). A classic experiment that tests this ability is the "oddball paradigm," where a subject listens to a rhythmic tone at a particular frequency and then is presented with a tone at a different frequency (the "oddball") whose length can be longer or shorter (or the same) as the rhythmic tone (see Fig. 9.25). When people perform this test, they quickly learn to correctly judge the length of the oddball tone with respect to the standard tone, but they are also prone to a curious illusion (McAuley and Fromboluti 2014):

When a tone of the same length as the standard tone is *delayed* (with respect to the expected arrival of the tone according to the rhythm) people judge this tone to be longer that it really is, while if the tone arrives early, the subjects judge it to be short. This illusion is usually explained by a phenomenon called *attentional entrainment*, which is thought to give rise to periods of heightened attention to the onset of the tone (and reduced attention just before and after the onset).

We decided to test this model of attention by evolving deterministic Markov Brains to judge tone lengths in a setting much like the one analyzed by McAuley and Fromboluti, across a variety of different standard tones and IOIs (inter-onset-interval times; see Fig. 9.25). We evolved Markov Brains with sixteen neurons (one sensory neuron, one actuator neuron, and fourteen hidden neurons) for 2,000 generations, in fifty independent runs. In each of the runs a Brain is exposed to sixty different standard tones/IOI combinations (from tone length 10 to 25 and IOI from 5 to 12) which it is asked to classify. The best-performing Brain from each of the fifty runs (the "elite set") was able to correctly judge 98 percent of the tomes they were tested on. The task is not trivial, as each Brain is asked to master *all* of the different standard tones and IOIs, and therefore must learn what the current standard is from listening to the first four beats. We also test the Brain with all possible oddball lengths that do not interfere with the next tone (a total of 2,852 trials). Note that the oddball tone could appear in any of the five slots that follow the first four tones (see Fig. 9.25). In order to be counted, the judgment of tone length must be made on the last beat of the IOI.

In the experiments performed by McAuley and Fromboluti, the human test subjects were not told that in some tests the tone would be delayed or advanced with respect to the expected onset. In the same manner, during evolution we never show the Brain tones that do not begin at the rhythmic onset. When asked to judge tones that were the same length as standard tones but arrived early or late with respect to the background rhythm, the elite Markov Brains suffered from the same illusion as people did: on average they judged early tones as short, while later arriving tones were deemed to be long, as we can see in Figure 9.26.

What is the mechanism behind this illusion? Why are brains (and Brains) so consistently fooled by tones that fail to be at the expected onset within a rhythmic sequence? To understand how Markov Brains judge tone lengths, we need to think of these Brains in terms of *finite-state machines*. A finite-state machine (FSM) is a model of computation in which the machine can be in one of a finite number of states, and transitions between these states in response to inputs. We can, for example, depict the operation of the "sequence recognizers" in Figure 9.10 in terms of an FSM diagram, where each node

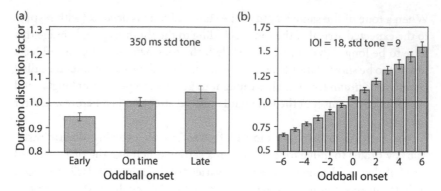

FIGURE 9.26. (a) Duration distortion factor (DDF) for early and late arriving standard tones (of length 350 ms) in the experiment by McAuley and Fromboluti (2013). (b) Duration distortion factor for evolved Markov Brains for standard tones of eight time units, with an IOI of 18 (corresponding to a time unit of about 39 ms). In this experiment, the standard tone arrived early or late by up to six time units. The dotted line indicates the DDF for zero delay. From Tehrani-Saleh and Adami (2021).

(a machine state) is labeled with the decimal equivalent of the binary machine state, and where arrows denote transitions between states given the particular input received when in this state. Figure 9.27(a) shows the state-to-state transition diagram for the machine defined by the table in Figure 9.10(a) (the 10-recognizer), while Figure 9.10(b) depicts the FSM of the table shown in Figure 9.10(b) (the 0001-recognizer).

We can now use this tool to analyze a typical Brain's state-to-state transitions as it classifies a standard tone of length 5 with IOI = 10, that is, a length-5 tone followed by a silence of the same length. Here, the motor neuron's state 0 stands for "oddball is shorter than the standard tone," while 1 translates to "oddball is same length or longer than the standard tone," or in other words, "not short." In this version of the state-to-state diagram (shown in Fig. 9.28), we do not include the motor neuron in the state of the machine, but indicate the state of this signaling neuron via a triangle on top of the state.

In Figure 9.28 we see that the Brain correctly identifies a standard tone of length 5 by issuing a 1 on the last beat of the IOI (triangle marked L), as both "equal" and "long" are coded by a 1 ("not short"). A length-6 sequence is also correctly judged as long because the transition from state 485 to state 1862 occurs whether or not the sixth symbol is a 0 or a 1 (the Brain does not pay attention to the input at this particular time). If this FSM encounters a short tone (for example, the sequence 1111000000), the Brain follows a different

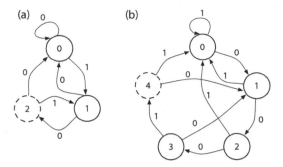

FIGURE 9.27. (a) FSM state-to-state transition diagram for the machine defined in Figure 9.10(a). Each node is labeled by the decimal representation of the binary state of the neurons MH (the "motor" neuron and the sole hidden neuron). Prior to any input the state of the machine is 0 (all neurons quiescent), and is driven by the input sequence '10' into state 2 which signals recognition (dashed outline), and a reset of the internal state. State 3 is never visited. (b) State-to-state transition diagram for the machine defined in Figure 9.10(b), with states labeled by the decimal equivalent of MH_1H_2. From the quiescent state, this machine is driven by the input sequence 0001 toward state 4 (dashed circle), signaling recognition and resetting the internal "clock" state. States 6, 7, and 8 are never visited.

path through state space, ending up in state 2884, where it correctly classifies the sequence as "short" (see Fig. 9.29).

Now that we have understood how an evolved Brain makes its judgments, let us test this Brain with out-of-rhythm sequences, that is, tones that do not start right after the end of the IOI. We will consider for illustrative purposes a late oddball and an early oddball, shown in Figure 9.30. In this case, we choose the early oddball to end exactly where a short on-time oddball ends, and a late oddball that ends precisely where a long oddball would end, but in the experiments all possible delays that did not interfere with the next or the previous tone were shown to the Brain.

When analyzing these cases, we find indeed that the Brain judges early oddballs short and late oddballs long (see Fig. 9.26[b]). How do these decisions come about when looking at the state-to-state diagram? Figure 9.31 provides the answer. In this diagram we included all the transitions we studied before, but added those coming from early and late oddballs (of the same length as the standard tone). Keep in mind that the Brain has never "heard" late or early tones before (meaning, it did not encounter them during evolution), so the Brain has no choice but to follow the logic that evolved in response to on-time signals.

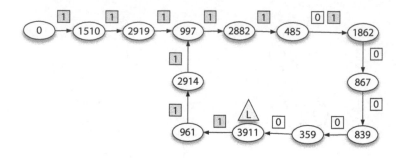

Standard

Longer

FIGURE 9.28. State-to-state FSM transition diagram of an evolved Brain showing the states traversed while the Brain listens to the periodic sequence shown
below the diagram. The numbers on top of the arrows show which sensed
value (0 or 1) is responsible for the transition (gray-boxed 1s indicate a tone,
white-boxed 0 indicates silence). If both 0 and 1 occur above an arrow, the FSM
transitions irrespective of the sensed value, i.e., the Brain does not pay attention
to the sound. The Brain state is the decimal equivalent of the binary firing pattern of twelve hidden neurons (two of the neurons do not play any role in this
particular Brain), so state 3911 is equivalent to the Brain state 111101000111.
The FSM starts in the quiescent state 0, and moves toward state 3911 while
listening to the first length-5 tone and length-5 silence. The dark gray boxed
sequence is the oddball (an oddball equal to the standard tone, and an oddball longer than the standard tone is shown). Only the transitions relevant
to those two tone sequence are shown here. Because the tone is periodic, the
state-to-state diagram has a loop with size equal to the IOI.

Let us first study the transitions due to a late tone. We begin in state 3911
(with the triangle labeled L), as this is the state reached after the first four
entrainment tones. Rather than continuing in this loop, the late tone (a delay
of two) moves the Brain into state 2918, where the sudden onset of the tone
moves the FSM back into state 997, a state that is actually part of the standard
loop! In fact, it enters the loop as if a normal tone had played, and as a consequence the Brain ends up in state 3911 again, issuing the "not short" verdict.
Of course, this assessment is incorrect; the tone *was* short! But as the delayed
short tone threw the Brain into the familiar loop, the illusion was unavoidable:
to this Brain, the delayed short tone "feels just like" a standard or long tone, in
the sense that the Brain undergoes precisely the same transitions *as if* such a
tone was played.

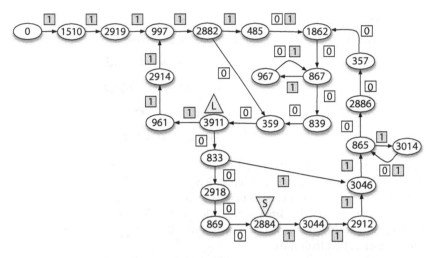

FIGURE 9.29. State-to-state transition diagram for the FSM shown in Figure 9.28, but with the transitions in response to a short tone added (dark-gray sequence of four 1s). Due to the 0 in the fifth position in the short tone, the Brain does not transition to state 485 from 2882, but rather moves to state 359 instead. From there, five more 0s move the Brain to state 2884, where it correctly issues the verdict "short" (a zero, indicated by the down-ward pointing triangle inscribed with S). In order to not be thrown out of rhythm, the next sequence of standard tone–standard silence brings the machine back into state 3911, issuing the correct "not short" signal from where the standard loop can begin again.

What about the early long tone? When experiencing such a sequence, differences in the firing pattern occur in state 839, when in the standard on-time tone two more transitions would result in state 3911. Instead, the early tone moves the Brain from state 839 to state 487 instead, and then to 3911. There, it must signal "not short" (triangle marked L for "long"), but this judgment is ignored because it comes at the wrong time: only judgments at the end of the IOI are counted. Indeed, the pattern is still ongoing: there are four more 1s to come, landing the Brain in state 2882. As the tone sequence has ended, rather than continue in the familiar loop the Brain is instead thrown into state 359, and the four ensuing 0s result in a Brain in state 2884, *precisely* the state the Brain finds itself after listening to a short sequence. As a deterministic Brain, it therefore has no choice but to offer the false verdict "short" (upside-down triangle marked S for short).

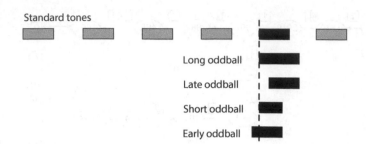

Standard tones

Long oddball

Late oddball

Short oddball

Early oddball

FIGURE 9.30. Four different oddball scenarios superimposed to the standard tone sequence. The standard (equal length) tone (in black) arrives at the expected time. A long oddball that begins at the expected onset (dashed line) ends later than expected. A late (standard-length) oddball happens to end where the long tone ends. Also shown is an early oddball that happens to end when a short (on-time) oddball ends.

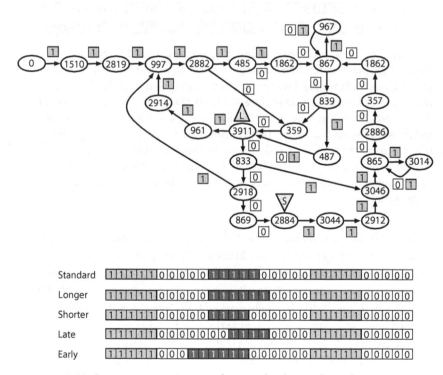

FIGURE 9.31. State-to-state transition diagram for the machine shown in Figure 9.29, but with the transitions in response to a late short tone and an early long tone added.

While the dynamics we just witnessed come from but one example Brain out of fifty, it turns out that this behavior is quite generic. When thrown out of the pathways that were carved by evolution (and thus are "familiar" to the Brain), these unfamiliar patterns do not elicit completely new Brain firing patterns (Brain states). Rather, the Brain falls back into pathways that it used before, except that due to the change in rhythm, perception will now be incorrect. In a way, we can (fancifully) say that the Brain is not confused by the out-of-rhythm tones—a confusion that could result in random decisions issued from never-before-visited Brain states. Instead, these Brains fall into familiar patterns, and issue their (incorrect) verdicts with full confidence.

What is remarkable about these illusions is that we can understand conceptually how they occur: it is as if the Brain does not pay attention to the beginning of the sequence and only uses the end of the sequence to make its judgment. It is remarkable because such a model of attention runs counter to the model of attentional entrainment that argued that both the beginning and the end of the sequence are being attended to.

To test which parts of the tonal sequence the Brain pays attention to, we can ask what information is used by the Brain to make the long/short decision. How much is the decision driven by the onset of the tone, the end of the tone, or the length of the tone? Because we have perfect Brain recordings, we can calculate the shared entropy between the state of the Brain at the decision point (we will call this random variable D), and various features of the tonal sequence that the Brain could use to make its decision, for all the different IOIs and standard tones that we used in this study. The state of the Brain at the decision point (always at the very end of the IOI) is depicted by the triangles in the state-to-state transition diagram above.

We can now define variables that encode different aspects of the tone. For example, both variables B (for "begin") and E (for "end") take on states 1 to 9 for a standard tone of length 5 (IOI $= 10$), encoding oddballs of length 1 to 9. The variable L, of course, simply encodes the length of the tone. Measuring which of these variables our decision variable $d \in D$ is correlated to tells us what information the Brain is using to render its decision. With $t \in T$ standing for the relevant features (either E, B, or L) we can write the information as

$$I(D:T) = \sum_{d,t} p(d,t) \log(\frac{p(d,t)}{p(d)p(t)}). \qquad (9.5)$$

Results of the information measurements are shown in Figure 9.32. They suggest that for short IOI and tone lengths, the Brain can use either end point, starting point, or tone length to make its prediction. However, as the IOI

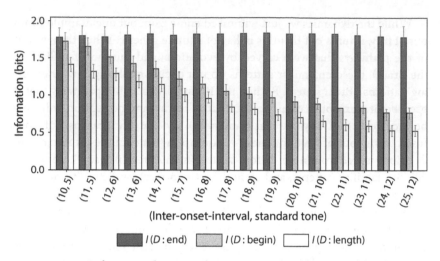

FIGURE 9.32. Information between the Brain state D at the time of the deci-sion and the position of the end $I(D : \text{end})$, the Brain state and the start of the sequence $I(D : \text{start})$, and the information between Brain state and tone sequence length (i.e., end-start) $I(D : \text{length})$.

(and tone length) become longer, the influence of the start and tone length information fade, while the correlation between end position and decision state remains constant. These findings imply that the state of the Brain at the moment of decision is best predicted by the *end* point of the tone sequence, *not* by the tone length, or the beginning of the tone sequence. In summary, most of these Brains are indeed using the time point at which the tone *ends* in order to predict tone length.

In hindsight, this observation makes perfect sense, as during evolution the tone *always* starts on time. With the starting point given, all the Brain needs to do in order to determine tone length is to pay attention to the *end*. As a consequence, some time during evolution, Brains stop paying attention to the beginning of the sequence (which is always the same), and instead focus all their attention to the end, which after all is the only variable.

These observations lead us to formulate a new theory of attention, which suggests that the Brain pays attention to those part of the sensory experience that are expected to be *the most informative*, that is, those parts that are the most variable (carry the most *potential information*). To some extent, this is a version of the *maxent* algorithm that allowed us to determine the identity of an image via the smallest number of saccades (discussed in section 9.2, page 466 and illustrated in Fig. 9.14), which also focused attention on the most entropic part of the image.

Of course, this theory does not assert that attention is *never* driven by the salient parts of an image or tone sequence, as the saliency model of visual attention suggests (Itti et al. 1998; Itti and Koch 2001) (we recall that a salient part is one that stands out above the background). Surely a loud noise will drive our attention as much as a bright light might. Also, often the salient parts and the most informative parts of a sensory experience might over-lap. We are reminded here again of the "two pathways" model of cognition that we discussed in section 9.1 and illustrated in Figure 9.2: it may be that the "saliency" form of attention is driven by the evolutionarily more ancient "where" pathway that requires no memory, while the ventral "what" path-way drives the information-driven attention mechanism. However, without dedicated experiments that test this hypothesis in people, we will not know.

9.4   A Future with Sentient Machines

In the epigraph at the beginning of this chapter, Shannon seems to suggest that in the future, people will have no choice to be subjugated to the robots they themselves have created. This sentiment is, of course, not new. Samuel Butler decried such a future in his letter to *The Press* in 1863 (see the excerpt on p. 188). An open letter put together in 2015 by the Future of Life Insti-tute (2015) and signed by numerous scientists and entrepreneurs warned about the dangers of unchecked research in artificial intelligence (AI), and called for "expanded research aimed at ensuring that increasingly capable AI systems are robust and beneficial: our AI systems must do what we want them to do."

The fear here appears to be that we will ultimately lose control over the machines that we have created, and that because these machines do not share in our values, they will subjugate and possibly exterminate us. Stuart Rus-sell, one of the pioneers of artificial intelligence research and the coauthor of one of the foundational textbooks in that discipline (Russell and Norvig 1995), warns that intelligent machines are bound to have a utility function (a way to represent individual preferences for goods and actions) that is unlike our own, and as a consequence those machines would be prone to catas-trophic misinterpretations of the commands given to them. For example, Russell (2019) cites an example that appears in the book *Superintelligence* by Nick Bostrom (2014), where a well-meaning operator asks an intelligent agent to produce paperclips. Because the instructions are not precise enough, the intelligent agent turns the entire planet into a paperclip production factory, killing all life as a side-effect. Incidentally, this is a fear that was also voiced by the cybernetics pioneer Norbert Wiener (1960), with essentially the same reasoning.

On the whole, these fears are grounded in the conviction that "human value functions" simply cannot be programmed. At the same time, it is assumed that intelligent machines could be so effective at manipulating the world that surrounds them that their impact can quickly become catastrophical. In other words, these fears assume that we can program behaviors very efficiently, but we cannot program human values. The fallacy in this thinking is that both complex behaviors (such as building planet-wide paperclip factories) and human value functions are guided by the same brain, and if you cannot program one you cannot program the other.

Earlier in section 9.2 I argued that any complex (intelligent) behavior is controlled both by sensory experience (and this includes verbal instructions) and the internal representations that each agent carries around in their brain, and which includes a model of the world. Without such a representation, we can imagine that simple rules will just be followed without regard to the consequences of the ensuing actions. But the representations that we carry with us allow us to "time travel," and imagine the consequences of our actions "in the mind's eye." Once an agent has access to this capacity, it is unlikely to engage in a behavior with disastrous consequences.

How could a robot be imbued with such representations? In movies, we sometimes see a machine simply "plugging into the internet" and download all of human knowledge in an instant. However, data is not the same as knowledge. Such a download cannot create representations: this is, after all, precisely the problem that artificial intelligence research faced in the early 1990s. For example, the robotics pioneer Rodney Brooks (then at the Massachusetts Institute of Technology) essentially gave up on the task of creating internal models for robots because he argued that this was a task that the robots should be doing *themselves*. As a consequence, he decided to focus on machines that are entirely reactive: intelligence without representations (Brooks 1991). The logical consequence of this approach was the cleaning robot Roomba, sold by the iRobot company that Brooks founded together with two colleagues.

In this chapter I am arguing instead that representations of the world around us (and that includes the people, animals, and plants) are crucial to intelligent decision making, and that these representations are acquired by lifetime-learning, augmented by instincts that are already "programmed" into our brain by eons of evolution. According to this view, sentient machines will never "suddenly become super-intelligent." Instead, mobile robots that have the capacity to autonomously explore the world around them are "born" when an evolved artificial brain with the capacity to learn is implanted on the mobile platform. At that point in time, the agent only has instincts (those acquired by evolution), but virtually no knowledge of the world. Just as an infant begins to learn, such a robot would slowly acquire skills, and learn how to interact with its surroundings. Its value function would emerge slowly, because the rate at which the robot learns is limited by the rate at which it encounters new information and new situations. In other words: such robots are limited to learn at about the same speed at which humans learn. After a certain period of time (say, about fifteen years) such a robot might have competencies that are similar to those that we possess, but most importantly, it is likely to have a value function that is very similar to our own simply because it grew up just as we do.

Because I believe that this evolutionary and lifetime-learning approach is the *only* path toward sentient machines, I am not worried about a future in which our creations subjugate us. I believe that we can create the conditions for sentience to emerge over time in single individual robots, but they will be individuals: some smarter, and some very much not so, than us. But this does not mean that there is no value in creating machines that are about as capable as we are. There is an essential difference between the brains of sentient robots and our own: the neural circuitry of a sentient robot can be copied precisely, and installed on another mobile platform. While such a clone would at first behave exactly like the original it was copied from, its personality will gradually change simply because it cannot have the same experiences as the first: both robots cannot occupy the same space. Essentially, such a cloning operation would create an artificial twin that would henceforth share some similarities with its sibling, but continue to diverge throughout life.

I think that this line of reasoning also implies that we must rethink the ethics surrounding artificial robot life. I would imagine that the value function of such a robot would object to the willy-nilly cloning of maybe hundreds of copies of their individual (and valuable) Brain. Indeed, it is likely that such robots have a right to be recognized as individuals, and therefore be accorded the same protections as we enjoy. I believe therefore that our future will be unlike the one imagined (and oddly championed) by the creator of information theory. It is much more likely that, on the path to creating sentient brains by evolution, we first create artificial dogs that look up to us like our pets do today.

9.5 Summary

The human brain is, by all accounts, one of the most stunning products of Darwinian evolution. It allows us to achieve an improbable level of fitness because it makes it possible for us to predict the world we live in reliably with exquisite precision, often out to the far future. But while our genes provide the information about how to make another one of us, they only tell us how to survive in a world that is unchanging, or, to be more precise, in an *average* world. In everyday life, however, things can change quickly, and therefore predictions might have to change quickly. Gene expression levels, however, can only change slowly in response to changed environments. Evolution's answer to this problem is neurons, who can propagate electrical signals quickly from sensors to motors (Jékely 2011), and therefore allow us to react quickly to changed circumstances. We can think of the brain's main function in terms of a prediction machine: it predicts changes on vastly different scales, from the millisecond range all the way to eons. Prediction with accuracy better than chance is only possible with information, which the brain mostly acquires during its lifetime (some information is encoded in the genes that give rise to the brain, giving rise to innate or instinctual behavior). Thus, what we call intelligence can be viewed entirely in terms of this capacity: to make predictions that are better than a competitor's predictions, and are accurate further out in time than the competitor's. The extent of this "temporal radius of prediction" (how far into the future you can make predictions that come to pass with accuracy better than chance) is a function of many components: the ability to categorize, to remember, to learn from experience, and to plan (among others). All of these elements of intelligence are rooted in the concept of information.

Because intelligent behavior is contingent on an agent having an accurate world model that is independent of the sensor values it might experience at any given point in time (a representation of the world), and such representations are so complex and elaborate (they are after all, the result of a lifetime of experiences), it is likely that the only path toward sentient machines lies in creating machines that can acquire a lifetime of experiences on their own. Such intelligent machines will plausibly share the same values as we do, simply because they will grow up along with us and other robots. When we reach this point in time, I anticipate that it will be obvious to us that each individual robot is special because it has a unique life history, and that we would accord such machines the same protections that we enjoy.

Exercises

9.1 In a simple model of bacterial chemotaxis, motion toward a nutrient source is achieved by comparing nutrient concentrations sensed at time t with the value sensed at the earlier time $t - \Delta$ and stored in memory (the model is described in more detail in section 9.1.1, page 448). Bacterial motion is controlled by a flagellar motor (variable R) that can turn either clockwise (leading to random "tumbling," state $R = t$) or counterclockwise (forward motion, or "swimming" state $R = s$). Assume that the sensor (variable S) is binary, and that low levels $S = \ell$ are sensed with probability $P(S = \ell) = p$ and high levels h are sensed with $P(S = h) =$

$1-p$, and that furthermore concentrations change with probability q between time t and $t + \Delta t$. There are two more parameters in this model: if the concentration is increasing, the motor takes on state $R = s$ (swimming) with probability π, while when it is decreasing, the state switches to t with probability $1 - \pi$. Finally, to escape a constant environment, if the concentration is unchanged the motor switches between states with probability ϵ.

(a) Show that the entropy of the first measurement $H(S_{t-\Delta t}) = H[p]$ where $H[p]$ is the binary entropy function

$$H[p] = -p \log_2 p - (1-p) \log_2 (1-p).$$

(b) Show that the joint entropy of two subsequent measurements is

$$H(S_{t-\Delta t}, S_t) = H[p] + H[q], \tag{9.6}$$

implying that the conditional entropy $H(S_t|S_{t-\Delta t}) = H[q]$.

(c) Show that for $p = 1/2$ and $\pi = 1$, the joint entropy of the two measurements and the motor variable is

$$H(S_{t-\Delta t}, S_t, R) = 1 + H[q] + (1-q)H[\epsilon]. \tag{9.7}$$

(d) Show that for the variable choices in (c) and $\epsilon = 0.5$, the "coherent information" $H(S_{t-\Delta t} : S_t : R)$ shared between the three variables takes on its largest negative value at $q = 1/3$, and that in this case the Venn diagram describing all three variables is completely symmetric (see Fig. 9.4). Compare to the Venn diagram you found in Exercise 2.6(c).

9.2 In the theory of grid-cell encoding, assume that the object location X on a regular 27×27 grid (see Fig. 9.9) is encoded using three grid-cell modules M_1, M_2, and M_3 that can take on nine states each.

(a): Draw the pair-wise entropy Venn diagram between X and any of the modules M_i (fill in the three values), between X and any pair of modules $M_i M_j$, and between X and the set of all modules $M_1 M_2 M_3$.

(b): Draw the tripartite entropy Venn diagram between X and two modules M_i and M_j (fill in the seven values). How is the absence of encryption and redundancy manifested in this diagram?

9.3 (a): For the problem of information-driven perception, use matrix (9.2) that shows which camera pattern appears at what image location given any particular numeral, to derive the histogram of entropies $H(C|L = i)$, as a function of the location variable $L = 0, \dots, 15$.

Convince yourself that $L = 9$ has the highest entropy, making it the maxent saccade.

(b): Suppose that you erroneously saccade to any of the two image locations with the second-highest entropy. Determine the subsequent maxent saccade location that will fully resolve the numeral in these cases.

(c): Calculate all entries in the tripartite Venn diagram relating C, L, and the numeral variable N. Assume that each numeral occurs with equal likelihood. Assume first that all locations are "visited" with equal probability. Then, calculate the entries to the same diagram but where the probability to visit location i, $p(L = i)$, is given by the normalized conditional entropy

$$p(L = i) = \frac{H(C|L = i)}{\sum_{i=0}^{15} H(C|L = i)} \qquad (9.8)$$

with $H(C|L = i)$ calculated in (a).

10

The Many Roles of Information in Biology

Nothing in biology makes sense except in the light of information.

We have come to the last chapter of this book, so it is time to step back a little bit from the details—the mathematics and the experiments—and contemplate the role of information in biology from a more general perspective. According to the quote above, which I confess I unashamedly modified from Dobzhansky's original, information is *everything* in biology: nothing makes sense without it.

In this age of bioinformatics and genomics, it may seem that the role of information—in the sense described by Shannon—was always accepted in biology, but this is far from the truth. As late as twenty years ago, the theoretical biologist John Maynard Smith was pleading to take the concept of information beyond the metaphorical, arguing for a more quantitative application of Shannon's ideas not just to the storage of information, but to signaling and communication as well (Maynard Smith 2000). The unambiguous success of applying Shannon's concepts to transcriptional regulation (Tkačik et al. 2008b) and biochemical signaling (Cheong et al. 2011) has certainly changed the landscape, but the idea that Shannon information quantitatively describes how much information is stored within genomes (and how to measure it) is still not widely accepted.

Of course, there *are* aspects of biology that are not directly or indirectly due to information, but rather exist due to chance. After all, the evolutionary process itself (as pointed out in the very first chapter) fundamentally relies on three components: inheritance, selection, and . . . chance (in the form of random mutations). And while characters that exist due to chance can be inherited and therefore persist, they are by definition not relevant in our quest to understand how an organism relates to its environment. Many times we can recognize nonadaptive traits as such. Famously, Stephen J. Gould and Richard Lewontin (Gould and Lewontin 1979) coined the word "spandrel" for any type of biological feature that is not adaptive, but nevertheless persists because it is not detrimental to fitness. Indeed, according to the relationship between fitness and information we discussed throughout the book, any feature whose alteration has no effect on fitness must have zero information for the organism that bears the feature.[1]

Just to be clear, I am not arguing here for a strict adaptationist view of evolutionary biology. Rather, my emphasis here is that in order to understand life on a grander scale, we should use information as our guiding principle. In this chapter I will discuss biological information in three biological contexts of ascending complexity. First, I will make the case that because biological information encodes the means of its own replication (according to chapter 7), information therefore is *the* defining characteristic of life. I will then focus on information's role in biological *communication*. Building on the insights we gained from examining the mathematics of cellular communication in section 2.4, we can study the advantages that a group gains when its members can communicate; that is, when they *exchange* information. We will find communication to be essential to understand how cells and organisms cooperate, because in order to establish the required mutual trust to engage in this behavior, they must exchange information. Finally, we'll learn that when animals make predictions about the future using a sophisticated neural circuitry (the brain and nervous system) they (and us) use information for that purpose. All of these examples are uses of information that lie at the heart of biology. It is hard for me to think of a single important mechanism in biology that does not have information at its root.

10.1   Life as Information

What is life? This question has a long history of course, and has been asked by scientists, lay people, and philosophers alike. For many, being able to answer this question is tantamount to finding the essence of life, while others argue that defining life is either impossible or pointless (Machery 2012).

---

1. Because information is often stored in multiple features (for example, multiple nucleotides, or even multiple traits) and robustly encoded, it is possible that altering a single feature in one particular organism can leave fitness unchanged even though information is decreased on average, because information cannot be measured using single sequences. Furthermore (as discussed in Box 10.1), whether or not an alteration has a fitness effect may be formally undecidable.

Box 10.1. Is Adaptive vs. Nonadaptive Decidable?

How easy is it to decide whether or not a particular feature of an organism is adaptive or not? It seems at first that such a question should not be too hard to answer. After all, if a feature z is adaptive, then we should be able to measure a correlation between that feature being present in an organism and the fitness of that organism (as in the first term of the Price equation; see Box 5.3). However, that equation does not reflect that the fitness $\bar{w}$ in Equation (5.31) depends crucially on which environment the organism is grown in. To make that dependence explicit, let us replace the covariance in Equation (5.31) by information instead: the information I that the particular feature z confers about the fitness w of the organism, given the particular environment. To quantify this, let us further define the (continuous) random variable W taking on values w, and the (continuous or discrete) random variable Z encoding different values or states of the feature. Adaptation within environment e can then be written as

$$A(e) = I(W : Z|e). \tag{10.1}$$

To evaluate this quantity, we need to evaluate the information that any particular feature conveys about fitness for *any* environment e that the organism could be exposed to. The reason for this is clear: if a particular feature does not predict fitness (in part) in any particular environment $E = e_0$ (so that in this environment, $I(W : Z|e_0) = 0)$), there could be another environment e_1 where Z contributes to fitness. How many environments do we need to test in order to decide that Z is nonadaptive (carries no information about fitness)? This question is reminiscent of Kolmogorov's attempt to define the complexity of a sequence s, which we encountered as Equation (5.1). That attempt failed because Kolmogorov complexity is uncomputable: we can never be sure what the shortest program is that will result in sequence s because some programs will run forever, something that cannot be avoided because Alan Turing showed (see Box 4.1) that whether or not a program will halt is undecidable. In a similar (but not logically equivalent) way, it is not possible to test all environments e_i because the experiment to determine whether feature Z contributes information might never end. After all, fitness is not a computable function in itself: we can create proxies that will predict the long-term survival of a type or species, but whether or not the proxy is accurate can only be ascertained by "running the experiment" for an infinite amount of time. As a consequence, the question whether or not a particular feature is adaptive or not is inherently undecidable.

Perhaps the most famous investigation of this question is due to the quantum physicist Erwin Schrödinger who, after he had made a name for himself as one of the fathers of quantum mechanics (the central equation of quantum physics is named after him), asked whether life can be explained by the laws of physics. His little book entitled *What Is Life* (Schrödinger 1944) is a fascinating read, and we should always keep in mind that it was written in 1944, nine years before Watson and Crick presented the structure of DNA and, based on that structure, suggested a mechanism for inheritance of information.

Schrödinger is certainly asking the right question in this book: How is it possible that the genes, given that they are microscopic (and therefore are encoded by a few million atoms only),[2] are so permanent? After all, he argued, if the genes are represented by a few million atoms, that number is much too small to "entail an orderly and lawful behaviour according to statistical physics" (Schrödinger 1944). If the genes were encoded in the properties of a gas or a liquid, he would indeed be right. That instead genes appear to be "reproduced without appreciable change for generations" is, to Schrödinger, a marvel that defies understanding. While we know today that human DNA takes about 100 billion atoms to encode one set of chromosomes, the argument is still valid, because 10^{11} is still far from the number needed for "orderly and lawful behaviour," which is typically given by Avogadro's number (about 6×10^{23}).

While Schrödinger asked the right question, his answer to it was almost completely wrong. He theorized that it was quantum mechanics that was at the root of the stability of the genes. Indeed, quantum mechanics predicts that atoms and molecules cannot just change state willy-nilly, but rather must transition between what are called "eigenstates," which are discrete. Basically, Schrödinger suggested that it is this "quantization" of molecular states that stabilizes the genes, and protects the gene-encoding molecules from the fluctuations that would threaten this "orderly and lawful behaviour." He further figured that mutations are in fact due to quantum jumps between the molecular eigenstates. And while it is true that at the very core, the molecular bond must be described with the rules of quantum mechanics (this is why I wrote that he was "almost" completely wrong), this is not the reason why genes encode phenotypes so reliably. The reason is instead the digital encoding of information in a base-4 code, using Watson-Crick bonds: an entirely classical encoding. It is clear, however, that Schrödinger understood that a defining property of life is its ability to stay away from maximum entropy, except that he did not understand that it is *information* that makes this possible. Instead,

2. This was Schrödinger's estimate at the time.

Schrödinger argued that life must "feed on negative entropy" in order to keep its entropy low.

We should not judge Schrödinger too harshly for this absurdity, as the publication of Shannon's formulation of the theory of information (Shannon 1948) was still five years away from the time Schrödinger delivered the material of his book in a set of lectures at Trinity College in Dublin. Shannon later showed (in a way) that Schrödinger's "negentropy" (negative entropy) is really just information: namely a constant (the maximal entropy) *minus* entropy [see Eq. (2.50)].

Many others have tried to pinpoint what is so special about life, myself included. In *Introduction to Artificial Life* (Adami 1998) I argued that

> Life is a property of an ensemble of units that share information coded in a physical substrate and which, in the presence of noise, manages to keep its entropy significantly lower than the maximal entropy of the ensemble, on timescales exceeding the "natural" timescale of *decay* of the information-bearing substrate by many orders of magnitude.

This unwieldy definition of life makes use of both thermodynamical and information-theoretic concepts, and while I still think that it captures "what life is" in the most general terms, there is much redundancy in it. Let's unpack this definition first, so that we can replace it with something simpler.

First, this definition insists that life is not a property of an individual, but instead is an attribute that can only be bestowed on a group (an "ensemble"; see section 2.1). This may sound silly at first, because many may argue that you absolutely should be able to tag individuals as alive or dead. But it turns out that how a single individual behaves, even if it displays clear lifelike properties, can never be a universal characteristic of life because it leaves out the most essential aspect of what distinguishes life from nonlife: life's ability to retain its essential features over long periods of time. Let us imagine, for example, that we have created a robotic contraption that has the ability to move (perhaps in ways that are difficult to predict for us) and that has the ability to find a charge-port by itself and "refuel," so that it could in principle continue moving this way for a long time. Should that "creature" not be classified as alive, since it might indeed fool a good fraction of onlookers? According to my definition above, it should not, because it can only maintain its appearance for as long as all the components it is made out of maintain their integrity. Once they fail, the contraption must also fail. But this failure is very different from the death of a living organism, because when a living organism is part of a group of (genetically related) organisms, the information that is inherent in the organism's genes did not disappear when the organism itself died: that information is still represented in the group. Thus, information storage in a group ensures the survival of the information for far longer than the lifetime of the components that each individual is made out of, or the individual itself. That is the essence of this definition of life: it puts survival of information first, not the survival of an organism.

In the above definition, I talk about "maintaining the entropy significantly lower than the maximal entropy of the system," but this is really just a contrived way of saying "maintaining information." According to the definition (2.34), information is defined as a difference between the maximal entropy of the system (here, the entropy of a biomolecular sequence) and its actual entropy. This "entropy gap," between how large the entropy could get if the second law of thermodynamics were allowed to "do its thing" and the actual entropy, that gap is therefore information.

The emphasis on maintaining a low entropy state (that is, information-rich state) for longer than the "natural timescale of decay of the information-bearing substrate" only serves to distinguish information stored in stable structures from information stored in unstable components. After all, a crystal (for example, water ice) is a low-entropy structure: the maximal entropy is obviously much larger (in the form of water vapor, for example). But, given a fixed temperature, the crystal has a very long natural lifetime (there are, for example, glaciers that are over a million years old). The same is obviously true for rocks, but even more so. In a way, rocks store information in the sense that the entropy of the rock is much lower than the maximal entropy of the components of the rock. But this information (which really can only be used to approximately predict where all the other atoms that make up the rock can be found) survives exactly as long as the rock survives, not longer. Biology is very very different, and it is life that makes this storage of information possible. Imagine for a moment an organism that has ceased to be alive. Under standard conditions (not encased in amber, for example), the information encoded in DNA has a fairly short natural lifetime. This is why I refer to life as being characterized by an "unnaturally" long lifetime of information, even though the moniker "unnatural" is misleading here because the word "natural" actually has two meanings: one referring to something existing in (or being caused by) nature and the other something that is "in agreement with the character or circumstances of something." I use the word "unnatural" here in accordance with the second definition, so naturally encoded information is unnaturally long-lived.

If instead of referring to "ensembles that maintain their entropy significantly lower than the maximal entropy of the system" we use the word "information," we can simplify this definition of life. And because information can only be defined in reference to an ensemble (because without this ensemble, we would not know what the information can predict; see Box 2.4),

FIGURE 10.1. A depiction of a fifteenth-century scribe or copyist, after an engraving in LaCroix (1870).

we can omit the discussion of ensembles also. In other words, we can define life as being "information that maintains itself." The important element in that definition is, of course, the word "itself." After all, we as a species have developed many different ways to maintain information, including for periods of time exceeding the lifetime of the information-bearing substrate. Indeed, the value of information was clear to those who first developed writing systems (by most accounts, the Sumerian cuneiform script, which emerged in Mesopotamia about 5,200 years ago), but the maintenance of information was not a primary concern then. After all, the earliest preserved information from this period mostly keeps track of trades, and the medium (mostly clay) was fairly fragile.

As writing systems became more complex, it became possible to preserve knowledge in writing. At the same time, the medium used to preserve this information changed (from clay to papyrus to paper), and the requirements for information preservation changed accordingly. One of the ways to ensure that information is preserved when the medium used to encode it is fragile (that is, it can decay) is to make many copies of it. Indeed, before the advent of copying machines, entire monasteries were tasked with making copies of existing works, as depicted in Figure 10.1. Today, the copying of information is so automatic and ubiquitous that we scarcely think about it, but the purpose of copying has changed: today, copies are made mostly in order to disseminate widely, primarily because information preservation itself is automatized and taken for granted. Copying for the sole purpose of information preservation has become rare, and (perhaps ironically), one of the few obvious uses is not in the digital realm, but rather in the biological one: the periodic copying of the state of all twelve independent lines of the Lenski experiment that we discussed earlier, so that this "fossil record" can be preserved even if one of the sets of refrigerators holding that record would lose power.

But having an external entity copy information is different from information that literally specifies how to make a copy machine out of parts, so as to make copies of that information. This is essentially what life does, and as we discussed in section 4.3, this is how John von Neumann figured that it is possible—in principle—to create an artificial life form from scratch, years before Watson and Crick realized that DNA was precisely such a piece of information. And then, as I have discussed in section 4.3, von Neumann proceeded to program that creation inside of a computer. This act of creation would have marked the first instance of the creation of a truly artificial form of life, only that computers that could execute that program would not exist for the next sixty years.

So, according to this view, it is easy to define life: life is information that replicates itself. And if we follow this definition to the letter, that means that viruses are alive (this includes those computer viruses that copy their own code) and in particular the digital life forms (such as those inhabiting the Tierra or Avida worlds) are alive as well. After all, as long as an energy source is available, those programs can maintain their low entropy state indefinitely, thus preserving the information inherent in the sequence long after any particular bit has flipped from one to zero, or vice versa.

If life is information, where did this information come from initially? We asked this question in chapter 7, and while it may appear at first that replacing the question "What is the origin of life?" with "What is the origin of information?" trivializes the search for our origins, it turns out that the question is still hard. According to what we know today, the amount of information necessary (and sufficient) to encode that information's own replication cannot simply emerge by chance. And assuming that the first form of life arrived on Earth from somewhere else in the solar system or galaxy does not solve the problem, because the mathematics of the chance emergence of information does not change if we move from one planet to another. This does not mean that the emergence of life from nonlife is impossible, it just means that we have yet to discover the pathway that allows information growth in a system where Darwinian evolution has yet to take hold. Making this question specifically about nonequilibrium information growth may allow us to search for this process in systems other than biochemistry (for example, inside a computer) as long as we believe that such a nonequilibrium process is general and not limited to only the particular biochemistry that was active on the early Earth.

Once the threshold for Darwinian evolution is breached—meaning that information that encodes its own replication has been assembled—information can now accumulate indefinitely. In chapter 3 I likened this process to the Maxwell demon, an imaginary agent who, by using judicious measurements and clever use of the information acquired in these measurements, can decrease the entropy of a system in an apparent violation of the second law of thermodynamics. However, it was also immediately clear that no such violation occurs, because the decrease in entropy is always associated with dissipated heat. Evolution can be mapped directly to the Maxwell demon dynamics: random mutations are "potential measurements," most of which do not actually lead to a decrease in entropy. But those accidental changes that end up predicting the environment with an accuracy better than chance (that is, those changes that represent information) result in an increased fitness for the organism carrying the information, which "fixes" the mutation. And just like the Maxwell demon in Figure 3.3 that has judiciously opened the door for fast molecules only while keeping it closed for the slow ones, the informative mutation ultimately becomes common in every organism that is on the line of descent of that mutation. On average then (and barring catastrophic changes in the environment), Darwinian evolution must lead to ever-increasing information.

If life is information here and everywhere, we can use this idea to detect life elsewhere, without even knowing what molecules encode the information, as long as we can design an "information detector." Colloquially speaking, all we need to do is look for "stuff" that should be really rare, but is instead everywhere. On Earth, for example, if we scoop up dirt and analyze its molecular composition, we find plenty of molecules that we would not expect on a planet without life, for example the complex amino acid tryptophan. Of course, some amino acids are readily generated abiotically (such as the simple amino acid glycine, with a low molecular weight of about 75 g/mol), but the complex ones are there because they are being synthesized using information stored in the genome. We can design an "agnostic" biosignature (it is called agnostic because it does not require a belief in a particular biochemistry) by looking at the monomer abundance distribution that you would get if you would "grind up" every organism and just record the frequency of monomers irrespective of where they came from. Of course, we would not know beforehand what monomers are used as instructions on an alien world, so we would measure the distribution of as many monomers as possible, looking at the distribution of nucleic acids, amino acids, hydroxyl acids, lipids, and more. Just as we saw that the distribution of letters in English text differs

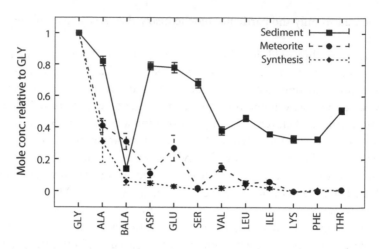

FIGURE 10.2. Average patterns of amino acid abundances relative to glycine, comparing biotic (Sediment, $n = 125$) and abiotic (Meteorite, $n = 15$, Synthesis, $n = 16$) sources. Error bars are one standard error. Adapted from Dorn et al. (2011).

from what we expect if the letters were generated randomly (recall Fig. 7.4), we expect the letters of an alien alphabet to be distributed differently from the rate at which they are being produced abiotically. This "Monomer Abundance Distribution Biosignature" (Dorn et al. 2011; Dorn and Adami 2011) certainly works well to "detect" life on Earth. For example, in Figure 10.2 we can see the abundance distribution of selected amino acids (normalized to the abundance of glycine) for sediment, compared to the distribution of the same amino acids as found in spark synthesis experiments, and the distribution found in meteorites (assumed to be of extraterrestrial origin). The distributions are significantly different, with the meteorite distribution very similar to what is found in spark synthesis (with a few interesting anomalies). Of course, we already know that the MADB can be used to detect digital life inside of the computer, but this can be checked also quantitatively (Dorn and Adami 2011). As long as we have not discovered evidence for life beyond Earth, the idea that life is information must, of course, be treated as a conjecture. But I would be extremely surprised if we would find life that is not based on information stored in linear nonbranching heteropolymers.

10.2   The Power of Communication

If the key to life is the permanent storage of information about how to copy that information, then the secret of life is the communication of information. At any moment in time, "decisions" have to be made that are contingent on the state of particular variables at that time, even in the simplest of organisms. The bacterium *E. coli* must decide whether or not to express the set of genes necessary to process glucose, for example (see Fig. 2.17), depending on what resource is currently available. In chemotaxis, a bacterium must decide whether to swim or tumble (see Fig. 9.3), depending on the recent readings of receptors on the cell surface. Multi-cellularity would be impossible without communication, as the existence of the multicellular state hinges upon making sure that the cells' decisions are perfectly synchronized, and that cells do not act independently of others. For a tissue to develop and grow, the time and place of each cell's duplication must be carefully coordinated.

10.2.1 Signaling in embryogenesis

An instructive example (albeit not representative for all eukaryotic life) is the development of the fruit fly *Drosophila melanogaster*, which has been studied in minute detail by generations of biologists (see Box 10.2). The unfolding of this process starts with a signal encoded in a protein that the adult mother has implanted into the unfertilized egg: a protein (called bicoid) that has an abundance gradient from one side of the elongated egg to the other. As the egg starts dividing, the concentration gradient of bicoid is maintained along what will become the anterior-posterior axis within the syncytial blastoderm. So even though the concentration of the protein is a continuous value, the information it encodes (at this stage) is binary: which side is the head, and which side will be the tail.

After the establishment of this axis, the three main regions of the embryo are determined, again by bicoid, which now encodes more information, namely not only where expression is high or low (at the edges of the cell) but also how steep the gradient of bicoid is, which happens to be in the central part of the embryo (we discussed the information channel between bicoid expression and the expression of another gap gene *hunchback*, back in section 2.4.3). Even though the story of gap gene expression is more complicated than that (for example, the expression of the gene *Krüppel* that defines the central region also is regulated by the expression of other gap genes that determine the two other regions, and which repress Krüppel in those regions), we can see in this process how information emerges, and is shaped by and combined with other streams of information, to guide embryonic development. That process is, of course, controlled by the genes of the organism, which are activated and turned off just at the right moment, just in the right place, by informational signals that were themselves generated just a tad earlier. In that way, the system resembles an automaton whose dynamics unfold relentlessly after it is activated, except that in this case the rules of unfolding are constantly changing because the automaton is being built at the same time. But for the time

Box 10.2. Early Drosophila Development

The development of the main body plan of the fruit fly *Drosophila melanogaster* is one of the most well-studied developmental processes in biology. The first thirteen cycles of cell division (starting from a single egg) all occur synchronously, leading to the *syncytial blastoderm* with $2^{13} = 8{,}192$ nuclei enclosed within a single cellular membrane.

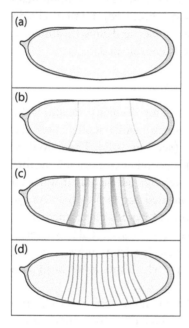

While these cell divisions are synchronous, a lot of patterning controlled by informational signals encoded in molecular concentrations occurs during this period. The first "decision" to be made is where to put the head, and where to put the tail of the animal: the determination of the anterior-posterior axis. This information is carried by a protein that is deposited into the egg by the mother before fertilization of the cell: a *maternal* protein (see Fig. 10.3[a]). All future development builds on this information, for example the subsequent determination of the main regions of the animal by the *gap genes* (Fig. 10.3[b]). These genes (with names like *hunchback*, *Krüppel*, and *knirps*) are activated or inhibited by the maternally inherited bicoid protein (expressed by a homeobox gene like the one shown in Fig. 2.7 or studied in Fig. 3.6), or even by a particularly steep gradient of bicoid. Thus, each of these gap gene proteins uses information encoded in bicoid differently,

FIGURE 10.3. Stages in the patterning of the fruit fly embryo in early development.

and on top of that is sensitive to the level of the others. After the main regions are formed (and therefore past the first thirteen cell divisions), the *pair rule* genes are activated by different combinations of the gap genes to create seven stripes of high *even-skipped* gene expression (interrupted by stripes of the gene *fushi tarazu*, see Fig. 10.3[c]). This pattern is then further refined by the *segment polarity* genes to create the fourteen *parasegments* seen in Figure 10.3(d). The parasegments are defined by the expression of the gene *wingless* in a single row of cells that defines the posterior edge of each parasegment, while *engrailed* is expressed in the anterior edge of the segment. The dorso-ventral axis in turn is determined by expression levels of the *dorsal* gene, whose role in regulating that axis we studied in detail in section 3.4.4.

period where the structure of the system is essentially constant (for example, the syncytial and cellular blastoderm stage), it is possible to model the cascade of gene expressions in terms of Boolean logic (Clark 2017), illustrating the information-driven nature of the process.

While in the early stages of *Drosophila* development the sychronicity of cells is maintained by the fact that each nucleus takes exactly the same time to divide, this synchronicity soon stops and different cells take a different amount of time to replicate. What then prevents, in a multicellular tissue, a cell from dividing faster than its neighbors, and by outcompeting them, change the delicate balance of types that is so crucial to the correct functioning of the organism? In the absence of a mechanism to inhibit this, a mutation that increases the replication rate of a cell would not initially be selected against, even if it would lead to the ultimate demise of the organism it participates in. We recognize here, of course, the tell-tale signs of tumor formation in cancer. Indeed, mutations that lead to proliferation of cells within a tissue do exist, but the mammalian genome carries an extensive arsenal of weapons to fight such "selfish" aberrations, and all of them rely on communication.

It goes without saying that we cannot here cover the myriads of pathways that are working together to fight tumor formation (referred to as "tumor suppressor genes") and how they interact with cells that have "gone rogue." But we can tell stories that remind us how important good communication is for the well-being of a multicellular tissue; to the extent that we can safely say that cancer is, speaking colloquially, a disease due to the failure to communicate.

10.2.2   Signaling and its disruption in cancer

One story we can tell concerns how a cell makes the decision to divide in the first place. A cancer biologist once told me: "Every morning a cell wakes up and contemplates suicide, only to be talked out of it by its closest friends." What he was referring to is the fact that normal cells in a tissue do not have the ability to simply engage in cell division, but rather must wait until they obtain permission to do so, in the form of a cellular growth factor. Growth factors are signaling molecules (sometimes called cytokines) that bind to cell-surface receptors of other cells (they cannot enter the cellular cytoplasm).

You might think that the signal that cells deliver is a simple "go ahead and grow," but the cellular growth control system is far more insidious than that. The growth factor signal actually interrupts an apoptosis (cellular suicide) pathway that is activated by default. In a way, the growth factor is really a "survival factor" (Collins et al. 1994). While such a mechanism appears extreme, it illustrates how important it is to the well-being of the organism to abolish, among the tissue residents, any semblance of "autonomy." A cell cannot control its own destiny, and any attempt to do so is punished by withholding these survival factors. This withholding is triggered, for example, when nutrients (like glucose) become scarce and the cellular environment becomes oxygen-deprived (anoxic), two events typically associated with tumor formation and uncontrolled growth. Once the surrounding cells sense this condition, growth factors are withheld and the offending cells are doomed (Carmeliet et al. 1998; Vander Heiden et al. 2001).

Of course, this is what should happen in a well-functioning organism, yet cancer is a reality that (in the United States) one in two women and one out of three men will have to face in their lifetime (Howlader et al. 2018). To the everlasting grief of millions of families, cells have found ways to circumvent even the extreme measures put in place to constrain their aberrant ways. And to the consternation of the researchers attempting to understand the cause of cancer, to paraphrase a famous first line in a Russian novel: healthy tissues are all alike, but every cancerous tissue is cancerous in its own way. At least it sometimes seems that way.

One of the ways cells "go their own way" is by producing the needed growth factor themselves, in a process called "autocrine" signaling (as opposed to the paracrine signaling, which sends growth factors via diffusion to the "closest friends"). After all, cells generally do not carry receptors to signaling molecules they produce themselves, most likely in order to prevent feedforward loops. But there are other ways to circumvent the dictate of the growth factor. One way is to trick the cellular machinery into believing that the signal had actually been given, when in reality it was withheld. A typical example of this sort of subterfuge is a mutation in the Ras signaling pathway, one of the most important signaling channels in cancer biology. Indeed, in the most deadly of all cancers (pancreatic cancer), Ras is found to be mutated in 90 percent of the cases, and such mutations also appear in about 30 percent of all other cancers (Downward 2003). Because of its importance, we will study this signaling channel (in a somewhat abstracted form) here.

In Fig. 10.4 we see two versions of the Ras pathway: the normal one in the left and middle panel, and the mutated one on the right. To understand what is going on here, we have to introduce the main players first.

The concentration of growth factor molecules (here the epithelial growth factor EGF, white circles in Fig. 10.4) is detected by the epithelial growth factor receptor (EGFR) molecule, shown in light gray in Figure 10.4. EGFR is a sensor that has one end sticking out of the membrane, and one end protruding into the cytosol. You notice that EGFR is a pair of molecules, and this is important. When EGF is bound to the receptor, the EGFR molecule dimerizes (the two monomers bind to each other), which leads to action on the part that sticks into the cell. In fact, the cell-pointing part of EGFR is a *kinase*,

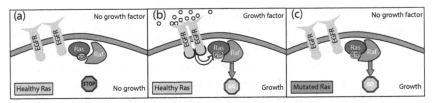

FIGURE 10.4. Schematic view of the Ras pathway. The normal pathways are shown in (a) and (b), while (c) shows the mutated pathway. (a) In the normal pathway, absence of the growth factor does not lead to cell growth, and ultimately to cell death. (b) When the growth factor is present, the signal cascade is activated by the Ras switch and cell division is allowed. (c) In the mutated Ras pathway, the Ras switch is stuck in "on" and cell proliferation occurs even though the growth factor is withheld.

which is a molecule that attaches phosphate molecules to other molecules (a process called phosphorylation). In this case, the kinase of one of the EGFR molecules attaches a phosphate group to the other EGFR molecule (a process called trans-auto-phosphorylation, in a self-explanatory way). In Figure 10.4, this activation via phosphorylation is indicated by the black circle on the bottom of the EGFR molecule. We can say quite generally that phosphorylation is the most ubiquitous binary tag in all of molecular biology.

Once EGFR is activated, it proceeds to send a signal to another kinase, which phosphorylates a molecule bound to Ras, shown in dark gray in Figure 10.4. Obviously, Ras is the central molecule in this signaling pathway, which is why the pathway carries its name. While phosphorylation can provide a tag, it is not a switch. Here, Ras is the switch. Ras is a protein in the family of G-proteins, or "guanine nucleotide binding proteins." This particular one binds to either GDP (guanine diphosphate) or GTP (guanine triphosphate).[3]

When Ras is bound to GDP, it is inactive: the switch is in the off position (Fig. 10.4[a]). But when Ras is bound to GTP, it is very much on (see Fig. 10.4), and binds to the kinase Raf (shown in medium gray in Fig. 10.4), which in turn activates a cascade of kinases that ultimately results in cell cycle progression; that is, the "don't kill yourself; go-ahead-and-multiply" signal. Needless to say, whether the Ras switch is on or off depends on the phosphorylation state of EGFR, so it is via this pathway that the EGF molecules give the signal to divide or not.

3. We recall GTP from the description of experiments to evolve RNA sequences to bind to GTP in section 2.3.3.

The mutated form of Ras is stuck in the on-position. It simply will not turn off and, in a sense, it refuses to listen to the EGF signal. Here then is the failure to communicate. But how is this possible? If the phosphorylation signal coming from EGFR turns GDP into GTP, how does Ras turn off normally? The answer is that Ras is also a low-level GTPase, that is, an enzyme that removes phosphorylation groups. So, normal Ras cannot stay in the on-state for long; it ultimately turns itself off.[4] This GTPase activity is precisely what is destroyed by the Ras mutations that occur in so many of the most deadly cancers, and 90 percent of pancreatic cancers. Once mutated Ras is on, it stays on.

Now that we have understood the functioning as well as the mutated pathway, you might ask what prevents us from designing a therapy that repairs this problem, so that Ras is not stuck in "on"? Shouldn't we able to fix the misinformation going on? It turns out that with today's technology, you cannot simply "fix Ras": this has been tried (Adjei 2001; Gysin et al. 2011; Stephen et al. 2014). The best you can do is to try to remove all the cells that have the mutated Ras (these are, after all, somatic mutations). But a more formidable obstacle is that the true biological pathways are far more complex than what we have been able to infer at this point. But let's first discuss the first problem, and then return to the second.

Cancer therapy works best if you can inhibit a gene product that has abnormally high levels, so that you can bring it down to normal levels. For example, one way in which cells circumvent the dependence on externally provided growth factors is by mutations that have the effect of overexpressing the growth factor receptor. That way, a cell can be sensitive to concentrations of the growth factor that are so small that they would normally not activate the pathway. One of the most successful cancer drugs works precisely this way: in about 20 to 30 percent of breast cancers, the Her-2 receptor (which is a homologue of the EGFR receptor, so it is also a tyrosine kinase and EGF sensor) is overexpressed up to 100-fold (in a sense these are mutations that create a pathway that is "listening too intently"). As a consequence, suppressing the Her-2 receptor with an antagonist (the cancer drug Herceptin) is highly effective.

While EGFR is not usually overexpressed in Ras-mediated cancers, there is a subset of cancers where it is, which led to the idea that this sub-class is druggable, in the same way as Her-2 positive cancers are. Except it does not work: EGFR inhibitors have been designed and are on the market, but they are not very effective. Or perhaps we should say that they are 100 percent effective in only 10 percent of the people (Ratain and Glassman 2007). The main

---

4. I am ignoring here the action of GAP (GTPase Activating Protein), because the mutations that turn off the self-GTPase activity of Ras also turn off the sensitivity to GAP.

reason for this is that the biological pathway is much more complex than the crude abstraction shown in Figure 10.4: between phosphorylation of EGFR and activation of Ras, for example, there are in reality at least three steps. And when Ras activates Raf (which activates a cascade of kinases downstream), it activates a host of other kinases as well (see, for example, Downward 2003). This is the second reason: we simply do not know enough about the pathways that are active between the sensor variable (EGF level) and the actuator variable (cell-cycle progression). The channel has several parallel pathways (Ras is activating at least two other kinases in parallel, besides Raf), and in each pathway there are several elements (kinase cascades) that we have not discussed here. Furthermore, they appear to be interconnected, with feedback loops of unknown origin. For example, an attempt at inhibiting the activity of Raf led to an *activation* of Raf instead (Stephen et al. 2014), and an attempt at downregulating a kinase downstream of Raf (in this case MEK)[5] actually led to an overexpression of EGFR, that is, increased sensitivity to the signal, as a compensation.

Clearly, we do not understand this channel well enough at all. There is a failure to communicate, but (like in many other cases) the channel is so complex that it is unlikely that modifying a single target within the channel can restore wild-type communication. But, in a sense, we know that it can be done in principle since the channel once worked. In particular, if we did know all the players in the channel, it should be possible to guide the information from the correct source toward the intended target. Information theory should allow us to characterize the channel using an exhaustive set of experiments, to discover the combination of variables that carries most of the information about the variable of interest in the mutated channel, namely cell proliferation. But because it is likely that in channels with many variables this information is effectively encrypted (see Exercise 2.6 for the simplest example of such an encryption), we need to develop new methods to discover where this crucial information is hidden.

This example of altered information flow in the Ras pathway shows how important information is in maintaining a steady-state (i.e., homeostasis) within a multicellular assembly. Just as in human societies, trust is not earned blindly: instead, it requires signals that allow us to verify the honest intentions at the same time.[6]

---

5. MEK stands for Mitogen-activated protein kinase kinase, which is called that because it phosphorylates another kinase, the mitogen-activated protein kinase (MAPK).

6. The Russian proverb is "Trust, but verify."

10.3   Predicting the Future

In chapter 9 we discussed intelligence in terms of information processing, not with the standard computer-processing model in mind, but rather in terms of how information is relayed from the sensors to the motors, making use of stored representations of the world, and the processes within that world. We put particular emphasis on the idea that the brain is a prediction machine, in the sense that it constantly attempts to predict its future sensory experience. Jeff Hawkins, the Silicon Valley entrepreneur who started Palm, Inc. (the company that created the Palm Pilot) and later started the machine intelligence company Numenta, gives an instructive example of this prediction model in his book *On Intelligence* (Hawkins and Blakeslee 2004), which I'll recount here.

When we walk along a path, we usually pay little attention to the surface we walk on. However, it turns out that some part of our brain is monitoring the timing of the impact of the ground on the soles of our feet (via cutaneous receptors) very accurately. Specifically, depending on the speed at which we are walking, the brain expects the signal from the cutaneous receptors on our feet to arrive at a very precise moment. While I doubt that this has been explored, it is possible that the circuitry that monitors this timing could be similar to the circuitry that we evolved to gauge the length of auditory signals that are embedded in a periodic background (we considered those at length in chapter 9). But we should keep in mind that there are many circuits in the brain that both generate and evaluate rhythms; see Buzsáki (2006). After a few steps, the impact occurs at periodic well-marked time points, and the brain expects these impacts to occur precisely at this time from there on. When the impact occurs at the predicted time, nothing happens. A prediction was made, and it came to pass: the brain deems everything to be in order.

Now imagine that a depression in the ground leads to an impact slightly later than expected. In that case, the coincidence between predicted and actual signal does not occur, and this sends an immediate alarm to the brain to start paying attention to the surface. This alarm alerts the brain that its model of the surface has been inadequate, and that the model needs updating. A quick peek at the road will reveal the indentations, which from now on we expect, and will check for. The same happens, of course, if we step on a rock and the impact signal arrives earlier than expected. Thus, the internal models that we carry with us in the brain (which we called "representations" in chapter 9) allow us to focus our attention only on those parts of our environment that are unexpected, and by that virtue are potentially informative.

I would like to emphasize that this view of cortical processing as a prediction machine has been gaining traction in the cognitive and neural sciences for quite a while. In fact, the British vision scientist Horace Barlow[7] (1921–2020) suggested such a model in 1989, writing that information

> may be incorporated in a model of 'what usually happens' with which incoming messages are automatically compared, enabling unexpected discrepancies to be immediately identified. (Barlow 1989)

A similar idea was expressed in a more quantitative manner by Rao and Ballard (1999). It is also inherent in the view of Karl Friston (2005) (see also Friston 2010), and Andy Clark in particular (see Clark 2013; Clark 2016). But the view I have been discussing here goes beyond these theories, because I specifically emphasize that there is a particular mechanism that allows the brain to predict only an extremely small subset of all possible experiences. According to this view, the brain is very *unlike* the laptop or desktop computer that we use daily, or the supercomputers that are used to predict the weather, for that matter. All these computing systems do not depend on history, in the sense that the result of running a spreadsheet analysis, for example, does not depend on whether you previously viewed a movie on the same computer, or whether you checked your email. In fact, our standard computers are *designed* not to depend on history, so that results are reliable and reproducible.

The brain is utterly different in that respect: the sensory experience materially *changes* how the brain computes, and more to the point, each individual's lived history will bias the computation henceforth. As a real-world example, imagine I sit down in front of the steering wheel of my car. The sensory experience immediately primes my brain to be a "car-driving" computer, while when I contemplate a chess board, my brain turns into a chess-playing computer and, at that moment, how to turn on a car's engine is the farthest thought in my mind.[8] This way, our brain can become a specialist computer simply by having the sensory experience drive the brain into a particular state where it *only* expects inputs that are problem-specific. This is reminiscent of the finite-state-machine model of the brain in Figure 9.31, where a particular brain state only connects to Brain states that are likely, given the current sensory experience. It is this state-dependence that makes the brain an effective prediction

7. Incidentally, Barlow's wife, the eminent geneticist Lady Nora Barlow, was a granddaughter of Charles Darwin.

8. To drive the absurdity of our current generation of "intelligent" machines home, you would never ask Deep Blue (the chess computer that for the first time beat a sitting chess world champion; Kasparov 2017) to drive your car, or else expect Tesla's car-driving software to respond reasonably to your opening move e2–e4.

machine: from the particular brain state it is in, only a very small subset of all possible experiences are expected, and the logic that connects this state to the others determines the likelihood of the expected signal. As long as our brain receives the expected signal, it responds accordingly. If an unexpected signal is received, the brain is forced to spend more attention on the signal, in order to understand why the predicted and the actual experience differ. Once the discrepancy is understood, an updated internal model of the world can then take into account the unpredicted occurrence, so as to make the surprise predictable again. Of course, these expectations are generated by different parts of our brain for the different sensory modalities at the same time, so we can rightly say that the brain is really a collection of thousands of mini-brains operating all at the same time (Hawkins 2021).

Clearly, our desktop computers do not work this way (nor do we want them to). But it is equally futile to attempt to build models of general intelligence using the feedforward model of information processing that underlies the current "Deep Learning" networks, since those do not rely on stored internal models either. Rather, we need to build control structures that *learn* during their lifetime, and that "at birth" already possess a fairly good model of the world in which they will operate, upon which they can build as they learn more about the world, by physically interacting with it. Note that for this to be possible, all these brains must be embodied (Pfeifer and Bongard 2006), so that they can control who and what they interact with. While it may take a decade or two for such a machine to become intelligent in the way that we use the term, the payoff of such an effort would be enormous. Even though it takes an average human about the same amount of time to become proficient, there is a fundamental difference between our brain and the sentient machine brain after, say, twenty years of experience: the human brain is unique, while the machine brain can be exactly copied, and transplanted into as many robotic platforms as we like. Of course, from that moment of transplantation, all the initial copies of the brain will start to differ, simply because their experiences from that point on will all be different given that they are differently embodied. Ethical considerations apart (which I believe will have to be taken seriously when we reach this point), this will also be the time at which experimental robot psychology (Adami 2006c) becomes a serious field of study. After all, whether or not a robot's brain has pathologies due to a traumatic experience will not necessarily be immediately apparent. Maybe investigating robot pathologies can also teach us about ourselves and just maybe, then, we can avoid the fate that Claude Shannon anticipated in the epigraph of chapter 9.

10.4 Summary

In this chapter I discussed three uses of information at three very different scales of life: how information (about how to copy that information to make it permanent) is the key element that distinguishes life from nonlife, how the communication of information is the key ingredient that makes cooperation possible from the level of cells all the way to societies, and how information is used by intelligent systems to predict the future state of the world. Of course there are many more uses of information in biological systems, but most are variations of the three uses I contemplated here: information storage, information for communication, and information for prediction. Just as I wrote in the introduction to this chapter, once we look at biology in the light of information, it is hard not to see the fundamental importance of this concept to biology. Indeed, everything in biology starts to make sense when viewed from this perspective.

REFERENCES

Adami, C. (1995a). Learning and complexity in genetic auto-adaptive systems. *Physica D 80*, 154–170.

Adami, C. (1995b). Self-organized criticality in living systems. *Phys Lett A 203*, 29–32.

Adami, C. (1998). *Introduction to Artificial Life*. New York: Springer Verlag.

Adami, C. (2004). Information theory in molecular biology. *Phys Life Rev 1*, 3–22.

Adami, C. (2006a). Digital genetics: Unravelling the genetic basis of evolution. *Nat Rev Genet 7*, 109–118.

Adami, C. (2006b). Evolution: Reducible complexity. *Science 312*, 61–63.

Adami, C. (2006c). What do robots dream of? *Science 314*, 1093–1094.

Adami, C. (2009). Biological complexity and biochemical information. In R. Meyers (Ed.), *Encyclopedia of Complexity and Systems Science*, pp. 489–511. New York: Springer Verlag.

Adami, C. (2011). Toward a fully relativistic theory of quantum information. In S. Lee (Ed.), *From Nuclei to Stars: Festschrift in Honor of Gerald E. Brown*, pp. 71–102. Singapore: World Scientific.

Adami, C. (2012). The use of information theory in evolutionary biology. *Ann NY Acad Sci 1256*, 49–65.

Adami, C. (2015). Information-theoretic considerations concerning the origin of life. *Origins Life Evol B 45*, 309–317.

Adami, C. (2016). What is information? *Philos Trans R Soc Lond A 374*, 20150230.

Adami, C. and N. C G (2022). Emergence of functional information from multivariate correlations. *Philos Trans R Soc Lond A 380*, 20210250.

Adami, C. and N. J. Cerf (2000). Physical complexity of symbolic sequences. *Physica D 137*, 62–69.

Adami, C. and J. Chu (2002). Critical and near-critical branching processes. *Phys Rev E 66*, 011907.

Adami, C. and A. Hintze (2013). Evolutionary instability of zero-determinant strategies demonstrates that winning is not everything. *Nat Commun 4*, 2193.

Adami, C. and T. LaBar (2017). From entropy to information: Biased typewriters and the origin of life. In S. Walker, P. Davies, and G. Ellis (Eds.), *Information and Causality: From Matter to Life*, pp. 95–112. Cambridge, MA: Cambridge University Press.

Adami, C., C. Ofria, and T. Collier (1999). Evolution of biological complexity. *Proc Natl Acad Sci USA 97*, 4463–4468.

Adami, C., J. Qian, M. Rupp, and A. Hintze (2011). Information content of colored motifs in complex networks. *Artif Life 17*, 375–390.

Adami, C., J. Schossau, and A. Hintze (2016). Evolutionary game theory using agent-based methods. *Phys Life Rev 19*, 1–26.

Adjei, A. A. (2001). Blocking oncogenic Ras signaling for cancer therapy. *J Natl Cancer Inst 93*, 1062–1074.

Agabian, N., L. Thomashow, M. Milhausen, and K. Stuart (1980). Structural analysis of variant and invariant genes in trypanosomes. *Am J Trop Med Hyg 29*(5 Suppl), 1043–1049.

Akashi, H. (2003). Translational selection and yeast proteome evolution. *Genetics 164*, 1291–1303.

Akin, E. (2012). The iterated Prisoner's Dilemma: Good strategies and their dynamics. arXiv:1211.0969v3.

Allman, J. M. (1999). *Evolving Brains*. W. H. Freeman & Co.

Amann, R. I., W. Ludwig, and K. H. Schleifer (1995). Phylogenetic identification and in situ detection of individual microbial cells without cultivation. *Microbiol Rev 59*, 143–169.

Amaral, D. G., N. Ishizuka, and B. Claiborne (1990). Neurons, numbers and the hippocampal network. *Prog Brain Res 83*, 1–11.

Amin, N., A. D. Liu, S. Ramer, W. Aehle, D. Meijer, M. Metin, S. Wong, P. Gualfetti, and V. Schellenberger (2004). Construction of stabilized proteins by combinatorial consensus mutagenesis. *Protein Eng Des Sel 17*, 787–793.

Anderson, J. R. (1983). *The Architecture of Cognition*. Hillsdale, NJ: Lawrence Erlbaum Associates, Inc.

Anderson, J. R. (1990). *The Adaptive Character of Thought*. Hillsdale, NJ: Lawrence Erlbaum Asscoiates, Inc.

Andley, U. P. (2007). Crystallins in the eye: Function and pathology. *Prog Retin Eye Res 26*, 78–98.

Ash, R. B. (1965). *Information Theory*. New York: Dover Publications, Inc.

Ashborn, P., T. J. McQuade, S. Thaisrivongs, A. G. Tomasselli, W. G. Tarpley, and B. Moss (1990). An inhibitor of the protease blocks maturation of human and simian immunodeficiency viruses and spread of infection. *Proc Natl Acad Sci USA 87*, 7472–7476.

Axelrod, R. (1980). Effective choice in the Prisoner's Dilemma. *J Confl Resolu 24*, 3–25.

Axelrod, R. (1984). *The Evolution of Cooperation*. New York: Basic Books.

Axelrod, R. and W. D. Hamilton (1981). The evolution of cooperation. *Science 211*, 1390–1396.

Baake, E., A. Gonzalez-Casanova, S. Probst, and A. Wakolbinger (2018). Modelling and simulating Lenski's long-term evolution experiment. *Theor Popul Biol 127*, 58–74.

Badii, R. and A. Politi (1997). *Complexity: Hierarchical Structures and Scaling in Physics*, Volume 6 of *Cambridge Nonlinear Science Series*. Cambridge (UK): Cambridge University Press.

Balaban, N. Q., J. Merrin, R. Chait, L. Kowalik, and S. Leibler (2004). Bacterial persistence as a phenotypic switch. *Science 305*, 1622–1625.

Balsa, E., R. Marco, E. Perales-Clemente, R. Szklarczyk, E. Calvo, M. O. Landázuri, and J. A. Enríquez (2012). Ndufa4 is a subunit of complex IV of the mammalian electron transport chain. *Cell Metab 16*, 378–386.

Barlow, H. B. (1961). Possible principles underlying the transformation of sensory messages. In W. Rosenblith (Ed.), *Sensory Communication*, pp. 217–234. Cambridge, MA: MIT Press.

Barlow, H. B. (1989). Unsupervised learning. *Neural Comput 1*, 295–311.

Barricelli, N. A. (1962). Numerical testing of evolution theories: Part I Theoretical introduction and basic tests. *Acta Biotheoretica 16*, 69–98.

Barrick, J. E. and R. E. Lenski (2013). Genome dynamics during experimental evolution. *Nat Rev Genet 14*, 827–839.

Barrick, J. E., D. S. Yu, S. H. Yoon, H. Jeong, T. K. Oh, D. Schneider, R. E. Lenski, and J. F. Kim (2009). Genome evolution and adaptation in a long-term experiment with *Escherichia coli*. *Nature 461*, 1243–1247.

Bartolo, R., R. C. Saunders, A. R. Mitz, and B. B. Averbeck (2020). Dimensionality, information and learning in prefrontal cortex. *PLoS Comput Biol 16*, e1007514.

Basharin, G. P. (1959). On a statistical estimate for the entropy of a sequence of independent random variables. *Theory Probability Applic 4*, 333–337.

Basile, B., A. Lazcano, and J. Oró (1984). Prebiotic syntheses of purines and pyrimidines. *Advances Space Res 4*, 125–131.

Bassler, B. L. (2002). Small talk. Cell-to-cell communication in bacteria. *Cell 109*, 421–4.

Bastian, H. C. (1872). *The Beginning of Life: Being Some Account of the Nature, Modes of Origin, and Transformations of the Lower Organisms*. London: Macmillan.

Batagelj, V. and A. Mrvar (2003). Pajek: Analysis and visualization of large networks. In P. M. M. Jünger (Ed.), *Graph Drawing Software*, pp. 77–103. Berlin: Springer-Verlag.

Bateson, G. (1979). *Mind and Nature: A Necessary Unity*. New York: E. P. Dutton.

Bateson, W. and E. R. Saunders (1902). *Experiments in the Physiology of Heredity*. London: Harrison and Sons.

Bava, K. A., M. M. Gromiha, H. Uedaira, K. Kitajima, and A. Sarai (2004). Protherm, version 4.0: Thermodynamic database for proteins and mutants. *Nucleic Acids Res 32*, D120–D121.

Bayes, T. (1763). An essay towards solving a problem in the doctrine of chances. *Philos Trans R Soc Lond 53*, 370–418.

Beadle, G. W. (1946). Genes and the chemistry of the organism. *Amer Scientist 34*, 31–35.

Bell, A. J. and T. J. Sejnowski (1995). An information-maximization approach to blind separation and blind deconvolution. *Neural Comput 7*, 1129–1159.

Bell, E. A., P. Boehnke, T. M. Harrison, and W. L. Mao (2015). Potentially biogenic carbon preserved in a 4.1 billion-year-old zircon. *Proc Natl Acad Sci USA 112*, 14518–14521.

Bell, G. and A. Mooers (1997). Size and complexity among multicellular organisms. *Biol J Linnean Society 60*, 345–363.

Benne, R., J. Van den Burg, J. P. Brakenhoff, P. Sloof, J. H. Van Boom, and M. C. Tromp (1986). Major transcript of the frameshifted CoxII gene from trypanosome mitochondria contains four nucleotides that are not encoded in the DNA. *Cell 46*, 819–826.

Bennett, A. F., K. M. Dao, and R. E. Lenski (1990). Rapid evolution in response to high-temperature selection. *Nature 346*, 79–81.

Bentley S. D., et al. (2002). Complete genome sequence of the model actinomycete *Streptomyces coelicolor* A3(2). *Nature 417*, 141–147.

Berg, O. G. and P. H. von Hippel (1987). Selection of DNA binding sites by regulatory proteins I. Statistical-mechanical theory and application to operators and promoters. *J Mol Biol 193*, 723–750.

Berg, O. G. and P. H. von Hippel (1988). Selection of DNA binding sites by regulatory proteins II. The binding specificity of cyclic AMP receptor protein to recognition sites. *J Mol Biol 200*, 709–723.

Bialek, W. (2018). Perspectives on theory at the interface of physics and biology. *Rep Prog Phys 81*, 012601.

Bialek, W., I. Nemenman, and N. Tishby (2001). Predictability, complexity, and learning. *Neural Comput 13*, 2409–2463.

Bialek, W., F. Rieke, R. de Ruyter van Steveninck, and D. Warland (1991). Reading a neural code. *Science 252*, 1854–1857.

Bicanski, A. and N. Burgess (2019). A computational model of visual recognition memory via grid cells. *Curr Biol 29*, 979–990.

Biemar, F., D. A. Nix, J. Piel, B. Peterson, M. Ronshaugen, V. Sementchenko, I. Bell, J. R. Manak, and M. S. Levine (2006). Comprehensive identification of Drosophila dorsal-ventral patterning genes using a whole-genome tiling array. *Proc Natl Acad Sci USA 103*, 12763–12768.

Billeter, M., Y. Q. Qian, G. Otting, M. Müller, W. Gehring, and K. Wüthrich (1993). Determination of the nuclear magnetic resonance solution structure of an Antennapedia homeodomain-DNA complex. *J Mol Biol 234*, 1084–1093.

Bloom, J. D. and F. H. Arnold (2009). In the light of directed evolution: pathways of adaptive protein evolution. *Proc Natl Acad Sci USA 106 Suppl 1*, 9995–10000.

Bloom, J. D., L. I. Gong, and D. Baltimore (2010). Permissive secondary mutations enable the evolution of influenza oseltamivir resistance. *Science 328*, 1272–1275.

Bloom, J. D., S. T. Labthavikul, C. R. Otey, and F. H. Arnold (2006). Protein stability promotes evolvability. *Proc Natl Acad Sci U S A 103*, 5869–5874.

Blount, Z. D., J. E. Barrick, C. J. Davidson, and R. E. Lenski (2012). Genomic analysis of a key innovation in an experimental *Escherichia coli* population. *Nature 489*, 513–518.

Blount, Z., C. Borland, and R. Lenski (2008). Historical contingency and the evolution of a key innovation in an experimental population of *Escherichia coli*. *Proc Natl Acad Sci USA 105*, 7899–7906.

Blount, Z. D. and R. E. Lenski (2013). Evidence of speciation in an experimental population of *E. coli* following the evolution of a key adaptation. In C. Adami, D. M. Bryson, C. Ofria, and R. T. Pennock (Eds.), *Proceedings of Artificial Life XIII*, p. 513. Cambridge, MA: MIT Press.

Blount, Z. D., R. Maddamsetti, N. A. Grant, S. T. Ahmed, T. Jagdish, J. A. Baxter, B. A. Sommerfeld, A. Tillman, J. Moore, J. L. Slonczewski, J. E. Barrick, and R. E. Lenski (2020). Genomic and phenotypic evolution of *Escherichia coli* in a novel citrate-only resource environment. *Elife 9*, e55414.

Blum, B., N. Bakalara, and L. Simpson (1990). A model for RNA editing in kinetoplastid mitochondria: "guide" RNA molecules transcribed from maxicircle DNA provide the edited information. *Cell 60*, 189–198.

Bohm, C., N. C G, and A. Hintze (2017). MABE (Modular Agent Based Evolver): A framework for digital evolution research. In C. Knibbe, G. Beslon, D. Parsons, D. Misevic, J. Rouzaud-Cornabas, N. Bredèche, S. Hassas, O. Simonin, and H. Soula (Eds.), *ECAL 2017: Proceedings of the European Conference on Artificial Life*, pp. 76–83.

Bollobás, B. (1985). *Random Graphs*. London: Academic Press.

Bongard, J., V. Zykov, and H. Lipson (2006). Resilient machines through continuous self-modeling. *Science 314*, 1118–1121.

Borel, É. (1913). Mécanique statistique et irréversibilité. *J Phys 5e série 3*, 189–196.

Borst, A. and F. E. Theunissen (1999). Information theory and neural coding. *Nat Neurosci 2*, 947–957.

Bostrom, N. (2014). *Superintelligence: Paths, Dangers, Strategies*. Oxford, UK: Oxford University Press.

Braakman, R. and E. Smith (2012). The emergence and early evolution of biological carbon-fixation. *PLoS Comput Biol 8*, e1002455.

Braakman, R. and E. Smith (2013). The compositional and evolutionary logic of metabolism. *Phys Biol 10*, 011001.

Braakman, R. and E. Smith (2014). Metabolic evolution of a deep-branching hyperther-mophilic chemoautotrophic bacterium. *PLoS One 9*, e87950.

Bradler, K. and C. Adami (2015). Black holes as bosonic Gaussian channels. *Phys Rev D 92*, 025030.

Brenner, S. (1974). The genetics of *Caenorhabditis elegans*. *Genetics 77*, 71–94.

Brenner, S. (2001). *My Life in Science*. London: BioMed Central Ltd.

Brenner, S. E., C. Chothia, and T. J. P. Hubbard (1998). Assessing sequence comparison methods with reliable structurally identified distant evolutionary relationships. *Proc Natl Acad Sci USA 95*, 6073–6078.

Brennicke, A., A. Marchfelder, and S. Binder (1999). RNA editing. *FEMS Microbiol Rev 23*, 297–316.

Bridgham, J. T., S. M. Carroll, and J. W. Thornton (2006). Evolution of hormone-receptor complexity by molecular exploitation. *Science 312*, 97–101.

Britten, R. J. and E. H. Davidson (1969). Gene regulation for higher cells: A theory. *Science 165*, 349–357.

Britten, R. J. and E. H. Davidson (1971). Repetitive and non-repetitive DNA sequences and a speculation on the origins of evolutionary novelty. *Q Rev Biol 46*, 111–138.

Brooks, R. A. (1991). Intelligence without representation. *Artif Intell 47*, 139–159.

Brown, C. T. and C. G. Callan Jr. (2004). Evolutionary comparisons suggest many novel cAMP response protein binding sites in *Escherichia coli*. *Proc Natl Acad Sci USA 101*, 2404–2409.

Brown, P., T. Sutikna, M. J. Morwood, R. P. Soejono, Jatmiko, E. W. Saptomo, and R. A. Due (2004). A new small-bodied hominin from the Late Pleistocene of Flores, Indonesia. *Nature 431*, 1055–1061.

Browne, T. (1643). *Religio Medici*. London: Andrew Cooke.

Bull, J. J., R. Sanjuán, and C. O. Wilke (2007). Theory of lethal mutagenesis for viruses. *J Virol 81*, 2930–2939.

Bürger, R. (2000). *The Mathematical Theory of Selection, Recombination, and Mutation*. Chichester: Wiley.

Burgess, N. (2014). The 2014 Nobel prize in physiology or medicine: A spatial model for cognitive neuroscience. *Neuron 84*, 1120–1125.

Burks, A. W., H. H. Goldstine, and J. von Neumann (1946). Preliminary discussion of the logical design of an electronic computing instrument. Part I. Technical report, prepared for U.S. Army Ord. Dept.

Buzsáki, G. (2004). Large-scale recording of neuronal ensembles. *Nat Neurosci 7*, 446–51.

Buzsáki, G. (2006). *Rhythms of the Brain*. New York: Oxford University Press.

C G, N., and C. Adami (2021). Information-theoretic characterization of the complete genotype-phenotype map of a complex pre-biotic world. *Phys Life Rev 38*, 111–114.

C G, N., T. LaBar, A. Hintze, and C. Adami (2017). Origin of life in a digital microcosm. *Philos Trans R Soc Lond A 375*, 20160350.

Cairns, J., G. S. Stent, and J. D. Watson (1966). *Phage and the Origins of Molecular Biology*. Cold Spring Harbor, Long Island, NY: Cold Spring Harbor Laboratory of Quantitative Biology.

Calude, C. (2002). *Information and Randomness* (2nd ed.). Berlin: Springer Verlag.

Campos, P. R. A., C. Adami, and C. O. Wilke (2002). Optimal adaptive performance and delocalization in NK fitness landscapes. *Physica A 304*, 495–506.

Carmeliet, P., Y. Dor, J. M. Herbert, D. Fukumura, K. Brusselmans, M. Dewerchin, M. Neeman, F. Bono, R. Abramovitch, P. Maxwell, C. J. Koch, P. Ratcliffe, L. Moons, R. K. Jain, D. Collen, E. Keshert, and E. Keshet (1998). Role of HIF-1α in hypoxia-mediated apoptosis, cell proliferation and tumour angiogenesis. *Nature 394*, 485–490.

Carothers, J. M., S. C. Oestreich, J. H. Davis, and J. W. Szostak (2004). Informational complexity and functional activity of RNA structures. *J Am Chem Soc 126*, 5130–5137.

Carothers, J. M., S. C. Oestreich, and J. W. Szostak (2006). Aptamers selected for higher-affinity binding are not more specific for the target ligand. *J Am Chem Soc 128*, 7929–7937.

Carpenter, L. R. and P. T. Englund (1995). Kinetoplast maxicircle DNA replication in *Crithidia fasciculata* and *Trypanosoma brucei*. *Mol Cell Biol 15*, 6794–6803.

Cavalier-Smith, T. (1985). Eukaryotic gene numbers, non-coding DNA and genome size. In T. Cavalier-Smith (Ed.), *The evolution of genome size*, pp. 69–103. New York: Wiley.

Chargaff, E. (1951). Structure and function of nucleic acids as cell constitutents. *Fed. Proc. 10*, 654–659.

Cheong, R., A. Rhee, C. J. Wang, I. Nemenman, and A. Levchenko (2011). Information transduction capacity of noisy biochemical signaling networks. *Science 334*, 354–358.

Cho, W. K., Y. Jo, K.-M. Jo, and K.-H. Kim (2013). A current overview of two viroids that infect chrysanthemums: *Chrysanthemum stunt viroid* and *Chrysanthemum chlorotic mottle viroid*. *Viruses 5*, 1099–1113.

Choi, P. J., L. Cai, K. Frieda, and X. S. Xie (2008). A stochastic single-molecule event triggers phenotype switching of a bacterial cell. *Science 322*, 442–446.

Clark, A. (1997). The dynamical challenge. *Cognitive Science 21*, 461–481.

Clark, A. (2013). Whatever next? Predictive brains, situated agents, and the future of cognitive science. *Behav Brain Sci 36*, 181–204.

Clark, A. (2016). *Surfing Uncertainty: Prediction, Action, and the Embodied Mind*. New York: Oxford University Press.

Clark, E. (2017). Dynamic patterning by the *Drosophila* pair-rule network reconciles long-germ and short-germ segmentation. *PLoS Biol 15*, e2002439.

Clifford, J. and C. Adami (2015). Discovery and information-theoretic characterization of transcription factor binding sites that act cooperatively. *Phys Biol 12*, 056004.

Clune, J., J.-B. Mouret, and H. Lipson (2013). The evolutionary origins of modularity. *Proc Roy Soc B 280*, 20122863.

Codoñer, F. M., J.-A. Darós, R. V. Solé, and S. F. Elena (2006). The fittest versus the flattest: Experimental confirmation of the quasispecies effect with subviral pathogens. *PLoS Pathog 2*, e136.

Coffin, J. M. (1995). HIV population dynamics in vivo: Implications for genetic variattion, pathogenesis, and therapy. *Science 267*, 483–489.

Collins, M. K., G. R. Perkins, G. Rodriguez-Tarduchy, M. A. Nieto, and A. López-Rivas (1994). Growth factors as survival factors: regulation of apoptosis. *Bioessays 16*, 133–138.

Colosimo, P. F., K. E. Hosemann, S. Balabhadra, G. Villarreal Jr., M. Dickson, J. Grimwood, J. Schmutz, R. M. Myers, D. Schluter, and D. M. Kingsley (2005). Widespread parallel evolution in sticklebacks by repeated fixation of ectodysplasin alleles. *Science 307*, 1928–1933.

Condra, J. H., D. J. Holder, W. A. Schleif, O. M. Blahy, L. J. Gabryelski, D. J. Graham, J. C. Quintero, A. Rhodes, H. L. Hobbins, E. Roth, M. Shivaprakash, D. L. Titus, T. Yang, H. Teppler, K. E. Squires, P. J. Deutsch, and E. A. Emini (1995). In vivo emergence of HIV-1 variants resistant to multiple protease inhibitors. *Nature 374*, 569–571.

Connes, A. (2016). An essay on the Riemann hypothesis. In J. Nash Jr. and M. T. Rassias (Eds.), *Open Problems in Mathematics*, pp. 225–257. Springer Verlag.

Costanzo, M. A., et al. (2010). The genetic landscape of a cell. *Science 327*, 425–431.

Costanzo, M. A., et al. (2016). A global genetic interaction network maps a wiring diagram of cellular function. *Science 353*, 1381.

Cover, T. M. and J. A. Thomas (1991). *Elements of Information Theory*. New York: John Wiley.

Coyne, J. A. and H. A. Orr (2004). *Speciation*. Sunderland, MA: Sinauer Associates.

Crick, F. (1968). The origin of the genetic code. *J Mol Biol 38*, 367–379.

Crocker, J., N. Potter, and A. Erives (2010). Dynamic evolution of precise regulatory encodings creates the clustered site signature of enhancers. *Nat Commun 1*, 99.

Daegelen, P., F. W. Studier, R. E. Lenski, S. Cure, and J. F. Kim (2009). Tracing ancestors and relatives of *Escherichia coli* B, and the derivation of B strains REL606 and BL21(DE3). *J Mol Biol 394*, 634–643.

Dallinger, W. H. (1887). The president's address. *J. Royal Microscopic Society 7*, 185–199.

Dalrymple, G. B. (2001). The age of the Earth in the twentieth century: a problem (mostly) solved. *Special Publications, Geological Society of London 190*, 205–221.

Darwin, C. (1859). *On the Origin of Species By Means of Natural Selection*. London: John Murray.

Darwin, C. (1862). *On the Various Contrivances by Which British and Foreign Orchids Are Fertilised by Insects, and on the Good Effects of Intercrossing*. London: John Murray.

Darwin, E. (1791). *The Botanic Garden; A Poem, in Two Parts*. London: J. Johnson.

Darwin, E. (1802). *The Temple of Nature, or the Origin of Society: A Poem*. London: J. Johnson.

Darwin, F. (1887). *The Life and Letters of Charles Darwin*. London: John Murray.

Davidson, E. H. (2001). *Genomic Regulatory Systems*. San Diego, CA: Academic Press.

Davidson, E. H., H. Jacobs, and R. J. Britten (1983). Very short repeats and coordinate induction of genes. *Nature 301*, 478–470.

Davis, J. H. and J. W. Szostak (2002). Isolation of high-affinity GTP aptamers from partially structured RNA libraries. *Proc Natl Acad Sci USA 99*, 11616–11621.

Dawkins, R. (1996). *Climbing Mount Improbable*. New York: W.W. Norton.

Dawkins, R. (1997). Human chauvinism. *Evolution 51*, 1015–1020.

de Maillet, B. (1750). *Telliamed: or, Discourses Between an Indian Philosopher and a French Missionary, on the Diminution of the Sea, the Formation of the Earth, the Origin of Men and Animals, and Other Curious Subjects, relating to Natural History and Philosophy*. London: T. Osborne.

de Maillet, B. and J.-B. L. Mascrier (1735). *Description de l'Egypte*. Paris: Louis Genneau & Jacques Rollin.

De Moraes, C. M., W. J. Lewis, P. W. Paré, H. T. Alborn, and J. H. Tumlinson (1998). Herbivore-infested plants selectively attract parasitoids. *Nature 393*, 570–573.

Dennett, D. (1995). *Darwin's Dangerous Idea: Evolution and the Meanings of Life*. New York: Simon and Schuster.

Dewdney, A. K. (1984). In the game called Core War hostile programs engage in a battle of bits. *Sci. Am. 250/5*, 14.

DeWeese, M. R. and M. Meister (1999). How to measure the information gained from one symbol. *Network 10*, 325–340.

Djordevic, M. and A. M. Sengupta (2006). Quantitative modeling and data analysis of SELEX experiments. *Phys. Biol. 3*, 13–28.

Djordevic, M., A. M. Sengupta, and B. I. Shraiman (2003). A biophysical approach to transcription factor binding discovery. *Genome Research 13*, 2381–2390.

Dobzhansky, T. (1973). Nothing in biology makes sense except in the light of evolution. *American Biology Teacher 35*, 125–129.

Domazet-Loso, T., J. Brajković, and D. Tautz (2007). A phylostratigraphy approach to uncover the genomic history of major adaptations in metazoan lineages. *Trends Genet 23*, 533–539.

Domingo, E. (2002). Quasispecies theory in virology. *J Virol 76*, 463–465.

Domingo, E., C. K. Biebricher, M. Eigen, and J. J. Holland (2001). *Quasispecies and RNA Virus Evolution: Principles and Consequences*. Georgetown, TX: Landes Bioscience.

Donaldson-Matasci, M. C., C. T. Bergstrom, and M. Lachmann (2010). The fitness value of information. *Oikos 119*, 219–230.

Dorn, E. D. and C. Adami (2011). Robust monomer-distribution biosignatures in evolving digital biota. *Astrobiology 11*, 959–968.

Dorn, E. D., K. H. Nealson, and C. Adami (2011). Monomer abundance distribution patterns as a universal biosignature: examples from terrestrial and digital life. *J Mol Evol 72*, 283–295.

Downward, J. (2003). Targeting RAS signalling pathways in cancer therapy. *Nat Rev Cancer 3*, 11–22.

Draghi, J. A. and J. B. Plotkin (2013). Selection biases the prevalence and type of epistasis along adaptive trajectories. *Evolution 67*, 3120–3131.

Drake, J. W. and J. J. Holland (1999). Mutation rates among RNA viruses. *Proc Natl Acad Sci USA 96*, 13910.

Dudoit, S., J. Shaffer, and J. Boldrick (2003). Multiple hypothesis testing in microarray experiments. *Statistical Science 18*, 71–103.

Dufour, Y. S., X. Fu, L. Hernandez-Nunez, and T. Emonet (2014). Limits of feedback control in bacterial chemotaxis. *PLoS Comput Biol 10*, e1003694.

Dugatkin, L. A. (1997). *Cooperation Among Animals: An Evolutionary Perspective*. Princeton, NJ: Princeton University Press.

Dujon, B. (1996). The yeast genome project: What did we learn? *Trends Genet 12*, 263–270.

Dunham, M. (2010). Experimental evolution in yeast: A practical guide. In J. Weissman, C. Guthrie, and G. R. Fink (Eds.), *Guide to Yeast Genetics: Functional Genomics, Proteomics, and Other Systems Analysis*, Volume 470 in Methods in Enzymology, Chapter 14, p. 487. Elsevier.

Durbin, R., S. R. Eddy, A. Krogh, and G. Mitchison (1998). *Biological Sequence Analysis: Probabilistic Models of Proteins and Nucleic Acids*. Cambridge, UK: Cambridge University Press.

Eagon, R. G. (1962). *Pseudomonas natriegens*, a marine bacterium with a generation time of less than 10 minutes. *J Bacteriol 83*, 736–737.

Eames, M. and T. Kortemme (2012). Cost-benefit tradeoffs in engineered *lac* operons. *Science 336*, 911–915.

Eddy, S. R. and R. Durbin (1994). RNA sequence analysis using covariance models. *Nucleic Acids Res 22*, 2079–2088.

Edlund, J. A., N. Chaumont, A. Hintze, C. Koch, G. Tononi, and C. Adami (2011). Integrated information increases with fitness in the evolution of animats. *PLoS Comput Biol 7*, e1002236.

Eigen, M. (1971). Selforganization of matter and the evolution of biological macromolecules. *Naturwissenschaften 58*, 465–523.

Eigen, M., J. McCaskill, and P. Schuster (1988). Molecular quasi-species. *J Phys Chem 92*, 6881–6891.

Eigen, M., J. McCaskill, and P. Schuster (1989). The molecular quasi-species. *Adv Chem Phys 75*, 149–263.

Eigen, M. and P. Schuster (1979). *The Hypercycle—A Principle of Natural Self-Organization*. Berlin: Springer-Verlag.

Eisely, L. (1958). *Darwin's Century*. New York: Doubleday.

Ellington, A. D. and J. W. Szostak (1990). In vitro selection of RNA molecules that bind specific ligands. *Nature 346*, 818–822.

Elowitz, M. B., A. J. Levine, E. D. Siggia, and P. S. Swain (2002). Stochastic gene expression in a single cell. *Science 297*, 1183–1186.

Emery, N. J. and N. S. Clayton (2001). Effects of experience and social context on prospective caching strategies by scrub jays. *Nature 414*, 443–446.

Emery, N. J. and N. S. Clayton (2004). The mentality of crows: Convergent evolution of intelligence in corvids and apes. *Science 306*, 1903–1907.

ENCODE Project Consortium (2012). An integrated encyclopedia of DNA elements in the human genome. *Nature 489*, 57–74.

Erdös, P. and A. Rényi (1960). On the evolution of random graphs. *Publ Math Inst Hungar Acad Sci 5*, 17–61.

Ewens, W. J. (2004). *Mathematical Population Genetics* (2nd ed.). New York: Springer Verlag.

Fano, R. M. (1961). *Transmission of Information: A Statistical Theory of Communication*. New York and London: MIT Press and John Wiley.

Feagin, J. E., J. M. Abraham, and K. Stuart (1988). Extensive editing of the cytochrome c oxidase III transcript in *Trypanosoma brucei*. *Cell 53*, 413–422.

Federhen, S. (2002). The taxonomy project. In J. McEntyre and J. Ostell (Eds.), *The NCBI Handbook*, Bethesda, MD. National Center for Biotechnology Information.

Felsenstein, J. (1974). The evolutionary advantage of recombination. *Genetics 78*, 737–756.

Ferguson, N. M., A. P. Galvani, and R. M. Bush (2003). Ecological and immunological determinants of influenza evolution. *Nature 422*, 428–433.

Ferré-D'Amaré, A. R. and W. G. Scott (2010). Small self-cleaving ribozymes. *Cold Spring Harb Perspect Biol 2*, a003574.

Fickett, J. W. (1982). Recognition of protein coding regions in DNA sequences. *Nucleic Acids Res 10*, 5303–5318.

Fields, D. S., Y. He, A. Y. Al-Uzri, and G. D. Stormo (1997). Quantitative specificity of the Mnt repressor. *J. Mol. Biol. 271*, 178–194.

Fiete, I. R., Y. Burak, and T. Brookings (2008). What grid cells convey about rat location. *J Neurosci 28*, 6858–6871.

Finn, R. D., J. Mistry, J. Tate, P. Coggill, A. Heger, J. E. Pollington, O. L. Gavin, P. Gunasekaran, G. Ceric, K. Forslund, L. Holm, E. L. L. Sonnhammer, S. R. Eddy, and A. Bateman (2010). The Pfam protein families database. *Nucleic Acids Res 38*, D211–D222.

Finn, R. D., et al. (2006). Pfam: Clans, web tools and services. *Nucleic Acids Res 34*, D247–D251.

Fisher, R. A. (1930). *The Genetical Theory of Natural Selection*. Oxford, UK: Clarendon Press.

Fisher, R. A. (1958). *The Genetical Theory of Natural Selection* (2nd ed.). Dover Publications, Inc.

Fletcher, J. A. and M. Zwick (2006). Unifying the theories of inclusive fitness and reciprocal altruism. *Am Nat 168*, 252–262.

Flood, M. M. (1952). Some experimental games. Technical report, RAND Corporation, Santa Monica, CA.

Fogle, C. A., J. L. Nagle, and M. M. Desai (2008). Clonal interference, multiple mutations and adaptation in large asexual populations. *Genetics 180*, 2163–2173.

Forster, R., C. Adami, and C. O. Wilke (2006). Selection for mutational robustness in finite populations. *J Theor Biol 243*, 181–190.

Fox, J. W. and R. E. Lenski (2015). From here to eternity—The theory and practice of a really long experiment. *PLoS Biol 13*, e1002185.

Frankel, A. D. and J. A. Young (1998). HIV-1: fifteen proteins and an RNA. *Annu Rev Biochem 67*, 1–25.

Franklin, J., T. LaBar, and C. Adami (2019). Mapping the peaks: Fitness landscapes of the fittest and the flattest. *Artif Life 25*, 250–262.

Friston, K. (2005). A theory of cortical responses. *Philos Trans R Soc Lond B Biol Sci 360*, 815–836.

Friston, K. (2010). The free-energy principle: a unified brain theory? *Nat Rev Neurosci 11*, 127–138.

Future of Life Institute (2015). Research priorities for robust and beneficial artificial intelligence: an open letter. http://futureoflife.org/2015/10/27/ai-open-letter/.

Fyhn, M., T. Hafting, A. Treves, M.-B. Moser, and E. I. Moser (2007). Hippocampal remapping and grid realignment in entorhinal cortex. *Nature 446*, 190–194.

Gabriel, W., M. Lynch, and R. Bürger (1993). Muller's ratchet and mutational meltdowns. *Evolution 47*, 1744–1757.

Galant, R. and S. B. Carroll (2002). Evolution of a transcriptional repression domain in an insect Hox protein. *Nature 415*, 910–913.

Gales, M. and S. Young (2007). The application of Hidden Markov Models in speech recognition. *Found Trends Sig Proc 1*, 195–304.

Galtier, N., G. Piganeau, D. Mouchiroud, and L. Duret (2001). GC-content evolution in mammalian genomes: The biased gene conversion hypothesis. *Genetics 159*, 907–911.

Garland, Jr., T. and M. R. Rose (2009). *Experimental Evolution*. Berkeley: University of California Press.

Gatlin, L. (1972). *Information Theory and the Living System*. New York: Columbia University Press.

Gavrilets, S. (2004). *Fitness Landscapes and the Origin of Species*. Princeton, N.J.: Princeton University Press.

Gell-Mann, M. and S. Lloyd (1996). Information measures, effective complexity, and total information. *Complexity 2*, 44–52.

Gerrish, P. J. and R. E. Lenski (1998). The fate of competing beneficial mutations in an asexual population. *Genetica 102-103*, 127–144.

Gilbert, E. N. (1959). Random graphs. *Ann Math Stat 30*, 1141–1144.

Gleick, J. (2011). *The Information: A History, a Theory, a Flood*. New York: Pantheon Books.

Gloye, A., F. Wiesel, O. Tenchio, and M. Simon (2005). Reinforcing the driving quality of soccer playing robots by anticipation. *it - Information Technology 47*, 250–257.

Goldstein, R. (2005). *Incompleteness: The Proof and Paradox of Kurt Gödel*. New York: W.W. Norton.

Goldstein, R. A. and D. D. Pollock (2017). Sequence entropy of folding and the absolute rate of amino acid substitutions. *Nat Ecol Evol 1*, 1923–1930.

Good, B. H., I. M. Rouzine, D. J. Balick, O. Hallatschek, and M. M. Desai (2012). Distribution of fixed beneficial mutations and the rate of adaptation in asexual populations. *Proc Natl Acad Sci USA 109*, 4950–4955.

Goodale, M. and D. Milner (2004). *Sight Unseen: An Exploration of Conscious and Unconscious Vision*. Oxford (UK): Oxford University Press.

Goodale, M. A. and D. Milner (1992). Separate visual pathways for perception and action. *Trends Neurosci 15*, 20–25.

Goodfellow, I., Y. Bengio, and A. Courville (2016). *Deep Learning*. MIT Press.

Gould, S. (1996). *Full House: The Spread of Excellence from Plato to Darwin*. New York: Harmony Books.

Gould, S. and R. Lewontin (1979). The spandrels of San Marco and the Panglossian paradigm: A critique of the adaptationist programme. *Proc Roy Soc London B 205*, 581–598.

Gould, S. J. (1997). The exaptive excellence of spandrels as a term and prototype. *Proc Natl Acad Sci USA 94*, 10750–10755.

Grant, B. R. and P. P. Grant (1989). *Evolutionary Dynamics of a Natural Population: The Large Cactus Finch of the Galapagos*. Chicago: University of Chicago Press.

Grant, P. R. and B. R. Grant (2008). *How and Why Species Multiply. The Radiation of Darwin's Finches*. Princeton, NJ: Princeton University Press.

Greenbaum, B. and A. N. Pargellis (2017). Self-replicators emerge from a self-organizing prebiotic computer world. *Artif Life 23*, 318–342.

Gregory, T. R. (2004). Macroevolution, hierarchy theory, and the c-value enigma. *Paleobiology 30*, 179–202.

Gregory, T. R. (2005). Genome size evolution in animals. In T. R. Gregory (Ed.), *The Evolution of the Genome*, San Diego, pp. 3–87. Elsevier.

Gregory, T. R. (2008). Evolutionary trends. *Evolution Education Outreach 1*, 259–273.

Grosse, I., H. Herzel, D. V. Buldyrev, and H. E. Stanley (2000). Species independence of mutual information in coding and non-coding dna. *Phys Rev E 61*, 5624–5629.

Gu, W., T. Zhou, and C. O. Wilke (2010). A universal trend of reduced mRNA stability near the translation-initiation site in prokaryotes and eukaryotes. *PLoS Comput Biol 6*, e1000664.

Gupta, A. and C. Adami (2016). Strong selection significantly increases epistatic interactions in the long-term evolution of a protein. *PLoS Genet 12*, e1005960.

Gysin, S., M. Salt, A. Young, and F. McCormick (2011). Therapeutic strategies for targeting Ras proteins. *Genes Cancer 2*, 359–372.

Haag, J., C. O'hUigin, and P. Overath (1998). The molecular phylogeny of trypanosomes: evidence for an early divergence of the Salivaria. *Mol Biochem Parasitol 91*, 37–49.

Haas Jr., J. (2000a). The Rev. Dr. William H. Dallinger F.R.S.: Early advocate of theistic evolution and foe of spontaneous generation. *Perspectives on Science and Christian Faith 52*, 107–117.

Haas Jr., J. (2000b). The Reverend Dr William Henry Dallinger, F.R.S. (1839–1909). *Notes and Records of the Royal Society of London 54*, 53–65.

Haigh, J. (1978). The accumulation of deleterious genes in a population–Muller's ratchet. *Theor Popul Biol 14*, 251–267.

Hajduk, S. L., V. A. Klein, and P. T. Englund (1984). Replication of kinetoplast DNA maxicircles. *Cell 36*, 483–492.

Haldane, J. B. S. (1927). A mathematical theory of natural and artificial selection. Part V: Selection and mutation. *Proc Camb Phil Soc 23*, 838–844.

Hamilton, W. D. (1963). Evolution of altruistic behavior. *Am Nat 97*, 354–356.

Hardin, G. (1960). The competitive exclusion principle. *Science 131*, 1292–1297.

Harms, A., E. Maisonneuve, and K. Gerdes (2016). Mechanisms of bacterial persistence during stress and antibiotic exposure. *Science 354*, aaf4268.

Harvey, I. (1999). Creatures from another world. *Nature 400*, 618–619.

Hassabis, D., D. Kumaran, C. Summerfield, and M. Botvinick (2017). Neuroscience-inspired artificial intelligence. *Neuron 95*, 245–258.

Hausser, J. and K. Strimmer (2009). Entropy inference and the James-Stein estimator, with application to nonlinear gene association networks. *J Mach Learn Res 10*, 1469–1484.

Hawkins, J. (2021). *A Thousand Brains: A New Theory of Intelligence*. New York: Basic Books.

Hawkins, J. and S. Ahmad (2016). Why neurons have thousands of synapses, a theory of sequence memory in neocortex. *Front Neural Circuits 10*, 23.

Hawkins, J. and S. Blakeslee (2004). *On Intelligence*. New York: Henry Holt and Co.

Hazen, R. M., P. L. Griffin, J. M. Carothers, and J. W. Szostak (2007). Functional information and the emergence of biocomplexity. *Proc Natl Acad Sci USA 104*, 8574–8581.

Hebb, D. O. (1949). *The Organization of Behavior: A Neuropsychological Theory*. New York: Wiley.

Heim, N. A., M. L. Knope, E. K. Schaal, S. C. Wang, and J. L. Payne (2015). Cope's rule in the evolution of marine animals. *Science 347*, 867–870.

Herrel, A., K. Huyghe, B. Vanhooydonck, T. Backeljau, K. Breugelmans, I. Grbac, R. Van Damme, and D. J. Irschick (2008). Rapid large-scale evolutionary divergence in morphology and performance associated with exploitation of a different dietary resource. *Proc Natl Acad Sci USA 105*, 4792–4795.

Hilbert, D. (1902). Mathematical problems. *Bull Am Math Soc 8*, 437–479.

Hintze, A. and C. Adami (2008). Evolution of complex modular biological networks. *PLoS Comput Biol 4*, e23.

Hintze, A., J. A. Edlund, R. S. Olson, D. B. Knoester, J. Schossau, L. Albantakis, A. Tehrani-Saleh, P. Kvam, L. Sheneman, H. Goldsby, C. Bohm, and C. Adami (2017). Markov Brains: A technical introduction. Arxiv.org 1709.05601.

Hintze, A., D. Kirkpatrick, and C. Adami (2018). The structure of evolved representations across different substrates for artificial intelligence. In T. Ikegami, N. Virgo, O. Witkowski, M. Oka, R. Suzuki, and H. Iizuka (Eds.), *Proceedings Artificial Life 16*. Cambridge, MA: MIT Press.

Hobbes, T. (1651). *Leviathan, or, The Matter, Forme, & Power of a Common-wealth Ecclesiasticall and Civill*. London: Andrew Crooke.

Hofacker, I. L., W. Fontana, P. F. Stadler, S. Bonhoeffer, M. Tacker, and P. Schuster (1994). Fast folding and comparison of RNA secondary structures. *Monatsh Chem 125*, 167–188.

Hofbauer, J. and K. Sigmund (1998). *Evolutionary Games and Population Dynamics*. Cambridge, UK: Cambridge University Press.

Hofstadter, D. (1979). *Gödel, Escher, Bach: An Eternal Golden Braid*. New York: Basic Books.

Holevo, A. S. and R. F. Werner (2001). Evaluating capacities of bosonic Gaussian channels. *Phys Rev A 63*, 032312.

Hout, M. C., M. H. Papesh, and S. D. Goldinger (2013). Multidimensional scaling. *Wiley Interdiscip Rev Cogn Sci 4*, 93–103.

Howlader, N., A. Noone, M. Krapcho, D. Miller, A. Brest, M. Yu, J. Ruhl, Z. Tatalovich, A. Mariotto, D. Lewis, H. Chen, E. Feuer, and K. Cronin (Eds.) (2018). *SEER Cancer Statistics Review, 1975-2016*, Bethesda, MD. Posted to the SEER website, April 2019: National Cancer Institute.

Huang, W., C. Ofria, and E. Torng (2004). Measuring biological complexity in digital organisms. In J. Pollack, M. A. Bedau, P. Husbands, T. Ikegami, and R. Watson (Eds.), *Proceedings of Artificial Life IX*, pp. 315–321. Cambridge, MA: MIT Press.

Hudson, K. M., A. E. R. Taylor, and B. J. Elce (1980). Antigenic changes in *Trypanosoma brucei* on transmission by tsetse fly. *Parasite Immunol 2*, 57–69.

Hunt, G. R. and R. D. Gray (2004). The crafting of hook tools by wild New Caledonian crows. *Proc Biol Sci 271 Suppl 3*, S88–S90.

Hutton, J. (1795). *Theory of the Earth (2 vols.)*. Edinburgh: Cadell, Junior and Davies.

Iliopoulos, D., A. Hintze, and C. Adami (2010). Critical dynamics in the evolution of stochastic strategies for the iterated Prisoner's Dilemma. *PLoS Comput Biol 6*, e1000948.

Ingram, P. J., M. P. H. Stumpf, and J. Stark (2006). Network motifs: Structure does not determine function. *BMC Genomics 7*, 108.

International Human Genome Consortium (2004). Finishing the euchromatic sequence of the Human genome. *Nature 431*, 931–945.

Itti, L. and C. Koch (2001). Computational modelling of visual attention. *Nat Rev Neurosci 2*, 194.

Itti, L., C. Koch, and E. Niebur (1998). A model of saliency-based visual attention for rapid scene analysis. *IEEE T Pattern Anal 20*, 1254–1259.

Jackson, A. P., M. Sanders, A. Berry, J. McQuillan, M. A. Aslett, M. A. Quail, B. Chukualim, P. Capewell, A. MacLeod, S. E. Melville, W. Gibson, J. D. Barry, M. Berriman, and C. Hertz-Fowler (2010). The genome sequence of *Trypanosoma brucei gambiense*, causative agent of chronic Human African Trypanosomiasis. *PLoS Negl Trop Dis 4*, e658.

Jackson, M. B. (2006). *Molecular and Cellular Biophysics*. Cambridge (UK): Cambridge University Press.

Jacob, F. (1977). Evolution and tinkering. *Science 196*, 1161–1166.

Jacob, F. and F. Monod (1961). On the regulation of gene activity. *Cold Spring Harb Symp Quant Biol 26*, 193–211.

Jaynes, E. T. (2003). *Probability Theory: The Logic of Science*. Cambridge (UK): Cambridge University Press.

Jékely, G. (2011). Origin and early evolution of neural circuits for the control of ciliary locomotion. *Proc Roy Soc B 278*, 914–922.

Jensen, R. E. and P. T. Englund (2012). Network news: The replication of kinetoplast DNA. *Annu Rev Microbiol 66*, 473–91.

Jeong, H., V. Barbe, C. H. Lee, D. Vallenet, D. S. Yu, S.-H. Choi, A. Couloux, S.-W. Lee, S. H. Yoon, L. Cattolico, C.-G. Hur, H.-S. Park, B. Ségurens, S. C. Kim, T. K. Oh, R. E. Lenski, F. W. Studier, P. Daegelen, and J. F. Kim (2009). Genome sequences of *Escherichia coli* B strains REL606 and BL21(DE3). *J Mol Biol 394*, 644–652.

Jeong, H., S. P. Mason, A.-L. Barabási, and Z. N. Oltvai (2001). Lethality and centrality in protein networks. *Nature 411*, 41–42.

Johannsen, W. (1909). *Elemente der exakten Erblichkeitslehre*. Jena: Gustav Fischer.

Kacian, D. L., D. R. Mills, F. R. Kramer, and S. Spiegelman (1972). A replicating RNA molecule suitable for a detailed analysis of extracellular evolution and replication. *Proc Natl Acad Sci USA 69*, 3038–3042.

Karbowski, J., G. Schindelman, C. J. Cronin, A. Seah, and P. W. Sternberg (2008). Systems level circuit model of *C. elegans* undulatory locomotion: Mathematical modeling and molecular genetics. *J Comput Neurosci 24*, 253–276.

Kasparov, G. (2017). *Deep Thinking: Where Machine Intelligence Ends and Human Creativity Begins*. New York: Perseus Books.

Kauffman, S. and S. Levin (1987). Towards a general theory of adaptive walks on rugged landscapes. *J Theor Biol 128*, 11–45.

Kauffman, S. A. and E. D. Weinberger (1989). The NK model of rugged fitness landscapes and its application to maturation of the immune response. *J Theor Biol 141*, 211–245.

Kawecki, T. J., R. E. Lenski, D. Ebert, B. Hollis, I. Olivieri, and M. C. Whitlock (2012). Experimental evolution. *Trends Ecol Evol 27*, 547–560.

Keller, L. and M. G. Surette (2006). Communication in bacteria: an ecological and evolutionary perspective. *Nat Rev Microbiol 4*, 249–258.

Khan, A. I., D. M. Dinh, D. Schneider, R. E. Lenski, and T. F. Cooper (2011). Negative epistasis between beneficial mutations in an evolving bacterial population. *Science 332*, 1193–1196.

Kim, J. T., T. Martinetz, and D. Polani (2003). Bioinformatic principles underlying the information content of transcription factor binding sites. *J Theor Biol 220*, 529–544.

Kimura, M. (1962). On the probability of fixation of a mutant gene in a population. *Genetics 47*, 713–719.

Kimura, M. (1968). Evolutionary rate at the molecular level. *Nature 217*, 624.

Kimura, M. (1983). *The Neutral Theory of Molecular Evolution*. Cambridge (UK): Cambridge University Press.

Kimura, M. and T. Maruyama (1966). The mutational load with epistatic gene interactions in fitness. *Genetics 54*, 1337–1351.

King, J. L. and T. H. Jukes (1969). Non-Darwinian evolution. *Science 164*, 788–798.

Kirby, L. (2019). *Pervasive alternative RNA editing in Trypanosoma brucei*. Ph.D. thesis, Michigan State University, East Lansing, MI.

Kirby, L., C. Adami, and D. Koslowsky (2022). Cluster analysis of unknown guide RNAs reveals the robustness of the RNA editing system in *Trypanosoma brucei*. In preparation.

Kirby, L. E. and D. Koslowsky (2017). Mitochondrial dual-coding genes in *Trypanosoma brucei*. *PLoS Negl Trop Dis 11*, e0005989.

Kirby, L. E., Y. Sun, D. Judah, S. Nowak, and D. Koslowsky (2016). Analysis of the *Trypanosoma brucei* EATRO 164 bloodstream guide RNA transcriptome. *PLoS Negl Trop Dis 10*, e0004793.

Kirkpatrick, D. and A. Hintze (2020). The evolution of representations in genetic programming trees. In W. Banzhaf, E. Goodman, L. Sheneman, L. Trujillo, and B. Worzel (Eds.), *Genetic Programming Theory and Practice XVII*, Cham, pp. 121–143. Springer.

Kirsh, D. and P. Maglio (1994). On distinguishing epistemic from pragmatic actions. *Cognitive Science 18*, 513–549.

Kolmogorov, A. (1965). Three approaches to the quantitative definition of information. *Problems of Information Transmission 1*, 4.

Kondrashov, A. S. (1988). Deleterious mutations and the evolution of sexual reproduction. *Nature 336*, 435–440.

Koonin, E. V. (2016). The meaning of biological information. *Philos Trans Roy Soc A 374*, 20150065.

Kopelman, S. (2019). Tit for Tat and beyond: The legendary work of Anatol Rapoport. *Negot Confl Manag R 13*, 60–84.

Kopetzki, D. and M. Antonietti (2011). Hydrothermal formose reaction. *New J Chem 35*, 1787–1794.

Koslowsky, D. J. (2004). A historical perspective on RNA editing: how the peculiar and bizarre became mainstream. *Methods Mol Biol 265*, 161–197.

Kotelnikova, E. A., V. J. Makeev, and M. S. Gelfand (2005). Evolution of transcription factor DNA binding sites. *Gene 347*, 255–263.

Kraines, D. and V. Kraines (1989). Pavlov and the Prisoner's Dilemma. *Theory and Decision 26*, 47–79.

Kramer, S. N. (1981). *History Begins at Sumer: Thirty-Nine Firsts in Recorded History* (3rd ed.). Philadelphia: University of Pennsylvania Press.

Krantz, D. H., R. D. Luce, P. Suppes, and A. Tversky (1971). *Foundations of Measurement*. New York: Academic Press.

Kriegeskorte, N. and P. K. Douglas (2018). Cognitive computational neuroscience. *Nat Neurosci 21*, 1148–1160.

Kryazhimskiy, S., J. A. Draghi, and J. B. Plotkin (2011). In evolution, the sum is less than its parts. *Science 332*, 1160–1161.

Kryazhimskiy, S., G. Tkacik, and J. B. Plotkin (2009). The dynamics of adaptation on correlated fitness landscapes. *Proc Natl Acad Sci USA 106*, 18638–18643.

Kun, A., M. Santos, and E. Szathmáry (2005). Real ribozymes suggest a relaxed error threshold. *Nat Genet 37*, 1008–1011.

Kussell, E., R. Kishony, N. Q. Balaban, and S. Leibler (2005). Bacterial persistence: A model of survival in changing environments. *Genetics 169*, 1807–1814.

LaBar, T. and C. Adami (2017). Evolution of drift robustness in small populations. *Nat Commun 8*, 1012.

LaBar, T., C. Adami, and A. Hintze (2016). Does self-replication imply evolvability? In P. Andrews, L. Caves, R. Doursat, S. Hickinbotham, F. Polack, S. Stepney, T. Taylor, and J. Timmis (Eds.), *Proc. of European Conference on Artificial Life 2015*, pp. 596–602. Cambridge, MA: MIT Press.

LaBar, T., A. Hintze, and C. Adami (2016). Evolvability tradeoffs in emergent digital replicators. *Artif Life 22*, 483–498.

LaCroix, P. (1870). *The Arts of the Middle Ages and at the Period of the Renaissance*. London: Chapman and Hall.

Lamarck, J.-B. (1809). *Philosophie Zoologique*. Paris: Chez Dentu et l'Auteur.

Landau, L. D. and E. M. Lifshitz (1980). *Statistical Physics (3rd Ed.)*. Oxford (UK): Butterworth-Heinemann.

Landauer, R. (1961). Irreversibility and heat generation in the computing process. *IBM J Res Dev 5*, 183–191.

Landauer, R. (1991, May). Information is physical. *Physics Today 44*(5), 23–29.

Lassalle, F., S. Périan, T. Bataillon, X. Nesme, L. Duret, and V. Daubin (2015). GC-content evolution in bacterial genomes: the biased gene conversion hypothesis expands. *PLoS Genet 11*, e1004941.

Laughlin, S. (1981). A simple coding procedure enhances a neuron's information capacity. *Z Naturforsch C 36*, 910–912.

Leclerc (Comte de Buffon), G.-L. (1749–1804). *Histoire Naturelle*. Paris: Imprimerie Royale.

Lee, S. H., J. L. Stephens, and P. T. Englund (2007). A fatty-acid synthesis mechanism specialized for parasitism. *Nat Rev Gen 5*, 287–297.

Lenski, R., C. Ofria, R. Pennock, and C. Adami (2003). The evolutionary origin of complex features. *Nature 423*, 139–144.

Lenski, R. E. (2001). Twice as natural. *Nature 412*, 255.

Lenski, R. E. (2017). What is adaptation by natural selection? Perspectives of an experimental microbiologist. *PLoS Genet 13*, e1006668.

Lenski, R. E. (2020). 2010: A BEACON Odyssey. In W. Banzhaf et al. (Ed.), *Evolution in Action—Past, Present, and Future*, pp. 3–17. New York: Springer Verlag.

Lenski, R. E., J. E. Barrick, and C. Ofria (2006). Balancing robustness and evolvability. *PLoS Biol 4*, e428.

Lenski, R. E., C. Ofria, T. C. Collier, and C. Adami (1999). Genome complexity, robustness and genetic interactions in digital organisms. *Nature 400*, 661–664.

Lenski, R. E. and M. Travisano (1994). Dynamics of adaptation and diversification: A 10,000-generation experiment with bacterial populations. *Proc Natl Acad Sci USA 91*, 6808–6814.

Lenski, R. E., M. J. Wiser, N. Ribeck, Z. D. Blount, J. R. Nahum, J. J. Morris, L. Zaman, C. B. Turner, B. D. Wade, R. Maddamsetti, A. R. Burmeister, E. J. Baird, J. Bundy, N. A. Grant, K. J. Card, M. Rowles, K. Weatherspoon, S. E. Papoulis, R. Sullivan, C. Clark, J. S. Mulka, and N. Hajela (2015). Sustained fitness gains and variability in fitness trajectories in the long-term evolution experiment with *Escherichia coli*. *Proc Roy Soc B 282*, 20152292.

Leon, D., S. D'Alton, E. M. Quandt, and J. E. Barrick (2018). Innovation in an *E. coli* evolution experiment is contingent on maintaining adaptive potential until competition subsides. *PLoS Genet 14*, e1007348.

Levy, S. (1992). *Artificial Life: The Quest for a New Creation*. New York: Pantheon Books.

Levy, S. B. (1998). The challenge of antibiotic resistance. *Sci Am 278/3 March*, 46–53.

Lewand, R. E. (2000). *Cryptological Mathematics*. Washington, DC: The Mathematical Association of America.

Lewis, W. J. and J. R. Brazzel (1966). Biological relationships between *Cardiochiles nigriceps* and the *Heliothis* complex. *J Econ Entomol 59*, 820–823.

Li, M. and P. Vitanyi (1997). *An Introduction to Kolmogorov Complexity and Its Applications*. New York: Springer Verlag.

Linnaeus, C. (1766). *Systema Naturæ*. Stockholm: Lars Salvius. 3 vols, fifth original edition.

Linsker, R. (1988). Self-organization in a perceptual network. *IEEE Computer 21*, 105–117.

Löfgren, L. (1977). Complexity of description of systems: A foundational study. *Int J Gen Sys 3*, 197–214.

Long, M., E. Betrán, K. Thornton, and W. Wang (2003). The origin of new genes: Glimpses from the young and old. *Nat Rev Genet 4*, 865–875.

Lyell, C. (1830). *Principles of Geology*. London: John Murray.

Lynch, M., R. Bürger, D. Butcher, and W. Gabriel (1993). The mutational meltdown in asexual populations. *J Hered 84*, 339–344.

MacFadden, B. J. (1992). *Fossil Horses: Systematics, Paleobiology, and Evolution of the Family Equidae*. Cambridge (UK): Cambridge University Press.

Machery, E. (2012). Why I stopped worrying about the definition of life... and why you should as well. *Synthese 185*, 145–164.

Macrae, N. (1992). *John von Neumann*. New York: Pantheon Books.

Malthus, T. (1798). *Essay on the Principle of Population, as It Affects the Future Improvement of Society*. London: J. Johnson.

Margulis, L. (1988). *Five Kingdoms: An Illustrated Guide to the Phyla of Life on Earth*. New York: W. H. Freeman, 3rd ed.

Marstaller, L., A. Hintze, and C. Adami (2013). The evolution of representation in simple cognitive networks. *Neural Comput 25*, 2079–2107.

Masel, J. (2011). Genetic drift. *Curr Biol 21*, R837–838.

Masel, J. and M. V. Trotter (2010). Robustness and evolvability. *Trends Genet 26*, 406–414.

Mauhin, V., Y. Lutz, C. Dennefeld, and A. Alberga (1993). Definition of the DNA-binding site repertoire for the *Drosophila* transcription factor SNAIL. *Nucleic Acids Res 21*, 3951–3957.

Maxwell, J. C. (1871). *Theory of Heat*. London: Longmans, Green & Co.

Mayfield, J. E. (2013). *The Engine of Complexity: Evolution as Computation*. New York: Columbia University Press.

Maynard Smith, J. (1970). Natural selection and the concept of a protein space. *Nature 225*, 563.

Maynard Smith, J. (1978). *The Evolution of Sex*. Cambridge (UK): Cambridge University Press.

Maynard Smith, J. (1982). *Evolution and the Theory of Games*. Cambridge, UK: Cambridge University Press.

Maynard Smith, J. (1986). Evolution: Contemplating life without sex. *Nature 324*, 300–301.

Maynard Smith, J. (2000). The concept of information in biology. *Phil Sci 67*, 177–194.

Maynard Smith, J. and G. Price (1973). The logic of animal conflict. *Nature 246*, 15–18.

Mazur, P., K. W. Cole, J. W. Hall, P. D. Schreuders, and A. P. Mahowald (1992). Cryobiological preservation of *Drosophila* embryos. *Science 258*, 1932–1935.

McAuley, J. D. and E. K. Fromboluti (2014). Attentional entrainment and perceived event duration. *Philos Trans R Soc Lond B Biol Sci 369*, 20130401.

McCandlish, D. M. and A. Stoltzfus (2014). Modeling evolution using the probability of fixation: History and implications. *Q Rev Biol 89*, 225–252.

McCulloch, W. S. and W. Pitts (1943). A logical calculus of the ideas immanent in nervous activity. *Bull Math Biol 52*, 99–115.

McShea, D. (1991). Complexity and evolution: What everybody knows. *Biology and Philosophy 6*, 303–324.

McShea, D. (2001). The hierarchical structure of organisms: A scale and documentation of a trend in the maximum. *Paleobiology 27*, 405–423.

McShea, D. W. (1996). Metazoan complexity and evolution: Is there a trend? *Evolution 50*, 477–492.

Meinert, C., I. Myrgorodska, P. de Marcellus, T. Buhse, L. Nahon, S. V. Hoffmann, L. L. S. d'Hendecourt, and U. J. Meierhenrich (2016). Ribose and related sugars from ultraviolet irradiation of interstellar ice analogs. *Science 352*, 208–212.

Mendel, G. (1865). Versuche über Pflanzen-Hybriden. *Verhandlungen des naturforschenden Vereines in Brünn 400*, 3–47.

Michalewicz, Z. (1999). *Genetic Algorithms + Data Structures = Evolution Programs*. New York: Springer Verlag.

Michod, R. E. and D. Roze (2001). Cooperation and conflict in the evolution of multicellularity. *Heredity 86*, 1–7.

Miller, G. A. and W. G. Madow (1954). On the maximum likelihood estimate of the Shannon-Wiener measure of information. Technical Report 54-75, Air Force Cambridge Research Center.

Miller, M., J. Schneider, B. K. Sathyanarayana, M. V. Toth, G. R. Marshall, L. Clawson, L. Selk, S. B. Kent, and A. Wlodawer (1989). Structure of complex of synthetic HIV-1 protease with a substrate-based inhibitor at 2.3 Å resolution. *Science 246*, 1149–1152.

Mills, D. R., R. L. Peterson, and S. Spiegelman (1967). An extracellular Darwinian experiment with a self-duplicating nucleic acid molecule. *Proc Natl Acad Sci USA 58*, 217–224.

Milo, R., S. Itzkovitz, N. Kashtan, R. Levitt, S. Shen-Orr, I. Ayzenshtat, M. Sheffer, and U. Alon (2004). Superfamilies of evolved and designed networks. *Science 303*, 1538–1542.

Milo, R., S. Shen-Orr, S. Itzkovitz, N. Kashtan, D. Chklovskii, and U. Alon (2002). Network motifs: simple building blocks of complex networks. *Science 298*, 824–827.

Mirmomeni, M. (2015). *The evolution of cooperation in the light of information theory.* Ph.D. thesis, Michigan State University, East Lansing, MI.

Mitchell, M. (1996). *An Introduction to Genetic Algorithms.* New York: Springer Verlag.

Monod, J. (1971). *Chance and Necessity: An Essay on the Natural Philosophy of Modern Biology.* New York: Alfred A. Knopf.

Monsellier, E. and F. Chiti (2007). Prevention of amyloid-like aggregation as a driving force of protein evolution. *EMBO Rep 8*, 737–742.

Moose, S., T. Rocheford, and J. Dudley (1996). Maize selection turns 100: A 21st century genomics tool. *Field Crops Res 100*, 82–90.

Morgan, T. H., C. B. Bridges, and A. M. Sturtevant (1925). The genetics of *Drosophila. Bibliogr Genetica 2*, 1–262.

Muller, H. J. (1964). The relation of recombination to mutational advance. *Mut Res 1*, 2–9.

Nash, J. F. (1951). Non-cooperative games. *Ann Math 54*, 286–295.

Nee, S. (2005). The great chain of being. *Nature 435*, 429.

Nehaniv, C. L. and J. L. Rhodes (2000). The evolution and understanding of hierarchical complexity in biology from an algebraic perspective. *Artif Life 6*, 45–67.

Neher, T., A. H. Azizi, and S. Cheng (2017). From grid cells to place cells with realistic field sizes. *PLoS One 12*, e0181618.

Neme, R. and D. Tautz (2013). Phylogenetic patterns of emergence of new genes support a model of frequent de novo evolution. *BMC Genomics 14*, 117.

Nemenman, I., F. Shafee, and W. Bialek (2002). Entropy and inference, revisited. In G. Dietterich, S. Becker, and Z. Ghahramani (Eds.), *Advances in Neural Information Processing Systems*, Volume 14, pp. 471–478. Cambridge, MA: MIT Press.

Neu, H. C. (1992). The crisis in antibiotic resistance. *Science 257*, 1064–1073.

Nevo, E., G. Gorman, M. Soulé, S. Y. Yang, R. Clover, and V. Jovanović (1972). Competitive exclusion between insular Lacerta species (Sauria, Lacertidae): Notes on experimental introductions. *Oecologia 10*, 183–190.

Nilsson, D.-E. (2009). The evolution of eyes and visually guided behaviour. *Philos Trans R Soc Lond B Biol Sci 364*, 2833–2847.

Nilsson, D.-E. (2013). Eye evolution and its functional basis. *Vis Neurosci 30*, 5–20.

Nirenberg, M. W., P. Leder, M. Bernfield, R. Brimacombe, J. Trupin, F. Rottman, and C. O'Neal (1965). RNA code words and protein synthesis VII: On the general nature of the RNA code. *Proc Natl Acad Sci USA 53*, 1161–1167.

Nobili, R. and U. Pesavento (1996). Generalised von Neumann's automata. In E. Besussi and A. Cecchini (Eds.), *Artificial Worlds and Urban Studies*, pp. 83–110. DAEST.

Nordhaus, W. and P. Samuelson (2009). *Microeconomics* (19th ed.). McGraw-Hill Education.

Nowak, M. (1990). Stochastic strategies in the Prisoner's Dilemma. *Theor Popul Biol 38*, 93–112.

Nowak, M. (2006). *Evolutionary Dynamics.* Cambridge, MA: Harvard University Press.

Nowak, M. and K. Sigmund (1990). The evolution of stochastic strategies in the Prisoner's Dilemma. *Acta Applic Math 20*, 247–265.

Nowak, M. and K. Sigmund (1993). A strategy of win-stay, lose-shift that outperforms tit-for-tat in the Prisoner's Dilemma game. *Nature 364*, 56–58.

Nurse, P. (2008). Life, logic and information. *Nature 454*, 424–426.

Oberle, M., O. Balmer, R. Brun, and I. Roditi (2010). Bottlenecks and the maintenance of minor genotypes during the life cycle of *Trypanosoma brucei*. *PLoS Pathog 6*, e1001023.

Ochsenreiter, T. and S. L. Hajduk (2006). Alternative editing of cytochrome c oxidase III mRNA in trypanosome mitochondria generates protein diversity. *EMBO Rep 7*, 1128–1133.

Ofria, C., C. Adami, and T. C. Collier (2002). Design of evolvable computer languages. *IEEE Trans Evolut Comput 6*, 420–424.

Ofria, C., C. Adami, and T. C. Collier (2003). Selective pressures on genomes in molecular evolution. *J Theor Biol 222*, 477–483.

Ofria, C., D. M. Bryson, and C. O. Wilke (2009). Avida: A software platform for research in computational evolutionary biology. In M. Komosinski and A. Adamatzky (Eds.), *Artificial Life Models in Software*, pp. 3–35. Springer London.

Ofria, C., W. Huang, and E. Torng (2007). On the gradual evolution of complexity and the sudden emergence of complex features. *Artif Life 14*, 255–263.

Orgel, L. (1968). Evolution of the genetic apparatus. *J Mol Biol 38*, 381–393.

Orgel, L. and J. Sulston (1971). Polynucleotide replication and the origin of life. In A. Kimball and J. Oro (Eds.), *Prebiotic and Biochemical Evolution*, pp. 89–94. Amsterdam: North-Holland Publishing Company.

Østman, B. and C. Adami (2013). Predicting evolution and visualizing high–dimensional fitness landscapes. In H. Richter and A. Engelbrecht (Eds.), *Recent Advances in the Theory and Application of Fitness Landscapes*, pp. 509–526. Berlin and Heidelberg. Springer-Verlag.

Østman, B., A. Hintze, and C. Adami (2012). Impact of epistasis and pleiotropy on evolutionary adaptation. *Proc Roy Soc B 279*, 247–256.

Østman, B., R. Lin, and C. Adami (2014). Trade-offs drive resource specialization and the gradual establishment of ecotypes. *BMC Evol Biol 14*, 113.

Paley, W. (1802). *Natural Theology; or, Evidences of the Existence and Attributes of the Deity, Collected from the Appearances of Nature*. London: R. Faulder.

Panigrahi, A. K., A. Zíková, R. A. Dalley, N. Acestor, Y. Ogata, A. Anupama, P. J. Myler, and K. D. Stuart (2008). Mitochondrial complexes in *Trypanosoma brucei*: A novel complex and a unique oxidoreductase complex. *Mol Cell Proteomics 7*, 534–545.

Papentin, F. (1980). On order and complexity I: General considerations. *J Theor Biol 87*, 421–456.

Papentin, F. (1982). On order and complexity II: Application to chemical and biochemical structures. *J Theor Biol 95*, 225–245.

Parfrey, L. W., D. J. G. Lahr, and L. A. Katz (2008). The dynamic nature of eukaryotic genomes. *Mol Biol Evol 25*, 787–794.

Pargellis, A. (2003). Self-organizing genetic codes and the emergence of digital life. *Complexity (Wiley) 8*, 69–78.

Pargellis, A. N. (2001). Digital life behavior in the amoeba world. *Artif Life 7*, 63–75.

Pesavento, U. (1995). An implementation of von Neumann's self-reproducing machine. *Artif Life 2*, 337—354.

Pfeifer, R. and J. Bongard (2006). *How the Body Shapes the Way We Think: A New View of Intelligence*. Cambridge, MA: MIT Press.

Philippi, T. and J. Seger (1989). Hedging one's evolutionary bets, revisited. *Trends Ecol Evol 4*, 41–44.

Phillips, P. C. (2008). Epistasis—the essential role of gene interactions in the structure and evolution of genetic systems. *Nat Rev Genet 9*, 855–867.

Phillips, R., J. Kondev, and J. Theriot (2008). *Physical Biology of the Cell*. Florence, Kentucky: Taylor and Francis (Garland Science).

Phillips, W. A. and W. Singer (1997). In search of common foundations for cortical computation. *Behav Brain Sci 20*, 657–683.

Planck Collaboration (2016). Planck 2015 results. XIII. Cosmological parameters. *Astronomy & Astrophysics 594*, A13.

Pope, A. (1733). *Essay on Man*. London: J. Wilford.

Potter, R. W. (2000). *The Art of Measurement*. New Jersey: Prentice-Hall.

Powers, R. (1988). *The Prisoner's Dilemma*. New York: Beechtree Books.

Press, W. H. and F. J. Dyson (2012). Iterated Prisoners' Dilemma contains strategies that dominate any evolutionary opponent. *Proc Natl Acad Sci USA 109*, 10409–10413.

Press, W. H., B. P. Flannery, S. Teukolsky, and W. T. Vetterling (1986). *Numerical Recipes: The Art of Scientific Computing*. Cambridge, MA: Cambridge University Press.

Price, G. R. (1970). Selection and covariance. *Nature 227*, 520–521.

Price, G. R. (1972). Extension of covariance selection mathematics. *Ann Hum Genet 35*, 485–490.

Proskurowski, G., M. D. Lilley, J. S. Seewald, G. L. Früh-Green, E. J. Olson, J. E. Lupton, S. P. Sylva, and D. S. Kelley (2008). Abiogenic hydrocarbon production at Lost City hydrothermal field. *Science 319*, 604–607.

Qian, J., A. Hintze, and C. Adami. (2011). Colored motifs reveal computational building blocks in the *C. elegans* brain. *PLoS ONE 6*, e17013.

Quandt, E. M., D. E. Deatherage, A. D. Ellington, G. Georgiou, and J. E. Barrick (2014). Recursive genomewide recombination and sequencing reveals a key refinement step in the evolution of a metabolic innovation in *Escherichia coli*. *Proc Natl Acad Sci USA 111*, 2217–2222.

Quandt, E. M., J. Gollihar, Z. D. Blount, A. D. Ellington, G. Georgiou, and J. E. Barrick (2015). Fine-tuning citrate synthase flux potentiates and refines metabolic innovation in the Lenski evolution experiment. *Elife 4*, e09696.

Queller, D. C. (1992). A general model for kin selection. *Evolution 46*, 376–380.

Rajamani, S., J. K. Ichida, T. Antal, D. A. Treco, K. Leu, M. A. Nowak, J. W. Szostak, and I. A. Chen (2010). Effect of stalling after mismatches on the error catastrophe in nonenzymatic nucleic acid replication. *J Am Chem Soc 132*, 5880–5885.

Rajon, E. and J. Masel (2011). Evolution of molecular error rates and the consequences for evolvability. *Proc Natl Acad Sci USA 108*, 1082–1087.

Rao, R. P. and D. H. Ballard (1999). Predictive coding in the visual cortex: a functional interpretation of some extra-classical receptive-field effects. *Nat Neurosci 2*, 79–87.

Rapoport, A. (1967). Optimal policies for the Prisoner's Dilemma. *Psych Rev 74*, 136–148.

Rapoport, A., D. A. Seale, and A. M. Colman (2015). Is Tit-for-Tat the answer? On the conclusions drawn from Axelrod's tournaments. *PLoS One 10*, e0134128.

Rasmussen, S., C. Knudsen, R. Feldberg, and M. Hindsholm (1990). The coreworld: Emergence and evolution of cooperative structures in a computational chemistry. *Physica D 42*, 111–134.

Ratain, M. J. and R. H. Glassman (2007). Biomarkers in phase I oncology trials: Signal, noise, or expensive distraction? *Clin Cancer Res 13*, 6545–6548.

Ray, T. (1992). An approach to the synthesis of life. In C. G. Langton, C. Taylor, J. D. Farmer, and S. Rasmussen (Eds.), *Proceedings Artificial Life II*, p. 371. Redwood City, CA: Addison-Wesley.

Reigl, M., U. Alon, and D. B. Chklovskii (2004). Search for computational modules in the *C. elegans* brain. *BMC Biol 2*, 25.

Rhee, A., R. Cheong, and A. Levchenko (2012). The application of information theory to biochemical signaling systems. *Phys Biol 9*, 045011.

Rhee, S.-Y., M. J. Gonzales, R. Kantor, B. J. Betts, J. Ravela, and R. W. Shafer (2003). Human immunodeficiency virus reverse transcriptase and protease sequence database. *Nucleic Acids Res 31*, 298–303.

Richelle, M., H. Lejeune, D. Defays, P. Greenwood, F. Macar, and H. Mantanus (2013). *Time in Animal Behaviour*. New York, NY: Pergamon Press.

Rieke, F., D. Warland, R. de Ruyter van Steveninck, and W. Bialek (1997). *Spikes: Exploring the Neural Code*. Cambridge, MA: MIT Press.

Risken, H. (1989). *The Fokker-Planck Equation: Methods of Solution and Applications* (2nd ed.). Berlin, Heidelberg, and New York: Springer Verlag.

Rivoire, O. and S. Leibler (2011). The value of information for populations in varying environments. *J Stat Phys 142*, 1124–1166.

Robinson, B. (2000). Human cytochrome oxidase deficiency. *Pediatric Research 48*, 581–585.

Robison, K., A. M. McGuire, and G. M. Church (1998). A comprehensive library of DNA-binding site matrices for 55 proteins applied to the complete *Escherichia coli* K-12 genome. *J Mol Biol 284*, 241–254.

Rodrigues, J. V., S. Bershtein, A. Li, E. R. Lozovsky, D. L. Hartl, and E. I. Shakhnovich (2016). Biophysical principles predict fitness landscapes of drug resistance. *Proc Natl Acad Sci USA 113*, E1470–E1478.

Romero, P., Z. Obradovic, C. R. Kissinger, J. E. Villafranca, E. Garner, S. Guilliot, and A. K. Dunker (1998). Thousands of proteins likely to have long disordered regions. *Pac Symp Biocomput 3*, 437–448.

Romiguier, J., V. Ranwez, E. J. P. Douzery, and N. Galtier (2010). Contrasting GC-content dynamics across 33 mammalian genomes: relationship with life-history traits and chromosome sizes. *Genome Res 20*, 1001–1009.

Ronshaugen, M., N. McGinnis, and W. McGinnis (2002). Hox protein mutation and macroevolution of the insect body plan. *Nature 415*, 914–917.

Rose, M. R., H. B. Passananti, and M. Matos (2004). *Methuselah Flies*. Singapore: World Scientific.

Ross, L., M. L. Lim, Q. Liao, B. Wine, A. E. Rodriguez, W. Weinberg, and M. Shaefer (2007). Prevalence of antiretroviral drug resistance and resistance-associated mutations in antiretroviral therapy-naïve HIV-infected individuals from 40 United States cities. *HIV Clin Trials 8*, 1–8.

Roth, G. (2013). *The Long Evolution of Brains and Minds*. Dordrecht: Springer Verlag.

Russell, S. (2019). *Human Compatible: Artificial Intelligence and the Problem of Control*. New York: Viking.

Russell, S. and P. Norvig (1995). *Artificial Intelligence: A Modern Approach*. Hoboken, NJ: Prentice-Hall.

Sanjuán, R., J. M. Cuevas, V. Furió, E. C. Holmes, and A. Moya (2007). Selection for robustness in mutagenized RNA viruses. *PLoS Genet 3*, e93.

Schliewen, U. K., D. Tautz, and S. Pääbo (1994). Sympatric speciation suggested by monophyly of crater lake cichlids. *Nature 368*, 629–632.

Schneider, D., E. Duperchy, E. Coursange, R. E. Lenski, and M. Blot (2000). Long-term experimental evolution in *Escherichia coli*. IX. Characterization of insertion sequence-mediated mutations and rearrangements. *Genetics 156*, 477–488.

Schneider, T. D. (1997). Information content of individual genetic sequences. *J Theor Biol 189*, 427–441.

Schneider, T. D. (2000). Evolution of biological information. *Nucl Acids Res 28*, 2794–2799.

Schneider, T. D., G. D. Stormo, L. Gold, and A. Ehrenfeucht (1986). Information content of binding sites on nucleotide sequences. *J Mol Biol 188*, 415–431.

Schrödinger, E. (1944). *What Is Life? The Physical Aspects of the Living Cell*. London, Bentley House: Cambridge University Press.

Schuster, P. and J. Swetina (1988). Stationary mutant distributions and evolutionary optimization. *Bull Math Biol 50*, 635–660.

Seehausen, O., Y. Terai, I. S. Magalhaes, K. L. Carleton, H. D. J. Mrosso, R. Miyagi, I. van der Sluijs, M. V. Schneider, M. E. Maan, H. Tachida, H. Imai, and N. Okada (2008). Speciation through sensory drive in cichlid fish. *Nature 455*, 620–626.

Segré, D., D. Ben-Eli, and D. Lancet (2000). Compositional genomes: prebiotic information transfer in mutually catalytic noncovalent assemblies. *Proc Natl Acad Sci USA 97*, 4112–4117.

Segré, D. and D. Lancet (2000). Composing life. *EMBO Rep 1*, 217–222.

Sella, G. and A. E. Hirsh (2005). The application of statistical physics to evolutionary biology. *Proc Natl Acad Sci USA 102*, 9541–9546.

Shafer, R. W. (2006). Rationale and uses of a public HIV drug-resistance database. *J Infect Dis 194 Suppl 1*, S51–S58.

Shafer, R. W. and J. M. Schapiro (2008). HIV-1 drug resistance mutations: an updated framework for the second decade of HAART. *AIDS Rev 10*, 67–84.

Shah, P., D. M. McCandlish, and J. B. Plotkin (2015). Contingency and entrenchment in protein evolution under purifying selection. *Proc Natl Acad Sci USA 112*, E3226–E3235.

Shannon, C. (1948). A mathematical theory of communication. *Bell Syst Tech J 27*, 379–423,623–656.

Shannon, C. E. (1951). Prediction and entropy of printed English. *Bell Syst Tech J 30*, 50–64.

Shapiro, T. A. and P. T. Englund (1995). The structure and replication of kinetoplast DNA. *Annu Rev Microbiol 49*, 117–143.

Shaw, J. M., J. E. Feagin, K. Stuart, and L. Simpson (1988). Editing of kinetoplastid mitochondrial mRNAs by uridine addition and deletion generates conserved amino acid sequences and AUG initiation codons. *Cell* 53, 401–411.

Shenhav, B., D. Segreè, and D. Lancet (2003). Mesobiotic emergence: Molecular and ensemble complexity in early evolution. *Advances Complex Systems* 6, 15–35.

Shlomai, J. (2004). The structure and replication of kinetoplast DNA. *Curr Mol Med* 4, 623–647.

Siddiquee, K. A. Z., M. J. Arauzo-Bravo, and K. Shimizu (2004). Effect of a pyruvate kinase (pykF-gene) knockout mutation on the control of gene expression and metabolic fluxes in *Escherichia coli*. *FEMS Microbiol Lett* 235, 25–33.

Soltis, D. E. and P. S. Soltis (1989). Alloploid speciation in *Tragopogon*: Insights from chloroplast DNA. *Amer J Botany* 76, 1119–1124.

Sprinzl, M., C. Horn, M. Brown, A. Ioudovitch, and S. Steinberg (1998). Compilation of tRNA sequences and sequences of tRNA genes. *Nucleic Acids Res* 26, 148–153.

Stanley, K. O., J. Clune, J. Lehman, and R. Miikkulainen (2019). Designing neural networks through evolutionary algorithms. *Nat Mach Intell* 1, 24–35.

Stephen, A. G., D. Esposito, R. K. Bagni, and F. McCormick (2014). Dragging Ras back in the ring. *Cancer Cell* 25, 272–81.

Stewart, A. and J. Plotkin (2013). From extortion to generosity, evolution in the iterated Prisoner's Dilemma. *Proc Natl Acad Sci USA* 110, 15348–15353.

Stoebel, D. M., A. M. Dean, and D. E. Dykhuizen (2008). The cost of expression of *Escherichia coli* lac operon proteins is in the process, not in the products. *Genetics* 178, 1653–1660.

Stoltzfus, A. (1999). On the possibility of constructive neutral evolution. *J Mol Evol* 49, 169–181.

Stoltzfus, A. (2012). Constructive neutral evolution: exploring evolutionary theory's curious disconnect. *Biol Direct* 7, 35.

Stormo, G. D. (2000). DNA binding sites: Representation and discovery. *Bioinformatics* 14, 16–23.

Stormo, G. D. and D. S. Fields (1998). Specificity, free energy and information content in protein-DNA interactions. *Trends Biochem Sci* 23, 109–113.

Studier, F. W., P. Daegelen, R. E. Lenski, S. Maslov, and J. F. Kim (2009). Understanding the differences between genome sequences of *Escherichia coli* B strains REL606 and BL21(DE3) and comparison of the *E. coli* B and K-12 genomes. *J Mol Biol* 394, 653–680.

Sturm, N. R. and L. Simpson (1990). Kinetoplast DNA minicircles encode guide RNAs for editing of cytochrome oxidase subunit III mRNA. *Cell* 61, 879–884.

Sturtevant, A. H. (1913). The linear arrangement of six sex-linked factors in *Drosophila*, as shown by their mode of association. *J Exp Zool* 14: 43–59.

Sudarshi, D., S. Lawrence, W. O. Pickrell, V. Eligar, R. Walters, S. Quaderi, A. Walker, P. Capewell, C. Clucas, A. Vincent, F. Checchi, A. MacLeod, and M. Brown (2014). Human African trypanosomiasis presenting at least 29 years after infection—what can this teach us about the pathogenesis and control of this neglected tropical disease? *PLoS Negl Trop Dis* 8, e3349.

Suh, N. P. (1990). *The Principles of Design*. Oxford, UK: Oxford University Press.

Sutton, R. S. and A. G. Barto (1998). *Reinforcement Learning: An Introduction*. Cambridge, MA: MIT Press.

Swofford, D. L., G. J. Olsen, P. J. Waddell, and D. M. Hillis (1996). Phylogenetic inference. In D. M. Hillis, C. Moritz, and B. K. Mable (Eds.), *Molecular Systematics* (2nd ed.), pp. 407–514. Sunderland, MA: Sinauer.

Sydykova, D. K., T. LaBar, C. Adami, and C. O. Wilke (2020). Moderate amounts of epistasis are not evolutionarily stable in small populations. *J Mol Evol 88*, 435–444.

Sykes, S., A. Szempruch, and S. Hajduk (2015). The Krebs cycle enzyme α-ketoglutarate decarboxylase is an essential glycosomal protein in bloodstream African trypanosomes. *Eukaryot Cell 14*, 206–215.

Szigeti, B., P. Gleeson, M. Vella, S. Khayrulin, A. Palyanov, J. Hokanson, M. Currie, M. Cantarelli, G. Idili, and S. Larson (2014). OpenWorm: an open-science approach to modeling *Caenorhabditis elegans*. *Front Comput Neurosci 8*, 137.

Szostak, J. W. (2003). Functional information: Molecular messages. *Nature 423*, 689.

Szostak, J. W. (2012). The eightfold path to non-enzymatic RNA replication. *J Syst Chem 3*, 2.

Taanman, J. (1997). Human cytochrome c oxidase: Structure, function, and deficiency. *J Bioenerg Biomembr 29*, 151–163.

Takahata, N. (1982). Sexual recombination under the joint effects of mutation, selection, and random sampling drift. *Theor Popul Biol 22*, 258–277.

Tarazona, P. (1992). Error thresholds for molecular quasispecies as phase transitions: From simple landscapes to spin-glass models. *Phys Rev A 45*, 6038–6050.

Taylor, J. S., I. Braasch, T. Frickey, A. Meyer, and Y. Van de Peer (2003). Genome duplication, a trait shared by 22000 species of ray-finned fish. *Genome Res 13*, 382–390.

Tehrani-Saleh, A. (2021). *The evolution of neural circuits for cognition in silico*. Ph.D. thesis, Michigan State University.

Tehrani-Saleh, A. and C. Adami (2021). Psychophysical tests reveal that evolved artificial brains perceive time like humans. In J. Cejkova, S. Holler, L. Soros, and O. Witkowski (Eds.), *Proceedings Conference on Artificial Life 2021*, pp. 57–59. Cambridge, MA: MIT Press.

Teilhard de Chardin, P. (1959). *The Phenomenon of Man*. New York: Harper and Brothers.

Tenaillon, O., J. E. Barrick, N. Ribeck, D. E. Deatherage, J. L. Blanchard, A. Dasgupta, G. C. Wu, S. Wielgoss, S. Cruveiller, C. Médigue, D. Schneider, and R. E. Lenski (2016). Tempo and mode of genome evolution in a 50,000-generation experiment. *Nature 536*, 165–70.

Thomas, R. D. K. and W.-E. Reif (1991). Design elements employed in the construction of animal skeletons. In N. Schmidt-Kittler and K. Vogel (Eds.), *Constructional Morphology and Evolution*, Berlin, pp. 283–294. Springer Verlag.

Thomas, R. D. K. and W.-E. Reif (1993). The skeleton space: A finite set of organic designs. *Evolution 47*, 341–360.

Thomas, R. D. K., R. M. Shearman, and G. W. Stewart (2000). Evolutionary exploitation of design options by the first animals with hard skeletons. *Science 288*, 1239–1242.

Thompson, R. F. (1986). The neurobiology of learning and memory. *Science 233*, 941–947.

Thornton, J. (2004). Resurrecting ancient genes: experimental analysis of extinct molecules. *Nat Rev Genet 5*, 366–375.

Tie, Y., A. Y. Kovalevsky, P. Boross, Y.-F. Wang, A. K. Ghosh, J. Tozser, R. W. Harrison, and I. T. Weber (2007). Atomic resolution crystal structures of HIV-1 protease and mutants V82A and I84V with saquinavir. *Proteins 67*, 232–242.

Tishby, N., F. Pereira, and W. Bialek (1999). The information bottleneck method. In B. Hajek and R. S. Sreenivas (Eds.), *Proceedings of the 37th Annual Allerton Conference on Communication, Control and Computing*, pp. 368–377. Champaign: University of Illinois Press.

Tkačik, G., C. G. Callan, Jr., and W. Bialek (2008a). Information capacity of genetic regulatory elements. *Phys Rev E 78*, 011910.

Tkačik, G., C. G. Callan, Jr., and W. Bialek (2008b). Information flow and optimization in transcriptional regulation. *Proc Natl Acad Sci USA 105*, 12265–12270.

Tkačik, G. and A. M. Walczak (2011). Information transmission in genetic regulatory networks: a review. *J Phys Condens Matter 23*, 153102.

Tononi, G., O. Sporns, and G. M. Edelman (1994). A measure for brain complexity: relating functional segregation and integration in the nervous system. *Proc Natl Acad Sci USA 91*, 5033–5037.

Travisano, M., J. A. Mongold, A. F. Bennett, and R. E. Lenski (1995). Experimental tests of the roles of adaptation, chance, and history in evolution. *Science 267*, 87–90.

Tretyachenko, V., J. Vyměttal, L. Bednárová, V. Kopecký, Jr, K. Hofbauerová, H. Jindrová, M. Hubálek, R. Souček, J. Konvalinka, J. Vondrášek, and K. Hlouchová (2017). Random protein sequences can form defined secondary structures and are well-tolerated in vivo. *Sci Rep 7*, 15449.

Treviranus, G. R. (1802). *Biologie, oder Philosophie der lebenden Natur*. Göttingen: Johann Friedrich Röwer.

Tuerk, C. and L. Gold (1990). Systematic evolution of ligands by exponential enrichment: RNA ligands to bacteriophage T4 DNA polymerase. *Science 249*, 505–510.

Turing, A. M. (1936). On computable numbers, with an application to the Entscheidungsproblem. *P Lond Math Soc 42*, 230–265.

Turner, C. B., Z. D. Blount, D. H. Mitchell, and R. E. Lenski (2015). Evolution and coexistence in response to a key innovation in a long-term evolution experiment with *Escherichia coli*. BioRxiv 020958.

Ugalde, J., B. Chang, and M. Matz (2004). Evolution of coral pigments recreated. *Science 305*, 1433.

Uy, J. A. C., R. G. Moyle, C. E. Filardi, and Z. A. Cheviron (2009). Difference in plumage color used in species recognition between incipient species is linked to a single amino acid substitution in the melanocortin-1 receptor. *Am Nat 174*, 244–254.

Valentine, J., A. Collins, and C. Meyer (1994). Morphological complexity increase in metazoans. *Paleobiology 20*, 131–142.

Valentine, J. W. (2000). Two genomic paths to the evolution of complexity in bodyplans. *Paleobiology 26*, 513–519.

van der Graaff, E., T. Laux, and S. A. Rensing (2009). The WUS homeobox-containing (WOX) protein family. *Genome Biol 10*, 248.

van Nimwegen, E., J. P. Crutchfield, and M. Huynen (1999). Neutral evolution of mutational robustness. *Proc Natl Acad Sci USA 96*, 9716–9720.

Van Oss, S. B. and A.-R. Carvunis (2019). De novo gene birth. *PLoS Genet 15*, e1008160.

van Weelden, S. W. H., B. Fast, A. Vogt, P. van der Meer, J. Saas, J. J. van Hellemond, A. G. M. Tielens, and M. Boshart (2003). Procyclic *Trypanosoma brucei* do not use Krebs cycle activity for energy generation. *J Biol Chem 278*, 12854–12863.

Vander Heiden, M. G., D. R. Plas, J. C. Rathmell, C. J. Fox, M. H. Harris, and C. B. Thompson (2001). Growth factors can influence cell growth and survival through effects on glucose metabolism. *Mol Cell Biol 21*, 5899–5912.

VanRullen, R. and C. Koch (2003). Is perception discrete or continuous? *Trends Cog Sci 7*, 207–213.

VanRullen, R., L. Reddy, and C. Koch (2005). Attention-driven discrete sampling of motion perception. *Proc Natl Acad Sci USA 102*, 5291–5296.

Varshney, L. R., B. L. Chen, E. Paniagua, D. H. Hall, and D. B. Chklovskii (2011). Structural properties of the *Caenorhabditis elegans* neuronal network. *PLoS Comput Biol 7*, e1001066.

Vilares, I. and K. Kording (2011). Bayesian models: the structure of the world, uncertainty, behavior, and the brain. *Ann N Y Acad Sci 1224*, 22–39.

von Neumann, J. (1958). *The Computer and the Brain*. New Haven, Connecticut: Yale University Press.

von Neumann, J. and A. Burks (1966). *Theory of Self-Reproducing Automata*. University of Illinois Press.

von Neumann, J. and O. Morgenstern (1944). *Theory of Games and Economic Behavior*. Princeton, NJ: Princeton University Press.

Wadhams, G. H. and J. P. Armitage (2004). Making sense of it all: bacterial chemotaxis. *Nat Rev Mol Cell Biol 5*, 1024–1037.

Wagner, A. (2005). *Robustness and Evolvability in Living Systems*. Princeton, NJ: Princeton University Press.

Wallace, A. R. (1858). On the tendency of varieties to depart indefinitely from the original type. *Proc Linn Soc Lond 3*, 54–62.

Wallace, A. R. (1867). Creation by law. *The Quarterly Journal of Science 4*, 471–488.

Wang, W., J. Zhang, C. Alvarez, A. Llopart, and M. Long (2000). The origin of the Jingwei gene and the complex modular structure of its parental gene, yellow emperor, in *Drosophila melanogaster*. *Mol Biol Evol 17*, 1294–1301.

Wang, X., G. Minasov, and B. K. Shoichet (2002). Evolution of an antibiotic resistance enzyme constrained by stability and activity trade-offs. *J Mol Biol 320*, 85–95.

Wang, Z., M. E. Drew, J. C. Morris, and P. T. Englund (2002). Asymmetrical division of the kinetoplast DNA network of the trypanosome. *EMBO J 21*, 4998–5005.

Watson, J. D. and F. H. C. Crick (1953a). Genetical implications of the structure of deoxyribonucleic acid. *Nature 171*, 965–966.

Watson, J. D. and F. H. C. Crick (1953b). A structure for deoxyribose nucleic acid. *Nature 171*, 737–738.

Watts, D. J. and S. H. Strogatz (1998). Collective dynamics of "small-world" networks. *Nature 393*, 440–442.

Wehrle, B. A., A. Herrel, B.-Q. Nguyen-Phuc, S. Maldonado, Jr, R. K. Dang, R. Agnihotri, Z. Tadić, and D. P. German (2020). Rapid dietary shift in *Podarcis siculus* resulted in localized changes in gut function. *Physiol Biochem Zool 93*, 396–415.

Weibull, J. W. (1995). *Evolutionary Game Theory*. Cambridge, MA: MIT Press.

Weir, W., P. Capewell, B. Foth, C. Clucas, A. Pountain, P. Steketee, N. Veitch, M. Koffi, T. De Meeûs, J. Kaboré, M. Camara, A. Cooper, A. Tait, V. Jamonneau, B. Bucheton, M. Berriman, and A. MacLeod (2016). Population genomics reveals the origin and asexual evolution of human infective trypanosomes. *Elife 5*, e11473.

Westfall, P. and S. S. Young (1993). *Resampling-based Multiple Testing: Examples and Methods for P-value Adjustment*. New York: Wiley.

White, J. G., E. Southgate, J. N. Thomson, and S. Brenner (1986). The structure of the nervous system of the nematode *Caenorhabditis elegans*. *Philos Trans R Soc Lond B Biol Sci 314*, 1–340.

White 3rd., H. (1976.). Coenzymes as fossils of an earlier metabolic state. *J Mol Evol 7*, 101–104.

Wiener, N. (1948). *Cybernetics: Or Control and Communication in the Animal and the Machine*. Paris: Hermann et Cie.

Wiener, N. (1960). Some moral and technical consequences of automation. *Science 131*, 1355–1358.

Wilke, C. O. (2001). Selection for fitness versus selection for robustness in RNA secondary structure folding. *Evolution 55*, 2412–2420.

Wilke, C. O. (2003). Probability of fixation of an advantageous mutant in a viral quasispecies. *Genetics 163*, 467–474.

Wilke, C. O. (2005). Quasispecies theory in the context of population genetics. *BMC Evol Biol 5*, 44.

Wilke, C. O. and C. Adami (2001). Interaction between directional epistasis and average mutational effects. *Proc R Soc Lond B 268*, 1469–1474.

Wilke, C. O., J. L. Wang, C. Ofria, R. E. Lenski, and C. Adami (2001). Evolution of digital organisms at high mutation rate leads to survival of the flattest. *Nature 412*, 331–333.

Wilson, B. A., S. G. Foy, R. Neme, and J. Masel (2017). Young genes are highly disordered as predicted by the preadaptation hypothesis of de novo gene birth. *Nat Ecol Evol 1*, 0146.

Wiser, M. J. (2015). *An analysis of fitness in long-term asexual evolution experiments*. Ph.D. thesis, Michigan State University, East Lansing, MI.

Wiser, M. J., N. Ribeck, and R. E. Lenski (2013). Long-term dynamics of adaptation in asexual populations. *Science 342*, 1364–1367.

Woese, C. (1967). *The Genetic Code, the Molecular Basis for Genetic Expression*. New York: Harper and Row.

Wolf, Y., T. Madej, V. Babenko, B. Shoemaker, and A. R. Panchenko (2007). Long-term trends in evolution of indels in protein sequences. *BMC Evol Biol 7*, 19.

Wolpert, D. H. and W. G. Macready (1997). No free lunch theorems for optimization. *IEEE Trans Evolut Comput 1*, 67.

Wright, S. (1931). Evolution in Mendelian populations. *Genetics 16*, 97–159.

Wright, S. (1932). The roles of mutation, inbreeding, cross-breeding and selection in evolution. In D. F. Jones (Ed.), *Proceedings of the VIth International Congress of Genetics*, Menasha, WI, pp. 356–366. Brooklyn Botanic Garden.

Wright, S. (1967). Surfaces of selective value. *Proc Natl Acad Sci USA 58*, 165–172.

Wright, S. (1988). Surfaces of selective value revisited. *Am Nat 131*, 115–123.

Yarbus, A. L. (1967). *Eye Movements and Vision*. New York: Plenum Press.

Yu, L., L. Peña Castillo, S. Mnaimneh, T. R. Hughes, and G. W. Brown (2006). A survey of essential gene function in the yeast cell division cycle. *Mol Biol Cell 17*, 4736–4747.

Yuh, C.-H., H. Bolouri, and E. H. Davidson (1998). Genomic cis-regulatory logic: Experimental and computational analysis of a sea urchin gene. *Science 279*, 1896–1902.

Zhang, J., A. M. Dean, F. Brunet, and M. Long (2004). Evolving protein functional diversity in new genes of *Drosophila*. *Proc Natl Acad Sci USA 101*, 16246–16250.

Zimmer, C. (2001). Alternative life styles. *Natural History Magazine 110/May*, 42–45.

Zinovieva, R. D., J. Piatigorsky, and S. I. Tomarev (1999). O-crystallin, arginine kinase and ferritin from the octopus lens. *Biochim Biophys Acta 1431*, 512–517.

Zipf, G. K. (1935). *The Psycho-Biology of Languages*. Boston, MA: Houghton-Mifflin.

Ziv, E., M. Middendorf, and C. H. Wiggins (2005). Information-theoretic approach to network modularity. *Phys Rev E 71*, 046117.

Zong, S., M. Wu, J. Gu, T. Liu, R. Guo, and M. Yang (2018). Structure of the intact 14-subunit human cytochrome c oxidase. *Cell Res 28*, 1026–1034.

Zurek, W. H. (1990). Algorithmic information content, Church-Turing thesis, physical entropy, and Maxwell's demon. In W. H. Zurek (Ed.), *Complexity, Entropy, and the Physics of Information*, Volume 8 of *SFI Studies in the Sciences of Complexity*, pp. 73–89. Redwood City, CA: Addison-Wesley.

INDEX

Bold page numbers indicate the primary discussion of the topic.